T0342293

Semiconductor Laser Engineering, Reliability and Diagnostics

Semiconductor Laser Engineering, Reliability and Diagnostics

A Practical Approach to High Power and Single Mode Devices

Peter W. Epperlein

A John Wiley & Sons, Ltd., Publication

This edition first published 2013
© 2013, John Wiley & Sons Ltd

Registered office

John Wiley & Sons Ltd, The Atrium, Southern Gate, Chichester, West Sussex, PO19 8SQ,
United Kingdom

For details of our global editorial offices, for customer services and for information about how to apply
for permission to reuse the copyright material in this book please see our website at www.wiley.com.

Library of Congress Cataloging-in-Publication Data

Epperlein, Peter W.
    Semiconductor laser engineering, reliability, and diagnostics : a practical approach to high power and
single mode devices / Peter W. Epperlein.
        pages cm
    Includes bibliographical references and index.
    ISBN 978-1-119-99033-8 (hardback)
  1. Semiconductor lasers.   I. Title.
    TA1700.E67 2013
    621.36'61–dc23

                                                                                        2012025789

A catalogue record for this book is available from the British Library.

Print ISBN: 978-1-119-99033-8

Typeset in 10/12pt Times by Aptara Inc., New Delhi, India

To Eleonore

My beloved wife and closest friend

With deep gratitude and affection

Contents

Preface

Scope and purpose

Semiconductor diode lasers have developed dramatically in the last decade as key components in a host of new applications, with optical fibre communications and data storage devices as the original and main driving forces behind the enormous progress in diode laser technologies. The increase of laser output power, accompanied by improved laser reliability and widened laser wavelength range in all single-emitter and multi-element emitter devices, gave rise to the penetration of diode lasers into other mass-markets and emerging applications, such as laser pumping, reprographics, data recording, displays, metrology, medical therapy, materials processing, sophisticated weaponry, and free-space communications. As a consequence, diode lasers continue to represent a high percentage of the worldwide commercial laser revenues, 51% of the $6.4B in 2010 with 10% growth forecasted for 2011 (Laser Focus World, 2011)[1]. Huge progress has been made in high power, single transverse mode lasers over recent years, followed by new applications and along with increased requirements for device engineering, reliability engineering and device diagnostics.

This book is a fully integrated novel approach, covering the three closely connected fields of diode laser engineering, reliability engineering and diagnostics in their development context, correlation and interdependence. It is exactly the blend of the underlying basic physics and practical realization, with its all-embracing, complementary issues and topics that has not been dealt with so far in the current book literature in this unique way. This includes practical, problem-related design guidelines as well as degradation-, reliability- and diagnostic-related aspects and issues for developing diode laser products operating in single transverse mode with high power and reliability. And it is this gap in the existing book literature, that is, the gap between device physics in the all-embracing context, and the practical issues of real device exploitation, which is going to be filled by the publication on hand. Research and practical experience gained in industry and higher level education have provided a lot of empirical evidence that the market is in need of a book to fill this gap.

---

[1] Laser Focus World, (2011). *Laser Markets - Annual Review and Forecast*: in *eNewsletter*, February 8, 2011

The book provides a novel approach to the development of high power, single transverse mode, edge-emitting (in-plane) diode lasers, through addressing the complementary topics of device engineering (Part I), reliability engineering (Part II) and device diagnostics (Part III) in altogether nine chapters. Diode laser fundamentals and standard material, fabrication and packaging issues are discussed first. In a subsequent section a comprehensive and elaborate account is given on approaches and techniques for designing diode lasers, emitting high optical power in single transverse mode or diffraction limited beams. This is followed by a detailed treatment of the origins of laser degradation including catastrophic optical damage and an exploration of the engineering means to address for effective remedies and enhanced optical strength. The discussion covers also stability criteria of critical diode laser characteristics and key laser robustness factors. Clear design considerations are discussed in great detail in the context of reliability-related concepts and models, and along with typical programs for reliability tests and growth. A final extended third part of advanced diagnostic methods covers in depth and breadth, for the first time in book literature, functionality-impacting factors such as temperature, stress and material instabilities. It also presents the basics of those diagnostic approaches and techniques and discusses the diagnostic results in conjunction with laser product improvement procedures.

Main features

Among the main features characterizing this book are, that it is:

1. Providing a novel approach of high power, single transverse mode, in-plane diode laser development by addressing the three complementary areas of device engineering, reliability engineering and device diagnostics in the same book and thus closes the gap in the current book literature.

2. Addressing not only narrow stripe lasers, but also other single-element and multi-element diode laser devices, such as broad area lasers, unstable resonator lasers, tapered amplifier lasers, phase-locked coherent linear laser arrays and high power incoherent standard 1 cm laser bars, designed by applying the various known principles to achieve high power emission in a single transverse mode or diffraction-limited beam.

3. Furnishing comprehensive practical, problem-oriented guidelines and design considerations by taking into account also reliability related effects, key laser robustness factors, and functionality impacting factors such as temperature, stress and material instabilities, and dealing with issues of fabrication and packaging technologies.

4. Discussing for the first time in depth and breadth diagnostic investigations of diode lasers, and using the results for improving design, growth and processing of the laser device in the development phase.

5. Covering in detail the basics of the diagnostic approaches and techniques, many of which pioneered by the author to be fit-for-purpose, and indicating the applicability of these techniques and approaches to other optical and electrical devices.

6. Demonstrating significance of correlations between laser operating characteristics and material parameters, and showing how to investigate and resolve effectively thermal management issues in laser cavities and mirrors.

7. Providing in-depth insight into laser degradation modes including catastrophic optical damage, and covering a wide range of concepts and technologies to increase the optical robustness of diode lasers.

8. Discussing extensively fundamental concepts and techniques of laser reliability engineering, and providing for the first time in a book details on setting up and operating a typical diode laser reliability test program used in industry for product qualification.

9. Representing an invaluable resource for professionals in industry and academia engaged in diode laser product R&D, for academics, teachers and post-graduates for higher educational purposes, and for interested undergraduates to gain first insights into the aspects and issues of diode laser technologies.

10. Featuring two hundred figures and tables illustrating numerous aspects of diode laser engineering, fabrication, packaging, reliability, performance, diagnostics and applications, and an extensive list of references to all addressed technical topics at the end of each of the nine chapters.

Addressed niche markets

The underlying synergetic laser development approach will make this much needed guidebook, a kind of vade mecum of high practical relevance, a great benefit to a broad worldwide readership in industry, higher education, and academic research. Professionals including, researchers and engineers in optoelectronics industries who work on the development of high quality, diode laser products, operating in single transverse mode with high optical output power and high reliability, will regard this book as an invaluable reference and essential source of information. The book will also be extremely useful for academics, teachers and post-graduates for higher educational purposes or satisfying their requirements, if they are just interested in gaining first insights into the aspects and issues associated with the optimization of these diode laser products.

Book context

The book is based primarily on the author's many years of extensive and complex experience in diode laser engineering, reliability and diagnostics. The author accumulated his highly specialized knowledge and skills in hands-on and managerial roles both in global and start-up companies in cutting-edge optoelectronics industries, including IBM, Hewlett-Packard, Agilent Technologies, and IBM/JDSU Laser Enterprise (today part of Oclaro) – starting in the early nineties with his decisive and formative collaboration, as core member of the Laser Enterprise team, the spin-out of IBM Research, pioneering and commercializing its pre-eminent 980-nm pump laser technology for applications in terrestrial and submarine optical communications networks.

The inspiration to write exactly this book has come from the author's extensive semiconductor consulting experience, providing a realistic insight into the very obvious need for a practical, synergetic approach to diode laser development, along with the realization that there has not been any such publication available yet to meet these needs - both at industry and higher educational level. The author is confident, therefore, that the book on hand will be welcomed worldwide by the addressed, specialized readership with high, and growing demand, so that further editions are required much earlier than expected.

Acknowledgments

I would like to thank my former colleagues in the various semiconductor laser development departments for many thought-provoking discussions and helpful support, especially to Drs. Hans Brugger, Dan Clark, Dan Guidotti, Andrew Harker, Tony Hawkridge, Amr Helmy, Dan Mars, Andy McKee, Heinz Meier, Pat Mooney, John Oberstar, Mike Parry, Julia Shaw, Simon Stacey and Steve Wang.

Thanks equally go to my customers worldwide for their ongoing, encouraging requests in the past years for writing exactly this all-embracing book.

Special thanks for useful discussions and supportive communication to: Prof. Dan Botez, University of Wisconsin, USA; Prof. Dieter Bimberg, Technical University Berlin, Germany; Prof. Petr Eliseev, University of New Mexico, USA and Lebedev Physics Institute, Russia; Prof. Charlie Ironside, University of Glasgow, UK; Dr. Bob Herrick, JDSU Inc., USA; and Dr. David Parker, SPI Lasers, UK.

Special thanks also to my production editor Gill Whitley for all her cooperation and support, and for shepherding this book to publication with undiminished commitment and reliability. Lastly, I would like to express my deepest thanks to Ashley Gasque, a very experienced, most perceptive and resourceful acquisitions editor with CRC Press, USA, whose idea of a book based on my full-day short course at the SPIE Photonics West 2010, triggered off this publication.

Peter W. Epperlein
Colchester, Essex, UK
May 2012

About the author

Dr. Epperlein is currently Technology Consultant with his own semiconductor technology consulting business, Pwe-PhotonicsElectronics-IssueResolution, and residence in the UK. He provides technical consulting services worldwide to companies in photonics and electronics industries, as well as expert assistance to European institutions through evaluations and reviews of novel optoelectronics R&D projects for their innovative capacities including competitiveness, disruptive abilities, and proper project execution to pre-determined schedules.

He looks back at a thirty year career in cutting-edge photonics and electronics industries with focus on emerging technologies, both in global and start-up companies, including IBM, Hewlett-Packard, Agilent Technologies, Philips/NXP, Essient Photonics and IBM/JDSU Laser Enterprise. He holds Pre-Dipl. (B.Sc.), Dipl. Phys. (M.Sc.) and Dr. rer. nat. (Ph.D.) degrees in physics, magna cum laude, from the University of Stuttgart, Germany.

Dr. Epperlein is a well-recognized authority in compound semiconductor and diode laser technologies. He accumulated the broad spectrum of his professional competencies in most different hands-on and managerial roles, involving design and fabrication of many different optical and electrical devices, and sophisticated diagnostic research with focus on the resolution of issues in design, materials, fabrication and reliability, and including almost every aspect of product and process development from concept to technology transfer and commercialization. He has a proven track record of hands-on experience and accomplishments in research and development of optical and electrical semiconductor devices, including semiconductor diode lasers, light-emitting diodes, optical modulators, quantum well devices, resonant tunneling devices, field-effect transistors, and superconducting tunneling devices and integrated circuits.

His extensive investigations of semiconductor materials and diode laser devices have led to numerous world-first reports on special effects in laser device functionality. Key achievements and important contributions to the improvement of development processes in emerging semiconductor technologies include his pioneering development and introduction of novel diagnostic techniques and approaches. Many have been adopted by other researchers in academia and industry, and his publications of these pioneering experiments received international recognition, as demonstrated by thousands of references, for example, in Science Citation Index and Google, advanced search exact phrase for 'PW or Peter W Epperlein'. Many of those unique results added high value to the progress of new product or emerging technology development processes.

Dr. Epperlein authored or co-authored more than seventy peer-reviewed journal and conference technical papers, has given more than thirty invited talks at international conferences and workshops, and published more than ten invention disclosures in the IBM Technical Disclosure Bulletin. He has served as reviewer of numerous proposals for publication in technical journals and he was awarded five IBM Research Division Awards for achievements in diode laser technology, quality management and laser commercialization.

Dr. Epperlein started his career in emerging superconductor technologies in the late seventies, with sophisticated design, modelling and measurements on superconducting materials, tunneling effects, devices and integrated circuits in his more than five years collaboration in the then revolutionary IBM Josephson Junction Superconducting Computer Project (dropped by IBM end of 1983), which included a two-year International Assignment from the IBM Zurich Research Laboratory to the IBM Watson Research Center, N.Y., USA until the mid-eighties.

This term was followed by a fundamental career re-orientation from emerging superconductor to emerging semiconductor technologies, comprising more than twenty-five years in the fields of semiconductor technologies, optoelectronics, fibre-optic communications, and with his first role to start as core member of the pioneering IBM Laser Enterprise (LE) Team, to become a spinout of IBM Research in the early nineties. He contributed significantly to research, development and commercialization of the pre-eminent pump diode laser technology for applications in optical communication networks in the early nineties along with the transition of the LE-Research Team into a competitive market leader IBM/JDSU LE some five years later.

Part I

DIODE LASER ENGINEERING

Overview

The impressive technological advances that resulted in semiconductor diode laser technologies in the last decade can be grouped roughly into four areas: higher optical output power, higher single transverse mode and diffraction-limited output, increased range of lasing wavelengths, and significantly improved reliability (see Part II). Most noteworthy commercial demonstrations in high-power continuous wave (cw) outputs of single-emitter and multi-element emitter laser products, for example, in the 980 nm band, include 0.75 W ex-fiber for single spatial mode, narrow-stripe emitters, 12 W for tapered master oscillator power amplifier emitters with single-mode, diffraction-limited operation, 25 W for standard 100 μm wide aperture single-emitter devices, and 1000 W quasi-cw for standard 1 cm multi-element linear laser arrays with nearly diffraction-limited beams.

The development of novel design approaches including strained quantum wells and quantum cascade structures, as well as the advanced maturity of material systems such as compounds based on GaN, CdS, and GaSb, have significantly extended the operating wavelength range of semiconductor lasers throughout the visible spectrum into the ultraviolet regime down to about 0.375 μm on the short-wavelength side and far into the infrared regime with cw operation within 3–10 μm at room temperature, and beyond 10 μm up to 300 μm at operating temperatures around 77 K on the very long wavelength side. Compressively-strained InGaAs/AlGaAs quantum well lasers emitting in the 980 nm band are typical examples of lasers with wavelengths, which lattice-matched quantum well structures cannot deliver.

This part consists of two chapters. Chapter 1, on basic diode laser engineering principles, includes elaborate descriptions of relevant basic diode laser elements, parameters, and characteristics, aspects of high-power laser design, diode laser structures, materials, fabrication, and packaging technologies, and practical laser performance figures. Chapter 2 is on the design considerations for high-power single spatial mode operation. It provides an extensive account of various approaches and techniques for the development of high-power semiconductor lasers emitting in a single spatial mode or diffraction-limited beam. The discussion is mainly on design issues and operating parameter dependencies of narrow-stripe, in-plane lasers, but also on other single- and multi-element diode laser devices. This includes broad-area lasers, unstable resonator lasers, tapered amplifier lasers, phase-locked coherent linear laser arrays, and high-power incoherent standard 1 cm laser bars designed by applying the various known principles to realize high-power emission in a single transverse mode or diffraction-limited beam.

Chapter 1

Basic diode laser engineering principles

Semiconductor Laser Engineering, Reliability and Diagnostics: A Practical Approach to High Power and Single Mode Devices, First Edition. Peter W. Epperlein.
© 2013 John Wiley & Sons, Ltd. Published 2013 by John Wiley & Sons, Ltd.

Introduction

This chapter starts with a brief recap of the fundamental aspects and elements of diode lasers, including relevant features of the standard device types, with an emphasis on the advantages of quantum heterostructures for their effective use as active regions in the lasers. Common laser material systems are then discussed, along with lasing wavelength-dependent applications and best output power levels achieved in each individual high-power diode laser category for illustration and comparison. Various aspects of high-power issues are presented, including power-limiting factors and reliability tradeoffs. To develop a good understanding of diode laser operation, key electrical, optical and thermal parameters and characteristics are described. The chapter concludes with a description of the basic aspects of diode laser fabrication and packaging technologies.

1.1   Brief recapitulation

1.1.1   Key features of a diode laser

The basic device structure consists of a rectangular parallelepiped of a direct bandgap semiconductor, usually a III–V compound semiconductor such as GaAs, incorporating a forward-biased, heavily doped p–n junction to provide the optical gain medium in a resonant optical cavity, as illustrated schematically in Figure 1.1.

    Further basic elements include the optical con nement in the transverse vertical direction perpendicular to the active region and transverse lateral con nement of injected current, carriers and photons parallel to the active layer. Further details of these features will be illustrated below.

1.1.1.1   Carrier population inversion

The operating principle of a semiconductor laser requires the gain medium to be pumped with some external energy source, either electrical or optical, to build up and maintain a nonequilibrium distribution of charge carriers, which has to be large enough to enable a population inversion for the generation of optical gain. Pumping realized by optical excitation of electron–hole pairs is usually only important for the rapid characterization of the quality of the laser material without electrical contacts. The more technologically important technique, however, is direct electrical pumping using a forward-biased semiconductor diode with a heavily doped p–n junction at the center of all state-of-the-art semiconductor injection lasers, that is, diode lasers. The

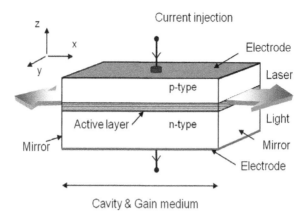

Figure 1.1 Illustration of a very basic diode laser chip. Typical dimensions in x direction are approximately 500 μm for the laser cavity length, in y direction 100 μm for the transverse lateral coordinate without lateral con nement structure, and in z direction a few micrometers for the transverse vertical extent of the p–n layer stack only (substrate not shown). The active layer structure depends on the transverse vertical layer con guration, which will be discussed along with different lateral con nement structures in some of the sections further below.

Fermi levels in these heavily doped and therefore degenerate n- and p-type materials lie in the conduction and valence band, respectively. With no bias voltage applied, the quasi-Fermi levels are identical across the p–n junction at thermal equilibrium with the conduction and valence bands bent, as shown in Figure 1.2a. In this steady state, further diffusion of electrons and holes across the p–n boundary is opposed by the built-in potential (diffusion potential) resulting from the depletion layer or space-charge region formed by the negatively charged acceptors and positively charged donors on the p- and n-sides, respectively. Simpli ed expressions for the depletion layer width W and built-in potential V_{bi} can be written as follows:

$$W = \left[\frac{2\varepsilon}{q} \left(\frac{N_a + N_d}{N_a N_d} \right) V_{bi} \right]^{1/2} \tag{1.1}$$

$$V_{bi} = \frac{k_B T}{q} \ln \frac{N_d N_a}{n_i^2} \tag{1.2}$$

where $\varepsilon = \varepsilon_r \varepsilon_0$ is the permittivity, q is the electron charge, T is the absolute temperature, k_B is the Boltzmann constant, N_a and N_d are the densities of acceptors and donors, respectively, and $n_i^2 = np$ is the square of the intrinsic carrier concentration, with n the electron density in the conduction band and p the hole density in the valence band, and with all shallow donors and acceptors fully ionized at room temperature (Sze, 1981).

When the p–n junction is forward biased by applying an external voltage nearly equal to the energy bandgap voltage, the built-in electric eld is reduced, and

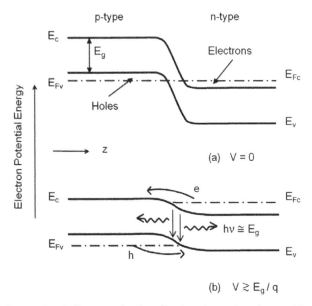

Figure 1.2 Energy band diagram of a heavily doped p–n junction at (a) zero bias and (b) forward bias. Dashed lines represent the quasi-Fermi levels. Schematic representation of densities and strong injections of electrons (e) and holes (h) under forward bias close to the energy gap voltage causing population inversion and radiative recombinations of electrons and holes in the narrow depletion zone.

free electrons and holes can diffuse across the junction into the p- and n-regions, respectively.

Figure 1.2b shows that electrons are then injected into the conduction band and holes into the valence band and for suf cient numbers can create a population inversion in a very narrow zone. In this so-called active region, electrons and holes can recombine, and photons, generated in the radiative recombination process with energies of about the bandgap energy E_g, can be reabsorbed or can induce stimulated emission. The following section treats the optical processes responsible for the generation of optical gain. Issues linked to the simple p–n junction approach and more ef cient p–n junction con gurations will be discussed in those sections dealing with different diode laser structures.

1.1.1.2 Net gain mechanism

Figure 1.3 shows a simpli ed illustration of different electronic transitions between the conduction and valence bands, which are important for establishing an optical gain mechanism. There are three basic types of radiative band-to-band transitions: (i) spontaneous emission; (ii) photon absorption, also called stimulated absorption; and (iii) stimulated emission.

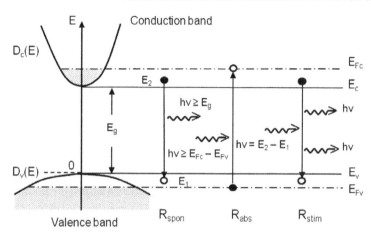

Figure 1.3   Energy level diagram illustrating electronic transitions between the conduction and valence bands of a nonequilibrium, degenerate, direct semiconductor at different rates R: R_{spon} represents the spontaneous emission with an electron (occupied state, solid circles) in the conduction band recombining with a hole (unoccupied state, open circles) in the valence band. R_{abs} depicts the photon absorption process from an occupied state in the valence band to an unoccupied state in the conduction band by generating an electron–hole pair leaving an electron in the conduction band and a hole in the valence band. The required incident photon energy has to be $\geq E_{Fc} - E_{Fv}$ where E_{Fv} is negative with the zero-point of energy at the top edge of the valence band. R_{stim} shows the stimulated emission process where an incident photon with suf cient energy stimulates the recombination of a conduction band electron with a hole in the valence band by producing a second photon, which has the same phase, wavelength, and direction of propagation as the incoming photon. This photon multiplication process is the basis for generating optical gain and coherent emission.

Most transitions of interest involve electrons and holes close to the bottom of the conduction band and to the top of the valence band, respectively, because carrier densities are highest there. Assuming the energy of an electron is E_2, and of a hole E_1, then the energy of a photon emitted in a recombination process is $E_{21} = h\nu \sim E_g$, which can be slightly higher than the bandgap energy; here h is the Planck constant and ν the photon frequency.

In the spontaneous emission process, the radiative recombination of an electron–hole pair generates a photon. This process is random in direction, time and phase, but does not lead to a coherent radiation when averaged over a large ensemble of emission processes. The spontaneous emission does not interact with photons through the recombination process, in contrast to the absorption and stimulated emission processes, which will be discussed next. Therefore, the transition rate for spontaneous emission, R_{spon}, is not dependent on the photon density, but is proportional to the product of the electron concentration at E_2 and hole concentration at E_1. The former is the product of the density of electronic states $D_c(E_2 - E_c)$ and the probability that these states are occupied by electrons given by the Fermi–Dirac distribution function f_2, whereas the latter is the product of the density of states $D_v(E_v - E_1)$ and the probability $(1 - f_1)$ that these states are not occupied by electrons. E_c and E_v are the energies at the conduction band minimum and valence band maximum, respectively. Thus, we can write

$$R_{spon} = A_{21} D_c (E_2 - E_c) f_2 D_v (E_v - E_1)(1 - f_1). \tag{1.3}$$

Here A_{21} is the transition probability for spontaneous emission from the energy state E_2 to E_1.

In the absorption process, a photon with energy $E_2 - E_1 \geq E_g$ is absorbed and excites an electron in the conduction band while leaving a hole in the valence band. Therefore, absorption is a three-particle process and its transition rate R_{abs} from E_1 to E_2 is proportional to the product of three particle concentrations: the photon density $n_{phot}(E_2 - E_1)$; the density of unoccupied states $D_c(E_2 - E_c)(1 - f_2)$ in the conduction band; and the density of occupied states $D_v(E_v - E_1)f_1$ in the valence band. With B_{12}, the transition probability constant, we obtain

$$R_{abs} = B_{12} n_{phot} (E_2 - E_1) D_v (E_v - E_1) f_1 D_c (E_2 - E_c)(1 - f_2). \tag{1.4}$$

In contrast to stimulated absorption discussed above, the various interactions involved in the stimulated emission process are reversed, that is, an incident photon stimulates the recombination of an electron–hole pair by simultaneously generating the emission of a new photon. This is a positive gain mechanism leading to the ampli cation of radiation, because the stimulated photons, which are aligned in direction and phase to the incident photons, are emitted into the incident radiation eld resulting in a strong coherent optical emission. Of course, the effective net gain is dependent on the difference between the transition rates of stimulated emission and absorption. The rate of stimulated emission is proportional to the incident photon density $n_{phot}(E_2 - E_1)$, the density of occupied states $D_c(E_2 - E_c)f_2$ in the conduction band, and the density of unoccupied states $D_v(E_v - E_1)(1 - f_1)$ in the valence band. It can be written as

$$R_{stim} = B_{21} n_{phot} (E_2 - E_1) D_c (E_2 - E_c) f_2 D_v (E_v - E_1)(1 - f_1) \tag{1.5}$$

where B_{21} is the probability of the stimulated emission transition.

At thermal equilibrium of the semiconductor, there is no net energy transfer to the optical eld, that is, the transition rates for spontaneous emission, stimulated emission, and absorption have to obey the following equation:

$$R_{abs} = R_{stim} + R_{spon}. \tag{1.6}$$

By using Equations (1.3) to (1.6), and the well-known expressions for the Fermi–Dirac distribution function f and the black-body radiation according to Planck's theory, we obtain after some simple algebraic manipulations

$$A_{21} = \frac{n_r^3}{\pi^2 \hbar^3 c^3} (E_2 - E_1)^2 B_{21} \tag{1.7}$$

and

$$B_{21} = B_{12} \tag{1.8}$$

where n_r is the refractive index of the medium, c is the velocity of light in vacuum, $\hbar = h/2\pi$, and $(E_2 - E_1) = h\nu$ is the photon energy. Equations (1.7) and (1.8) are the Einstein relations for radiative transitions.

In the following, we derive the necessary condition for net gain under nonequilibrium conditions, that is, strong carrier injections, by calculating the ratio between the transition rates for absorption and stimulated emission. We use Equation (1.4) for R_{abs} and (1.5) for R_{stim}, and the well-known expressions for the separate quasi-Fermi functions f_1 and f_2 in the valence and conduction bands, respectively. We obtain

$$\frac{R_{abs}}{R_{stim}} = \frac{f_1(1 - f_2)}{f_2(1 - f_1)} = \frac{\exp\{(E_2 - E_{Fc})/k_B T\}}{\exp\{(E_1 - E_{Fv})/k_B T\}} = \exp\{[h\nu - (E_{Fc} - E_{Fv})]/k_B T\} \tag{1.9}$$

with the quasi-Fermi level energies E_{Fc} for the conduction band and E_{Fv} for the valence band. The exponential function is greater than one for $E_{Fc} = E_{Fv}$ at thermal equilibrium and hence $R_{abs} > R_{stim}$ always. However, Equation (1.9) clearly shows that lasing operation can be achieved for the condition

$$(E_{Fc} - E_{Fv}) > h\nu > E_g \tag{1.10}$$

involving photons with energies larger than the bandgap energy. A semiconductor meeting the condition in Equation (1.10) is in the state of population inversion resulting in $R_{stim} > R_{abs}$, or, in other words, a photon is ampli ed rather than absorbed. Equation (1.10) establishes the gain bandwidth of the medium including the requirement $(E_{Fc} - E_{Fv}) > E_g$ for having gain at any frequency. The limiting case

$$E_{Fc} - E_{Fv} = E_g \tag{1.11}$$

is called the transparency condition, where the gain is zero for a photon frequency $\nu = E_g/h$. When the density of injected carriers is larger than the transparency density, then $(E_{Fc} - E_{Fv}) > E_g$ and a net gain develops for photon energies between E_g and $(E_{Fc} - E_{Fv})$ according to Equation (1.10). We will resume the discussion of gain-related issues in Section 1.3.

1.1.1.3 Optical resonator

To provide positive feedback for laser action, edge-emitting diode lasers usually employ a Fabry–Pérot resonator comprising two parallel, high-quality plane mirrors as shown in Figure 1.4. In semiconductors, this is easily achieved by cleaving the crystal perpendicular to the cavity at the end faces along well-de ned crystal planes. For GaAs the cleaved facets are (110) planes and the junction plane comprising the active layer is (100). Since the refractive index of the medium is very large (e.g., 3.6

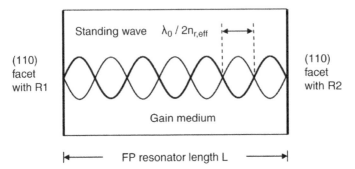

Figure 1.4   Schematics of a semiconductor diode laser illustrating the formation of a standing wave or longitudinal mode inside the gain medium of a Fabry–Pérot optical cavity with cleaved, uncoated facets at the end faces, which act as partially re ecting mirrors. The mode propagates in a dielectric waveguide with an effective refractive index determined by the indices of the waveguide core and claddings. The laser is likely to oscillate at a frequency that matches a longitudinal mode supported by the resonator.

for GaAs), the re ectivity of uncoated facet mirrors is already suf ciently high to produce a resonant cavity and provide the feedback for the onset of laser oscillations. Typical (power) re ectivities R of 32% for GaAs can be calculated from the Fresnel re ection at normal incidence at the GaAs/air interface

$$R = \frac{(n_{rs} - n_{ra})^2 + \kappa^2}{(n_{rs} + n_{ra})^2 + \kappa^2} \simeq \frac{(n_{rs} - n_{ra})^2}{(n_{rs} + n_{ra})^2} \tag{1.12}$$

where n_{rs} is the refractive index of the semiconductor, n_{ra} the index of the ambient air in this case, and κ the extinction coef cient. For $\kappa^2 \ll (n_{rs} - n_{ra})^2$, Equation (1.12) simpli es and R depends only on the refractive indices.

To maximize laser functionality, mirrors are coated with appropriate dielectric layers to adjust the re ectivities, which are usually very high (>90%) at the rear mirror and lower (<10%) at the output front mirror of the laser cavity. Section 1.3 gives a detailed account of this topic.

The optical modes of this resonator can be considered as the superposition of two plane light waves propagating normal to the mirror surfaces in opposite directions along the resonator axis in a laser active material of length L and refractive index $n_{r,eff}$. A standing wave develops between the mirrors when the cavity length is an integral number of half wavelengths

$$L = m\frac{\lambda}{2} = m\frac{\lambda_0}{2n_{r,eff}} \tag{1.13}$$

where m is a positive integer and the number of nodes of the standing wave, λ_0 is the wavelength in vacuum. Equation (1.13) implies that the electric eld is zero at both mirror surfaces. In Section 1.3, we will use the phase condition expressed in Equation (1.13) to describe the development of longitudinal laser modes.

1.1.1.4 Transverse vertical confinement

In order to achieve high gain, the photon density in the active region, where the gain-generating recombination processes are occurring, has to be maximized. This means that the all-important stimulated emission has to dominate the spontaneous emission, which can be evaluated from the transition rate ratio R_{stim}/R_{spon}. Using Equations (1.5), (1.3) and (1.7), (1.8), we obtain

$$\frac{R_{stim}}{R_{spon}} = \frac{\pi^2 \hbar^3 c^3}{n_r^3 (hv)^2} n_{phot} (hv). \tag{1.14}$$

Here we have again used $hv = E_2 - E_1$ for the photon energy. This equation clearly demonstrates the dominance of the stimulated emission over the spontaneous emission, increasing with increasing density of photons $n_{phot}(hv)$ of energy hv. However, it also shows that this photon density has to be much higher for lasers operating at higher energies hv to compensate for the inverse squared photon energy dependence in Equation (1.14).

In the previous section, we showed how to increase the photon density by optical feedback in a resonator cavity. By bandgap engineering, effective structures perpendicular to the active layer have been developed to con ne photons and charge carriers in the laser active region. Such structures include double-heterostructures (DHs) and quantum wells (QWs), schematically shown in Figure 1.5. Aspects of these structures can be found in Sections 1.1.3, 1.1.4 and 1.3.

Electrons and holes are con ned in the thin slab of undoped active material sandwiched between n- and p-doped cladding layers and are then forced to recombine

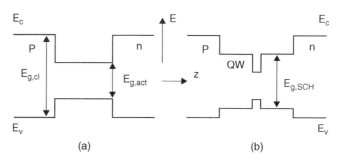

(a)   (b)

Figure 1.5   Schematic illustration of two examples for vertical transverse con nement structures of an edge-emitting (in junction plane) diode laser: (a) double heterostructure and (b) quantum well (QW) embedded in separate-con nement heterostructure (SCH). Simpli ed energy band diagrams (effects of band bending in the heterojunctions are ignored) versus vertical direction z for a thin slab of undoped active material sandwiched between p- and n-doped cladding layers.

inside this thin active layer. Typical thicknesses for DH and QW carrier-con ning active layers are about 0.1–0.2 µm and about 10 nm, respectively. The cladding layers have a higher conduction–valence bandgap energy, which leads to a couple of positive effects.

First, the three-layer structure acts like an optical waveguide in a vertical direction since usually semiconductors with higher bandgap energy have a lower refractive index than semiconductors with a low bandgap energy. Second, light generated in the low-bandgap active layer will not be lost by reabsorption in the cladding layers.

The optical con nement in the transverse vertical direction is low because of the thin active layers, and typical con nement factors can be as low as 20% and far below 1% for DH and QW structures, respectively, for refractive index differences between cladding and active layers of about 5%. However, there are structures available for improving the optical con nement in QW structures, and include surrounding the active layer with a thicker separate-con nement heterostructure (SCH) region with higher bandgap and lower refractive index to enhance the con nement of photons (and electrons). The advantages of such structures will be discussed in Section 1.3 and the fundamental transverse vertical mode in Chapter 2.

1.1.1.5   Transverse lateral confinement

In transverse lateral con nement, not only are carriers and photons con ned in a direction parallel to the active layer, but so too is the current owing through the device in a vertical direction. This triple con nement scheme is designed to deliver edge-emitting (in the plane of the active layer) diode lasers operating in fundamental single-mode with high external ef ciency and output power and at low threshold current. These topics and parameters will be described in detail in Section 1.3 and Chapter 2.

It is important to maximize the current injected into the active region by mini- mizing the leakage current, which is the difference between the total current injected into the laser device and the current passing through the active region. The lower the leakage current, the lower the probability that the output power saturates at high currents. Also, under continuous wave (cw) operation of the laser there is then only a small or negligible additional heating effect that can lead to a rollover of the light– current characteristic, resulting in a decrease of optical output power with increasing current (see Section 1.2.3.2). Current con nement can simply be realized by reduc- ing the area of current injection just by limiting the contact area. However, the most effective current con nement is combined with schemes to laterally con ne carriers and photons in the same con guration.

As discussed in the previous section, optical wave con nement in the vertical di- rection perpendicular to the active junction layer is formed by dielectric waveguiding. This is also known as index-guiding because the refractive index of the active layer is higher than that of the surrounding cladding layers, resulting in mode con nement through total internal re ection at the active layer/cladding interfaces.

However, optical eld con nement in the lateral direction can be grouped into two classes, gain-guiding and index-guiding, dependent on whether the mode is

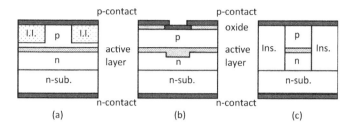

Figure 1.6   Schematic cross-section of lateral con nement structures. (a) Proton stripe realized by ion implantation (I.I.) provides gain guiding; region implanted by protons is of high resistivity restricting the current ow to an opening in the implanted region. (b) Simple rib waveguide provides weak index-guiding. (c) Etched-mesa (ideal) buried-heterostructure sandwiched laterally between insulating (Ins.), highly resistive current blocking layers with higher bandgap energies and lower refractive indices than those of the active layer provides strong index-guiding. Alternatively, semi-insulating materials or reverse-biased p–n junctions formed by regrowth can also be used as current blocking layers.

con ned by the lateral variation of the optical gain or the refractive index, respectively. Index-guided devices are further subdivided into weakly and strongly index-guided, depending on the strength of the lateral index step.

Figure 1.6 shows schematically the three classes. Each class is represented by a typical design selected from the numerous different approaches available to realize the lateral con nement of current, carriers and photons. As will be shown in detail in Section 1.3, gain-guiding provides current con nement, weak index-guiding current and photon con nement, and strong index-guiding provides current, carrier and photon con nement.

1.1.2   Homojunction diode laser

As described in Sections 1.1.1.1 to 1.1.1.3, the key parts and processes of a homojunction laser include:

- direct electrical pumping by strong electron and hole injection of a forward-biased ($V \cong E_g/q$) p–n junction with both p- and n-type regions of the same material heavily doped to $\gtrsim 10^{18}$ atoms/cm^3;

- carrier recombination and population inversion in the narrow active region; and

- optical feedback of the stimulated photons for coherent emission in an optical resonator.

The rst semiconductor lasers employed homojunctions and were plagued by serious problems. One of the main constraints of this laser type results from the very small potential barrier that an electron encounters when it is injected into the p-region and from its high mobility μ_n. The penetration depth d of the electron is then given by the diffusion length L_n

$$d = L_n = \sqrt{D_n \tau_n} \qquad (1.15)$$

where D_n is the diffusion coefficient and τ_n is the lifetime of the electron established by electron–hole recombination. D_n can be calculated from the mobility by using the Einstein relation

$$D_n = \frac{k_B T}{q} \mu_n. \qquad (1.16)$$

In GaAs, doped to $n \cong 2 \times 10^{18}$ cm^{-3}, the electron mobility is about 300 cm^2/(V s) at room temperature (Yu and Cardona, 2001). This corresponds to a diffusion coefficient of $D_n \cong 8$ cm^2/s. On the other hand, relevant mobility and diffusion coefficient values for holes are lower by a factor of about 20, and hence have only a negligible effect on the thickness of the effective recombination region. The electrons injected into the p-region become minority carriers in a sea of holes with a density of p_p and their lifetimes can be approximated by

$$\tau_n \approx \frac{1}{B p_p}. \qquad (1.17)$$

With typical values for the bimolecular recombination coefficient of $B \approx 2 \times 10^{-10}$ cm^3/s, which is valid for most III–V compound semiconductors, and for $p_p = 2 \times 10^{18}$ cm^{-3}, the effective lifetime of an injected electron is $\tau_n = 2.5$ ns. Using these values for D_n and τ_n in GaAs we get a diffusion length L_n of about 1.5 μm. This shows that the active region is quite thick compared to the thickness of the depletion layer of typically 0.1 μm. The region of inversion is located approximately 1.5 μm on the p-side of the junction. Since the diffusion length of the injected charge carriers is controlled by recombination, there is little control over the extent of the gain region. One of the problems with the homojunction laser is the poor overlap of the small gain region with the spatial laser mode. However, a much bigger problem is the very high threshold current density that usually requires the laser to be operated pulsed at room temperature or cw at cryogenic temperatures. Actually, there are two main reasons for the high threshold current.

The first reason can be evaluated from the current density necessary to generate gain. To establish this carrier density in a depth d, regardless of the dimensions of the diode laser, a current density of $J = qdG$ is required, where d is the electron penetration depth equal to the diffusion length L_n in a homojunction and G is the necessary pumping rate to maintain a steady state population inversion. Under steady state conditions, the injection rate must equal the recombination rate. G can be evaluated from $G = Bnp$ and results in approximately 8×10^{26} electron–hole pairs/(cm^3 s) to establish a carrier density of 2×10^{18} cm^{-3} for both electrons and holes usually

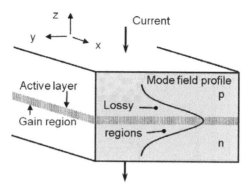

Figure 1.7   Schematic illustration of the poor overlap between optical mode and gain region, and the high absorption losses of the laser beam in a homojunction device.

required to form a population inversion, and by using for the bimolecular recombination coef cient $B \approx 2 \times 10^{-10}$ cm^3/s and for the diffusion length $L_n = d = 1.5$ μm. Finally, we get $J \cong 19 \times 10^3$ A/cm^2 for the rst contribution to the threshold current density.

The second reason lies in the fact that, due to the large transverse dimensions, the laser beam extends considerably into the n- and p-regions where it is strongly absorbed; in other words, the region outside the inversion introduces a strong absorption loss (Figure 1.7).

As will be discussed in Section 1.3, gain exists only for light with energy slightly greater than the bandgap energy. Moreover, there is no mode con nement or waveguiding of the laser mode having a size of some micrometers, with the consequence that the beam will spread to larger sizes in a short distance due to strong diffraction effects leading to even larger absorption losses and a smaller overlap with the gain region. These excessive losses due to poor optical and carrier con nement further increase the current density evaluated above, resulting in typical thresholds of approximately 10^5 A/cm^2.

1.1.3   Double-heterostructure diode laser

The double-heterostructure (DH) concept solves most of the problems associated with homojunction lasers. As illustrated in Figure 1.8, a thin active layer of undoped, direct lower bandgap semiconductor is sandwiched between thicker n- and p-type semiconductor cladding layers with higher bandgap energies and at the same time lower refractive indices.

Figure 1.8 shows the structure under full forward bias leading to the injection of large equal densities of electrons and holes into the active layer enclosed at both sides by heterobarriers. Electrons and holes are con ned in the potential well of the active layer by the conduction band offset ΔE_c and valence band offset ΔE_v, respectively. These two offsets share the difference ΔE_g between the energy gaps of the cladding layer and active layer: $\Delta E_c = x\Delta E_g$, $\Delta E_v = y\Delta E_g$, $x + y = 1$.

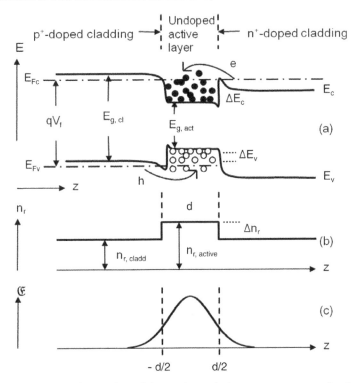

Figure 1.8   Schematic illustration of the carrier and photon con nement of a double heterostructure in the transverse vertical direction z of an edge-emitting diode laser. (a) Schematic
energy band edges of a strongly forward-biased laser (nearly at-band conditions). Con nement of injected electrons (e, solid circles) and holes (h, open circles) in the potential well
of the thin, undoped active layer formed by the conduction and valence band edge energy
discontinuities ΔE_c and ΔE_v, respectively. (b) Refractive index pro le $n_r(z)$ perpendicular to
the active layer determines the waveguiding strength of the mode in the transverse vertical
direction. (c) Field intensity pro le of the fundamental optical mode in the transverse vertical
direction traveling in the axial direction x of the optical cavity.

Consequently, there are three key advantages as shown in Figure 1.8. First, the
con ned carriers are forced to recombine in the thin, low-bandgap active layer to
produce optical gain. Second, the refractive index difference con nes the optical
mode close to the active layer, which acts as an optical waveguide and enables a
strong overlap between the optical mode and gain region. Third, the optical mode
con nement strongly reduces the reabsorption of the laser beam in the cladding
layers and hence the intrinsic optical losses that would otherwise be very high in
the absence of index-guiding as in the homojunction laser structure. However, the
injected carriers may over ow from the active region into the surrounding cladding
layers under very high injection conditions. As we will discuss in Section 1.3.7.3,
such carrier loss via leakage over the heterobarriers will have negative effects on,

for example, the internal quantum ef ciency and the characteristic temperature T_0, which characterizes the temperature dependence of the threshold current.

Typical thicknesses d of the active layer are in the range of 0.1 to 0.2 μm, so carrier densities of $\gtrsim 2 \times 10^{18}$ cm^{-3} required for population inversion can be achieved at much lower currents than in homojunction lasers. The gain scales as $1/d$; however, as d decreases, the optical con nement decreases and eventually dominates over the $1/d$ dependence, which is typically for $d < 0.1$ μm, and the gain then starts to decrease with decreasing d. Properly optimized DH lasers have a threshold current density of $\lesssim 10^3$ A/cm^2, that is, more than two orders of magnitude lower compared to corresponding homojunction lasers and therefore making cw room temperature operation feasible. The dependence of the threshold gain on the active layer thickness d will be discussed in Section 1.3.

The realization of a DH requires careful matching of the lattice constants of the active layer and the cladding layer materials to avoid the formation of mis t dislocations at the two interfaces due to the buildup of strong strain elds in the case of non-matching conditions. Mis t dislocations and other defects generally become nonradiative recombination centers with the detrimental effect of reducing the number of injected carriers effective in the gain-building stimulated emission process. There are two classical material systems that meet the lattice-matching condition (within ~0.1%) over some range of composition in each case: the GaAs active layer sandwiched between $Al_xGa_{1-x}As$ claddings; and $In_{1-x}Ga_xAs_yP_{1-y}$ embedded in InP. In addition, these material systems maintain a direct gap and enable a range of doping levels over a certain range of composition.

The following crucial material parameters, for example, for the GaAs/AlGaAs system, demonstrate its suitability for realizing a DH laser:

- The refractive index $n_r(GaAs) = 3.6$ is much larger than $n_r(Al_{0.3}Ga_{0.7}As) = 3.4$, thus the big index step of 0.2 provides strong *photon confinement*.

- The bandgap energy $E_g(GaAs) = 1.42$ eV is much smaller than $E_g(Al_{0.3}Ga_{0.7}As) \cong 1.79$ eV, which provides *reduced absorption* of the laser beam in the lossy claddings.

- The bandgap energy difference $\Delta E_g \cong 0.37$ eV leads to a conduction band discontinuity $\Delta E_c = x\Delta E_g \cong 0.60 \times 0.37$ eV $\cong 0.22$ eV and valence band discontinuity $\Delta E_v = y\Delta E_g \cong 0.40 \times 0.37$ eV $\cong 0.15$ eV, thus providing *effective carrier confinement*. The 60%/40% division of the bandgap difference is a typical ratio for the GaAs/AlGaAs system accepted in relevant technical publications (Yariv, 1997).

Details of these material systems and others will be described in subsequent sections.

1.1.4   Quantum well diode laser

As we saw in the previous section, DH lasers consist of an active layer sandwiched between two cladding layers with higher bandgap energy. The laser characteristics deteriorate with decreasing thickness of the narrow-bandgap active layer, for example, the threshold current increases because of the reduced con nement for injected carriers and photons. However, when the thickness becomes comparable to the de Broglie wavelength ($\lambda \cong h/p$ where p is the momentum of the electron or hole) quantum size effects can occur resulting in profound differences in material properties and laser performance, which will be discussed here and also in Sections 1.1.4.1 and 1.3.

Modern growth techniques, such as molecular beam epitaxy (MBE) and metal–organic chemical vapor deposition (MOCVD), have enabled the growth of re-producible layers as thin as a few monolayers of atoms. Most high-performance semiconductor diode lasers nowadays are based on quantum well (QW) designs consisting of the potential well sandwiched between potential barriers with wide bandgap energies.

Single-QW (SQW) and multi-QW (MQW) structures are employed, each with its own speci c characteristics, which will be discussed below. Lasers using QWs as active layers with typical thicknesses of 10 nm have very reduced threshold currents. On the one hand, this is due to the very small active volume resulting in a reduced injection current to reach threshold (transparency); that is, a thinning of the active layer reduces the active volume and transparency current proportionately. On the other hand, a signi cant contribution to the threshold reduction can be expected from the density of electronic states in the gain region of a QW, which is modi ed from the parabolic density of states of the bulk semiconductor. We know that the density of states for electrons of energy E in the conduction band in the three-dimensional case is

$$D_c(E) = \frac{1}{2\pi^2} \left(\frac{2m_c^*}{\hbar^2} \right)^{3/2} \sqrt{E} \tag{1.18}$$

where m_c^* is the conduction band effective mass, and energy E is measured from the band edge.

In a QW with ultrathin dimension L_z and otherwise macroscopic dimensions L_x and L_y in the QW plane, carrier motion perpendicular to the well layer is restricted and the kinetic energy of the carriers moving in that direction is quantized into discrete energy levels. The allowed electron wavevectors then have components $k_x = l\pi/L_x$, $k_y = m\pi/L_y$, and $k_z = n\pi/L_z$, where l, m, and n are integers greater than or equal to one. The small value of L_z requires the magnitude of k_z to be large. As k_z can never equal zero, there are no allowed states until $k_z \geq \pi/L_z$, or until the energy is at least $E = \hbar^2 k_z^2 / 2m_c^* = \hbar^2 \pi^2 / (2m_c^* L_z^2)$. This is in contrast to bulk material where there are available states beginning at the band edges E_c or E_v up to E. For a QW with suf ciently high and wide barriers the quantized electron energy levels measured from the conduction band edge can be approximated by

$$E_n = n^2 \frac{\hbar^2 \pi^2}{2m_c^* L_z^2} \tag{1.19}$$

where n is a positive integer marking the number of energy levels (Yariv, 1997). Equation (1.19) gives the energy levels from the conduction band edge (bottom of well), increasing inversely proportional to the square of the well thickness L_z. The ground state of the QW is located above the band edge. A similar relation is valid for heavy and light holes by using the relevant valence band effective masses. Figure 1.9a shows a simpli ed band diagram of a SQW structure illustrating the location of the energy levels in the conduction and valence bands of the well with band edge energy discontinuities ΔE_c and ΔE_v, respectively. Quantization of the energy levels occurs only perpendicular to the well layer. In contrast, carriers are free to move parallel to the well in the x and y directions, which leads to the known energy versus wavevector dependencies with the bottom of the subbands corresponding to the quantized levels. The total energy of the carriers in the well is equal to the amount calculated according to Equation (1.19) added by the kinetic energy $\hbar^2/2m^*(k_x^2 + k_y^2)$ of the electrons and holes, with m^* the relevant effective mass in the conduction and valence band, respectively.

By calculating the number of allowed electron wavevectors per unit area in the two-dimensional x–y plane for the nth energy level, the electron density of states in a QW per unit volume and energy can be derived as

$$D_{c,n}(E) = \frac{1}{2\pi^2} \left(\frac{2m_{c,n}^*}{\hbar^2} \right) \left(\frac{\pi}{L_z} \right). \qquad (1.20)$$

The density of states is independent of energy while at a speci c value of n. In fact, the total density of states increases by the constant amount according to Equation (1.20) at each of the energy levels E_n because an electron of a given total energy $E = E_n$ can be found either in the state n or in one of the states below n, which leads to a multiplication of the density of states, that is, a doubling at $n = 2$, tripling at $n = 3$, and so forth. This results in the well-known staircase density of states function. A similar treatment holds for holes in the valence bands. A plot of the density of states comparing a SQW to a bulk semiconductor is shown in Figure 1.9a.

This staircase-like density of states signi cantly modi es the gain in a QW compared to a bulk semiconductor laser. In a bulk semiconductor the states at band edge have the highest probability of being occupied but the lowest density, which leads to a spreading of the occupied carrier states over a large range of energies. Consequently, the gain curve is wide and highly energy dependent, and requires a large concentration of carriers to generate a signi cant population inversion. On the other hand, in a QW the ground state already has a higher density of states, which results in a carrier distribution with a signi cantly higher maximum value and a smaller energetic width than in a bulk material. The consequences are the following: rst, a signi cant gain at a given wavelength can be created by a small number of carriers; and, second, band- lling effects cause a much smaller spectral shift of the gain curve.

The optical properties of QWs are quite different from those in a bulk semiconductor because of the quantized energy levels formed in the well. The states $n = 1$ in the conduction band and valence band have the highest carrier population, because

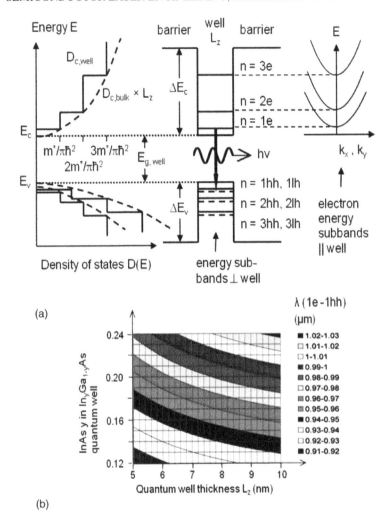

Figure 1.9   (a) Simpli ed illustration of a single quantum well structure including the density of states functions (solid lines) in the conduction and valence bands and the density of states (dashed curves) in a bulk semiconductor for comparison. Schematic representation of the corresponding quantized energy levels perpendicular to the well and the two-dimensional parabolic bands parallel to the well. Quantized electron and heavy-hole (hh) energy levels are shown in solid lines, light-hole (lh) energy levels in dashed lines. The most important transition from $n = 1e$ to $n = 1hh$ is highlighted. (b) Calculated wavelengths of the fundamental transition (1e–1hh) as a function of the InAs y mole fraction and well thickness L_z of a strained $In_yGa_{1-y}As/GaAs$ quantum well. Larger wavelengths are in the upper right and lower wavelengths in the lower left of the greyscale contrast image.

the respective quasi-Fermi function gives the highest probability of occupancy for these states. In addition, the density of states of the heavy-hole band is larger than that of the light-hole band. Transitions between two QW states are governed by the overlap of the wavefunctions in these quantized states. It has been shown in relevant textbooks (e.g., Coldren and Corzine, 1995) that, due to the orthogonality between the QW wavefunction solutions, the overlap integral reduces such that allowed transitions can only occur between energy levels for electrons and holes with the same quantum number n. Transitions where $|\Delta n| \geq 1$ are usually forbidden. From the above follows, that the highest optical gain will result from an $n = 1$ electron ($1e$) to $n = 1$ heavy-hole ($1hh$) transition. The total photon energy emitted in the transition is, to a good approximation,

$$h\nu \cong E_{g,well} + E_{1e} + E_{1hh} \qquad (1.21)$$

where $E_{g,well}$ is the bandgap energy of the well material, E_{1e} is the electron energy of the   rst quantized level with $n = 1$ (ground level) and E_{1hh} is the lowest heavy-hole energy level with $n = 1$.

Using the advanced commercial laser technology integrated program LASTIP (Crosslight Software Inc., 2010), we have calculated the energy subbands of a strained QW (cf. Section 1.1.4.1). Figure 1.9b shows the fundamental transition wavelength $\lambda(1e{-}1hh)$ as a function of the InAs y mole fraction and well thickness L_z of a strained In$_y$Ga$_{1-y}$As/GaAs QW. The long wavelength $\lambda > 1$ μm regime is located in the upper right corner of the contrast image and the wavelengths around 0.91 μm in the lower left corner. A 980 nm transition can be achieved typically for $y = 0.2$ and $L_z = 7$ nm, which is well below the critical thickness at this composition (see the next section). Replacing the GaAs barrier with a higher barrier material, for example, Al$_{0.1}$Ga$_{0.9}$As, would require a larger L_z by about 1 nm in order to achieve the same wavelength for the same well composition.

The thickness of the active layer of a QW laser is very small, typically between 5 and 10 nm. This implies that the con nement of radiation and charge capture within a single well is very poor. Optical con nement factors in thin QWs are below 1% and the ballistic travel distance of an injected electron is about 100 nm, which is at least 10 times larger than a QW 10 nm thick. A large injection current would then be required to achieve suf cient optical gain for lasing. However, these effects would counteract the intrinsic advantage of a QW structure (high gain at low carrier density leading to low threshold current density) resulting from the staircase density of states function, as discussed above.

Therefore, appropriate con nement structures are used to improve carrier capture and optical con nement to the well. Usually such structures consist of layers with higher bandgap energy and lower refractive index sandwiching the well layer. This structure is then surrounded by the cladding layers with higher bandgap energy and lower refractive index than the separate-con nement heterostructure (SCH) layers. One approach is shown in Figure 1.5b. In the wider rectangular potential well of the SCH structure, electrons are   rst trapped in the SCH layer where they form a standing wave before they cool down by phonon scattering and then drop into the well. The optical mode is con ned in the SCH layer of typically 0.2 μm thickness with low leakage into the adjacent cladding layers and ideally negligible absorption losses in these layers. A variation of the SCH approach is the graded refractive index separate-con nement heterostructure (GRIN-SCH) where the index decreases and the energy gap increases gradually within the SCH from the well to the claddings. In addition, this structure traps the carriers, and the electric eld, caused by the gradient, drives the carriers into the well. However, compared to the SCH, this structure has a lower density of states in the "funnel" region and therefore the carrier capture is less effective (see Sections 1.4.1.3 and 2.1.3.3 for details in fabrication).

A second approach to resolving the problem of poor con nement in SQW structures, and which has proven to be very effective, is to use MQWs separated by thin barrier layers, which have a lower bandgap than that of the cladding layers. Typical barrier widths are, for example, 5 nm for well thicknesses of 7 nm in a MQW GaAs/AlGaAs structure providing substantial improvement of con nement.

The extremely large gain of QWs together with the described con nement structures has led to the achievement of very low threshold current densities. For example, threshold current densities as low as 200 A/cm^2 have been recognized as typical values for AlGaAs/GaAs GRIN-SCH SQW lasers.

More details on gain, losses, con nement and threshold current in QW lasers can be found in Section 1.3 along with a schematic representation of the most important QW and con nement structures shown in Figure 1.24, including also some selected, qualitative density of states versus energy plots.

1.1.4.1   Advantages of quantum well heterostructures for diode lasers

The use of very thin active layers in QW lasers has a series of signi cant consequences leading to excellent operational features of these lasers. Some of the cw features of these lasers will be summarized brie y as follows.

Wavelength adjustment and tunability

The wavelength can be adjusted easily by changing the well thickness L_z. It can be evaluated, for example, for the dominating transition from the $n = 1$ quantized electron state to the $n = 1$ heavy-hole quantized state from $\lambda = 1.24/(E_{g,well} + E_{1e} + E_{1hh})$ by using Equations (1.19) and (1.21), where the units of the energy levels are electronvolts and those of the wavelength micrometers.

An example may demonstrate the effect: an InGaAs/InP SQW with different L_z values of 4, 8, and 12 nm will emit typically at wavelengths of 1.3, 1.48, and 1.55 μm, respectively (Asada *et al.*, 1984).

As we will show in Section 1.3, a QW laser can provide a wide and at gain spectrum due to the onset of the $n = 2$ quantized state in addition to the rst $n = 1$ state (Mittelstein *et al.*, 1989). Thus, lasing at shorter wavelengths of the $n = 2$ quantized state was observed by decreasing the laser cavity length to increase the threshold modal gain in a GaAs/AlGaAs SQW laser (Mittelstein *et al.*, 1986). Also, the wide and   at gain spectrum enables a large wavelength tuning range. With a grating-coupled external cavity con guration, tuning ranges of 105 nm in GaAs/AlGaAs SQW lasers (Mehuys *et al.*, 1989) and 170 nm in strained InGaAs/AlGaAs SQW (Eng *et al.*, 1990) lasers have been reported.

The wavelength range below 0.88 μm is well supported by lattice-matched DH and QW lasers, for example, in the AlGaAs and AlGaInP material systems. Lattice-matched InGaAsP lasers cover the range from about 1.1 to 1.6 μm and AlGaAsSb up to 2 μm. The range between 0.88 and 1.1 μm cannot be covered by any lattice-matched III–V compound semiconductor system. The gap in this important wave-length range can be lled, however, by using appropriate, strained, lattice-mismatched InGaAs/AlGaAs QW laser systems, which are capable of accommodating elastically the strain associated with the mismatch without forming detrimental mis t disloca-tions. Some details on this topic follow next.

Strained quantum well lasers

In a GaAs/AlGaAs QW the lattice constant a of the well layer GaAs is matched to better than 0.1% that of the AlGaAs barrier. If there is a lattice mismatch $\Delta a/a$ between the two materials of a few percent, however, such as in InGaAs/AlGaAs, then the introduction of mis t dislocations at the interfaces severely impacts the proper operation of the laser. The lattice constant, for example, of $In_{0.5}Ga_{0.5}As$, is larger than that of $Al_{0.2}Ga_{0.8}As$ by as much as 3.6%. It has been shown theoretically (Matthews and Blakeslee, 1974; People and Bean, 1985) and in many experiments (Anderson *et al.*, 1987; Hwang *et al.*, 1991) that usually no defects are generated as long as the well layer is thinner than the so-called *critical layer thickness* $L_{z,crit}$. The critical thickness for $In_{0.5}Ga_{0.5}As$ with 3.6% mismatch amounts to $\cong 5$ nm and for $In_{0.2}Ga_{0.8}As$ with 1.2% mismatch, it is $\cong 15$ nm (Anderson *et al.*, 1987).

Figure 1.10 shows schematically the crystal lattice deformation of a well layer with a larger lattice constant sandwiched between barrier layers with smaller lattice constants. To accommodate the lattice mismatch and to retain approximately the same unit cell volume, the two lattice constants must become equal in the well plane but the lattice constant of the well must also distort in the direction perpendicular to the well plane. This produces a biaxial compression in the well layer and a uniaxial tension along the orthogonal direction. The well layer then loses its cubic symmetry, which results in the removal of the degeneracy in the valence band edge and in changes in the energies and effective masses of both heavy- and light-hole valence bands relative to the conduction band edge (Adams, 1986; Yablonovitch and Kane, 1988) as illustrated in Figure 1.10. The strain can be increased, for example, by increasing the InAs mole fraction in an $In_xGa_{1-x}As/GaAs$ strained-layer system, and it changes at a rate of about 1% per $x = 0.15$ indium concentration in the range $0 < x < 0.5$ (Coleman, 1993).

Crucial for the laser action is that the heavy-hole mass is greatly reduced under compressive strain, with the consequence that the density of states in the valence band becomes comparable to that in the conduction band. Both the reduced carrier mass and the density of states in the valence band lead to a lower transparency carrier

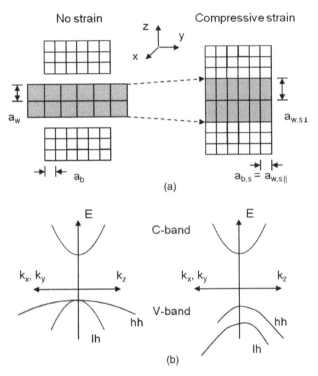

Figure 1.10   (a) Simpli ed illustration of the crystal lattice deformation resulting from sandwiching a thin quantum well layer with an original lattice constant a_w between thick barrier layers with a lattice constant $a_b < a_w$ to provide compressive strain in the well layer with $a_{w,s\parallel} = a_{b,s} < a_w < a_{w,s\perp}$. (b) Schematic energy versus wavenumber E–k band diagram for an unstrained direct semiconductor showing the approximately parabolic conduction band and degenerate heavy-hole (hh) and light-hole (lh) valence bands. The same material under biaxial compression includes (i) the split of the degeneracy of the valence band maximum leading to changes of the valence band masses, and (ii) the increases in the energies of both hh- and lh-valence bands relative to the conduction band.

density and higher differential gain compared to an unstrained QW. These values are most pronounced in the case when the effective masses in the conduction and valence bands are equal. Ultimately, the various effects resulting from the modi ed band structure will signi cantly raise the level of inversion in the valence band and approach that in the conduction band, which will lead to higher gain.

This effect made possible the development of the technologically important, strained-layer, pseudomorphic InGaAs/AlGaAs QW lasers emitting in the 900–1100 nm band. These highly developed lasers have demonstrated excellent performance and reliability, and are mainly applied for pumping optical ber ampli ers and ber lasers but are also used in various consumer applications. The 980 nm laser is the pre-eminent component in an erbium-doped ber ampli er, which is mainly deployed in long-distance optical communication systems. Chapter 2 discusses this laser in detail.

Strained QW lasers show many desirable properties (Adams and Cassidy, 1988; Vahala and Zah, 1988; Temkin *et al.*, 1991; Zah *et al.*, 1991; Tanbun-Ek *et al.*, 1991; Dutta, 1984) including:

- very low threshold current density due to a reduced transparency carrier density and increased electron–hole recombination time as a consequence of the reduced transparency density;

- higher differential material gain and therefore higher differential modal gain, which not only decreases the threshold current density, but also increases the laser ef ciency;

- lower laser linewidth;

- higher characteristic temperature T_0 due to better con ning potential barriers and the lower band lling;

- mode selectivity, that is, transverse electric (TE) (electric eld in the well plane) for compressive strain and transverse magnetic (TM) (electric eld perpendicular to the well plane) for tensile strain.

Finally, it has been shown that the threshold current and lasing wavelength in a GaAs/AlGaAs QW laser increase monotonically from compressive to tensile strain (Tiwari *et al.*, 1992).

In conclusion, to pro t from these useful features one could design, if technically possible, the unstrained active region of a laser such that it becomes a strained-layer QW in the new laser device without introducing any defects affecting proper operation.

Optical power supply

Many industrial applications of diode lasers such as pumping solid state lasers require extremely high optical power levels, which must be available with high ef ciency and reliability. These laser sources come in single-emitter and multi-emitter devices and are all based on QW structures. Reasons for the success of QW lasers as high-power generators are mainly threefold:

- First, the differential gain is higher, which leads to a lower transparency current (smaller density of states to be inverted) and hence lower threshold current.

- Second, the quantum ef ciency is very high because the transparency current (unused carrier injection) is lower in QW lasers.

- Third, the internal optical losses are much smaller in QW lasers because of the smaller optical con nement factor than in a bulk DH laser.

These effects lead to an improved *wall-plug efficiency*, which is the conversion ratio of the total electrical input power to optical output power (cf. Sections 1.3.7.1

and 2.4.5.3). The consequence is less heating of the device, which improves the laser performance by effectively increasing the optical output power and laser reliability by increasing the catastrophic optical damage level (see e.g. Chapters 3 and 4).

Temperature characteristics

From an application point of view, the temperature dependence of the laser characteristics is very important. The optical output power of a diode laser decreases at constant current operation with increasing temperature. Temperature characteristics determine the performance and reliability of a diode laser. The optical output power gradually decreases with temperature because of the increase in threshold current and decrease in laser ef ciency, which can be expressed empirically by separate, similar exponential dependencies on temperature (Hayashi *et al.*, 1971), see Section 1.3 for details. The characteristic temperature T_0 is a measure for the temperature dependence of the lasing characteristics. This characteristic temperature for the threshold current usually decreases as the ambient temperature increases and a large T_0 indicates a small change in lasing characteristics with temperature (the latter is highly desired for most applications).

In general, T_0 of QW lasers is higher than that of bulk active layer lasers, which is caused by the lower threshold current density and reduced carrier over ow from the wells into the adjacent layers (O'Gorman *et al.*, 1992). At higher threshold currents, higher lasing energy levels are populated due to the increased injected carrier density, which leads to an increase in the quasi-Fermi levels and, consequently, an increased electron escape rate.

More details on the physical mechanisms responsible for the temperature characteristics will be given in Section 1.3.7.3 along with T_0 data for different QW structures and material systems.

1.1.5 Common compounds for semiconductor lasers

The performance of a semiconductor laser depends sensitively on the intrinsic properties of the materials used in the design and fabrication. The most critical parameters to be considered in the selection of appropriate materials are both the bandgap energy, which determines the lasing wavelength, and the lattice constant, which is responsible for a defect-free interface between two semiconductors with different bandgaps. To reduce the formation of lattice defects, which can strongly impact the performance and reliability of a laser, the lattice constants of the different layers in the vertical structure should typically match to better than 0.1%. As we saw in the last section, the lattice constant requirement is slightly relaxed in a strained-layer system in that a small lattice mismatch of $\lesssim 3\%$ can be tolerated up to a certain critical layer thickness of typically <15 nm without the formation of any defects.

Considering the bandgap and lattice constant conditions to be met by the many layers involved in a structure – at least three for a DH device and, in more complex structures such as SCH QWs, at least ve layers – it seems that there is only a very limited set of semiconductors available that meet all the material speci cations required to make a high-quality diode laser according to the application requirements.

However, as Figure 1.11 shows, potential semiconductor materials for optoelectronic devices comprise binary, ternary, and quaternary III–V compound semiconductors. Mixing different kinds of binary compounds creates a new compound with properties intermediate between those of the original ones. The properties of the new compound vary linearly in proportion to the alloy composition according to Vegard's law (Vegard, 1921; the paper is discussed in many semiconductor textbooks). This procedure works well for lattice constants as long as the linear interpolation occurs

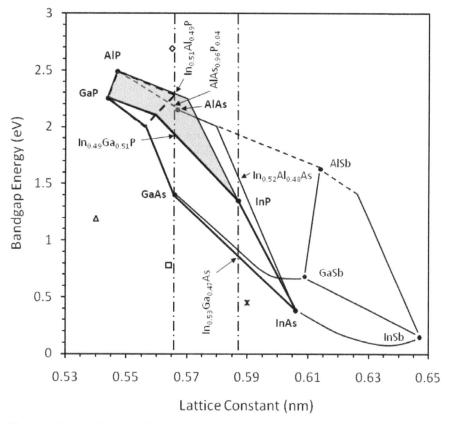

Figure 1.11 Bandgap energies and lattice constants of common III–V compound semiconductors used for high-power diode lasers. Compound binaries are represented as solid dots and ternaries as solid lines. Dashed lines represent regions of indirect gap. $In_{1-x}Ga_xAs_yP_{1-y}$ (clear region with thick solid lines) and $(Al_xGa_{1-x})_yIn_{1-y}P$ (light grey region) are obtained by varying compositions x and y. The two dot–dashed vertical lines show the range of bandgap energies of ternaries and quaternaries that can be grown lattice-matched on the binary substrates GaAs and InP. Some speci c ternaries are depicted. For comparison ZnSe ($\Diamond$), Si (Δ), Ge ($\square$), and PbS ($\times$) are also shown.

in the direct bandgap regime of the new compound. For the evaluation of other parameters, such as the bandgap, a second-order bowing parameter must be added to improve the value of the weighted average. Examples for a ternary are $Al_xGa_{1-x}As$ made up of the binaries GaAs and AlAs, and for a quaternary $In_{1-x}Ga_xAs_yP_{1-y}$ with GaAs, GaP, InAs, and InP as the constituents.

Figure 1.11 illustrates the correlation between bandgap energy and lattice constant for different binary, ternary, and quaternary compound semiconductors (Kressel and Butler, 1977; Casey and Panish, 1978). The emphasis here is on materials enabling laser emission in the wavelength range of approximately 0.63–1.55 μm where the highest optical output powers have been demonstrated. This range includes important wavelengths, such as at 1.31 and 1.55 μm used in optical ber communications, 0.98 and 1.48 μm for pumping ber ampli ers, 0.81 μm for pumping Nd:YAG due to the excellent match to its absorption spectrum, and wavelengths between 0.63 and 0.78 μm for optical data storage, laser printers, and machining. Most of these materials have a direct bandgap in $E–k$ space, which is a decisive condition for using them in the development of high-power diode lasers. Relevant features of these materials are discussed below. Regarding relevant laser power values, see e.g., Tables 1.2 to 1.5, Sections 1.2.5 and 2.1.3.3.

- $Al_xGa_{1-x}As$ can be grown on GaAs lattice-matched to better than 0.1% and it can be grown for any value of x without introducing defects due to lattice mismatch. However, only for $x < 0.45$ is it direct; for higher values it has an indirect gap. The x dependence of its energy gap can be approximated by $E_g(x < 0.45) = (1.424 + 1.247x)$ eV (Casey and Panish, 1978). The band offsets in the conduction and valence bands split to about 60% to 40% for AlGaAs/GaAs, respectively (Yariv, 1997). Lasing wavelengths are typically in the range of 700 to 900 nm for DH structures.

- The wavelength regime of 900–1100 nm can only be covered by the strained-layer system $In_xGa_{1-x}As$ sandwiched between AlGaAs or GaAs layers as discussed in the last section. The indium ion is larger than the gallium ion and therefore InGaAs has a higher lattice constant than GaAs with the consequence that InGaAs is under compressive strain. For layer thicknesses below a critical thickness the mismatch strain can be accommodated elastically without forming mis t dislocations, and the resulting biaxial compression modi es the valence band structure of InGaAs such that ef cient inversion now occurs also in the valence bands leading to increased gain. A typical band offset ratio is $\Delta E_c/\Delta E_v \approx 80\%/20\%$ (Arias et al., 2000). The lasing wavelength can be controlled by changing the thickness and indium mole fraction of the thin strained InGaAs QW layer (cf. Figure 1.9b). The same wavelength range of 900 to 1100 nm can also be obtained by using a strained InGaAs QW sandwiched between InGaAsP SCH and $In_{0.49}Ga_{0.51}P$ cladding layers. This system is aluminum-free and consequently has improved laser operation (Botez et al., 1996; Mawst et al., 1996) (see Section 2.1.3.3 for laser properties in this material system).

- $In_{1-x}Ga_xAs_yP_{1-y}$ grown on InP is the classical material system delivering wavelengths for long-distance ber optic communications in the range of 1.1 to 1.65 μm by changing the mole fractions x and y accordingly. Wavelengths at 1.3 and 1.55 μm are of particular interest, where the standard silica ber has minima in total dispersion and loss, respectively. A range of lattice-matched quaternaries extending from InP to the InGaAs ternary line can be grown by complying with the mole fraction condition $x = 0.4y + 0.067y^2$. Direct bandgap energies can be achieved ranging from 0.75 eV for the ternary end-point $In_{0.53}Ga_{0.47}As$ to 1.35 eV for InP. Compared to AlGaAs/GaAs, the band offsets in this system are quite different: only 40% of the band offset is in the conduction band, whereas the band offset in the valence band of 60% is much higher (Piprek $et\ al.$, 2000). In contrast, the InGaAlAs on InP material system delivers higher band offsets in the conduction band leading to lower electron leakage and improved laser characteristics (Zah $et\ al.$, 1992).

    There is an approach to extend GaAs-based lasers with all their positive characteristics, including high-temperature operation (high T_0), to 1.3 μm and to replace InGaAsP lasers with their thermal problems (low thermal conductivity, high electron escape due to low conduction band offset). This has been achieved by InGaAsN/GaAs QW structures with the incorporation of only a very low N content of <2% to avoid a strong degradation of luminescence ef ciency at higher concentrations (Kondow $et\ al.$, 1996, 1997). Excellent laser characteristics have been reported, including threshold current densities down to 500 A/cm^2 from narrow-stripe (∼4 μm) devices 350 μm long, lasing wavelengths between 1.24 and 1.3 μm, and reliable cw operation up to at least 100 °C with high characteristic temperatures T_0 of 110 K (Borchert $et\ al.$, 2000; Riechert $et\ al.$, 2000). The latter is caused by the high con nement energy for electrons in these structures, which is due to (i) the high conduction band offset of about 80%, and (ii) the increased electron mass compared to InGaAs (Hetterich $et\ al.$, 2000; Shan $et\ al.$, 1999). The higher electron mass, however, leads to transparency current densities roughly three times higher than for 980 nm InGaAs QW lasers.

- AlGaInP diode lasers are now well established to cover roughly the wavelength range of 600 to 700 nm in the visible red. $Ga_xIn_{1-x}P$ has a direct energy gap for x up to about 0.7 and an indirect gap at higher values of x. At $x = 0.51$ the material is lattice-matched to GaAs (Casey and Panish, 1978), which is therefore the preferred substrate. The substitution of Ga by Al produces hardly any change in the lattice constant because the ion sizes are approximately the same. Therefore, with the indium mole fraction xed at 0.49, Ga may be partially replaced by Al to provide various bandgap energies while maintaining a xed lattice constant matching GaAs. The resulting quaternary then has the composition $(Al_xGa_{1-x})_{0.51}In_{0.49}P$, which is a direct gap material for Al fractions x up to about 0.7. Strain can be produced in a thin GaInP layer by changing the In/Ga ratio, where compressive strain results from an increased In content and tensile strain from an increased Ga content. A higher In content leads to a reduced bandgap but increases both the optical and carrier con nement. A $Ga_yIn_{1-y}P$ QW layer sandwiched in $(Al_xGa_{1-x})_{0.51}In_{0.49}P$ separate con nement barrier layers is lattice-matched for $y = 0.51$ and compressively-strained for $y < 0.51$. Threshold current densities of about 200 and 700 A/cm^2 have been reported for compressively-strained SQW lasers emitting near 690 nm (Katsuyama et al., 1990) and 632 nm (Valster et al., 1992), respectively.

Appropriate materials to cover the wavelength regime $\lambda > 1.6$ µm are III–V, II–VI, and IV–VI (lead salts) semiconductor compounds and alloys (Eliseev, 1999). Typical representatives are InGaAsSb, HgCdTe, and PbSnTe, respectively. However, output powers of lasers made of these material systems are relatively low, and in the case of lead salts even low-temperature operation of the lasers emitting in the far-infrared region of 3–34 µm is required. Figure 1.11 shows the direct gap, binary compounds InSb and GaSb, which form together with InAs the alloy InGaAsSb successfully developed for commercial diode laser and photodiode applications.

Lattice-matched QW layers of $In_xGa_{1-x}As_ySb_{1-y}$ embedded in $Al_xGa_{1-x}As_ySb_{1-y}$ barriers have been grown on GaSb substrates and can cover the wavelength range of 1.7 to 4.4 µm. The lowest room temperature threshold current density was 260 A/cm^2 achieved for long (1000 µm), broad (100 µm), and low optical loss (10 cm^{-1}) lasers emitting a cw output power up to 190 mW/facet (Choi and Eglash, 1992). (AlGaIn)(AsSb) QW broad-area lasers and linear laser arrays (20 emitters on a 1 cm long bar) emitting in the range of 1.9 to 2.2 µm have achieved typical cw output powers of 2 and 20 W, respectively (Kelemen et al., 2008).

Material systems used for short wavelengths in the green, blue, and ultraviolet (UV) regime are II–VI compounds and III–V nitride systems, but they are not shown in the gure. Classical representatives in the former group are ZnS, ZnSe, and ZnSSe. However, heterojunction systems made from these materials suffer usually from high charge carrier escape, in particular electron loss, due to a much lower bandgap difference in the conduction band than in the valence band caused by the relatively high ionicity of these semiconductors. Doping with Cd can reduce the carrier loss by lowering the bandgap, which, however, increases the emitting wavelength. An important material, especially useful for the claddings, is MgZnSSe, which has a wide gap of about 3.5 eV and can be grown lattice-matched to GaAs or ZnSe. In general, the refractive indices and refractive index differences in these material systems are lower than in the standard III–V semiconductors, resulting in lower optical con nement.

The other group of materials comprises III–V nitrides with the basic binaries GaN and AlN, which are direct semiconductors of very large bandgap energies of 3.4 and 6.2 eV, respectively. In contrast to the AlGaAs system, the incorporation of Al to GaN causes a reduction in the lattice constant by about 2.4%. The addition of In to GaN shifts the wavelength from the UV to the visible and increases the lattice constant by about 11%. Because of the high ionicity of the nitrides, the refractive indices are even lower than in the previous II–VI materials. By adjusting the composition and doping, however, effective waveguiding can be achieved. For a long time effective technological progress was hampered by the mismatch in lattice constants and thermal conductivities between nitride layers and substrates. A recent breakthrough has been achieved by using a triple $In_{0.07}Ga_{0.93}N/In_{0.01}Ga_{0.99}N$ QW sandwiched in a GaN SCH and AlGaN claddings, which were deposited on an n-type GaN substrate. Ridge waveguide lasers 7 μm wide, 600 μm long, and with cleaved uncoated facets emitted with high ef ciency at room temperature record-high cw output powers up to 2 W at a wavelength of 405 nm and operating currents and voltages up to 1 A and 5.7 V, respectively (Saito *et al.*, 2008).

Table 1.1 summarizes the material systems described above, and includes common cladding/active layer structures grown on appropriate wafer substrates, typical lasing wavelength ranges for each material system, and wavelength-dependent applications grouped in four blocks.

1.2   Optical output power – diverse aspects

1.2.1   Approaches to high-power diode lasers

Edge- and surface-emitting lasers are the two fundamental concepts for implementing semiconductor diode lasers.

1.2.1.1   Edge-emitters

Edge-emitters have mirror facets that are perpendicular to the surface of the wafer substrate and the optical mode propagates parallel to the surface of the wafer (see Figure 1.1 and Section 1.1.1.3 above). These devices can be classi ed roughly into the following groups consisting of single-emitters and multi-emitters including:

- Narrow-stripe devices of typically 3–5 μm widths delivering kink-free, high output powers in the 1 W range in a single transverse vertical and lateral mode reliably up to high drive currents (see Sections 2.1 to 2.3).

- Broad-area lasers with typically 100 μm widths supplying output powers up to 25 W in a single spatial mode under certain design conditions (see Section 2.4).

- Tapered ampli er lasers including tapered unstable resonator lasers and monolithically integrated master oscillator power ampli ers (MOPAs), which comprise a single-mode laser and a ared-contact (with a typical width from 4 to 250 μm) power ampli er, are capable of delivering the highest diffraction-limited, single-mode output in the 12 W cw range (see Section 2.4).

- Laser array bars, which come in two forms. First, phase-locked, anti-guided arrays deliver coherent powers in narrow, diffraction-limited beams (cf. Section 1.2.2) in the 1 W cw range from, for example, 20 closely spaced, single emitters 120 μm wide in the bar. Second, spatially incoherent arrays with a laser bar 1 cm long as the basic building block achieve today the highest optical power

Table 1.1 Key diode laser material systems including typical cladding/active layer con gurations, substrates, lasing wavelength ranges and major applications in each of the four groups.

Cladding/active layer	Substrate	Wavelength [μm]	Major applications
GaN/In$_x$Ga$_{1-x}$N [1] typ. $x \cong 0.02$–0.30	Sapphire-GaN buffered	~ 0.35–0.50	Holographic storage. Image recorders, fax machines, printers.
Cd$_x$Zn$_{1-x}$S/CdS [2]	CdS	~ 0.35–0.50	Pumping solid state, e.g., Nd:YAG at 810 nm and fiber lasers.
(Al$_x$Ga$_{1-x}$)$_{0.5}$In$_{0.5}$P/Ga$_y$In$_{1-y}$P $x = 0.7$: T_0 high, J_{thr} low $y = 0.5$ latt. match. $y < 0.5$ comp. strain	GaAs	~ 0.60–0.70	Barcode readers. Blu-ray discs, HD DVD, opt. ROM (780 nm). Materials processing. Printing, graphics arts.
Al$_x$Ga$_{1-x}$As / GaAs $x < 0.45$ (direct gap)	GaAs	~ 0.70–0.92	Medical therapeutics. Aerospace, military systems, rangefinders
Al$_x$Ga$_{1-x}$As/In$_y$Ga$_{1-y}$As comp. strain Practical ranges: $y \sim 0.08$–0.42, $x \sim$ 0.20–0.85, $L_z \sim 7$–15 nm. J_{thr} minimum for $y \sim 0.2$, $x \sim 0.35$, $L_z \sim 8$ nm	GaAs	~ 0.90–1.10	Telecommunications. Fiber amplifier pumps. Materials processing. Medical therapeutics
InP/Ga$_x$In$_{1-x}$As$_y$P$_{1-y}$ $x \sim 0.47y$: l. m. to InP. Endpoint In$_{0.53}$Ga$_{0.47}$As. $\lambda = 1.3$ (1.55) μm: $x \sim$ 0.28 (0.37), $y \sim 0.6$ (0.8)	InP	~ 1.1–1.65	Fiber optics communications [silica fiber low dispersion (loss) at 1.3 (1.55) μm], transceivers, Raman amplifiers.
Al$_x$Ga$_{1-x}$As/In$_y$Ga$_{1-y}$As$_{1-x}$N$_x$ [3] $\lambda = 1.3$ μm for active: $x = 0.017$, $y = 0.35$; cladding: $x = 0.3$; $L_z \sim 7$ nm	GaAs	~ 1.24–1.30	Aerospace, military, rangefinders
Al$_x$Ga$_{1-x}$Sb$_{1-y}$As$_y$/ In$_x$Ga$_{1-x}$Sb$_{1-y}$As$_y$ [4] For $\lambda = 2.2$ μm (lattice-matched layers): Active: $x = 0.16$, $y = 0.14$ Cladding: $x = 0.75$, $y = 0.06$	GaSb or InAs	~ 1.70–4.40	Spectroscopic sensing of humidity, gas impurities, drugs. Next gen. of fiber-optics comms. based on novel fluoride, sulfide glasses with very low losses $\sim 10^{-3}$ dB/km at 2.4 μm range

[1], [2] Typ. blue/UV diode lasers. Record-high powers: 2 W of 7 μm wide lasers [1]; Saito *et al.*, 2008. [3] Potential alternative to InGaAsP: higher material gain and conduction band offset, higher T_0 (126 K) (Kondow *et al.*, 1996, 1997; Borchert *et al.*, 2000; Riechert *et al.*, 2000). [4] $\lambda = 1.9$–2.2 μm: 100 μm MQW laser with 190 mW cw/facet at 300 K (Choi and Eglash, 1992); 2 W cw for BA lasers, 20 W cw for 1 cm laser bars (Kelemen *et al.*, 2008).

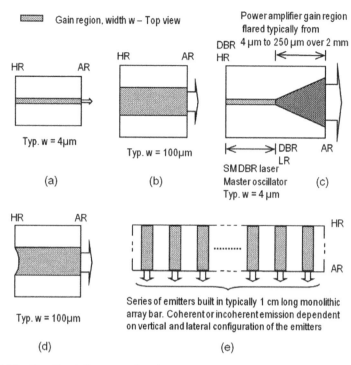

Figure 1.12   Simpli ed diagrams of basic edge-emitting diode laser structures (top views; effective active gain regions shown in hatched patterns; mirrors: high-re ectivity (HR), anti-re ectivity (AR), low-re ectivity (LR) distributed Bragg re ector (DBR)): (a) narrow-stripe device; (b) broad-area device with wide aperture; (c) tapered device utilizing a single-mode (SM) laser and a ared-contact power ampli er; (d) unstable resonator laser approach with curved mirror facet; (e) monolithic diode laser array bar structure.

up to 1 kW cw from a single monolithic laser chip having a lling factor of up to 80% (ratio between pumped and unpumped area of bar) and an electrical-to-optical power conversion ef ciency as high as 70% (see Section 2.4 for details).

Figure 1.12 shows simpli ed, schematic illustrations of these four laser types in top view. Speci c design issues and performance gures of these high-power, single-mode, diffraction-limited laser approaches and other conceptual techniques such as unstable resonator designs (see Figure 1.12) with large emitting curved apertures, and tapered lasers, will be discussed in detail in Section 2.4.

1.2.1.2   Surface-emitters

In surface-emitting devices, the laser output is normal to the surface of the wafer. These devices come in two categories. In one category there are lasers, which have their optical cavity normal to the wafer. These devices are known as vertical-cavity surface-emitting lasers (VCSELs). The other class comprises lasers where the active layer has the conventional waveguide structure but the laser beam is de ected normal to the substrate surface using (i) a 45° etched output mirror, (ii) a folded cavity realized with a 45° de ecting intracavity mirror or bent-waveguide structure, or (iii) a diffraction grating, which is etched on the top p-type cladding layer and couples the light out vertically to the surface (Iga and Koyama, 1999).

VCSEL devices have very short cavity lengths in the micrometer range, which would imply extremely high mirror losses. To counteract this negative effect and to decrease the threshold gain, which is inversely proportional to the cavity length, the re ectivity of the mirrors must be close to 100%. This can be realized by using distributed Bragg re ectors (DBRs) composed of a semiconductor multilayer structure consisting of a series of alternating, high and low refractive index layers a quarter of a wavelength thick (see also Section 1.3.8). Re ectivities of more than 99% can be achieved, for example, with 20 pairs of an epitaxially grown GaAs/AlAs DBR. The very short cavity leads to very large mode spacing and therefore only a single longitudinal mode is excited in the spectral gain regime. A cross-sectional view of the concept of a VCSEL is illustrated in Figure 1.13. VCSELs have become a maturing technology that has been successfully commercialized by many companies for applications such as parallel processing of information or parallel optical interconnection between computers. They have attractive features including a low-divergence circular output beam, temperature-insensitive output characteristics, high-speed modulation capability at low driving currents, submilliamp threshold currents, wafer-scale fabrication, and on-wafer testing. These high-performance devices have the desired circular output beam but their single-mode output power is limited to ~10 mW cw for device diameters <10 μm. For larger device diameters >100 μm

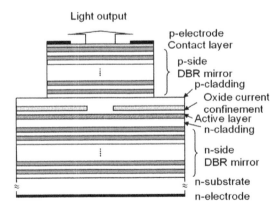

Figure 1.13 Schematic illustration of the vertical-cavity surface-emitting laser (VCSEL) approach with an oxide current con nement layer. The laser cavity is perpendicular to the substrate plane and the mirrors of the resonator consist of distributed Bragg re ectors (DBR) with high re ectivities close to 100%. The length of the laser cavity is short, typically in the range 1–3 μm.

the power can be around 200 mW cw; however, the output beam then has multiple transverse modes.

Vertical external cavity surface-emitting lasers (VECSELs) make use of the VCSEL design and offer a technologically important concept to increase output power and beam quality directly in a vertical resonator device. These devices use a three-mirror, coupled cavity design comprising the active region sandwiched between the p-doped, high re ectivity $R > 99\%$ Bragg re ector at the bottom, the n-doped Bragg re ector with a partial transparency at the top, and the curved external mirror. Electric current is injected through a circular p-aperture at the bottom, while the n-contact has an anti-re ection-coated circular aperture to allow the beam through to the external mirror. The p-mirror is soldered to the heat sink for ef cient heat removal. The shape and size of the external mirror control the power and transverse mode operation, whether the laser operates in multi-mode or in a single fundamental mode. Electrically pumped VECSELs emitting at 980 nm have generated 1 W cw multi-mode and 0.5 W cw in a fundamental TEM_{00} mode and single frequency with 90% coupling ef ciency into a single-mode ber (McInerney *et al.*, 2003).

Surface-emitting structures have also been used to fabricate high-power laser arrays, which can be designed such that neighboring emitters in the array oscillate in phase, that is, the phase of the optical eld of an emitter is synchronized with those of the other emitters in the array. This leads to a sharp emission including a single-lobe far- eld pattern (Chang-Hasnain, 1994).

1.2.2   High optical power considerations

There is no exact de nition of the power value necessary to call a diode laser a laser with high power, but it is generally accepted that levels above 100 mW for narrow-stripe, single-mode devices and above 1 W for all other single- and multi-emitter lasers can be considered to be high power. Lasers with low powers in the range below 10 mW serve many applications, including 780 nm GaAs lasers used in compact disc (CD) players, 670 nm AlGaInP lasers in barcode scanners, and 1.3 and 1.5 μm InGaAsP single-mode lasers in optical communication systems. In contrast, high-power lasers have different applications depending on the operation of the laser. For example, to pump solid state lasers the focus is only on the highest possible power, which can be achieved, for example, by a laser array.

High powers as well as near diffraction-limited, coherent laser beams are crucial for applications such as optical recording, printing, frequency doubling, free-space communication, laser tweezers, and ber ampli er pumping with the requirements of transmitting the light over long distances or focusing the beam to a very small spot of high-power density. Excellent examples of such application-tailored lasers are the kink-free, single-mode, ber-coupled power levels of 750 mW achieved from narrow, ridge-waveguide, strained-layer InGaAs/AlGaAs SQW lasers emitting in the 980 nm band for pumping erbium-doped ber ampli ers used in long-haul ber optic communication links (Bookham, Inc., 2009). In this context, it is important to note the signi cance of the brightness of a laser.

1.2.2.1   Laser brightness

The maximum peak intensity, which can be obtained by focusing a laser beam, is proportional to the beam brightness, and the brighter the laser, the further the distance that energy can be propagated. The brightness of a given laser source cannot be changed and is independent of the optical system that follows the source. Brightness B is given by the power P emitted per unit solid angle per unit area and is measured in units of W sr^{-1} cm^{-2}. B is de ned as $B = P/(d\theta_\perp D\theta_\parallel)$, where d is the maximum extent of the near- eld in the mirror plane normal to the active layer, D is the maximum extent parallel to the active layer, and $\theta_\perp$ and $\theta_\parallel$ are the full-width, half-maximum (FWHM) points of the far- eld divergence angles measured in radians in the direction normal (fast-axis) and parallel (slow-axis) to the active layer.

For a diffraction-limited laser output ($\theta_\perp = \lambda/d$, $\theta_\parallel = \lambda/D$) we obtain the brightness $B = P/\lambda^2$. The brightness of a laser is orders of magnitude higher than that of a spatially incoherent light source of similar power. The brightness of a 1 W, 980 nm laser can be calculated to be about 1×10^9 W sr^{-1} cm^{-2}, and hence is seven orders of magnitude larger than that of a high-pressure mercury-vapor lamp emitting about 10 W at 546 nm. The same ratio is also obtained for the two peak intensities by focusing the beams of the two sources. These gures demonstrate the importance of using focused, high-brightness laser beams in industrial material processing applications such as welding, cutting, marking, drilling, and so forth.

1.2.2.2   Laser beam quality factor M^2

The beam quality factor M^2, also called the beam propagation factor, characterizes the degree of imperfection and focusing ability of a laser beam. It compares the characteristics of a real beam to those of a pure fundamental TEM_{00} mode, that is, $M^2 = 1$ for a diffraction-limited Gaussian beam. The closer M^2 is to 1.0, the better the beam can be focused. For any other beam $M^2 > 1$. This factor relates the divergence of a real laser beam in the far- eld to its near- eld waist size (Siegmann, 1990, 1993).

According to the ISO 11146 standard (ISO, 2005), the relationship between M^2, wavelength λ, near- eld beam waist (minimum spot size) w_0, and beam divergence half-angle θ is given by $M^2 = \theta w_0/(\lambda/\pi)$. As $\lambda/\pi w_0$ is the diffraction-limited angular divergence of a Gaussian beam, therefore M^2 is a single number. The quality factor M^2 of a real laser beam describes the deviation from a theoretical diffraction-limited Gaussian beam with $M^2 = 1$ and is always ≥ 1.

M^2 is measured by focusing the laser beam and measuring several beam pro les along the beam caustic. The divergence angle is determined by $\theta = d_f/f$ where d_f is the beam diameter at the focal distance f of the lens used. Thus, to evaluate M^2 it is necessary to measure both the near- eld and far- eld beam waist. Commercial instruments are available for measuring M^2 in real time. The M^2 formalism offers a convenient way to de ne in one single number the quality of a laser beam, which can have, for example, distorted, multimode, or partially incoherent characteristics. The M^2 factor can be very different for elliptical laser beams such as in broad-area lasers or diode laser bars (Chapter 2). In these lasers, M^2 is much larger in the transverse lateral (slow-axis) than in the transverse vertical (fast-axis) direction.

1.2.3   Power limitations

As we will see in one of the following sections, the output power of a diode laser is measured by the optical power output (P) as a function of the drive current (I) input characteristic P/I. In an ideal case, the P/I curve is a straight line with a slope corresponding to an ideal ef ciency of one emitted photon per injected electron. However, in reality the characteristics look different and the useful power levels can be impacted by roughly four factors including (i) the appearance of *kinks* in the P/I, (ii) power rollover with increasing current, (iii) catastrophic optical damage, and (iv) aging effects. Figure 1.14 shows a schematic representation of these effects.

These performance-degrading effects will be described in detail in subsequent sections and chapters. In the following, however, a short summary should illustrate the physical processes leading to these effects.

1.2.3.1   Kinks

In general, multimode operation of the laser is the cause of the occurrence of a discontinuity (*kink*) in the P/I curve, which is the most common mechanism limiting the effective power for single spatial mode radiation. The excitation of higher-order spatial modes causes a distortion of the far- eld radiation pattern and a deviation from the ideal Gaussian-like shape, which is re ected at the onset of modes higher than the fundamental mode. The physical origin for the excitation of these additional modes can be seen in the perturbation of the designed waveguide refractive index or gain pro le, which is caused by nonuniformities in the local carrier population and/or

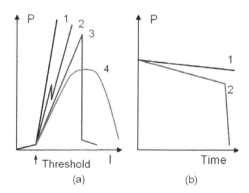

Figure 1.14   Limitations of useful diode laser optical output power illustrated schematically (a) in power (P) versus current (I) characteristics: (1) ideal output, (2) kink, (3) catastrophic optical damage (COD) at mirrors or in cavity, (4) thermal rollover; and (b) during aging power versus time: (1) gradual degradation, (2) sudden (COD) power degradation.

temperature. *P/I* curves with a *kink* are usually reversible. Details of the kink issue will be discussed in Section 1.3 and Chapter 2.

1.2.3.2 Rollover

The characteristic of this effect is the continuous decrease in laser ef ciency (power increase per current unit) with increasing current. This is due to Ohmic losses, which increase at higher injection currents and lead to higher heat powers dissipated in the proximity of the p–n junction. The resulting increase of the active layer temperature leads to a reduction of the internal quantum ef ciency (see Section 1.3) with the effect that the power level ultimately saturates, and eventually decreases (*rolls over*) with increasing current. The actual reason for this effect is carrier escape from the active region into the adjacent cladding layers where the carriers cannot contribute to population inversion and gain. Ineffective carrier con nement and insuf ciently high band offsets are responsible for this carrier loss. The optical output can nally diminish to close to zero for very high temperatures, leading to a collapse of the laser gain. Thermal rollover is usually a reversible event and it is the main power-limiting mechanism in: (i) 1.3–1.5 μm InGaAsP/InP lasers, which have poor temperature characteristics due to low band offsets and high nonradiative Auger recombination processes; and (ii) diode lasers with high thermal resistance. Section 1.3 presents more details on the thermal rollover effect.

1.2.3.3 Catastrophic optical damage

Catastrophic optical damage (COD) is an irreversible process, which can occur at laser mirror facets and in the bulk of the cavity of the laser by strong heating due to high optical power densities and/or nonradiative carrier recombinations. COD at mirrors, also called COMD, is the result of strong surface recombination via traps, which causes a depletion of charge carriers at the crystal surface. The depleted bands of the active region now become absorbing at the lasing wavelength. The heat generated in this process raises the local temperature very strongly. At a critically high optical ux density, the raised temperature causes a suf cient shrinkage of the local bandgap energy such that the optical absorption becomes even higher. This positive feedback can cause a thermal runaway with ultimately melting of the end facet of the laser, irreversibly diminishing any useful laser output power. COD is especially prevalent in aluminum-containing semiconductors and as a result strongly impacts the reliability of AlGaAs lasers. However, the application of appropriate facet passivations and coatings (see Chapters 3 and 4) greatly increased the COMD level by at least one order of magnitude compared to untreated mirrors and which is currently higher than 100 MW/cm^2 for the leading, narrow-stripe 980 nm pump laser on the market (Lichtenstein *et al.*, 2004; Bookham, Inc., 2009).

COD events in the cavity are due to the generation and growth of so-called dark-line defects (DLDs), which are regions with greatly reduced radiative ef ciency. The main origins of a DLD are threading dislocations originating from defects in the wafer substrate, and stacking faults introduced during the epitaxial growth of the active region layers. These dislocations can grow into networks by nonradiative recombination processes with a growth rate promoted by the presence of mechanical strain and thermal gradients. The formation of DLDs is also sensitively dependent on the material system, and is much more pronounced in AlGaAs/GaAs than in InGaAsP/InP lasers. A DLD can act as a sink for injected carriers and as an absorber of laser light in the cavity. These effects can cause a rapid degradation of laser performance with nal catastrophic failure in the active region due to strong Joule heating. COD-related phenomena and effects will be discussed in detail in Chapters 3–9.

1.2.3.4 Aging

The operation of a diode laser at constant drive current and temperature over a long period of time usually leads to a reduction in optical output power. The strength and type of the transition to lower powers depend sensitively on the laser structure, materials, fabrication technology, and operating conditions as well as the type of root causes involved. In brief, thermal dissipation, high current density, and rapid, catastrophic-like optical damage events are the major reliability-limiting factors. As we saw in the last paragraph, thermal and carrier-induced degradation mechanisms are linked to the formation of extended lattice defects, which can migrate and grow in the active region with the effect of reducing the optical power over time. Chapters 3 and 4 describe the fundamental diode laser degradation mechanisms. Chapters 5 and 6 discuss basic reliability engineering concepts and techniques, and describe a diode laser reliability test program required to achieve quali cation of a laser product.

1.2.4 High power versus reliability tradeoffs

Depending on the speci c laser application, there are certain tradeoffs to be considered to achieve laser operation at *high optical power* and *high reliability* (Mehuys, 1999). There are three different cases to be taken into account:

- Case 1: Reliability is limited by the optical power density at the output facet. Then a reduction in the optical overlap of the active region can be applied to increase the COMD level.

- Case 2: Reliability is limited by heat dissipation. Then an increase in the optical overlap of the active region may help to lower the threshold current density and increase the electrical-to-optical power conversion ef ciency.

- Case 3: Reliability is limited by the current density in the semiconductor. One solution would be to increase the effective active region area, which lowers the current density at a xed total drive current. A second solution could be to lower the drive current by adjusting the optical overlap of the active region. In a third solution the external differential quantum ef ciency could be increased by adjusting the mirror re ectivities (see Section 1.3.8).

1.2.5 Typical and record-high cw optical output powers

1.2.5.1 Narrow-stripe, single spatial mode lasers

Table 1.2 lists state-of-the-art, cw, optical output powers of narrow-stripe, single spatial mode diode lasers as a function of the lasing wavelength. The values given are for selected commercial laser products and research devices from the UV to the mid-IR. Figure 1.15 shows the power data of the table plotted as a function of the lasing wavelength. Two results stand out and are worth looking at closer regarding the design and technology used.

First, the Toshiba Research Group (Saito *et al.*, 2008) succeeded in fabricating GaN-based, ridge waveguide lasers 7 μm wide with the highest power characteristics at a wavelength of 405 nm and at room temperature. The devices emitted typical cw powers of 2 W at 1 A operating current and 5.7 V operating voltage with a high slope ef ciency of 2.6 W/A. This device is not included in Figure 1.15 because no direct or indirect information is available whether the laser emitted in single spatial mode or not; however, we consider the technology as suf ciently advanced to be reported here. The triple 3 nm $In_{0.07}Ga_{0.93}N$/10 nm $In_{0.01}Ga_{0.99}N$ QW/barrier stack is sandwiched in a 100 nm GaN SCH layer with adjacent AlGaN cladding layers grown practically lattice-matched on an n-type GaN substrate.

Table 1.2 State-of-the-art, continuous wave (cw), optical output powers of narrow-stripe, single spatial mode diode lasers as function of the lasing wavelength from the ultraviolet to mid-infrared regime. Selected products and research devices (status 2009).

Wavelength [nm]	Cw power [mW]	Company/institution
375	20	Nichia (2009)
405	470	Ryu *et al.* (2006)
405	2000 (7 μm ridge)	Saito *et al.* (2008)
450	80	OSRAM (2009)
488	60	Nichia (2009)
515	5	Nichia (2009)
640	250	opnext (2009)
660	130	Mitsubishi Electric (2009)
785	150	opnext (2009)
808	200	SANYO (2009)
850	200	Frankfurt Laser (2009)
915	300	opnext (2009)
980	750 (ex-fiber)	Bookham (2009)
1060	852	Zah *et al.* (2004)
1300	600 (ex-fiber)	Garbuzov *et al.* (2002)
1400–1480	710 (ex-fiber)	Garbuzov *et al.* (2001)
1510	320 (ex-fiber)	JDSU (2009)
2300	30	Frankfurt Laser (2009)
2500	20	Frankfurt Laser (2009)

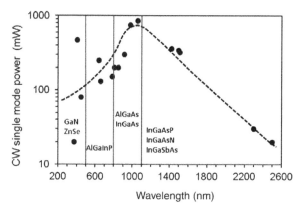

Figure 1.15   State-of-the-art, continuous wave, optical output powers of narrow-stripe, single transverse mode diode lasers as a function of the lasing wavelength. Data are from selected commercial laser products and research devices. Spectral regimes of relevant material systems are indicated. Dashed line is guide to the eye.

Two of the key issues, which hampered progress in GaN laser research for a long time were growth on a poorly lattice-matched substrate (e.g., sapphire) and the impossibility of realizing p-type conductivity of GaN and AlGaN. By growing on GaN the lattice mismatch problem has been resolved and it has been shown (Nakamura, 1991) that Mg-doped GaN grown by MOCVD after a suitable post-growth treatment by annealing turned out to be p-type. Annealing activates the Mg acceptors and yields a low-resistivity p-type material. AlGaN can be made n- and p-type with Al content no higher than 30%.

The second striking result is the power of 750 mW ex- ber at a wavelength of 980 nm (Bookham, Inc., 2009). The technology for this laser device was actually developed by IBM Laser Enterprise, Zurich, in the early 1990s and comprises a compressively-strained InGaAs/AlGaAs GRIN-SCH SQW structure grown lattice-matched on an n-GaAs substrate by MBE. A carefully designed and wet-etched, self-aligned, narrow ($\sim$3–4 µm) ridge waveguide structure enables the laser to operate in the fundamental spatial mode up to high currents by yielding linear, kink-free light output powers up to 1.4 W ex-facet (Lichtenstein *et al.*, 2004). The laser has a very high rollover power of 1.75 W at 25 °C heat sink temperature, and, taking into account a single-mode near- eld pattern of size $\sim$3 µm $\times$ 0.6 µm FWHM, the maximum power density is well above 100 MW/cm^2 without causing any COMD failures. The proprietary facet passivation technology includes facet cleaving in ultrahigh vacuum followed by depositing in situ a thin silicon passivation layer with a thickness $\lesssim$10 nm depending on the technique used for depositing the re ectivity modi cation coatings outside the high-vacuum chamber. This technique has proven to provide maximum protection from laser chip failures due to any gradual and sudden facet degradation mechanisms. The excellent and unique performance and reliability data of this laser comply fully with the stringent highest requirements established for pumping erbium-doped ber ampli ers (EDFAs) used in terrestrial and submerged optical communication systems.

We will learn more about this laser type when we discuss in Chapters 2 and 4 design considerations for high-power, single-mode emission and for optical strength enhancements at laser mirrors, respectively.

1.2.5.2 Standard 100 μm wide aperture single emitters

Table 1.3 lists the best wavelength-dependent optical power values found for 100 μm wide broad-area (BA) diode lasers in the range 405 to 2200 nm. These devices usually lase in the fundamental transverse vertical mode (fast axis), but not in the fundamental transverse lateral mode (slow axis). However, there are techniques available to improve the operation of BA lasers in the fundamental lateral mode, which will be discussed in Section 2.4. Nevertheless, the data in the table are meant for information and comparison.

There are two values which catch the eye, the 25.3 W cw at 940 nm (Petrescu-Prahova *et al.*, 2008) and 9 W cw at 1240 nm (Bisping *et al.*, 2008).

The former data are for InGaAs/GaAsP QW devices with optimized cavity lengths L and d/Γ_{tv} ratios aimed at maximizing the thermal rollover power, where d is the total active region width and Γ_{tv} the transverse vertical optical con nement factor (see Section 1.3 for basic relations). Crucial to the design is that L directly scales

Table 1.3 State-of-the-art, continuous wave (cw), optical output powers of broad-area commercial and research diode lasers with 100 μm wide apertures as function of the lasing wavelength from the blue to mid-infrared regime. Selected products and research devices (status 2009).

Wavelength [nm]	Cw power [W]	Company/institution
405	0.25	Nichia (2009)
450	0.50	Nichia (2009)
630	0.40	LDX Optronics (2009)
690	1.00	nLight (2009)
750	1.50	LDX Optronics (2009)
808	5.00	Li *et al.* (2008); Lumics (2009)
915	20.00	Bookham (2009)
940	25.30	Petrescu-Prahova *et al.* (2008)
960	20.00	Bookham (2009)
975	20.00	Bookham (2009)
1060	16.00	Tarasov *et al.* (2004)
1210	2.00	LDX Optronics (2009)
1240	9.00 (200 μm wide)	Bisping *et al.* (2008)
1300	8.00	Livshits *et al.* (2000)
1500	5.00	Garbuzov *et al.* (1996)
1850	1.00	LDX Optronics (2009)
1900–2200	2.00	Kelemen *et al.* (2008)

proportional to d/Γ_{tv}, that is, higher d/Γ_{tv} ratios can be obtained by lowering Γ_{tv} and keeping d constant at 6–7.5 nm (cf. also "Broad waveguides" in Section 2.1.3.5). This has two consequences. First, a low Γ_{tv} reduces the overlap at the output mirror between the high optical power generated in longer devices and the active QW, which is vulnerable to degradation. Second, the threshold current density of about 130 A/cm^2 is the same for all devices in the optimization process (see Section 1.3). The thermal rollover power increases with d/Γ_{tv} in the range of 0.78 to 1.17 μm and L in the range of 3.5 to 5 mm with a record maximum cw power of 25.3 W achieved for $d/\Gamma_{tv} = 1.17$ μm and $L = 5$ mm. The high d/Γ_{tv} values are obtained by using an asymmetric structure, which includes an optical trap on the n-side (cf. "Optical traps" in Section 2.1.3.5 and Chapter 4 for optical strength enhancement techniques). The optical trap pulls most of the guided optical eld to the n-side and reduces Γ_{tv} to typical values between 0.5 and 0.7%. The devices are mounted on water-cooled Cu carriers.

The 9 W at 1240 nm in the table is for laser structures based on a $Ga_{0.68}In_{0.32}N_{0.007}As_{0.993}$ QW 6.5 nm thick embedded in $GaAs_{0.99}N_{0.01}$ strain compensating layers 5 nm thick. This active region is centered in a GaAs SCH layer 1100 nm thick with adjacent $Al_{0.4}Ga_{0.6}As$ cladding layers 1200 nm thick. The laser devices have internal optical losses as low as 0.5 cm^{-1}, high internal quantum ef ciencies of 80%, and excellent temperature dependence of the P/I characteristic (see Section 1.3 for basic relations). These favorable data, which result from an optimized vertical structure and the high quality of grown materials, have led nally to the highest room temperature, cw output power ever reported for a high-power laser in the technologically important wavelength range around 1240 nm. The power of 9 W has been achieved for a device 200 μm wide and 2.5 mm long. To make it comparable to standard 100 μm wide devices, the power has to be scaled and a best-guess value may be around 5 W in the absence of any further information. This power, however, has to be compared to the 8 W cw obtained from an AR/HR-coated InGaAsN SQW laser 100 μm wide emitting at a wavelength of 1.3 μm (Livshits *et al.*, 2000).

1.2.5.3   Tapered amplifier lasers

A selection of the highest optical power levels for tapered ampli er lasers can be found in Table 1.4. The two high powers of 10 W (Ostendorf *et al.*, 2008) and 12 W (Paschke *et al.*, 2008) are for similar devices comprising InGaAs/AlGaAs QW vertical structures. Typical lengths for the ridge waveguide oscillator section are 2 mm and for the tapered ampli er section 4 mm. In the latter example with the higher power, a 1 mm long, sixth-order surface grating de ned by projection lithography and fabricated by reactive ion etching was implemented in the ridge waveguide. This device shows longitudinal single-mode emission at 980 nm over the full operating range up to record-high 12 W cw and nearly diffraction-limited optical power at 15 A drive current with a high average slope ef ciency of 0.85 W/A and conversion ef ciency of 44% (cf. Section 1.3.7.1 for de nitions).

Table 1.4 State-of-the-art, continuous wave (cw), optical output powers of selected tapered ampli er single-emitter lasers at various lasing wavelengths. Selected products and research devices (status 2009).

Wavelength [nm]	Cw power [W]	Company/institution
757	0.5	m2k/DILAS (2009)
770	1.0	m2k/DILAS (2009)
787	1.5	m2k/DILAS (2009)
855	0.5	m2k/DILAS (2009)
975	8.3	Michel *et al.* (2008)
976	10.0	Ostendorf *et al.* (2008)
980	12.0	Paschke *et al.* (2008)
982	2.0	m2k/DILAS (2009)
1037	1.0	m2k/DILAS (2009)
1064	4.0	Ostendorf *et al.* (2009)
1240	1.0	Bisping *et al.* (2008)
1550	1.0	QPC (2009)

1.2.5.4 Standard 1 cm diode laser bar arrays

Table 1.5 shows the highest cw power levels detected for incoherent monolithic diode laser bar arrays. Optimized material quality and innovative design of the vertical laser structure, mounting, and cooling techniques led to rst output powers beyond the 500 W barrier with 509 W cw at 540 A from 1 cm InGaAs/AlGaAs bars with a 50% lling factor and 3 mm cavity length. The high power is due to a low-loss waveguide structure with a low fast-axis, far- eld angle of 27°, high power conversion ef ciency

Table 1.5 State-of-the-art, continuous wave (cw) power levels detected for incoherent monolithic 1 cm diode laser bar arrays versus lasing wavelength from the red to mid-infrared regime. Selected products and research devices (status 2009).

Wavelength [nm]	Cw power [W]	Company/institution
638	8	Mitsubishi Electric (2009)
790	100	nLight (2009)
808	100	nLight (2009)
825	100	nLight (2009)
915	200	Bookham (2009)
940	200	Bookham (2009)
940	509	Jenoptik Diode Lab (2006); Sebastian *et al.* (2007)
940	950	Li *et al.* (2007)
980	200	Bookham (2009)
1030	110	Bookham (2009)
1870	10	m2k/DILAS (2009)
1900–2200	20	Kelemen *et al.* (2008)

>68%, optimized facet coatings, and ef cient active cooling (Jenoptik Diode Lab, 2006; Sebastian *et al.*, 2007). Signi cantly improved 980 nm laser bar structures with low internal loss $\alpha_i < 0.6$ cm^{-1}, high internal quantum ef ciency $\eta_i > 98\%$, high $T_0 \cong 200$ K, and $T_\eta \cong 700$ K (see Section 1.3 for de nitions of these various parameters) enable rst record-high cw output power of 950 W at 1120 A (supply limit) from a microchannel-cooled 1 cm bar (Li *et al.*, 2007). The bar consists of 65 emitters, each 115 μm wide and 5 mm long, has a 77% lling factor, and performs with a conversion ef ciency η_c of 70 and 67% for 4 and 5 mm cavity lengths, respectively. Far- eld beam divergence angles for laser bars are typically in the range of 5 to 10° FWHM and 20 to 35° FWHM in the slow-axis and fast-axis direction, respectively, depending on the technology and layout of the bar used.

1.3   Selected relevant basic diode laser characteristics

1.3.1   Threshold gain

Light generated in the active region of the laser device propagates along the optical waveguide of the active layer structure and is partially re ected at the mirror facets of the Fabry–Pérot resonator, which contains the laser active material. This propagation and repeated re ection of light cause a loss and gain of light. Loss is formed by two components: rst, mirror losses, which are caused by the nal mirror re ectivities; and, second, cavity losses, which are due to free-carrier absorption losses in the active layer and cladding layers as well as scattering losses at structural inhomogeneities of the heterointerfaces.

Gain is generated in the stimulated emission process, which is turned on by population-inverted energy levels due to heavy carrier injection. We have to distin-guish between *material gain*, which is the gain of the actual active material, and *modal gain*, which is determined by the ratio of the transverse dimension of the active layer to that of the cavity mode and depends on the details of the speci c laser con guration. Modal gain is always smaller than material gain. This topic will be described in detail in the following sections.

Figure 1.16 illustrates the loss and gain of power of a propagating optical mode in a cavity of length L and with mirror re ectivities R_1 and R_2. Internal optical losses of the unbounded active material per unit length are expressed by a loss coef cient α_i in units of cm^{-1} and volume gain per unit length of the active material is expressed by the gain coef cient g also in units of cm^{-1}. Thus, we obtain for the power P_{rt} of the optical mode after one roundtrip with P_0 the power at the start

$$P_{rt} = P_0 R_1 R_2 \exp\{2L(g - \alpha_i)\}. \tag{1.22}$$

The threshold condition states that the gain compensates the loss after one roundtrip. In this case, $P_{rt} = P_0$, and we obtain for the threshold gain

$$g_{th} = \alpha_i + \frac{1}{2L} \ln\left(\frac{1}{R_1 R_2}\right) \tag{1.23}$$

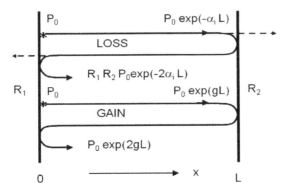

Figure 1.16   Schematic illustration of gain and loss of an optical wave during a roundtrip in a Fabry–Pérot resonator with initial power P_0, cavity length L, internal loss α_i, gain g, and mirror re ectivities R_1 and R_2.

which is the minimum gain for turning on the lasing operation. The rst term is the internal cavity loss as described above (see also the next sections). The second term characterizes the mirror loss α_m. Let us take a simple AlGaAs/GaAs laser with a bulk active layer, cleaved uncoated facets, and a length of 600 μm to illustrate the magnitude of the threshold gain. The mirror loss is calculated to be $\cong 20 \text{ cm}^{-1}$ by using $R \cong 0.32$ (see Equation 1.12) and for the internal loss in the bulk active layer $\cong 15 \text{ cm}^{-1}$ is a typical value which leads in total to a threshold gain of about 35 cm^{-1} required to start the lasing process.

1.3.2   Material gain spectra

1.3.2.1   Bulk double-heterostructure laser

In Section 1.1.1.2, "Net gain mechanism," we showed that the lasing operation can be achieved for the condition $E_{Fc} - E_{Fv} \geq h\nu \geq E_g$ (see Equation 1.10) involving photons with energies $h\nu$ larger than the bandgap energy. A semiconductor meeting this condition is in the state of population inversion, resulting in $R_{stim} > R_{abs}$ or, in other words, a photon is ampli ed rather than absorbed. The gain is zero for $h\nu < E_g$, since there are no carrier recombinations at these energies, and it becomes zero again at $h\nu = E_{Fc} - E_{Fv}$.

The limiting case $E_{Fc} - E_{Fv} = E_g$ is called the transparency condition, where the gain is zero for a photon energy $h\nu = E_g$. A typical minimum carrier density to achieve transparency in a bulk GaAs/AlGaAs laser is $N_{tr} \cong 1.5 \times 10^{18} \text{ cm}^{-3}$. At transparency, the material losses of the active material are compensated by the optical gain. The transparency density includes both radiative carrier loss such as spontaneous recombinations and nonradiative carrier loss including carrier escape. When the density of injected carriers N is larger than the transparency density, then $E_{Fc} - E_{Fv} > E_g$ and a net gain develops for photon energies between E_g and $E_{Fc} - E_{Fv}$.

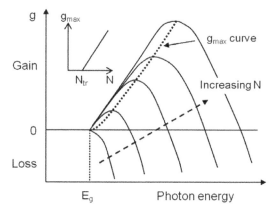

Figure 1.17   Schematics of the gain formation for various carrier densities N injected in a bulk DH laser. The material is transparent for photon energies smaller than the bandgap energy E_g of the semiconductor material. The inset shows the typical linear dependence of the maximum gain g_{max} on carrier density N for a bulk laser (3D). N_{tr} is the carrier density at transparency.

Typical plots of optical gain (or loss = negative gain) spectra as a function of photon energy are illustrated schematically in Figure 1.17 at different injection currents (carrier densities). Quantitative, calculated gain spectra can be found in various textbooks (e.g. Yariv, 1997).

One can observe two key features. First, with increasing carrier density, the difference in quasi-Fermi energies $E_{Fc} - E_{Fv}$ increases, and this results in a broadening of the spectra or increase in the gain bandwidth. The bandwidth increase is, however, small and was evaluated to about 3% of the bandgap energy at a carrier injection of 2.5×10^{18} cm^{-3}. Second, the gain peak shifts gradually to higher energies due to band- lling effects. At higher photon energies, the semiconductor absorbs.

The remarkably simplifying feature is that for typical gain coef cients of interest in a bulk DH laser ($20 < g < 80$ cm^{-1}), the variation of the volume material gain curve maximum versus the density of injected carriers can be well approximated by the linear relation

$$g_{max} = g = \sigma \, (N - N_{tr}) \tag{1.24}$$

where $\sigma = dg/dN$ is the differential gain with the dimension of an area. Typical values for σ are 1.5×10^{-16} cm^2 and 1×10^{-16} cm^2 for GaAs and In$_{0.58}$Ga$_{0.42}$As$_{0.9}$P$_{0.1}$, respectively (Yariv, 1997; Svelto, 1998).

1.3.2.2   Quantum well laser

Schematic material gain spectra for a QW laser at different injection currents are illustrated in Figure 1.18.

There are several interesting characteristics. Due to the high density of states and its narrow energetic distribution, the maximum of the gain curve shows nearly no

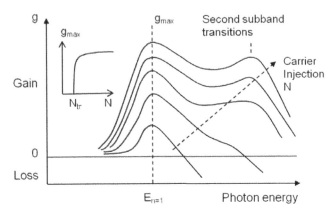

Figure 1.18   Material gain spectra $g(E)$ of a single quantum well laser are illustrated schematically as a function of injected carrier density N. The inset shows a larger initial slope of the maximum gain g_{max} versus N dependence than in 3D due to the much higher density of states in a quantum well (2D); however, g_{max} saturates when electron and hole states are fully inverted. N_{tr} is the carrier density at transparency. At higher carrier injections, the population of the next higher $n = 2$ quantized state contributes to the gain.

shift in energy with increasing current. For the same reasons the material gain is also higher than in bulk DH lasers according to calculations (Yariv, 1997). The same calculations showed also that equal increments of current will yield larger increments of gain in QWs, and that the energetic spread for effective gain is larger by a factor of about 2–3 than in DH lasers.

Due to the two-dimensional density of states in QWs, carriers are more ef cient than in a bulk laser, because added carriers contribute to the gain at its maximum. In contrast to DH lasers, carriers move the gain peak away from the bottom of the band, rendering all carriers at energies below that of g_{max} as useless (compare insets in Figure 1.17 and Figure 1.18). For the same reason the differential gain is higher at lower injection levels than in the bulk. However, the gain saturates at an injection when the electron and hole states are fully inverted, whereas the maximum never saturates in a bulk DH laser due to the   lling of an ever-increasing density of states. Also N_{tr} (QW) $\ll N_{tr}$ (bulk DH) because the density of states to be inverted is signi cantly smaller; for example, a GaAs QW has $\approx 10^{12}$ states/cm^2 in the minimal energy range of $k_B T$ to be inverted, compared to $\approx 10^{13}$ states/cm^2 in an active layer of a DH laser 100 nm thick (Weisbuch and Vinter, 1991).

A   nal interesting feature in Figure 1.18 is that the gain peak at the lower energy, which is due to the population of the $n = 1$ well ground state,   attens with increasing current and a second peak appears at higher currents due to the population of the next higher $n = 2$ state.

For QW lasers, the gain and injection carrier density relation is nonlinear, because the gain saturates in a g versus N plot at suf ciently high carrier injections due to the   at feature of the two-dimensional, step-like density of states (Zhao and Yariv, 1999;

Coldren and Corzine, 1995). We can approximate the material gain versus carrier injection density relation for a SQW laser by

$$g = g_0 \ln \frac{N}{N_{tr}} \tag{1.25}$$

where g_0 is the gain constant (which is proportional to the differential gain) for one QW. Equation (1.25) can be extended to a MQW laser by adding the factor n_{qw} for the number of QWs. For not very large n_{qw} values, the QWs are decoupled and the carrier injection is uniform within the QWs. Then the modal gain g_{mod}, which is the material gain g multiplied by the con nement factor Γ, which we will see in the next sections, is approximately proportional to n_{qw} because each QW contributes almost equally to the laser mode eld.

1.3.3   Optical confinement

The optical con nement factor Γ plays a key role in the design of any semiconductor diode laser. It is de ned as the degree of overlap of the optically guided wave with the gain region of thickness d and can be written as

$$\Gamma = \Gamma_{tv} = \frac{\displaystyle\int_{-d/2}^{+d/2} |E(z)|^2}{\displaystyle\int_{-\infty}^{+\infty} |E(z)|^2} dz \tag{1.26}$$

where $|E(z)^2|$ is the electric eld intensity pro le in the direction z perpendicular to the active layer. This is the de nition of the transverse vertical con nement factor Γ_{tv}; a similar expression to Equation (1.26) can be de ned for the transverse lateral con nement factor Γ_{tl} with the integration then over the width w of the lateral waveguide. A simpli ed illustration for the transverse vertical con nement factor can be found in Figure 1.19.

In general, thin active layers lead to a broad spreading of the optical mode resulting in a large near- eld spot size, which transforms into a narrow far- eld divergence angle. In contrast, thick active layers give a narrow, well-con ned mode with a small near- eld spot size and consequently a large far- eld angle. This information is used to optimize the performance of QW lasers, tune the laser beam divergence angles, and maximize the optical strength of the mirror facets at high optical power outputs. These aspects will be discussed in the sections and chapters that follow.

For the fundamental transverse vertical mode, Botez (1978, 1981) derived a remarkably simple expression accurate to within 1.5%

$$\Gamma_{tv} \cong \frac{D^2}{(2 + D^2)} \tag{1.27}$$

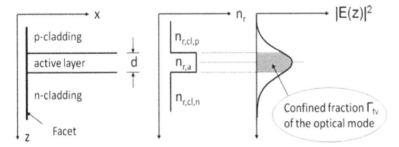

Figure 1.19   A simpli ed illustration of the transverse vertical optical con nement factor Γ_{tv}. $|E(z)|^2$ is the electric eld intensity pro le in the direction z perpendicular to the active layer, $n_{r,cl}$ and $n_{r,a}$ are the refractive indices of the cladding layers and active layer, respectively.

with

$$D = 2\pi \frac{d}{\lambda} \left(n_{r,a}^2 - n_{r,cl}^2\right)^{1/2} \qquad (1.28)$$

where $n_{r,a}$ and $n_{r,cl}$ are the refractive indices of the active layer and the claddings, respectively, and λ is the lasing wavelength. The effective index was also derived (Botez, 1981) and can be approximated by

$$n_{r,eff}^2 \cong n_{r,cl}^2 + \Gamma_{tv} \left(n_{r,a}^2 - n_{r,cl}^2\right). \qquad (1.29)$$

This will be required for the discussion on the lateral modes in the next chapter. To achieve a large modal gain

$$g_{mod} = \Gamma_{tv} g \qquad (1.30)$$

the normalized thickness D of the active layer has to be large. However, as we will see in the discussion on the single spatial mode in the next chapter, this will lead to the excitation of higher order modes and, in order to let the laser operate in the fundamental vertical mode, D has to be below a critical value. Similar expressions to Equations (1.27) to (1.29) can be written for the lateral con nement issue. These will be de ned and used in Chapter 2 also for establishing the conditions for fundamental lateral mode operation.

Figure 1.20a shows the dependence of Γ_{tv} on the active layer thickness d calculated for a 980 nm $In_{0.2}Ga_{0.8}As/Al_{0.3}Ga_{0.7}As$ DH laser. Typical values are 25% for an active layer 100 nm thick; however, for a 10 nm QW con nement factors are as low as a few percent. The con nement factor for a SCH SQW laser with thickness L_z can be evaluated by

$$\Gamma_{tv,qw} \approx \frac{L_z}{d_{mode}} \qquad (1.31)$$

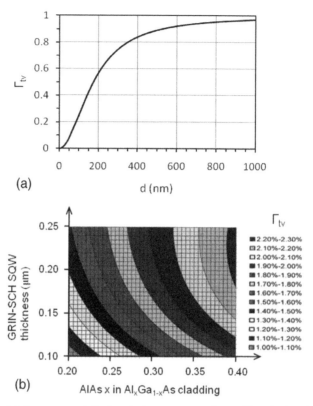

Figure 1.20   (a) Transverse vertical optical con nement factor Γ_{tv} calculated as a function of the active layer thickness d of a 980 nm $In_{0.2}Ga_{0.8}As/Al_{0.3}Ga_{0.7}As$ DH laser by using for the refractive indices $n_r(Al_{0.3}Ga_{0.7}As) = 3.3$ and $n_r(In_{0.2}Ga_{0.8}As) = 3.7$. (b) Calculated Γ_{tv} as a function of the GRIN-SCH layer thickness and AlAs x mole fraction in the $Al_xGa_{1-x}As$ cladding layers of $In_{0.2}Ga_{0.8}As/Al_xGa_{1-x}As$ 6 nm SQW lasers. High Γ_{tv} values are in the upper right corner and low values in the lower left corner of the greyscale contrast image.

where d_{mode} is the spread of the mode perpendicular to the active layer, which is roughly the thickness of the SCH layer. Detailed calculations (Nagarajan and Bowers, 1999) yielded con nement factors of typically 3% for a 5 nm InGaAs SQW embedded in different AlGaAs SCH energy pro les including step-SCH and GRIN-SCH structures.

Figure 1.20b shows transverse vertical con nement factors calculated as a function of the AlGaAs GRIN-SCH layer thickness and AlAs x mole fraction in the $Al_xGa_{1-x}As$ cladding layers of $In_{0.2}Ga_{0.8}As/Al_xGa_{1-x}As$ 6 nm SQW lasers. For the calculation we used the LASTIP simulation code (Crosslight Software Inc., 2010) already introduced in Section 1.1.4 above. High con nement factors above 2% can be found in the greyscale contrast image in the upper right corner and low con nement factors around 1% in the lower left corner. The data can be understood in terms of

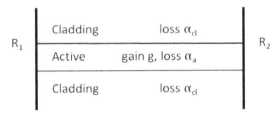

Figure 1.21 Simpli ed schematics of gain and loss in the individual vertical layers of a Fabry–Pérot cavity structure with the end mirror re ectivities R_1 and R_2.

high AlAs x values (low refractive index) leading to stronger mode con nement and therefore higher Γ_{tv} values dependent on the GRIN-SCH thickness, whereas low GRIN-SCH thicknesses and low AlAs x values (high index) lead to strong mode spreading and hence to low con nement factors (see also "Thin waveguides" in Section 2.1.3.5).

1.3.4   Threshold current

To derive an expression for the threshold current requires knowledge of the modal gain at threshold and combining it with the relevant modal gain versus injection carrier density relation. The optical eld distribution in the laser structure has to be considered, which includes the distribution of gain and loss in the individual layers as given by the con nement factor Γ_{tv}. This is illustrated for a simple Fabry–Pérot structure in Figure 1.21.

In the following sections, we will derive the threshold current for a DH and QW laser and discuss the variation of threshold current with active layer thickness and cavity length.

1.3.4.1   Double-heterostructure laser

From Equations (1.22) and (1.23) we obtain

$$\Gamma_{tv}g_{th} = \Gamma_{tv}\alpha_a + (1 - \Gamma_{tv})\alpha_{cl} + \frac{1}{2L}\ln\frac{1}{R_1 R_2} \tag{1.32a}$$

or

$$g_{th} = \alpha_a + \frac{(1 - \Gamma_{tv})}{\Gamma_{tv}}\alpha_{cl} + \frac{1}{2\Gamma_{tv}L}\ln\frac{1}{R_1 R_2} \tag{1.32b}$$

where α_a and α_{cl} denotes the loss in the active layer and cladding layers, and $\Gamma_{tv}g_{th}$ is the modal gain or ampli cation of the optical wave per unit length at threshold. $\Gamma_{tv}\alpha_a$ is the loss per unit length of the guided wave in the active layer. Therefore, $(1 - \Gamma_{tv})\alpha_{cl}$ is the loss of the optical mode per unit length outside of the active layer in the con ning material. Equation (1.32) simply states that the modal gain is exactly balanced by the total losses including active layer, cladding layer and mirror losses. Using Equation (1.24) we obtain for the threshold carrier density N_{th}

$$N_{th} = \frac{\alpha_a}{\sigma} + \frac{1 - \Gamma_{tv}}{\sigma \Gamma_{tv}} \alpha_{cl} + \frac{1}{2L\sigma \Gamma_{tv}} \ln \frac{1}{R_1 R_2} + N_{tr}. \tag{1.33}$$

The threshold current density J_{th} is related to the threshold carrier density N_{th} by

$$J_{th} = \frac{qd}{\eta_i \tau_r} N_{th} \tag{1.34}$$

where η_i is the internal quantum ef ciency, which is the fraction of carriers that recombine radiatively (see one of the next sections), τ_r is the radiative recombination time, q the electron charge and d the active layer thickness. Equation (1.34) was obtained from $R = \eta_i J_{th}/qd$ for the recombination rate of carriers injected with current density J_{th} and from the steady state condition for the carrier density $N_{th} = R\tau_r$. Finally, we obtain the desired expression for the current density at lasing threshold as

$$J_{th} = \frac{qd}{\eta_i \tau_r} \left\{ \frac{\alpha_a}{\sigma} + \frac{1 - \Gamma_{tv}}{\sigma \Gamma_{tv}} \alpha_{cl} + \frac{1}{2L\sigma \Gamma_{tv}} \ln \frac{1}{R_1 R_2} + N_{tr} \right\}. \tag{1.35}$$

An example may demonstrate the usefulness of these equations. We take an AlGaAs/GaAs DH laser with uncoated facets ($R_1 = R_2 \cong 0.32$), refractive indices $n_{r,a} = 3.6$, $n_{r,cl} = 3.4$, an active layer thickness of $d = 0.1$ μm, a cavity length of $L = 300$ μm, and a lasing wavelength of 850 nm. For simplicity reasons, we assume $\alpha_a = \alpha_{cl} = \alpha_i$ and take $10 \, \text{cm}^{-1}$ as a typical value for the loss coef cient in an AlGaAs/GaAs DH laser. The con nement factor is then calculated as $\Gamma_{tv} \cong 0.25$ by using Equations (1.27) and (1.28). Finally, by taking $3.6 \times 10^{-16} \, \text{cm}^2$ for the differential gain σ and 2×10^{18} carriers/cm³ for the transparency density N_{tr} (Svelto, 1998), we obtain for the threshold carrier density $N_{th} = (0.11 + 0.42 + 2) \times 10^{18}$ carriers/cm³. To evaluate the threshold current density, we take $\eta_i \cong 0.9$ and $\tau_r \cong 4$ ns as typical values and obtain $J_{th} \cong (4.45 \times 10^{-16} \, \text{A cm}) \times (2.53 \times 10^{18}$ carriers/cm³$) = 1125 \, \text{A/cm}^2$. This calculated value is in remarkably good agreement with experimental values.

If the mirror losses are much greater than the internal losses, the threshold current density dependence expressed in Equation (1.35) is then mainly determined by the ratio d/Γ_{tv}, which has an optimum value for a normalized thickness $D = 1.42$ (cf. Equations 1.27 and 1.28).

As can be seen from Equation (1.35), the transparency current J_{tr} can be determined from measurements on lasers of different cavity lengths L and by plotting J_{th} versus $1/L$. The intersection of the plot with the J_{th} axis yields $J_{th,1/L=0} = J_{tr} + (qd\alpha_i)/(\eta_i \tau_r \Gamma_{tv} \sigma)$ by again using $\alpha_a = \alpha_{cl} = \alpha_i$, that is, not splitting the intrinsic loss into losses in the active and passive regions of the optical waveguide for simplicity reasons. J_{tr} can be extracted from the intersection point value by subtracting the second term, which requires knowledge of the values of the individual parameters α_i, η_i, τ_r, Γ_{tv}, σ. The dependence of the threshold current on cavity length and active region thickness will be discussed in Sections 1.3.4.3 and 1.3.4.4, below.

1.3.4.2   Quantum well laser

A procedure similar to that described in Section 1.3.4.1 can be applied to a QW laser by rewriting Equation (1.32b) as

$$g_{th} = \alpha_a + \frac{\left(1 - n_{qw}\Gamma_{tv,qw}\right)}{n_{qw}\Gamma_{tv,qw}}\alpha_{cl} + \frac{1}{2n_{qw}\Gamma_{tv,qw}L}\ln\frac{1}{R_1 R_2} \qquad (1.36)$$

where n_{qw} is the number of QWs in the active region and $\Gamma_{tv,qw}$ is the transverse vertical optical con nement factor of one QW.

The threshold carrier density is then obtained with the help of Equation (1.25) as

$$N_{th} = N_{tr}\exp\left\{\left[n_{qw}\Gamma_{tv,qw}\alpha_a + \left(1 - n_{qw}\Gamma_{tv,qw}\right)\alpha_{cl} + \frac{1}{2L}\ln\frac{1}{R_1 R_2}\right] / \left[n_{qw}\Gamma_{tv,qw}g_0\right]\right\}.$$
$$(1.37)$$

We select as an example an AlGaAs/GaAs SQW ($n_{qw} = 1$) laser with similar parameters and simpli cations as in the above DH laser, but with a well thickness $L_z = 10$ nm, a con nement factor $\Gamma_{tv,qw} = 0.03$, a gain constant $g_0 \cong 2400$ cm^{-1} and a transparency density $N_{tr} \cong 2.5 \times 10^{18}$ cm^{-3} (Coldren and Corzine, 1995). Equation (1.37) then yields $N_{th} \cong 2 \times N_{tr} \cong 5 \times 10^{18}$ carriers/cm^3. With this value and Equation (1.34) where d is the well thickness L_z, the threshold current density becomes $J_{th} \cong (4.5 \times 10^{-17} \text{ A cm}) \times (5 \times 10^{18} \text{ carriers/cm}^3) \cong 220$ A/cm^2, which is in good agreement with experimental values.

1.3.4.3   Cavity length dependence

We take a QW laser for the discussion and rewrite Equation (1.37) as the current at lasing threshold

$$I_{th} = wLJ_{tr}\exp\left\{\left(\alpha_i + \frac{1}{2L}\ln\frac{1}{R_1 R_2}\right) / \left(n_{qw}\Gamma_{tv,qw}g_0\right)\right\}. \qquad (1.38)$$

Here w is the active layer width, L is the laser length, and $J_{tr} = (qL_z/\eta_i\tau_r)N_{tr}$ is the transparency current, which was obtained from Equation (1.34) for the relation between current density and carrier density. For simplicity, we do not distinguish between active and passive losses and again set $\alpha_a = \alpha_{cl} = \alpha_i$. The threshold current dependence on cavity length of a QW laser is different from that of a bulk DH laser. It can be shown from Equation (1.38) that there is a minimum threshold current for a QW laser at

$$I_{th,min} = \frac{1}{2}\ln\frac{1}{R_1 R_2} \times \frac{wJ_{tr}}{n_{qw}\Gamma_{tv,qw}g_0} \times \exp\left\{\left(\frac{\alpha_i}{n_{qw}\Gamma_{tv,qw}g_0}\right) + 1\right\} \qquad (1.39)$$

which is achieved at a length

$$L_{min} = \frac{1}{2} \ln \frac{1}{R_1 R_2} \times \frac{1}{n_{qw} \Gamma_{tv,qw} g_0}. \tag{1.40}$$

From the last three equations, we can evaluate how the number of QWs determines the threshold current. Both the transparency current and internal loss coefficient consist of a part, which scales with the number of QWs and a part, which is independent of it. For the transparency current, the radiative loss of carriers within the active QW is proportional to the number of QWs. In contrast, the carrier loss in the SCH region, which embeds the QW, and the carrier leakage are actually independent of the number of QWs. The internal loss coefficient consists of free-carrier absorption in the active QW, which scales with the number of QWs. It also consists of free-carrier absorption in the region adjacent to the QW, waveguide scattering, and impurity absorption effects, all of them are independent of the number of QWs. There is also a strong influence on the threshold current versus number of QWs dependence by the threshold modal gain. At low modal gain, the exponential term in Equation (1.38) tends to one, with the consequence that the threshold current of SQW lasers becomes lower due to a lowering of the transparency current density. At high threshold modal gain, the exponential term in Equation (1.38) dominates the threshold current. The use of MQWs will significantly reduce this term leading together with a higher gain coefficient to lower threshold currents despite the fact that the transparency current density is larger than in SQW lasers. We can see from Equations (1.39) and (1.40) that lower minimum threshold currents can be obtained at shorter laser lengths in MQW lasers (Zhao and Yariv, 1999).

Using the data of the example discussed in the last part of Section 1.3.4.2 we can get for a laser with a width of 4 µm and length 300 µm a threshold current of about 3 mA. In addition, from Equations (1.39) and (1.40) the threshold current minimum becomes about 2 mA at a cavity length of 160 µm for a SQW laser 4 µm wide with the same data as used in the example in the last section. These results are in excellent agreement with experimental data (Chen et al., 1992).

Figure 1.22 shows schematically the cavity length dependence of the threshold current of a QW laser. There are typically three different ranges. The linear curve for lengths $L > L_{min}$ represents the fact that the threshold current increases with cavity length. In that regime the slope of the curve increases with the number n_{qw} of QWs and also increases with the interface recombination velocity v_s, which is a measure for the loss of carriers due to nonradiative recombination via defect states in the bandgap of the active QW.

The regime around the minimum is characterized by a threshold, which decreases with n_{qw}, increases with v_s, and decreases with mirror reflectivity R_1, R_2 at a cavity length L_{min} decreasing with increasing n_{qw} and R_1, R_2 (Engelmann et al., 1993). A high reflectivity, however, leads to a low external differential quantum efficiency, as we will see in one of the sections below, because it is basically the ratio between mirror loss and total cavity loss.

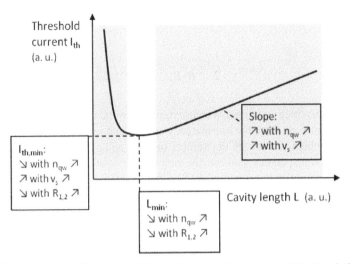

Figure 1.22 Schematic illustration of the cavity length dependence of the threshold current of a quantum well laser characterized by three different ranges: (i) threshold $I_{th,min}$ and cavity length L_{min} at the minimum, (ii) strong increase of the threshold below the minimum, (iii) weaker increase of threshold with a certain slope for longer cavities above the minimum. Dependence ($\searrow$ denotes a decrease, $\nearrow$ denotes an increase) of $I_{th,min}$, L_{min}, and slope on increasing number of quantum wells n_{qw}, interface/surface recombination velocity v_s, and re ectivities R_1, R_2 is indicated.

In the regime $L < L_{min}$ the threshold current shows a striking increase toward short cavity lengths, which is a direct consequence of the gain saturation in a QW laser at high carrier injections and the dominating mirror losses (see Equation 1.38). A large gain is required to overcome the high mirror losses in this *short-cavity effect* regime. According to Engelmann *et al.* (1993), this anomalous increase of I_{th} shifts to smaller L values with increasing n_{qw}, which multiplies the fraction of the optical mode subject to the gain due to an increased Γ_{tv} and therefore suppresses the gain saturation problem even at very short cavities. The short-cavity effect is particularly pronounced in SQW lasers due to a low gain con nement Γ_{tv}. Its location is affected by Auger recombination and carrier leakage but to a lesser extent by interface recombination.

1.3.4.4 Active layer thickness dependence

The variation of the threshold current density J_{th} with active layer thickness d is plotted in Figure 1.23 for an AlGaAs/GaAs DH and a GRIN-SCH SQW laser. In the case of the DH laser, we can understand and analyze the dependence with the help of Equation (1.35).

For suf ciently large values of d above the minimum, the threshold carrier density N_{th} is nearly the same as the transparency density N_{tr}, as can be seen from the example in Section 1.3.4.1, and N_{tr} increases with d like the number of states to be inverted. The differential gain σ decreases because fewer states per unit volume are populated by the same drive current density and the slight increase of the con nement factor

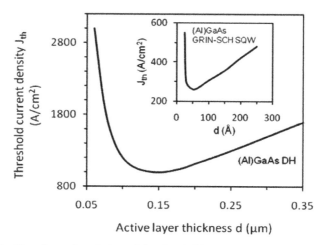

Figure 1.23 Experimental variation of the threshold current density J_{th} with active layer thickness d plotted for an AlGaAs/GaAs DH and a GRIN-SCH SQW (inset) laser. Small current steps at wider quantum well widths indicate the population of higher subbands.

Γ_{tv} can be considered as negligible. Therefore, considering all effects, J_{th} increases linearly with d in the region of larger active layer thicknesses.

However, for suf ciently small d values the $1/\Gamma_{tv}$ terms in Equation (1.35) increase dramatically because: (i) the con nement factor Γ_{tv} diminishes very quickly in proportion to d^2 (see Equations 1.27 and 1.28); (ii) the mode extends considerably into the doped cladding layers; (iii) the transparency density decreases only proportional to d; and (iv) the differential gain increases only in proportion to $1/d$. Overall, including a reduced effective gain and increased losses leads to a strongly increased threshold carrier density resulting in a dramatic increase in the threshold current density as the active layer thickness decreases to small values.

In the case of the GRIN-SCH SQW laser, we use the results of detailed calculations carried out on the volume gain versus injection current dependence with the well thickness as parameter (Weisbuch and Vinter, 1991).

These calculations showed a continuous decrease in the transparency current with decreasing well thickness from high, quasi-3D values to smaller values and an increase in the volume gain with diminishing d as the volume diminishes. The con nement factor decreases proportional to d with decreasing d (see Equation 1.31). Using Equations (1.34) and (1.37) we conclude therefore that the transparency current is the main parameter responsible for the decrease in the threshold current density with decreasing d in the region of larger d values (see inset of Figure 1.23). In this regime, the population of higher lying quantum states may also play a role, in particular for wider wells, as the subband spacing becomes smaller. This is indicated by the appearance of small current steps in the J_{th} versus d plot shown in the inset of Figure 1.23.

In very thin wells carriers escape into the GRIN-SCH optical con nement region leading to a large transparency current, which is the main source for the strongly increasing threshold current density in very thin well layer lasers. The carrier escape is caused by the large con ning energy, which drives the quasi-Fermi level high in the well. The effects at both larger and smaller well thicknesses account for the existence of a minimum in the J_{th} versus d dependence.

1.3.5   Transverse vertical and transverse lateral modes

The distribution of the propagating optical fields in the laser cavity can generally be characterized by two independent sets of modes, transverse electric (TE) and transverse magnetic (TM) ones. As already discussed above, these are subdivided into transverse vertical modes, which are perpendicular to the active layer, and transverse lateral modes, which are parallel to the active layer. TE modes have polarizations with the electrical field vector parallel to the active layer as opposed to TM modes where the light is polarized perpendicular to the active layer. Semiconductor diode lasers usually operate in the TE mode due to the lower threshold gain g_{th} (TE) $< g_{th}$ (TM), which is also true for the technologically important, pseudomorphic, compressively strained InGaAs/AlGaAs QW lasers (Coleman, 1993).

The transverse vertical mode is formed by the standing wave between the heterojunctions of the active layer structure, which determines the confinement strength of the optical field, and the conditions for fundamental transverse vertical mode operation. The latter will be discussed in detail in Chapter 2, whereas the optical confinement issue was briefly discussed in Sections 1.1.1.4 and 1.3.3. The following section gives a rundown of possible vertical confinement structures along with their key features.

The transverse lateral mode is determined by the standing wave in the direction parallel to the active layer. Fundamental mode operation is strongly determined by the effective, lateral width and structure of the active region and the change of refractive index from the active region to adjacent layers. Structures for stabilizing the transverse lateral mode have already been briefly dealt with in Section 1.1.1.5. Sections 1.3.5.2 and 2.1.4 will discuss this topic in more detail.

1.3.5.1   Vertical confinement structures – summary

Double-heterostructure

- Example: n-AlGaAs/GaAs/p-AlGaAs. $d_{GaAs} \approx 0.08$–0.2 μm.

- Efficient carrier confinement due to band discontinuities.

- Optical confinement factor Γ_{tv} large ≈ 0.6 (0.9) at $d = 0.2$ (0.6) μm.

- Threshold current density large, typically 1 kA/cm^2.

Single quantum well

- Well thickness L_z typically in range of 5 to 10 nm.

- Lasing wavelength adjustable by changing L_z and/or barrier height.

- $\Gamma_{tv,qw} \propto \Delta n_r L_z^2$ is very low. Example: $\Gamma_{tv,qw} \lesssim 1\%$ for an InGaAsP/InP SQW with $L_z = 10$ nm.

- Without the optical con nement structure, threshold current density is high caused by high losses due to spreading of the mode into the lossy cladding layers and gain saturation at high carrier densities due to the constant density of states of the $n = 1$ subband. This makes the threshold current more sensitive to increased cladding losses, increasing with temperature because of higher free-carrier absorption at high temperatures.

Strained quantum well

- Lattice-mismatch between well and barrier. Generated strain is elastically re-laxed by deformation of the well material if $L_z < L_{z,crit}$.

- Access to certain wavelengths not available from any lattice-matched III–V compound semiconductor system. Example: $\sim$900–1100 nm band wave-lengths only accessible by pseudomorphic, compressively-strained InGaAs/AlGaAs QW lasers.

- Threshold current density is reduced and slope ef ciency of power output versus drive current characteristic is increased by a reduced density of states and thus of the reduced hole effective mass due to a strain-induced separation of light-hole and heavy-hole valence bands.

- Temperature dependence of lasing characteristics is improved with higher characteristic temperatures.

- High mode selectivity: TE polarized light emission for electron to heavy-hole recombinations for compressive strain as opposed to TM polarized light emission for electron to light-hole recombinations for tensile strain.

Separate confinement heterostructure SCH and graded-index SCH (GRIN-SCH)

- To improve signi cantly photon and carrier con nement of SQW.

- To counteract carrier over ow from well under high injection.

- To fully exploit reduced density of states in QWs to achieve lasing at low carrier injection.

- Optimum SCH or GRIN-SCH thickness (waveguide layer) for a given com-position difference between cladding layer and waveguide layer is that, which maximizes the QW optical con nement factor $\Gamma_{tv,qw} \propto \Delta n_r L_z^2$. Example: $d_{SCH} \cong 170$ nm and $d_{GRIN-SCH} \cong 300$ nm for a maximum $\Gamma_{tv,qw} \cong 3.5\%$ for SQW ($L_z = 10$ nm) AlGaInP/GaInP lasers emitting in the 650 nm band.

Multiple quantum well (MQW)

- Low threshold current densities due to strong optical con nement.

- At high loss: MQW is always better than SQW due to the higher differential gain in the gain–current curve. In contrast, the saturated gain of the SQW is not large enough to reach threshold gain.

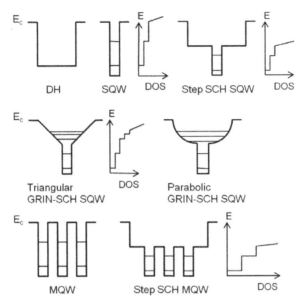

Figure 1.24 Schematic illustration of band structure (conduction band only shown for clarity) of various active region structures including DH, SQW, GRIN-SCH, and MQW and corresponding energy (E) dependent density of states (DOS) for some of them.

- At low loss: SQW is always better than MQW due to both its lower J_{tr} (only states of one QW have to be inverted) and lower internal loss ($\Gamma_{tv,qw}\alpha_i$ scales with number of wells)

Figure 1.24 shows the schematics of the various active layer structures along with some associated density of states (DOS).

1.3.5.2 Lateral confinement structures

As mentioned in Section 1.1.1.5, transverse lateral con nement has to be realized for current, carriers, and photons to achieve high performance of edge-emitting diode lasers. There are practically three types of implementation approaches: gain guiding provides current con nement, weak index-guiding current, and photon con nement, and strong index guiding provides current, carrier, and photon con nement. In the following we describe the concept and key features of each of the approaches.

Gain-guiding concept and key features

As illustrated in Figure 1.25, gain guiding is generated by injecting current through an aperture in a dielectric insulating layer. Other techniques include the formation of a current path by Zn diffusion or by restricting the current ow to an opening in high-resistivity areas created by ion implantation (see also Figure 1.6). The active region of all these structures is planar and continuous. The optical mode distribution

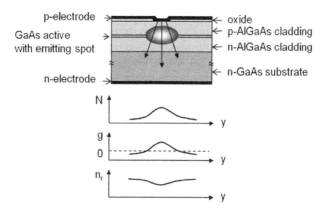

Figure 1.25   Schematic cross-section of a simple gain-guided laser structure with an oxide stripe for current con nement. The gain-guiding concept is shown for a GaAs/AlGaAs laser. The effective pro les for carrier density *N*, gain *g*, and refractive index n_r at high injection conditions are also shown qualitatively. Layer thicknesses not to scale.

along the active layer is determined by the optical gain and therefore these lasers are called gain-guided lasers.

Light ampli cation by stimulated emission occurs only in the gain region, which is pumped by the injection of carriers through the drive current. Outside this region high optical losses impact the mode. The optical gain is determined by the carrier distribution, which is impacted by lateral current spreading in the p-type cladding layer and carrier diffusion in the active layer. *Spatial hole burning* (holes burned in the spatial distribution of inversion within the active region) can occur under relatively high injection conditions in the center part where the high stimulated emission rate decreases the carrier density, which leads to an increase in the refractive index resulting in a contraction of the emission spot width, called *self-focusing* (Van der Ziel, 1981). The effective index reduction can be as strong as 5×10^{-3}.

The location of this transverse lateral mode with a highly contracted emission spot is, however, not stable and can move laterally to higher refractive index areas by any irregularity in the built-in index pro le. This is usually associated with a local nonlinearity (*kink*) appearing in the *P/I* characteristic, because of the spatial separation of the light con nement region and high-gain region leading to a movement of the mode along the active layer plane, an excitation of higher order lateral modes or a transition from the TE to TM mode. Index guiding along the active layer can mitigate the self-focusing effect by stabilizing the optical mode (Lang, 1980). Fundamental lateral mode operation can be achieved with narrow-stripe lasers exhibiting kinks at much higher operating powers than do lasers with wider stripes (waveguides). This operating mode is necessary for ber-coupled applications.

Other disadvantages include high threshold current densities and low differential quantum ef ciencies due to lossy waveguides and carrier-induced index suppression resulting in index *anti-guiding*. Signi cant *astigmatism* in the output beam and high susceptibility to *self-pulsations* are further undesirable features. Astigmatism is caused by the difference in the beam waist position in the direction parallel and perpendicular to the active layer. In the former, the phase front of the mode is curved since the mode leaks laterally into an absorbing medium where no current is injected, which locates the beam waist within the cavity; in the latter, the phase front is plane because of the index guiding locating the beam waist at the mirror surface. Self-pulsations are sustained oscillations where the emitted light pulsates at high frequencies of several hundred megahertz. A saturable absorption model can account for the pulsation phenomenon (Dixon and Joyce, 1979). Central to this model is a nonlinear gain versus carrier density dependence, which causes regions with smaller carrier density to act as saturable absorbers.

Weakly index-guiding concept and key features

A possible realization of a weakly index-guided structure was shown in Figure 1.6. Common to all such designs is that the thickness of at least one layer is laterally nonuniform. There are many possible structures, which can simply be grouped in rib and ridge waveguide-type lasers.

In both types of scheme, the lateral laser structure is modi ed such that an effective index step of $\lesssim 10^{-2}$ is generated in the rib or ridge zone, which is larger than the carrier-induced index suppression leading then to a relatively stable index guiding of the lateral mode. In rib lasers the thickness, for example, of the waveguide layer or the active layer (Figure 1.6), can be varied laterally; however, current spreading in the p-cladding layer can impact the threshold current density.

In ridge lasers (Figure 1.26) where the ridge is formed in the upper p-cladding by etching and embedded in a dielectric layer, the loss of effective current by current spreading is less pronounced. However, carrier diffusion in the active layer, which extends beyond the ridge impacts the threshold current but also produces a continuous lateral variation of gain and index. The partial overlap of the mode with the dielectric layer forms an effective index step with a height determined by the height of the ridge and the residual thickness to the active layer. A sensitive adjustment of the etch depth is required to provide enough effective lateral index step for single lateral mode operation. The beam quality and fundamental mode operation of ridge waveguide lasers are sensitively dependent on the design of critical ridge dimensions and their control during device fabrication.

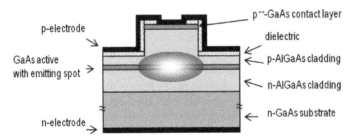

Figure 1.26 Schematic cross-section of a weakly index-guided ridge waveguide laser providing lateral current and photon con nement illustrated for a GaAs/AlGaAs laser. Layer thicknesses not to scale.

This topic will form one of the main discussion points in Chapter 2. Narrow ridge waveguide, single-emitter lasers have been extensively investigated and are widely employed in many key application areas because of:

- simple fabrication technology requiring only one single epitaxial growth step;

- low optical losses;

- low threshold currents;

- long lifetimes;

- easy integration; and, above all,

- record-high optical powers emitted in a single transverse vertical and lateral mode.

Strongly index-guiding concept and key features

The basic structure of a strongly index-guided laser combining all three lateral confinements for current, carriers, and photons was described in Section 1.1.1.5 and illustrated in its simplest form in Figure 1.6. The active layer is buried on all sides in higher bandgap materials with lower refractive indices creating a high lateral index step along the active layer of about 0.3 for InGaAsP/InP, which is at least higher by a factor of 100 than the index change induced by carriers. Current path and optical con nement are tightly kept within bounds including lateral carrier diffusion in these devices, called buried-heterostructure (BH) lasers, resulting in lasing characteristics determined mainly by the waveguide, which con nes the optical mode within the buried active area. BH lasers come in many different forms, which can be classi ed in two groups comprising structures with planar and nonplanar active layers.

We restrict our discussion to a planar-type device, shown in Figure 1.27. It is an etched mesa BH device, which is fabricated by rst growing a planar active waveguide con guration followed by etching a narrow mesa stripe down to the active layer and

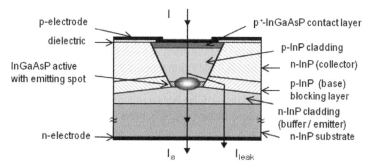

Figure 1.27   Schematic cross-section of an etched-mesa buried heterostructure providing current, carrier, and photon con nement shown for an InGaAsP/InP laser. The current leakage path through the mesa burying layers formed by regrowth is indicated in a simpli ed manner. Layer thicknesses not to scale.

subsequently regrowing additional lattice-matched layers with higher bandgap and lower index to embed this mesa. The purpose of the regrown material is twofold: to confine current, carriers, and photons to the center part; and to block current flow in these burying layers. This can be accomplished by reverse-biased homojunctions under forward-bias operation, whereas the heterojunctions in the center path are forward biased to inject the confined current into the active layer. Alternatively, highly resistive material or semi-insulating material such as Fe-doped InP for InGaAsP/InP lasers can also be used as current blocking layers.

Leakage current I_{leak} poses a potential issue in BH lasers. Its magnitude varies with the drive current I_a through the active layer and depends sensitively on the thickness and carrier concentration of the current-confining layers. The leakage current paths in a mesa etched BH laser (Figure 1.27) can be summarized as follows (Agrawal and Dutta, 1993):

- Diode leakage path: Formed by forward-biased homojunction p-InP blocking layer/n-InP buffer layer. The effective saturation parameter of this diode is determined by current spreading in the p-InP layer, which is smaller for thinner p-InP and consequently diode leakage tends to be lower for thin p-InP regrown layers.

- Transistor leakage path: Formed by n-InP buffer layer (emitter), p-InP blocking layer (base), and n-InP top regrown layer (collector). Increasing the width of the p-InP layer increases the base width of the transistor, which reduces the current gain of the transistor and hence the leakage through the transistor path.

- Thyristor leakage path: Formed by p-InP cladding, n-InP top regrown layer, p-InP blocking layer, and n-InP buffer layer. The thyristor is off at low currents and hence leakage through this path is low. At high currents the thyristor may turn on, leading to very high leakage currents.

BH diode lasers are characterized by positive and negative features. The former generally include low threshold currents in the range of 10 to 20 mA, bandwidths as high as 22 GHz have been demonstrated in 1.3 μm lasers with low-capacitance structures (Huang *et al.*, 1992), and operation to high output powers in stable, fundamental mode operation for active layer widths of 1.5 μm in the 1.55 μm band of InGaAsP/InP communications lasers. The latter include complex fabrication technology steps, relatively high threshold current densities, and the degradation of the buried heterointerface as a potential, additional degradation mechanism. This mode is associated with a breakdown or degradation of the active region due to a decrease of injected carriers. The degradation of the buried heterointerface can be classed as a wear-out failure and not a sudden failure (see Chapters 3 and 5).

1.3.5.3 Near-field and far-field pattern

The field profile of the optical wave traveling along the laser cavity manifests at the laser facet as a 2D field distribution in the transverse vertical and lateral direction, called the near-field pattern (NFP). The light emitted from the NFP spot propagates freely into space and the laser beam is strongly broadened in both directions by diffraction. According to diffraction theory, the light spot some distance away from the NFP is called the far-field pattern (FFP). The transition occurs at a distance $\approx w^2/\lambda_0$, where w is some characteristic full width of the NFP (Coldren and Corzine, 1995). The phase front of the mode is planar in the NF of index-guided waveguides but approaches a spherical shape toward the FFP.

The FFP is characterized by the beam divergence angles at FWHM of the intensity peak where $\theta_\perp$ is the angle in the direction perpendicular to the active layer (so-called *fast axis*) and $\theta_\parallel$ is the angle parallel to the active layer (so-called *slow axis*). For small angles the FFP is the Fourier transform of the NFP, or, in other words, the width (angle) of a FFP is inversely proportional to that of a NFP in a good analogy to the diffraction of light through a narrow slit.

Figure 1.28 illustrates these effects using two different NFPs of a simple ridge laser. It shows also slow- and fast-axis FF intensity profiles of an InGaAs/AlGaAs

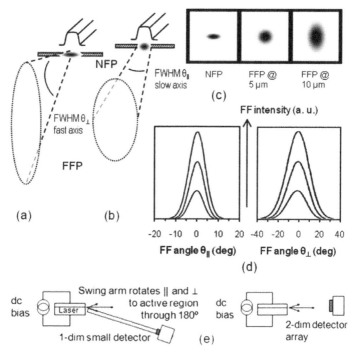

Figure 1.28   Schematic illustration of near-field (NFP) and far-field radiation patterns (FFP) for a ridge laser with high aspect ratio (a) and low aspect ratio (b) between vertical and horizontal beam divergence angle. (c) Radiation intensity distribution at different positions along the optical cavity axis of a laser. (d) Transverse lateral and vertical far-field intensity profiles of a single-mode InGaAs/AlGaAs laser measured at various optical output powers in the range of 100 to 300 mW. (e) Schematic diagrams for recording far-field intensity profiles with a rotating 1D small detector (left) and intensity maps with a 2D detector array (right).

laser measured at different laser powers in the range of 100 to 300 mW and con rms its fundamental transverse mode operation. Typical setups for measuring FFPs are shown in Figure 1.28e. The NFP width $w_\perp$ perpendicular to the junction plane depends on the thickness and composition of various layers including the active, con nement, and cladding layers. It can be approximated (Botez and Ettenberg, 1978; Agrawal and Dutta, 1993) by

$$w_\perp \cong d(2\ln 2)^{1/2} \left(0.321 + 2.1 D^{-3/2} + 4 D^{-6}\right) \tag{1.41}$$

where d is the active layer thickness and the normalized thickness D is given by Equation (1.28). Equation (1.41) yields fairly accurate results for $1.8 < D < 6$.

The NFP parallel to the active layer depends on the lateral guiding structure. For an index-guided laser Equation (1.41) can be applied for $w_\parallel$ after replacing D by W, where W is the normalized waveguide width given by (Agrawal and Dutta, 1993)

$$W = 2\pi \frac{w}{\lambda} \left(n^2_{r,eff,in} - n^2_{r,eff,out}\right)^{1/2}. \tag{1.42}$$

Here $n_{r,eff,in}$ and $n_{r,eff,out}$ are the effective refractive indices within and outside of the effective waveguide width w. The lateral NF for gain-guided lasers extends considerably beyond the stripe width in contrast to index-guided lasers, where a lateral index step as small as 0.005 is suf cient to con ne the NF within the effective active layer width.

The FFPs in the direction parallel and perpendicular to the active layer display the spread of the laser mode and are important parameters in many applications requiring well-de ned divergence angles $\theta_\perp$ and $\theta_\parallel$, for example, to couple laser light into bers. Also the FFP carries useful information because its shape and smoothness depend sensitively on the type and strength of transverse vertical and lateral wave-guiding, the type and number of excited spatial modes, and the quality of the wave-guide material and interfaces regarding uniformity and structural irregularities. In fundamental mode operation, the NFP and FFP display single-peak, smooth Gaussian pro les, which become distorted when higher order spatial modes are present in the laser operation.

The FF emission pattern for the fundamental mode of a symmetric vertical wave-guide laser has a fast-axis beam divergence angle $\theta_\perp$, which is given to a good approximation by the expression (Botez, 1982, 1981; Botez and Herskowitz, 1980)

$$\theta_\perp = \frac{0.65 D \left(n^2_{r,a} - n^2_{r,cl}\right)^{1/2}}{1 + 0.15 \left(1 + n_{r,a} - n_{r,cl}\right) D^2} \tag{1.43}$$

where $n_{r,a}$ and $n_{r,cl}$ are the refractive indices of the active and cladding layers, respectively, and D is given by Equation (1.28). It is claimed that Equation (1.43) supplies results accurate to within 3% for $D \leq 2$, which corresponds to an active layer thickness $d \leq 0.3$ μm for an InGaAsP/InP DH laser emitting at 1.55 μm.

An experimental value of 23° for a 1.3 μm InGaAsP/InP laser with $d = 0.05$ μm compares well to the 20° calculated according to Equation (1.43) (Itaya *et al.*, 1979).

Typical beam divergence angles of 980 nm, strained-layer InGaAs/AlGaAs GRIN-SCH SQW ridge waveguide lasers are in the range of 20 to 35° for the fast-axis and 5 to 10° for the slow-axis.

In Chapter 2, we will discuss the experimental and calculated dependencies of the FF angles of a ridge InGaAs/AlGaAs laser on the composition and dimensions of layers and structures including the AlAs mole fraction of the QW cladding, width of the GRIN-SCH layer, width and depth of the ridge, and distance from the bottom edge of the ridge to the active layer (called the residual waveguide thickness). The insertion of so-called mode puller layers, such as inverse refractive index (IRIS) or spread index (SPIN) structures, into the cladding layers to decrease the transverse vertical FF angles, but to keep simultaneously the threshold current density low, will also be discussed in Chapter 2.

We will also describe quantitatively the linking of FF angles to electrical and optical laser parameters such as threshold current, ef ciency, and maximum kink-free optical power. All the above relations are crucial for optimizing this important diode laser with respect to high-power and single-mode operation.

1.3.6   Fabry–Pérot longitudinal modes

To obtain the spectral separation between two adjacent longitudinal modes $\Delta\lambda_m$ we start from the phase condition $m\lambda_0 = 2Ln_{r,eff}$ expressed in Equation (1.13), perform the total differential, and then use the partial derivatives $\partial n_{r,eff} = (\partial n_{r,eff}/\partial\lambda_0)\partial\lambda_0$, $\partial\lambda_0 = \Delta\lambda_m$, and $\partial m = \Delta m = -1$, which leads to a positive change $\Delta\lambda_m$ of the mode wavelength. After some simple algebraic manipulations we nally nd for the wavelength separation between two modes

$$\Delta\lambda_m = \frac{\lambda_0^2}{2L\left(n_{r,eff} - \dfrac{\partial n_{r,eff}}{\partial\lambda_0}\lambda_0\right)} = \frac{\lambda_0^2}{2Ln_{r,gr.eff}} \tag{1.44}$$

where $n_{r,gr.eff} = n_{r,eff}[1 - (\lambda_0/n_{r,eff})(\partial n_{r,eff}/\partial\lambda_0)]$ is the effective index of the group velocity $v_{gr} = c/n_{r,gr.eff}$ of the longitudinal optical mode (c is the velocity of light in vacuum). The group effective index in semiconductors is typically 20–30% larger than the effective index, depending on the speci c wavelength relative to the band edge (Coldren and Corzine, 1995).

Figure 1.29 illustrates the formation of longitudinal modes along the cavity by considering the Fabry–Pérot (FP) mode spectrum of the cavity and the gain versus wavelength pro le. Whenever the modal gain condition, Equation (1.32) or Equation (1.36), is met at a certain FP wavelength λ_m, lasing starts at this wavelength and has to comply with the phase condition, Equation (1.13). Only modes closest to the peak gain are ampli ed, where the number of lasing modes depends on the width of the gain maximum, and other modes are not excited because their losses are higher than

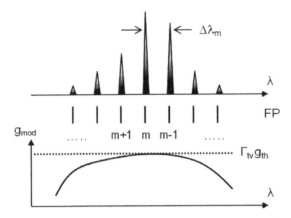

Figure 1.29   Schematic illustration of the formation of the laser longitudinal mode emission spectrum and its dependence on gain spectrum and longitudinal Fabry–Pérot (FP) cavity modes.

the available modal gain. Mode spacings are, for example, <1 nm for InGaAsP/InP lasers 300 μm long.

The wavelength of a lasing longitudinal mode shifts to shorter wavelengths by a reduction of the refractive index induced by the injected carriers at a density N (Casey and Panish, 1978; Sell *et al.*, 1974). The size of the shift due to this so-called plasma effect can be found by using the calculus of variation of Equation (1.13) which results in

$$\delta\lambda_m\,(N) = \frac{\lambda_0}{n_r}\frac{\mathrm{d}n_r}{\mathrm{d}N}\delta N. \qquad (1.45)$$

Typical values for $\mathrm{d}n_r/\mathrm{d}N$ range from -1×10^{-21} cm³ to -6×10^{-21} cm³ for 0.85 μm AlGaAs/GaAs and 1.55 μm InGaAsP/InP lasers, respectively (Casey and Panish, 1978; Nash, 1973).

On the other side, the lasing peak wavelength increases with increasing temperature caused by the Joule heating effects of the drive current, which include the temperature dependence of the refractive index, bandgap energy, and cavity length. Similar to Equation (1.45), we nd for the mode wavelength shift caused by the temperature dependence of the refractive index

$$\delta\lambda_m\,(T_j) = \frac{\lambda_0}{n_r}\frac{\mathrm{d}n_r}{\mathrm{d}T_j}\delta T_j. \qquad (1.46)$$

Here T_j is the junction temperature of the diode laser. Temperature coef cients of the refractive index $\mathrm{d}n_r/\mathrm{d}T_j$ are between 2×10^{-4} K^{-1} and 5×10^{-4} K^{-1} for common IR diode lasers. Table 1.6 summarizes the mode wavelength shifts caused by the temperature dependence of the refractive index and bandgap energy for some laser material systems.

Table 1.6   Lasing mode wavelength shifts caused by the temperature dependence of the refractive index and bandgap energy (in brackets) for some diode laser material systems in the infrared range.

$\delta\lambda_m/\delta T_j$ [nm/K]	Laser material	Lasing wavelength band [μm]
0.08 (0.25)	AlGaAs/GaAs	0.85
0.10 (0.40)	InGaAsP/InP	1.32
0.12 (0.60)	InGaAsP/InP	1.55

As can be seen from the table, the wavelength shift caused by the change of the bandgap energy with temperature is about 3–5 times higher compared to that caused by the refractive index change.

Finally, we want to mention the phenomenon of *mode hopping*. Steps can occur in the emission wavelength versus injected current (temperature) plot, which can be attributed to longitudinal mode hopping, at which the shift of peak gain causes the dominant mode to change to an adjacent, higher wavelength longitudinal mode with lower order number m, resulting in a wavelength shift of one FP mode spacing for each mode hop. Mode hopping is suppressed until the gain at the adjacent mode is higher than that at lasing. Mode hopping is manifested by small ripples or kinks in the P/I characteristic, which can be very detrimental for certain applications of single-mode diode lasers.

1.3.7   Operating characteristics

In this introductory section we give an overview of the electrical characteristics based on optical output power P, voltage drop V across the laser, and drive current I through the laser and comprising P/I and V/I curves, together with the derivative characterizations, which include the characteristics dP/dI, d^2P/dI^2, dV/dI, and $I\,dV/dI$. In the following four subsections, we then discuss laser ef ciency, optical power, and temperature characteristics along with the relevant expressions, and describe how to measure internal parameters such as the intrinsic optical loss α_i and quantum ef ciency η_i.

With increasing injection current, a diode laser is driven from a low bias state with spontaneous emission to a state where stimulated emission is dominant enough to generate suf cient gain to balance all losses for ef cient light ampli cation. This transition point occurs at laser threshold (Figure 1.30a). At threshold, the spontaneous emission clamps as the carrier density and thus the optical gain clamps because most injected carriers are immediately converted to lasing light. This is due to the stimulated emission process, which is much faster by a factor of 10^3–10^4 than the processes of spontaneous and nonradiative recombination. In fact, the carrier density above threshold still increases with current with a slope τ_{stim}/qd (see Equation 1.34) determined by the stimulated emission lifetime and which is lower

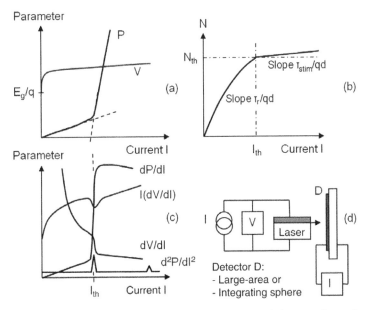

Figure 1.30 Schematic representation of diode laser characteristics: (a) direct characteristics including power P versus current I and voltage V versus I; (b) injected carrier density N versus input current I illustrating the clamping effect of the carrier density at threshold causing the gain to clamp too; (c) derivative characteristics including four frequently used derivative curves to investigate diode lasers; (d) simple setup for measuring diode laser optical power/voltage/current characteristics.

by the above factor for the regime below threshold. Figure 1.30b illustrates the carrier clamping effect. Above threshold, the coherent emission power grows steadily with current.

Below threshold, the V/I characteristic is similar to that of an ideal heterojunction diode with a parasitic series resistance R_s comprising the contact resistance and the resistance of the various layers. The voltage turns on when the applied voltage exceeds the bandgap voltage of the active layer and rapidly increases, but above threshold the voltage V_d across the diode laser saturates because the carrier density saturates and the measured voltage drop V across the laser then becomes $V = V_d + IR_s$ (Figure 1.30a).

The value of the measured threshold current depends on the applied evaluation method. The methods are as follows: (i) linear t: takes I_{th} at the point at which a straight-line t to the linear portion of the P/I curve above threshold intercepts the current axis; (ii) two-segment t: takes I_{th} at the point at which a straight-line t to the linear portion of the P/I curve above threshold intercepts the straight-line t to the linear portion of the P/I below threshold; (iii) rst derivative: takes I_{th} at the point at 50% of the maximum of the rising edge of the dP/dI curve; and (iv) second derivative: takes I_{th} at the point at which d^2P/dI^2 is a maximum (Hertsens, 2005).

For commercial applications, it is important to note that the last three methods are compatible with the standards Telcordia GR-468-CORE and GR-3010-CORE (Telcordia Technologies, 2010).

There are four frequently used derivative curves calculated from the P/I and V/I characteristics, which can be very useful in sorting out any nonlinearities showing up in the P/I and V/I. Figure 1.30c shows a schematic representation of these derivative characteristics, and Figure 1.30d shows a simple setup for measuring laser power, voltage, and current.

In addition to the threshold current, the dP/dI versus drive current characteristic also supplies the instantaneous slope ef ciency above threshold and is sensitive to nonlinearities and kinks in the P/I curve. A P/I kink is usually linked to a change in the optical mode parallel to the active layer in the form of mode movement, mode transition, or excitation of higher order modes. Also the turn-on of current leakage paths at higher currents, which may lead to a rollover of the P/I can be identi ed with the dP/dI.

The d^2P/dI^2 curve is mainly used for determining the threshold current, but diode laser manufacturers use it also to determine the exact location of the kink caused by the activation of higher order transverse lateral modes in addition to the fundamental mode. The location of the kink in the P/I gives the so-called kink-free optical power up to that where the laser operates in the fundamental single mode and is a key parameter for pumping ber ampli ers with high single-mode power.

The dV/dI versus I plot supplies the dynamic resistance curve and is the effective resistance to a change in current at a certain current. Typically, there is a downward shift in this curve at lasing threshold separating the spontaneous emission operation (light-emitting diode, LED) from the stimulated emission operation (diode laser).

The $I\,dV/dI$ versus I plot is a powerful tool to measure series and parallel, linear and nonlinear resistive circuit elements and to identify shunt current paths. As an example, the information contained in the dV/dI drop at threshold can be quanti ed by using the known V/I equation for an ideal diode with a series resistance R_s as described above and given by $I = I_s \exp\{(qV_d)/(nk_BT) - 1\}$, and by taking the derivative of the voltage across the laser $V = V_d + IR_s$. I_s is the saturation current and n is the diode ideality factor. By neglecting the second term in the rst expression because the exponential term is usually much greater than one, we nally obtain $I\,dV/dI = nk_BT/q + IR_s$ for below threshold and $I\,dV/dI = IR_s$ above threshold (voltage clamps because carrier density clamps). This means that the slope of $I\,dV/dI$ versus I curve is R_s above and below threshold, but there is a drop of nk_BT/q at threshold. In the presence of a shunt path across the diode laser, which can occur in strongly index-guided lasers, investigations showed that a peak in the measured $I\,dV/dI$ versus I curve can be present before threshold is reached (Agrawal and Dutta, 1993; Wright et al., 1982). From all these data, useful information about the junction characteristic of the diode laser can be calculated, including ideality factor and contact resistance, and laser manufacturers use these data for laser process control.

Finally, it should be mentioned that a kink in the P/I curve can generally also be observed as a kink in the $I\,dV/dI$ versus I characteristic. This can be understood from the fact that a change in the optical mode changes also the average carrier density in the active layer, which ultimately generates a change in voltage via the adjustment of the quasi-Fermi levels.

1.3.7.1   Optical output power and efficiency

The power emitted by stimulated emission into the modal volume due to a current density $J > J_{th}$ can be written as

$$P_{stim} = A\,(J - J_{th})\,\frac{\eta_i h\nu}{q} \tag{1.47}$$

where A is the junction area and η_i is the internal differential quantum ef ciency, which represents the fraction of injected carriers that recombine radiatively and generate photons (cf. Equation 1.34 and accompanying text). Equation (1.47) expresses the effective number of photons with energy $h\nu$ generated in the modal volume per second by the number of injected carriers per second. A fraction of this stimulated power is dissipated inside the cavity via the distributed losses α_i and the other fraction is coupled through the cavity end mirrors as useful laser output. These two power fractions are proportional to the effective internal loss α_i and mirror loss $\alpha_m = (1/2L)\ln(1/R_1 R_2)$, and the output power (in units of W or mW) through both mirrors can thus be written as a function of current above threshold

$$P_{out} = \frac{\eta_i h\nu}{q}\,(I - I_{th})\,\frac{\dfrac{1}{2L}\ln\dfrac{1}{R_1 R_2}}{\alpha_i + \dfrac{1}{2L}\ln\dfrac{1}{R_1 R_2}}\;[\text{W or mW}]. \tag{1.48}$$

The ratio between the optical powers emitted at the two mirrors can be derived as (Agrawal and Dutta, 1993)

$$\frac{P_{out,1}}{P_{out,2}} = \frac{(1 - R_1)}{(1 - R_2)}\sqrt{\frac{R_2}{R_1}}. \tag{1.49}$$

For $R_1 = R_2$ we get $P_{out,1} = P_{out,2} = {}^1\!/_2 P_{out}$, and for all values of R_1 and R_2 the sum of the two emitted powers is P_{out}, since a change in facet re ectivity does not affect the total power, only its partition between the two facets. The power re ectivity of a facet can be determined by using Equation (1.49).

The external differential quantum ef ciency η_d is de ned as the ratio of the photon output rate to the photon generation rate that results from an increase in the carrier injection rate

$$\eta_d = \frac{d\left(\dfrac{P_{out}}{h\nu}\right)}{d\left(\dfrac{I - I_{th}}{q}\right)} = \eta_i\,\frac{\dfrac{1}{2L}\ln\dfrac{1}{R_1 R_2}}{\alpha_i + \dfrac{1}{2L}\ln\dfrac{1}{R_1 R_2}} = \eta_i\,\frac{\alpha_m}{\alpha_i + \alpha_m}. \tag{1.50}$$

Inserting Equation (1.50) in Equation (1.48) we then get for the total output power

$$P_{out} = \frac{h\nu}{q}\eta_d\,(I - I_{th})\,[\text{W or mW}]\,. \tag{1.51}$$

While the external differential quantum efficiency is limited by energy conservation to $\eta_d < 1$, the slope of the P/I characteristic, called slope efficiency η_{sl}, depends on the lasing wavelength, and is given in units of W/A or mW/mA by

$$\eta_{sl} = \frac{\mathrm{d}P_{out}}{\mathrm{d}I} = \eta_d\frac{h\nu}{q} = 1.24\eta_d\frac{1}{\lambda\,[\mu\mathrm{m}]}\left[\frac{\mathrm{W}}{\mathrm{A}} \text{ or } \frac{\mathrm{mW}}{\mathrm{mA}}\right] \tag{1.52}$$

where the wavelength λ is measured in units of μm. Typical high slope efficiencies at room temperature are in the range of $\gtrsim 0.8$ W/A for high-quality strained 980 nm InGaAs/AlGaAs SQW lasers (Lichtenstein et al., 2004).

The overall net power conversion efficiency is an important parameter for determining the optical output power achievable from a given electrical input power. The size of electrical power required to achieve a certain optical power is important, because the dissipated power determines sensitively the strength of heating of the laser during operation, which in turn impacts laser performance and reliability (see Section 1.3.7.3 and Chapters 3–6). The electrical-to-optical conversion efficiency η_c, also called *wall-plug* efficiency, is simply the ratio between optical output power and electrical input power, given as

$$\eta_c = \frac{P_{out}}{VI} = \frac{\dfrac{h\nu}{q}\eta_d\,(I - I_{th})}{I\,(V_d + R_s I)} = \eta_i\frac{h\nu}{q}\frac{(I - I_{th})}{I\,(V_d + R_s I)}\frac{\alpha_m}{\alpha_i + \alpha_m} \tag{1.53}$$

where we used for the terminal voltage across the laser $V = V_d + R_s I$ with V_d being the diode voltage and R_s a series resistance (see Section 1.3.7 above). According to Equation (1.53), the series resistance is the main cause accounting for the small discrepancy between the energy qV furnished to each injected carrier and the photon energy $h\nu$. In practice, the applied voltage is slightly higher than the energy gap voltage of the active layer E_g/q and can be well approximated by $\approx 1.4E_g/q$. For optimum coupling of light out of the cavity, α_m can be made much larger than α_i, and under these conditions η_c approaches η_i with the consequence that, if the internal quantum efficiency is large, then the wall-plug efficiency can also be made large. Conversion efficiencies for diode lasers have been increased above the 70% level in recent years.

Government-funded programs such as the super high-efficiency diode sources (SHEDS) program administered by the Defense Advanced Research Projects Agency (DARPA, 2011) in the USA have set the goal to push the conversion efficiency to the 80% mark and beyond (Stickley et al., 2006).

In Section 2.4.5.3, we present a comprehensive discussion on diverse measures to maximize optical output power and power conversion efficiency.

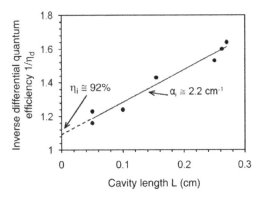

Figure 1.31   Experimental reciprocal differential quantum ef ciencies versus cavity lengths for compressively-strained $In_{0.2}Ga_{0.8}As/Al_{0.3}Ga_{0.7}As$ GRIN-SCH SQW ridge lasers. The internal quantum ef ciency η_i and internal optical loss α_i were derived from the plot according to Equation (1.54).

1.3.7.2   Internal efficiency and optical loss measurements

The external differential quantum ef ciency η_d increases with decreasing cavity length L and can be used to measure the internal quantum ef ciency η_i and internal optical loss α_i. Using Equation (1.50) we obtain

$$\frac{1}{\eta_d} = \frac{1}{\eta_i}\left\{1 + \frac{2\alpha_i}{\ln 1/R_1 R_2}L\right\} = \frac{1}{\eta_i}\left\{1 + \frac{\alpha_i}{\ln 1/R_m}L\right\} \qquad (1.54)$$

where $R_m = R_1 = R_2$ is the mirror re ectivity assumed for simplicity to be equal for both facets. Values for η_i and α_i can be extracted from experimental data by plotting measured values of $1/\eta_d$ versus L for lasers cleaved to different lengths, where the re ectivities are known or calculated for uncoated mirrors according to Equation (1.12). Values for η_d can be determined from Equation (1.52) by using experimental values for the slope ef ciency η_{sl} and lasing wavelength λ. According to Equation (1.54), the intercept of the plotted data with the $1/\eta_d$ axis gives $1/\eta_i$ and the slope of the plotted data yields $2\alpha_i/\eta_i \ln(1/R_1R_2)$ or $\alpha_i/\eta_i \ln(1/R_m)$, which can be used with η_i to get α_i.

Figure 1.31 plots a set of typical data taken from in-plane, 980 nm strained-layer InGaAs/AlGaAs GRIN-SCH SQW ridge lasers, from which a distributed internal optical loss coef cient $\alpha_i \cong 2.2$ cm^{-1} and internal quantum ef ciency $\eta_i \cong 92\%$ could be derived.

1.3.7.3   Temperature dependence of laser characteristics

Usually the threshold current increases and the slope ef ciency decreases with increasing device temperature caused by laser self-heating at high drive currents or ambient heating effects. Ideally, it is desirable to eliminate or at least minimize these detrimental temperature effects because they limit sensitively the application, in particular, of high power lasers, and impact not only laser performance but also laser reliability. The temperature dependence of the threshold current and the slope ef ciency can usually be described by the empirical expressions

$$I_{th,2} = I_{th,1} \exp\{(T_2 - T_1)/(T_0)\} \tag{1.55}$$

and

$$\Delta I_2 = \Delta I_1 \exp\{(T_2 - T_1)/(T_\eta)\}, \tag{1.56}$$

respectively. Here T_1 and T_2 are two closely spaced temperatures with $T_2 > T_1$, and T_0 and T_η are characteristic temperatures expressing the temperature sensitivity of the threshold current and slope ef ciency, respectively. The ΔI terms in Equation (1.56) denote the above-threshold-current increment required to obtain a desired output power at both temperatures. Note that both characteristic temperatures are sensitively related to laser materials, heterostructure design, and external parameters such as laser length, width, facet re ectivities, and device heat sink ef ciency. They usually decrease as the laser junction temperature increases and smaller values indicate a larger dependence of lasing characteristics on temperature.

In general, T_0 values are higher in wider bandgap materials as opposed to lower bandgap materials. Near room temperature, the experimental values are greater than 120 K for 850 nm GaAs/AlGaAs DH lasers compared to lower values in the range $\approx$50 to 70 K for InGaAsP/InP DH lasers emitting between 1.3 and 1.55 μm. GaAs/AlGaAs QW lasers have higher values in the range $\approx$150 to 180 K and strained InGaAs/AlGaAs QW devices furnished highest values of >200 K. Considering only the structure dependence of T_0, one can derive the general relationship T_0(MQW) > T_0(GRIN-SCH) > T_0(SCH) (Weisbuch and Vinter, 1991). Experimental values for T_η are typically higher by a factor of 2–3 than values for T_0: for example, $T_0 = 115$ K and $T_\eta = 285$ K for 730 nm emitting InGaAsP/InGaAlP SCH SQW lasers (Al-Muhanna et al., 1998). An example may illustrate the effect by comparing the ratios of I_{th} at 20 °C and 80 °C for a GaAs/AlGaAs QW laser with $T_0 = 160$ K, and an InGaAsP/InP QW laser with $T_0 = 55$ K. The threshold current ratio amounts to $I_{th}(80\ °C)/I_{th}(20\ °C) = 1.47$ for the former and 2.78 for the latter. This indicates a greater variation in threshold current with temperature for the InP laser.

The characteristic temperature is actually not a physical parameter but a tting parameter, which is linked to four main physical mechanisms, which can explain the temperature characteristics, and which include an increase in the following parameters with temperature:

- Gain width and hence a gain peak height decrease.

- Nonradiative recombination including Auger and processes via surface states and deep levels.

- Optical losses including free-carrier and intravalence band absorption effects.

- Thermal escape of injected carriers over con ning barriers into the con ne-ment and cladding layers, and resistive leakage currents through blocking layers.

Looking at Equation (1.35), we can identify three factors, namely, transparency density N_{tr}, differential gain σ, and optical loss coef cient α_a in the active layer, which show a strong temperature dependence. Thermal broadening of carrier distri-butions due to the temperature dependence of the Fermi–Dirac function introduces a temperature dependence in the parameters N_{tr} and σ, and can be written as $N_{tr} \propto T$ and $\sigma \propto 1/T$, which means an increase of N_{tr} and decrease of σ, because electron and hole energies are spread over a greater range at higher temperatures. Nonradia-tive recombination via temperature-activated defects and Auger processes will affect the internal quantum ef ciency η_i and cause a drop in η_i at higher temperatures. Intervalence band and free-carrier absorption give rise to a temperature dependence of both loss terms α_a and α_{cl}, which are proportional to temperature T to a good approximation.

Thermal spillover of carriers over the heterobarriers, while not included explicitly in Equation (1.35), has been found to be a substantial factor affecting the temperature characteristics. Carrier leakage also reduces the internal quantum ef ciency η_i and its strength is very dependent on the materials and heterostructure used. It is smaller in QW and strained QW than in bulk diode lasers due to the lower threshold current density and higher con ning barriers. With high threshold current devices the injected carrier density pushes the quasi-Fermi level up, leading to a population of higher order subbands in the well (*band filling* effect) with the consequence that the higher energy carriers, especially electrons, easily escape from the active well into adjacent layers. The occupancy of optical cavity states plays a decisive role in the difference between MQW, GRIN-SCH, and SCH structures. The goal therefore includes reducing the density of states at the energy levels most likely to be populated by *hot* carriers escaping the active layer at elevated temperatures.

Even in optimized structures, the quasi-Fermi level is so high in the conduction band that population of the cavity states occurs. It can be shown by calculation that, at threshold, the number of carriers is about the same in the SCH optical cavity and in the active layer, but only 20% of the carriers are in a GRIN-SCH structure due to the reduced density of states in the triangular optical cavity (Weisbuch and Vinter, 1991). It can also be shown that this detrimental population of optical cavity states is lowest with MQW structures. In a MQW, the bulk density of states (DOS) is the farthest away in energy and the high-energy tail of the Fermi–Dirac distribution populates mainly the 2D-DOS of the well structure. In evaluating the relative strength of carrier leakage, these DOS effects can be considered as the primary causes for the relationship given above for T_0 of MQW, GRIN-SCH, and SCH lasers.

Auger recombination and intravalence band absorption are the major factors for the reduced characteristic temperatures of low-bandgap InGaAsP/InP lasers. Carrier leakage also plays a role due to Auger recombination, because hot electrons with energies higher than the band discontinuities are generated in the transition. Auger recombination and intravalence band absorption can be mitigated by modifying the valence band structure in strained QW structures (cf. Section 1.1.4.1). The Auger recombination rate is also lower in a MQW than a SQW laser. This follows from the fact that the Auger rate varies as N^3 where the carrier density at threshold is $N_{th} \cong 2.5 \times 10^{18}$ cm^{-3} for a SQW and $\cong 8 \times 10^{17}$ cm^{-3} (Agrawal and Dutta, 1993) for a MQW laser, leading to a lower Auger rate by a factor of 30 and a corresponding improvement in T_0 for a MQW laser. The internal optical loss can be reduced by using QW structures due to the very low optical con nement and thus helps to improve the temperature characteristics.

The key factors affecting the temperature characteristics of the slope ef ciency include free-carrier absorption and intravalence band absorption in the active layer and con nement layers, with the effect of increasing the overall optical loss coef cient α_i and decreasing the internal quantum ef ciency η_i with increasing temperature. Other contributing factors such as Auger recombination and carrier over ow play a minor role, because the rate of these processes is practically constant and independent of temperature, since the carrier density is clamped at threshold and does not actually change in the slope ef ciency regime at current levels beyond threshold.

1.3.8   Mirror reflectivity modifications

Optimization of the diode laser performance requires modi cation of the intrinsic power re ectivity of cleaved facets, which is about 0.32 for GaAs (cf. Equation 1.12), by creating one end mirror with high re ectivity and the other facet with a low re ectivity for coupling out the laser power. However, there are tradeoffs to be considered, such as the threshold current decreases but the differential quantum ef ciency decreases too (cf. Equation 1.50) with increasing mean mirror re ectivity (see Figure 1.32). To maintain a reasonable differential quantum ef ciency one would need to reduce the internal loss of the waveguide.

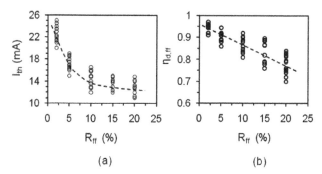

Figure 1.32   Plots of experimental threshold currents I_{th} (a) and front-facet differential quantum ef ciencies $\eta_{d,ff}$ (b) versus front-facet re ectivities R_{ff} for ridge waveguide compressively-strained InGaAs/AlGaAs GRIN-SCH SQW lasers 4 µm wide. Dashed lines are trendlines.

The re ectivity of cleaved facets can be modi ed by dielectric coatings. Coating the facets with appropriate thin dielectric layers has the additional advantage of passivating and protecting the sensitive mirror facets from degradation effects and thereby enhancing considerably the effective laser output and damage level. Chapters 3 and 4 will deal in detail with degradation modes and laser facet passivation technologies. Figure 1.33a shows a typical coating of the low-re ectivity front mirror and high-re ectivity back mirror for an edge-emitting, high-power FP diode laser. It also illustrates the re ection and transmission of light at the interfaces of a single dielectric lm on a semiconductor. Stringent requirements are imposed on the optical robustness of laser mirror coatings, which include:

- high transparency at the lasing wavelength;

- chemical, stoichiometric, and mechanical stability under high optical power exposure and extreme environmental conditions;

- excellent adhesion to facet surface;

- ideally zero mechanical stress;

- prevention or suppression of gradual and sudden laser facet degradation mechanisms;

- high reliability and long lifetimes under application-speci c conditions.

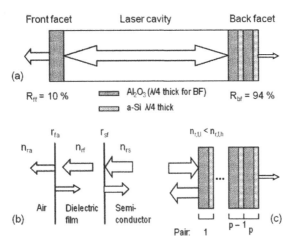

Figure 1.33 Schematic illustration of mirror coating approaches for high-power edge-emitting diode lasers. (a) Typical low-re ectivity front-facet single-layer coating and high-re ectivity back-facet (BF) Bragg stack coating. (b) Re ection and transmission of light at the surfaces of a semiconductor laser facet coated with a single dielectric lm where r_{fa} and r_{sf} denote the re ection coef cients at the surfaces of the lm/air and semiconductor/ lm, respectively. (c) Re ection and transmission at a Bragg mirror stack consisting of pairs ($p =$ number) of dielectric layers each a quarter-wavelength thick and each pair comprising a high refractive index $n_{r,f,h}$ and low refractive index $n_{r,f,l}$ lm.

If there are no optical absorption and scattering losses in the dielectric, the power reflectivity R_f of the single film with thickness d_f and refractive index $n_{r,f}$ can be expressed by the equation for normal incidence

$$R_f = \frac{r_{sf}^2 + r_{fa}^2 + 2r_{sf}r_{fa}\cos 2\beta}{1 + r_{sf}^2 r_{fa}^2 + 2r_{sf}r_{fa}\cos 2\beta} \tag{1.57}$$

where

$$r_{sf} = \frac{n_{r,s} - n_{r,f}}{n_{r,s} + n_{r,f}} \tag{1.58}$$

$$r_{fa} = \frac{n_{r,f} - n_{r,a}}{n_{r,f} + n_{r,a}} \tag{1.59}$$

$$\beta = \frac{2\pi}{\lambda_0} n_{r,f} d_f \tag{1.60}$$

and $n_{r,s}$ and $n_{r,a}$ are the refractive indices of the semiconductor and air, respectively, and λ_0 is the lasing wavelength in vacuum (Figure 1.33b). Equation (1.57) was derived by Born and Wolf (1999 [originally 1959]) on the basis of the characteristic (transfer) matrix for the film within the Fresnel reflectivity formalism for plane electromagnetic waves. Changes for the *modal* reflectivity of a diode injection laser are expected to be no greater than 15% (Ikegami, 1972). The reflectivity in Equation (1.57) changes periodically with β, that is, with a periodicity in thickness of $\lambda_0/(2n_{r,f})$ of the dielectric film. The maximum value of R_f can be found when $\cos 2\beta = 1$, that is, for thicknesses $d_f = m\lambda_0/2n_{r,f}$ (for $m = 1, 2, 3, \dots$) and becomes for normal incidence

$$R_{fmax} = \left(\frac{n_{r,s} - n_{r,a}}{n_{r,s} + n_{r,a}}\right)^2 \tag{1.61}$$

which is independent of $n_{r,f}$. R_{fmax} is the natural reflectivity for an uncoated mirror. The minimum value of Equation (1.57) is achieved at $\cos 2\beta = -1$, that is, for thicknesses $d_f = m\lambda_0/4n_{r,f}$ (for $m = 1, 3, 5, \dots$) and becomes for normal incidence

$$R_{fmin} = \left(\frac{n_{r,s}n_{r,a} - n_{r,f}^2}{n_{r,s}n_{r,a} + n_{r,f}^2}\right)^2. \tag{1.62}$$

This equation gives the condition for an anti-reflective coating, which would be strictly achieved at normal incidence for

$$n_{r,f} = \sqrt{n_{r,s}n_{r,a}}. \tag{1.63}$$

An effective anti-reflective (AR) coating on a 980 nm InGaAs/AlGaAs laser facet can be achieved by depositing a thin Al_2O_3 film of thickness $d_f = m\lambda_0/4n_{r,f}$

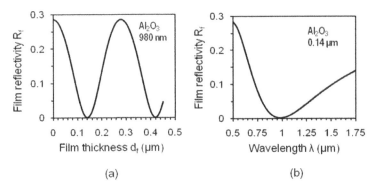

(a)   (b)

Figure 1.34 (a) Plot of the reflectivity R_f versus Al$_2$O$_3$ film thickness d_f calculated according to Equation (1.57) for the low-reflectivity front facet of a 980 nm InGaAs/AlGaAs laser. A 10% front-mirror reflectivity can be achieved at a thickness of 190 nm. (b) Calculated reflectivity R_f versus wavelength λ for 140 nm Al$_2$O$_3$ film thickness as found at the first minimum in the R_f versus d_f dependence in (a). The minimum in (b) is close to 980 nm as expected.

(for $m = 1, 3, 5, \ldots$), because Al$_2$O$_3$ with an index $n_{r,f} \cong 1.76$ at a wavelength of 980 nm meets the condition in Equation (1.63) well by using $n_{r,s} \cong 3.28$ for Al$_{0.3}$Ga$_{0.7}$As. Al$_2$O$_3$ has all the properties expected from an effective, efficient, and reliable mirror coating material (see above).

Figure 1.34a plots the reflectivity versus Al$_2$O$_3$ film thickness calculated according to Equation (1.57) for the low-reflectivity front facet of a 980 nm InGaAs/AlGaAs laser. As expected, the periodic dependence has a periodicity in thickness of $\lambda_0/(2n_{r,f})$ and the maxima and minima points are close to 29% and 0.1%, respectively. A 10% front-mirror reflectivity can be achieved at a thickness of 190 nm. The reflectivity versus wavelength dependence calculated for 140 nm Al$_2$O$_3$ thickness, as found at the first minimum in the R_f versus d_f dependence, is plotted in Figure 1.34b. The minimum in the figure is close to 980 nm as expected.

Periodically stratified films are highly reflective coatings. They are formed by so-called Bragg stacks, which are pairs of films with high $n_{r,f,h}$ and low refractive index $n_{r,f,l}$ and a thickness of each film set at a quarter-wavelength of $\lambda_0/4n_{r,f}$ (see Figure 1.33c). The constructive interference in the multilayer stack results in dramatically high reflectivity values, which can be increased close to unity by increasing the number p of periods of double films and the index ratio $n_{r,f,h}/n_{r,f,l}$ between the high and low refractive index films. For normal incidence the reflectivity of such a Bragg reflector has been derived by using the transfer matrix method (Born and Wolf, 1999) and is given by

$$R_{stack} = \left(\frac{1 - \left(\dfrac{n_{r,f,h}}{n_{r,s}} \right) \left(\dfrac{n_{r,f,h}}{n_{r,a}} \right) \left(\dfrac{n_{r,f,h}}{n_{r,f,l}} \right)^{2p}}{1 + \left(\dfrac{n_{r,f,h}}{n_{r,s}} \right) \left(\dfrac{n_{r,f,h}}{n_{r,a}} \right) \left(\dfrac{n_{r,f,h}}{n_{r,f,l}} \right)^{2p}} \right)^2 . \tag{1.64}$$

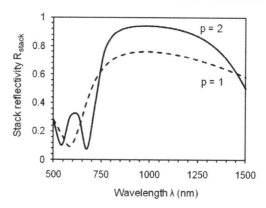

Figure 1.35 Spectral re ectivity of a periodic multilayer Bragg re ector calculated for the high-re ectivity back mirror of a 980 nm InGaAs/AlGaAs laser. Two pairs ($p = 2$) of quarter-wavelength-thick Al_2O_3 and a-Si layers provide a high re ectivity of >90% in a broad wavelength range ~850 to 1200 nm with a maximum of 94% at 980 nm.

The spectral re ectivity has been calculated with Equation (1.64) for the high-re ectivity (HR) back mirror of a 980 nm InGaAs/AlGaAs laser by using the standard coating materials of Al_2O_3 with $n_{r,f,l} = 1.76$ and a-Si (amorphous-Si) with $n_{r,f,h} = 3.69$ (Ioffe Physico-Technical Institute, 2006). Figure 1.35 shows a re ectivity of 0.76 for $p = 1$ and 0.94 for $p = 2$ at $\lambda = 980$ nm. A re ectivity of ≥ 0.9 is achieved in a rather broad wavelength range ~850 to 1200 nm. Alternative coating material combinations of TiO_2 ($n_{r,f,h} = 2.78$) with SiO_2 ($n_{r,f,l} = 1.45$) offer the advantage of developing reduced facet heating by laser light absorption due to the negligible absorption of TiO_2 at wavelengths $\gtrsim 450$ nm. However, due to the lower index of TiO_2, three periods, $p = 3$, would be required to obtain a re ectivity > 90%.

1.4   Laser fabrication technology

The purpose of this section is to give a brief overview of key steps for fabricating edge-emitting laser devices including laser wafer growth, processing, and packaging. While a variety of epitaxial growth techniques have been used to grow diode lasers, the two most important techniques for the growth of high-power diode lasers for both research and commercial applications are molecular beam epitaxy (MBE) and organometallic vapor-phase epitaxy (OMVPE), also referred to as metal–organic chemical vapor deposition (MOCVD). MOCVD, which comes in low-pressure (~0.1 atm) and atmospheric pressure systems, uses metal–organic precursors such as group-III alkyls and group-V hydrides as gas sources to react on a heated substrate to form the epitaxial lm. In contrast, MBE is carried out under ultrahigh vacuum (UHV; $<10^{-10}$ Torr) with thermal beams of atoms evaporated from heated effusion cells and directed to the atomically at and contaminant-free surface of a substrate held at high temperature to form the layer.

In the following, we focus on the MBE technique and describe some of the major issues, which have to be considered in the growth of a standard InGaAs/AlGaAs GRIN-SCH SQW laser structure on n-GaAs substrates. We also discuss the key steps to be taken in processing the laser wafer to fabricate a nished ridge waveguide diode laser chip (cf. Section 1.3.5.2 and Figure 1.26; see also Sections 2.1.3 to 2.1.6). Finally, we discuss diode laser packaging issues with an emphasis on single-mode, in-plane diode lasers along with associated materials, processes, and components, common package designs including optical alignment and coupling, thermal management, and atmosphere control requirements inside the package.

1.4.1   Laser wafer growth

UHV analysis techniques that can be applied include re ection high-energy electron diffraction (RHEED), Auger analysis, and mass spectrometry for monitoring the growth rate and studying the layer characteristics; the latter can also be achieved by optical techniques such as re ectometry. Usually the growth temperature is monitored by a single-wavelength optical pyrometer in conjunction with a thermocouple to provide a complementary temperature reading. The UHV conditions are maintained by ion pumps, cryopumps, and titanium sublimation pumps, and, in addition, liquid-nitrogen- lled cryoshrouds surround the effusion cells and the substrate to protect the substrate from any contamination by impurities.

1.4.1.1   Substrate specifications and preparation

Usually n-type GaAs substrates ≥ 2 inches with a (100) orientation are used. However, misorientations such as $2°$ toward the $\langle 110 \rangle$ direction reduce the formation of loop dislocations at the substrate/epitaxial layer interface, or $3–4°$ toward $\langle 111 \rangle$A yield a reduced incorporation of impurities like oxygen in AlGaAs, and smoother and sharper heterointerfaces (Chand *et al.*, 1994). The substrates are silicon doped, typically 1.5×10^{18} cm^{-3}, and have a low dislocation density measured by the density of etch pits of $<2 \times 10^3$ cm^{-2}. Slight substrate misorientations and lower defect densities result in improved laser performance and lifetimes.

The substrates are cleaned to remove any contamination layer on their surfaces. The process is performed in a class 100 clean-room environment and includes the following steps: cleaning in sulfuric acid, rinsing in owing deionized water, etching in a solution of deionized water with ammonia and hydrogen peroxide, and nal rinsing in deionized water. Each of these steps lasts between 1 and 2 min. The substrate is dried either by blowing dry with a nitrogen gas gun or by spinning in a spin processor, and then returned into the original container or loaded into the MBE system.

1.4.1.2   Substrate loading

Loading and unloading the growth chamber is achieved via a UHV buffer and load–lock chamber to minimize the introduction of detrimental impurities such as oxygen and water into the actual growth chamber. The load–lock is pumped to UHV conditions. A stress-free mounting of the substrate in the holder and a smooth transfer inside the system are required to ensure defect-free growth.

1.4.1.3   Growth

In a solid-source MBE, group-V elements are much more volatile than group-III elements. This effect is used for stoichiometry control where usually the substrate temperature is kept suf ciently low to increase the group-III element sticking coef cient (fraction of atoms in the beam sticking to the substrate) close to unity. The growth rate is then determined by the group-III element   ux and the group-V element   ux is adjusted to multiple times of that   ux. A typical V/III (e.g., As₄/Ga) beam equivalent pressure ratio p_{As_4}/p_{Ga} is 25. The sticking coef cients for all the components have to be determined empirically as a function of substrate temperature and beam   ux. Substrate temperatures T_{sub} for AlGaAs growth have to be high, in the range $\cong$700 to 720 °C, to avoid the formation of deep-level defects and nonradiative recombination centers (As $et\ al.$, 1988). In contrast, the growth temperature for the InGaAs QW is $\cong$510 °C, which requires a growth interruption just prior to the QW growth to allow the temperature to decrease from the higher temperature at AlGaAs growth to the lower value required for the QW growth. It is known that there is a tradeoff between the optimum V/III ratio for AlGaAs and InGaAs growth: InGaAs prefers a lower ratio (lower As₄   ux) whereas AlGaAs growth quality degrades if the   ux is too low. Useful As₄ beam equivalent pressure values are $\sim$0.8 × 10⁻⁵ Torr and $\sim$1.2 × 10⁻⁵ Torr for optimum growth of InGaAs and AlGaAs, respectively, which can be achieved, for example, by rapid switching between two separate As effusion cell sources for depositing the two materials at different As   uxes.

To establish a safe operating regime with optimum and stable laser parameters it is essential to know the sensitivity of key growth parameters such as temperature and As₄ pressure on laser performance parameters like threshold current density J_{th} and internal quantum ef ciency η_i. Experiments yielded the following results (Epperlein, 1999): roughly linear dependencies of

- J_{th} on T_{sub} with a slope of −40 A/cm² per 10 °C and η_i on T_{sub} with a slope +0.03 per 10 °C for ± 10 °C range of change in standard growth temperature; and

- J_{th} on p_{As_4} with a slope of +45 A/cm² per 1 × 10⁻⁵ Torr and η_i on p_{As_4} with a slope of −0.04 per 1 × 10⁻⁵ Torr for a range of change of twice the standard As₄ beam equivalent pressure.

By considering the above relationships and many other necessary dependencies and procedures such as the generation of growth parameter calibrations, effusion cell outgassing protocols, and initial growth conditions (details are beyond the scope of this book), the laser growth program has to be established and   nally run and controlled by computer. The program details the exact growth conditions, including growth times, temperature ramp functions, effusion cell temperatures, and shutter modulations used during the growth for each building block in the vertical structure to enable precisely controlled and reproducible layer thicknesses, compositions, and doping pro les. The layer sequence includes a thin GaAs buffer layer on top of the substrate surface, rst cladding grading, n-AlGaAs cladding layer, AlGaAs GRIN-SCH layer with embedded InGaAs QW, p-AlGaAs cladding layer, second cladding grading, and p^+-contact layer. Several different short AlGaAs/GaAs superlattices are positioned at speci c locations in the structure, each with its own purpose, such as to getter segregating surface impurities, to trap diffusing ions at heterojunction interfaces, to block threading dislocations, or just to serve as a useful marker in scanning electron microscopy (SEM) investigations to determine layer thicknesses.

1.4.2 Laser wafer processing

1.4.2.1 Ridge waveguide etching and embedding

One of the most critical processing steps is the formation of the ridge waveguide structure because it determines the nal optical laser parameters such as FF angles or the maximum kink-free optical output power and electrical parameters like the threshold current. The lateral waveguide pattern is de ned by a standard lithographic process and then transferred to the (Al)GaAs layers by wet chemical etching. The requirements for control over etch depth and ridge pro le are extremely stringent and include control of the former to within a few tens of nanometers (see Chapter 2). This can be achieved by suitable, epitaxially grown etch-stop layers or by precise adjustment of the etch solution and etching in several steps with intermediate depth control measurements. The etched ridge is embedded with a dielectric insulator such as Si_3N_4 deposited by a low-temperature plasma-enhanced chemical vapor deposition (PECVD) technique, but leaving uncovered the highly doped p^+-GaAs layer on top of the ridge used for the formation of the Ohmic contact. The dielectric layer is structured by a lift-off process. Ideally, waveguide etching and dielectric embedding are fully self-aligned processes. In Chapter 2, we will discuss within the single spatial mode issue other lateral waveguide formation technologies including QW intermixing, epitaxial regrowth, or the use of photonic crystals.

1.4.2.2 The p-type electrode

The structure of the p-electrode is formed by using another lithographic process, evaporative metallization and resist lift-off. A Ti/Pt/Au contact is evaporated at an angle to ensure good metal step coverage over the ridge and to minimize the stress level in the Pt lm. It is nonalloyed to avoid interdiffusion of the metallic species. A thin Ti layer acts as an adhesion promoter, Pt is used as a diffusion barrier for Au, and Au is used for the top contact layer. The barrier effectiveness depends strongly on the crystalline ne structure and stress in the Pt lm, which are strongly determined by the deposition method and process parameters of the contact. The p^+-GaAs contact layer, highly doped to $\geq 2 \times 10^{19}$ cm^{-3}, has to be kept free of damage and contamination

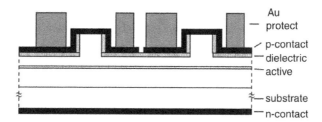

Figure 1.36   Schematic cross-section of a ridge waveguide structure illustrating the ridge protecting function of a thick patterned gold layer electroplated on the p-electrode of the wafer. Layer thicknesses and lateral dimensions not to scale.

before depositing the metal layers to ensure low speci c contact resistivity in the regime $\lesssim 1 \times 10^{-6}$ Ω cm^2. For further details, see Section 3.1.1.4 and Figure 3.2.

1.4.2.3   Ridge waveguide protection

It is advisable to deposit a thick ridge protective layer on the p-side of the wafer, which can be accomplished by electroplating Au using a patterned photoresist as a cast. Positive features of the some micrometer thick Au layers include protecting the ridge from mechanical damage during processing and handling, and providing more effective spreading of heat generated during laser operation. The waveguide protection layer also eases mounting of the laser chip p-side down onto a heat sink carrier. However, there are also negative effects, such as the generation of stress in the laser chip reaching typical levels of 2×10^8 dyn/cm^2 for as-plated Au due to the large thickness of the layer. Figure 1.36 shows schematically a cross-section of the ridge waveguide including the dielectric ridge embedding layer, n- and p-electrode layer, and thick Au ridge protective structure.

1.4.2.4   Wafer thinning and the n-type electrode

The back-side processing of the wafer comprises thinning of the wafer and deposition of the n-side electrode material. Thinning down to a thickness of about 100 μm is carried out by a lapping process in an Al$_2$O$_3$ slurry. The wafer is mounted with the protected p-side onto a glass lapping carrier by using a special wax. Protection of the p-side with the sensitive ridge structure is achieved by depositing before lapping a thin layer of polyimide, which is known for its excellent planarizing properties. Thinning is required mainly for two reasons, to facilitate a proper, defect-free cleaving of the laser facets and to reduce the thermal resistance of p-side-up mounted laser chips. After lapping, the n-side of the wafer is chemically etched to remove any lapping damage, yet not smooth enough to adversely affect adhesion of the n-metallization. The n-electrode consists of a standard AuGe/Ni/Au metal layer scheme and is evaporated onto the large substrate back-side area with subsequent alloying at high temperature below the Au/Ge eutectic point. This is to alloy the metal contact with the highly doped substrate material to deliver low-resistance metal contacts and to mitigate stress levels. According to standard practices, the n-contact is then strengthened by evaporating a second layer of metal to allow for good solderability of the laser die to a package submount.

1.4.2.5 Wafer cleaving; facet passivation and coating; laser optical inspection; and electrical testing

The nal back-end-of-line (BEOL) processing steps are as follows:

- Cleaving the wafer along crystallographic planes into bars with typical lengths of 10 mm comprising 40–80 laser chips with widths $\cong$ 100–250 μm and lengths $\cong$ 500–3000 μm (equal to bar width) depending on laser type and application.

- Passivating the cleaved facet surfaces immediately after cleaving to minimize detrimental corrosion effects. For details see Section 4.2.

- Coating the passivated facets to modify their re ectivity and to make them robust against any degradation mechanisms. The coating process is usually carried out in a PECVD or ion beam (IB) sputter deposition system depending on the actual coating material and passivation technique used. Selection of the material and thickness of the passivation layer is also determined by the fact that the PECVD process involves ions with much lower energies than the IB process. High-energy ions can damage irreversibly on impact a thin passivation layer a few nanometers thick. For details see Sections 1.3.8 and 4.2.3.

In Section 1.3.8, we discussed in detail re ectivity modi cation schemes including (i) the impact of re ectivity changes on major laser parameters, (ii) the requirements on the coating material, and (iii) quantitative expressions for calculating the required re ectivity values.

Chapters 3 and 4 will deal extensively with the physical effects linked to facet passivation and with potential passivation techniques including nonabsorbing mirror schemes. The nished devices are examined under an optical microscope for appropriate mechanical integrity and then electro-optically tested, usually on the bar level. Depending on the application, single laser chips cleaved from the bars or complete laser bars are mounted in appropriate packages t for use for electrical input/output and optical output via electrical terminals and the optical port.

1.4.3 Laser packaging

This is a comparatively brief section to conclude the chapter. In this section, we will look at some basic material, thermal, electro-optical, chemical, and geometrical requirements for packaging high-power diode lasers with a focus on single-emitter, single-mode, in-plane devices, and will refer to some current generic and new package designs for illustration, and address in particular soldering, optical alignment, coupling ef ciency, and temperature control issues. Packaging greatly in uences laser device performance and reliability.

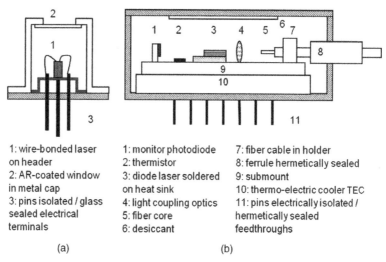

1: wire-bonded laser on header	1: monitor photodiode	7: fiber cable in holder
2: AR-coated window in metal cap	2: thermistor	8: ferrule hermetically sealed
3: pins isolated / glass sealed electrical terminals	3: diode laser soldered on heat sink	9: submount
	4: light coupling optics	10: thermo-electric cooler TEC
	5: fiber core	11: pins electrically isolated / hermetically sealed feedthroughs
	6: desiccant	
(a)	(b)	

Figure 1.37 Schematic cross-sectional view of two typical, hermetically sealable diode laser package formats for low to high output power applications. (a) Basic coaxial type TO can for devices up to 5 W with the option to attach and align a ber to the can and include a lens, TEC, photodiode, and thermistor (not shown) in multi-pin formats. (b) Typical rectangular butter y package, which is an industry standard 14-pin DIL bered package for applications up to about 5 W laser power by using a high-performance internal TEC.

1.4.3.1   Package formats

Commercially available diode lasers come in a wide variety of package formats. The vast majority of lower performance and cost-sensitive devices come in lower cost package formats, such as transistor outline (TO) can header packages. High-performance laser products including high-power pump lasers and most dense wavelength division multiplexing (DWDM) optical communications signal lasers are assembled and ber pigtailed in hermetically sealed 14-pin butter y packages or mini-dual in-line (mini-DIL) packages, just to mention the most common ones. An interesting new development is a low-pro le, uncooled, hermetically sealed, and low thermal resistance optical at package (OFP) for low-cost and high-reliability applications, which has a much smaller footprint and lower cost bill of materials (BOM) than the 14-pin butter y and can handle up to 6 W of optical output from the ber (Singh *et al.*, 2004). For illustration, cross-sectional schematic views of two typical packages are displayed in Figure 1.37. The following subsections describe the basic components and technologies related to diode laser packaging.

1.4.3.2   Device bonding

Device chip bonding, which means soldering the laser die on a monolithic planar heat sink substrate, is the rst step in assembly of the package. Usually single-mode laser chips are soldered p-side up directly to the metallized heat sink material, whereas

Table 1.7 Room temperature thermal conductivities and expansion coef cients of typical heat sink materials used for mounting optoelectronic devices. Data for the listed semiconductors are for comparison.

Material	Thermal conductivity at 300 K [W/(m K)]	Thermal expansion coefficient [ppm/K]
Diamond	2000	2.3
AlN	26	4.2
SiC	70	3.7
Al$_2$O$_3$	17	6.5
BN	600	3.7
GaAs	54	6.6
InP	70	4.5
Si	150	2.6
Ag	422	19.2
Cu	402	17.6
Al	226	23.4
W90Cu	185	5.9
W75Cu	230	9.8
Mo70Cu	195	8.1
Mo50Cu	235	10.4
C nanotubes	~3000	

high-power multimode lasers and laser bars are soldered p-side down onto the heat spreader. The heat sink material should have a thermal expansion coef cient close to that of the laser die and a high thermal conductivity; in addition, high electrical resistivity is required for high-frequency applications. Table 1.7 lists some common heat sink materials where AlN and CuW are the most popular ones for packaging diode lasers. Some data are from Fukuda (1999). As most heat sink materials are electrical insulators, they have to be metallized by depositing metals such as Ti/Pt/Au or Cr/Au to enable an effective and reproducible solder joint process.

The bonding con guration strongly in uences the temperature characteristics of the laser where a crucial factor is the wetting capability of the solder, which means the adhesion between the base metal and liquid solder. Normally, soldering is required to be ux free to prevent contamination and degradation of the facets and to assure long-term reliability. In addition, in order to prevent hot spots under high-power operation, because of the poor lateral heat conduction within the laser die, the soldering process has to be void free.

Two groups of solder metals are used: soft solders, which have low melting points (MPs), and hard solders with high melting points. Typical representatives for the rst group are In–52 wt% Sn (MP = 117 °C), Sn–60 wt% Pb (MP = 183 °C), and for the second group Sn–20 wt% Au (MP = 280 °C) and Au–88 wt% Ge (356 °C) (Fukuda, 1999). Hard solders generally cause a larger mechanical stress at the bonded part due to their higher bonding temperature, whereas soft solders absorb the mechanical stress built up at the interface between the laser die and heat sink during soldering, because they deform plastically under stress. However, soft solders become unstable during long-term laser operation, because of thermal fatigue and creep which increases in strength with decreasing melting point and when the mechanical stress is stronger than the solder's elasticity limit. This degradation and creep of the solder sensitively impacts the coupling ef ciency of the laser light into a ber. On the other hand, hard solders such as Au-rich AuSn enable long-term and reliable stability of the bonded laser die. However, AuSn shows very little creep, which may be a concern, because thermal expansion mismatches, for example, between the laser submount and the structure underneath, almost invariably lead to some degree of warpage, which cannot be allowed to vary with time in service. The mechanical stress per unit length, S_{bond}, can be approximated by (Fukuda, 1999)

$$S_{bond} = |\alpha_{HS} - \alpha_{die}| (T_{bond} - T_{amb}) E_{Ym} \qquad (1.65)$$

where α_{HS} and α_{die} are the thermal expansion coef cients of the heat sink and laser die, respectively, T_{bond} is the bonding temperature, which is close to the melting point of the solder, T_{amb} is the ambient temperature, and E_{Ym} is Young's modulus of the laser die.

An example should demonstrate the effect: bonding a GaAs laser die to an AlN material generates a compressive stress in the die at 25 °C of $\cong 6 \times 10^8$ dyn/cm^2 when a hard solder such as Au-rich AuSn ($T_{bond} \cong 280$ °C) is used and $\cong 2 \times 10^8$ dyn/cm^2 when the soft solder InSn ($T_{bond} \cong 120$ °C) is used. Here we also used 6.6×10^{-6} °C^{-1} and 4×10^{-6} °C^{-1} for the thermal expansion coef cient of GaAs and AlN, respectively, and 8.6×10^{11} dyn/cm^2 for Young's modulus of GaAs. The stress generated in the die with soft solders can easily be released via plastic deformation into the solder layer, whereas high mechanical stress levels in excess of 10^9 dyn/cm^2 caused by hard solders cannot be released into the hard solder, but will trigger the growth of slip dislocations in the laser chip, which will impact its performance and reliability.

1.4.3.3   Optical power coupling

Many diode laser applications require, but also prefer, ber delivery systems, where the laser beam is coupled into an optical ber to transport the light *to* the application.

The incident (acceptance) angle of light into a step index ber with a core refractive index $n_{r,co}$ larger than the index $n_{r,cl}$ of the cladding has to be such that the beam reaches the core–cladding interface at an angle $\geq \theta_{cr}$, the critical angle for total re ection, in order to be captured and propagated as a bound mode. The geometry of light coupling into an optical ber is illustrated in Figure 1.38.

The light-capturing capability of the ber can be expressed by the ber numerical aperture (NA), which is the sine of the largest angle θ_a contained within the cone of acceptance and is given as

$$NA = \sin\theta_a \qquad (1.66a)$$

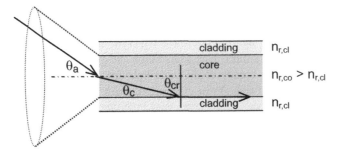

Figure 1.38 Schematics for calculating the numerical aperture of a step-index optical ber. Correlation between acceptance angle θ_a cone and critical angle θ_{cr} for total re ection at the ber core/cladding interface.

and using Snell's law we obtain after some manipulation

$$NA = \sqrt{\left(n_{r,co}^2 - n_{r,cl}^2\right)} \qquad (1.66b)$$

$$NA = n_{r,co}\sin\theta_c \qquad (1.66c)$$

where $\theta_c = 90° - \theta_{cr}$ is the supplementary angle for total re ection. Equation (1.66c) is a useful expression for the NA and relates it to the index of the core and the maximum angle at which a bound ray may propagate. The acceptance angle θ_a is then given by

$$\theta_a = \arcsin(NA) = \arcsin\sqrt{\left(n_{r,co}^2 - n_{r,cl}^2\right)}. \qquad (1.67)$$

Typical NA values for single-mode (SM) and multi-mode (MM) bers are 0.1 and 0.2–0.3, respectively. The higher the NA, the more modes in the ber, which means the larger the dispersion of this (MM) ber. The higher the NA of an SM ber, the higher its attenuation, because a signi cant proportion of optical power travels in the cladding, which is highly doped to achieve a high index contrast for SM.

In the following, we discuss the optical requirements to achieve the highest possible coupling of optical power from an SM diode laser into an SM optical ber. This can be achieved by matching both the amplitude and phase of the laser mode to the amplitude and phase of the ber mode. Technical realizations including quantitative coupling ef ciencies, dependencies, and alignment tolerances will also be discussed.

In Section 1.3.5.3, we discussed how light emitted from the usually elliptical NF spot propagates freely into space and broadens strongly in both directions by diffraction – stronger in the transverse vertical than transverse lateral direction. Important information about setting up quantitative conditions for mode matching can be obtained by determining the laser mode eld radii ($1/e^2$ intensity points) and wavefront

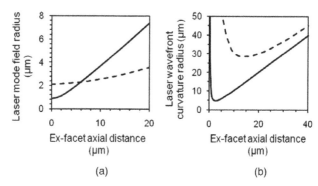

Figure 1.39 (a) Mode eld radii of an InGaAs/AlGaAs ridge laser calculated as a function of the axial distance from the facet for the transverse vertical (solid line) and transverse lateral (dashed line) direction. (b) Laser wavefront curvature radii calculated as a function of the axial distance from the facet for the transverse vertical (solid line) and transverse lateral (dashed line) direction.

radii of curvature (measure for phase) as a function of the axial distance from the laser facet for transverse vertical and transverse lateral directions.

Figure 1.39a shows the mode eld radii versus axial distance calculated for a narrow-stripe laser (Epperlein *et al.*, 2000). As expected, the mode eld radii increase with increasing axial distance, faster in the transverse vertical direction with a linear increase for $\gtrsim 5$ μm in this example. The FF angles can be taken from the asymptotic angles in the transverse vertical and lateral directions and are $\cong 22°$ and $\cong 9°$, respectively, in this case. Both curves intersect at $\cong 6$ μm where the transverse vertical and lateral mode eld radii are the same and hence the amplitude distribution is circular (see Figure 1.28c).

The shapes of the phase fronts also change in the axial direction with the phase constant at the facet, which is expressed by the large radii of the wavefront curvature in both directions (Figure 1.39b). At large distances the radii of curvature become equal to the axial distance, which means that the phase fronts are spheres centered at the facet. For both directions, the radii have minima at different locations in the case of an elliptical NF spot, whereas for circular intensity distributions the locations of the minima are the same. The distance from the facet to the minimum is called the Rayleigh range and marks the division between the NF and FF.

From Figure 1.39 we can see that there are two locations where matching the phases and amplitudes of the laser mode and ber mode (expressed by the mode eld radius, which is half the distance between the points in the ber where the electric eld amplitude decays to 1/e of its peak value) can be done best. At the facet, the phase of the laser mode is planar and therefore only the amplitude needs to be matched to the ber, which can be accomplished by using an elliptical core ber in case the NFP is elliptical. The laser phase is only planar very close to the facet, which requires the ber to be located very close to the facet to achieve high coupling. The other place is at the crossover point (Figure 1.39a) where the amplitude distribution is circular.

Mode amplitude matching can be achieved by selecting a fiber with a mode field radius, which equals that of the laser ($\cong 2$ μm in the example). Figure 1.39b shows that mode matching is not that straightforward at the mode field radii crossover point, because the radii of curvature in both directions are very different at this point ($\cong 6$ μm in the transverse vertical and >40 μm in the transverse lateral direction in the example). However, over the core of the fiber ($\cong 2 \times 2$ μm $= 4$ μm) the phase fronts are to a good approximation cylindrical and to match them a cylindrical lens is required, which is best realized by a fiber tipped with a wedge lens.

In summary, by placing a wedge-lensed fiber, which has a mode field diameter equal to the diameter of the laser intensity distribution at the position where the intensity distribution is circular, both the phase and amplitude of the modes can be matched. The best form for the wedge is a hyperbola; however, a simple wedge or double wedge is easier to fabricate and can match the phase sufficiently well to realize high ($>80\%$) coupling efficiencies, in case the mode amplitude is also well matched. If the laser mode is circular, the best form for the phase matching lens is a hyperboloid, which can be well approximated by a cone, equivalent to the simple wedge lens in the case of an elliptically shaped NF (laser mode). It should be pointed out that the lens only serves to match the phase.

Bulk optical systems can be used when the beam is allowed to expand to a size that is considerably larger than the fiber mode field diameter and is then focused with a discrete lens. The best lenses are aspheric lenses that can be designed to have no spherical aberration and hence phase matching should be good; however, they cannot match the elliptical spot of the laser to the standard circular core of the fiber, because they magnify equally in vertical and lateral planes and therefore a cylindrical or acylindrical lens would have to be added to correct this. There is, nevertheless, a very smart commercially available solution (Blue Sky Research, Inc., 2010). A diffraction-limited acylindrical μLens™ placed directly in front of the laser captures nearly 100% of the emitted power and converts the divergent elliptical output beam into a beam with a spherical wavefront and nearly circular profile. The beam behaves as if it was emitted from an ideal point source with a certain low divergence angle, and, if required, can then be precisely focused to a round spot using bulk optics.

To optimize coupling efficiencies, the optical coupling system comprising active and passive elements has to be aligned either passively or actively. In active alignment, the alignment between the components is performed under operation of the laser, whereas in passive alignment, the laser is not operated and the components are mounted on bonding pads patterned on the submount or heat sink.

Single-mode diode laser packaging invariably involves active alignment along a total of six axes, mainly due to manufacturing variances in the laser, laser assembly, and optical fiber core center. The various elements are aligned and fixed one by one and the monitored optical power is maximized. Hard soldering and laser welding are used for bonding and fixing the various components, including the diode laser on a heat sink, a photodiode, which monitors the laser output through the back facet, a thermistor that monitors the temperature, and various optical elements to the submount to achieve high reliability for a high-performance laser product in a butterfly package. The actual fiber is mounted inside a so-called fiber tube subassembly, which is welded to the wall of the package, and the fiber is threaded in through a hole and

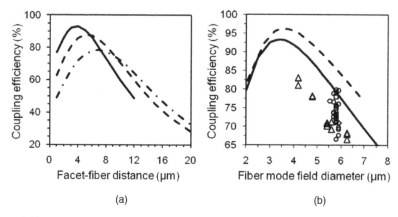

Figure 1.40 (a) Coupling efﬁciency versus laser facet–ﬁber tip distance calculated for laser light emitted under $11° \times 26°$ far-ﬁeld divergence angles into $24.5°$ wedge-lensed ﬁbers with mode ﬁeld diameters of 5.9 μm (solid line), 4.7 μm (dashed line), and 3.6 μm (dot–dashed line). (b) Coupling efﬁciency versus ﬁber mode ﬁeld diameter calculated for laser emission into $24.5°$ wedge-lensed ﬁbers (solid line) and ﬁbers with parabolically shaped lenses (dashed line). Experimental points from various different runs for coupling laser light into wedge-lensed ﬁbers.

attached to the submount or optical bench in front of the laser chip. Regarding loss of coupling due to misalignment, it is important to consider both mode amplitude and phase mismatching.

However, there is a conﬂict because systems that are more tolerant of amplitude mismatches are less tolerant of phase mismatches and vice versa. This can be illustrated by a simple example: butt coupling between two ﬁbers. For the same linear displacement, there will be a smaller mismatch of the amplitudes with a larger mode ﬁeld diameter (MFD). In contrast, for the same phase error, the smaller MFD allows a larger angular error. To achieve >90% coupling efﬁciency in a lensed ﬁber system, the lateral alignment accuracy is required to be better than 50 μm, something that takes very careful process optimization to maintain while ﬁxing the ﬁber by laser welding. The coupling efﬁciency drops from its maximum value toward larger facet–ﬁber distances with a typical rate of $\cong 5\%$ per micrometer (see Figure 1.40a).

Figures 1.40 and 1.41 show the dependence of the coupling efﬁciency on (i) the free-space distance between the laser facet and ﬁber tip, (ii) the ﬁber MFD, (iii) the FF angle of the laser in transverse lateral direction $\theta_\parallel$, and (iv) the FF angle in transverse vertical direction $\theta_\perp$ (Epperlein et al., 2000).

Key results are as follows:

- Calculated coupling efﬁciencies >90% at a facet–ﬁber gap of $\cong 4$ μm for a laser with FF angles $11° \times 26°$ and a $24.5°$ AR-coated wedge-lensed ﬁber with MFD $= 3.6$ μm.

- Calculated coupling efﬁciencies >90% at MFD $\cong 3.5$ μm for a $24.5°$ wedge-lensed ﬁber and for a laser with FF angles $10° \times 25°$ with experimental coupling values lower by $\lesssim 10\%$.

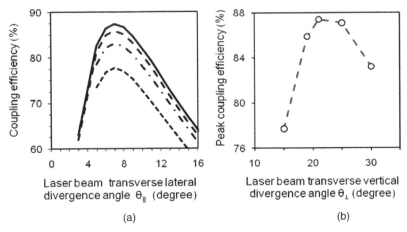

(a)   (b)

Figure 1.41   (a) Coupling ef ciency versus laser transverse lateral divergence angle cal- culated for laser light emitted at transverse vertical angles of 21° (solid line), 19° (long dashed line), 30° (dot–dashed line), and 15° (short-dashed line) into a 24.5° wedge-lensed single-mode ber with a mode eld diameter of 5.9 μm. (b) Peak coupling ef ciency versus laser transverse vertical divergence angle calculated for laser light emitted at a transverse lateral angle of 7° into a 24.5° wedge-lensed single-mode ber with a mode eld diameter of 5.9 μm.

- Calculated coupling ef ciencies >85% at $\theta_{\parallel} \cong 7°$ for a transverse vertical FF angle $\theta_{\perp} = 21°$ and a 24.5° wedge-lensed single-mode ber with MFD = 5.9 μm.

In the preceding subsections, we discussed optical coupling for single-mode diode lasers with diffraction-limited beams in both transverse directions. However, broad-area lasers typically 100 μm wide and 1 cm laser bars have a relatively poor beam quality, which is usually expressed by the beam parameter product (BPP), de ned as half the beam waist diameter in focus times half the FF divergence angle. The output beam of these lasers is characterized by a highly asymmetric pro le with regard to beam dimension and divergence angle. In the case of a laser bar, typical values for the source width are 10 mm in the slow-axis direction and 1 μm in the fast-axis direction with typical beam divergence angles of 5° and 35°, respectively. This means that the resulting BPPs are highly asymmetric: in the slow direction BPP $\cong$400 mm × mrad, which is far beyond the diffraction limit; and in the fast direction BPP $\cong$1 mm × mrad, which is nearly diffraction limited (Köhler et al., 2010). Ef cient ber coupling of such a diode laser is only possible if the different BPPs are adapted by shifting beam quality from one direction to the other and special symmetrization optics are applied for reshaping the beam (Bachmann et al., 2007). This symmetrization of the BPPs is equivalent to a minimization of the overall beam parameter product $BPP_{total} = (BPP_{slow}^2 + BPP_{fast}^2)^{1/2}$. The smallest possible BPP is achieved with a diffraction-limited Gaussian beam and is proportional to the wavelength, and consequently ber coupling becomes more dif cult at longer wavelengths.

1.4.3.4   Device operating temperature control

The strong temperature sensitivity of the diode laser characteristics including threshold current, output power, and lasing wavelength, but also of the laser long-term reliability and lifetime, requires operating the laser under temperature-controlled conditions. The stabilization of the operating laser temperature is usually realized by mounting the laser platform, including a temperature sensor, on a thermo-electric cooler (TEC) (Peltier). The sensor is typically a thermistor (thermally sensitive resistor; it has a large negative temperature coef cient for electrical resistance and is formed from a sintered alloy of oxides or carbonates of Fe, Ni, Mn, Mo, and Cu), which monitors the temperature and controls the cooling power of the TEC to achieve constant operating temperature. This is standard practice for diode lasers used as optical pumping sources for ber ampli ers and as transmitters for coherent communication systems. However, for high-power broad-area lasers and laser arrays water coolers have to be used, because of the large heat load these devices generate. For smaller heat loads, passive cooling is applied where the laser chip or bar on a heat sink is hard-soldered onto an expansion-matched substrate such as CuW, which is then clamped on a Cu heat exchanger. In the case of high heat loads, the device on an expansion-matched submount is hard-soldered onto an expansion-matched microchannel heat exchanger.

1.4.3.5   Hermetic sealing

Modules for high-performance diode lasers such as pump lasers and ber optic communication lasers are sealed hermetically to provide long-term stability and reliability.

To realize hermetically sealed packages the electrical and optical input and output terminals or ports have to be sealed hermetically:

- Electrical pins of TO headers are isolated from the package frame and sealed with glass by melting glass powder packed between the pin and frame with a subsequent oxidization to strengthen the seal.

- AR-coated optical windows partially metallized along the edge are joined to the cap of TO cans by brazing with silver solder.

- Electrical terminals of rectangular-type packages, such as butter y and mini-DIL, are hard-soldered onto a metallized layer in the ceramic frame.

- Fibers are inserted into a ferrule and xed with acrylate/epoxy or metallized bers are hard-soldered to the ferrule/frame.

The active components are wire-bonded each to a pin, the package is carefully cleaned of organic contamination, and a desiccant is included inside the package to getter moisture and organics. Before closing the package, an oxygen-rich atmosphere is established inside it. This suppresses so-called package-induced failures (PIFs), which are caused by a photothermal decomposition process of hydrocarbon traces in the atmosphere on a surface exposed to high optical intensity. The chemical reaction between hydrocarbons and oxygen burns off the hydrocarbons but leaves water as a by-product, which can be gettered by the desiccant. The coverlid is put on by hard soldering if the frame is metallized ceramic or by laser welding if the package frame is metallic.

References

Adams, A. R. (1986). *Electron. Lett.*, **22**, 250.

Adams, C. S. and Cassidy, D. T. (1988). *J. Appl. Phys.*, **64**, 6631.

Agrawal, G. P. and Dutta, N. K. (1993). *Semiconductor Lasers*, 2nd edn, Van Nostrand Reinhold, New York.

Al-Muhanna, A., Wade, J. K., Earles, T., Lopez, J., and Mawst, L. J. (1998). *Appl. Phys. Lett.*, **73**, 2869.

Anderson, T. G., Chen, Z. G., Kulakovskii, V. D., Uddin, A., and Vallin, J. T. (1987). *Appl. Phys. Lett.*, **51**, 752.

Arias, J., Esquivias, I., Larkins, E. C., Burkner, S., Weisser, S., and Rosenzweig, J. (2000). *Appl. Phys. Lett.*, **77**, 776.

As, D. J., Epperlein, P. W., and Mooney, P. M. (1988). *J. Appl. Phys.*, **64**, 2408.

Asada, M., Kameyama, A., and Suematsu, Y. (1984). *IEEE J. Quantum Electron.*, **QE-2**, 745.

Bachmann, F., Loosen, P., and Poprawe, R. (eds.) (2007). *High Power Diode Lasers: Technology and Applications*. Springer-Verlag, Berlin.

Bisping, D., Pucicki, D., Ho ing, S., Habermann, S., Ewert, D., Fischer, M., Koeth, J., Zimmermann, C., Weinmann, P., Kamp, M., and Forchel, A. (2008). *Proceedings of the IEEE 21st International Semiconductor Laser Conference, Sorrento, Italy, Conf. Dig.*, 51.

Blue Sky Research, Inc. (2010). *Products: laser components, virtual point source.* www.blueskyresearch.com.

Bookham, Inc. (now part of Oclaro, Inc.) (2009). *High power laser diodes.* www.oclaro.com.

Borchert, B., Egorov, A. Y., Illek, S., and Riechert, H. (2000). *IEEE Photonics Tech. Lett.*, **12**, 597.

Born, M. and Wolf, E. (1999). *Principles of Optics*, 7th (expanded) edn, Cambridge University Press, Cambridge.

Botez, D. (1978). *IEEE J. Quantum Electron.*, **QE-14**, 230.

Botez, D. (1981). *IEEE J. Quantum Electron.*, **QE-17**, 178.

Botez, D. (1982). *IEE Proc.*, **129**, Pt. 1, 237.

Botez, D. and Ettenberg, M. (1978). *IEEE J. Quantum Electron.*, **QE-14**, 827.

Botez, D. and Herskowitz, G. J. (1980). *Proc. IEEE*, **68**, 689.

Botez, D., Mawst, L. J., Bhattacharya, A., Lopez, J., Li, J., Kuech, T. F., Iakovlev, V. P., Suruceanu, G. I., Caliman, A., and Syrbu, A. V. (1996). *Electron. Lett.*, **32**, 2012.

Casey, Jr., H. C. and Panish, M. B. (1978). *Heterostructure Lasers, Part A: Fundamental Principles and Part B: Materials and Operating Characteristics*, Academic Press, New York.

Chand, N., Chu, S. N. G., Dutta, N. K., Lopata, J., Geva, M., Syrbu, A. V., Mereutza, A. Z., and Yakovlev, V. P. (1994). *IEEE J. Quantum Electron.*, **QE-30**, 424.

Chang-Hasnain, C. J. (1994). Chapter 9 in *Diode Laser Arrays* (eds. Botez, D. and Scifres, D. R.), Cambridge University Press, Cambridge.

Chen, T. R., Zhao, B., Zhuang, Y. H., Yariv, A., Ungar, J. E., and Oh, S. (1992). *Appl. Phys. Lett.*, **60**, 1782.

Choi, H. K. and Eglash, S. J. (1992). *Appl. Phys. Lett.*, **61**, 1154.

Coldren, L. A. and Corzine, S. W. (1995). *Diode Lasers and Photonic Integrated Circuits*, John Wiley & Sons, Inc., New York.

Coleman, J. J. (1993). Chapter 8 in *Quantum Well Lasers* (ed. Zory, P. S.), Academic Press, San Diego, CA.

Crosslight Software Inc. (2010). *Products: LASTIP*. www.crosslight.com.

Defence Advanced Research Projects Agency (DARPA) (2011). www.darpa.mil.

Dixon, R. W. and Joyce, W. B. (1979). *IEEE J. Quantum Electron.*, **QE-15**, 470.

Dutta, N. K. (1984). *J. Appl. Phys.*, **55**, 285.

Eliseev, P. G. (1999). Chapter 2 in *Semiconductor Lasers II: Materials and Structures* (ed. Kapon, E.), Academic Press, San Diego, CA.

Eng, L. E., Mehuys, D. G., Mittelstein, M., and Yariv, A. (1990). *Electron. Lett.*, **26**, 1675.

Engelmann, R. W. H., Shieh, C. L., and Shu, C. (1993). Chapter 3 in *Quantum Well Lasers* (ed. Zory, P. S.), Academic Press, San Diego, CA.

Epperlein, P. W. (1999). To be published. For further information see www.pwe-photonicselectronics-issueresolution.com.

Epperlein, P. W., Parry, M., Helmy, A., Drouot, V., Harrell, R., Moseley, R., Stacey, S., Clark, D., Harker, A., and Shaw, J. (2000). To be published. For further information see www.pwe-photonicselectronics-issueresolution.com.

Frankfurt Laser Company (2009). *Laser diode products*. www.frlaserco.com.

Fukuda, M. (1999). *Optical Semiconductor Devices*. John Wiley & Sons, Inc., New York.

Garbuzov, D. Z., Menna, R. J., Martinelli, R. U., DiMarco, L., Harvey, M. G., and Connolly, J. C. (1996). *Proceedings of the Conference on Lasers and Electro-Optics, CLEO'96, OSA Tech. Dig.*, postdeadline paper CPD10.

Garbuzov, D., Menna, R., Komissarov, A., Maiorov, M., Khal n, V., Tsekoun, A., Todorov, S., and Connolly, J. (2001). *Proceedings of the Optical Fiber Communication Conference, OSA Tech. Dig.*, **4**, PD18-1.

Garbuzov, D., Maiorov, M., Menna, R., Komissarov, A., Khal n, V. Kudryashov, I., Lunev, A., DiMarco, L., and Connolly, J. (2002). *Proc. SPIE*, **4651**, 92.

Hayashi, I., Panish, M. B., and Reinhart, F. K. (1971). *J. Appl. Phys.*, **42**, 1929.

Hertsens, T. (2005). *ILX Lightwave Application Note #5*. www.ilxlightwave.com.

Hetterich, M., Dawson, M. D., Egorov, A. Y., Bernklau, D., and Riechert, H. (2000). *Appl. Phys. Lett.*, **76**, 1030.

Huang, R. T., Wolf, D., Cheng, W. H., Jiang, C. L., Agarwal, R., Renner, D., Mar, A., and Bowers, J. E. (1992). *IEEE Photonics Technol. Lett.*, **4**, 293.

Hwang, D. M., Schwarz, S. A., Bhat, R., and Chen, C. Y. (1991). *Inst. Phys. Conf. Ser.*, **120**, 365.

Iga, K. and Koyama, F. (1999). Chapter 5 in *Semiconductor Lasers II: Materials and Structures* (ed. Kapon, E.), Academic Press, San Diego, CA.

Ikegami, T. (1972). *IEEE J. Quantum Electron.*, **QE-8**, 470.

Ioffe Physico-Technical Institute (2006). *New Semiconductor Materials, n-k data base.* www.ioffe.rssi.ru/SVA/NSM/.

ISO 11146 (2005). *International Organization for Standardization*, Products. www.iso.org/iso/store.htm.

Itaya, Y., Katayama, S., and Suematsu, Y. (1979). *Electron. Lett.*, **15**, 123.

JDSU, Corp. (2009). *Optical communication and laser products.* www.jdsu.com.

Jenoptik Diode Lab, GmbH (2006). *Press release Jan. 2006.* www.jenoptik.com.

Katsuyama, T., Yoshida, I., Shinkai, J., Hashimoto, J., and Hayashi, H. (1990). *Electron. Lett.*, **26**, 1375.

Kelemen, M., Gilly, J., Moritz, R., Rattunde, M., Schmitz, J., and Wagner, J. (2008). *Proc. SPIE*, **6876**, 68760E.

Köhler, B., Kissel, H., Flament, M., Wolf, P., Brand, T., and Biesenbach, J. (2010). *Proc. SPIE*, **7583**, 75830F.

Kondow, M., Uomi, K., Niwa, A., Kitatani, T., Watahiki, S. and Yazawa, Y. (1996). *Jpn. J. Appl. Phys.*, **35**, 1273.

Kondow, M., Kitatani, T., Nakatsuka, S., Larson, M. C., Nakahara, K., Yazawa, Y., Okai, M., and Uomi, K. (1997). *IEEE J. Sel. Top. Quantum Electron.*, **3**, 719.

Kressel, H. and Butler, J. K. (1977). *Semiconductor Lasers and Heterojunction LEDs*, Academic Press, New York.

Lang, R. (1980). *Jpn. J. Appl. Phys.*, **19**, L93.

LDX Optronics, Inc. (2009). *Laser products.* www.ldxoptronics.com.

Li, H., Towe, T., Chyr, I., Brown, D., Nguyen, T., Reinhardt, F., Jin, X., Srinivasan, R., Berube, M., Truchan, T., Bullock, R., and Harrison, J. (2007). *IEEE Photonics Technol. Lett.*, **19**, 960.

Li, L., Liu, G., Li, Z., Li, M., Li, H., Wang, X., and Wan, C. (2008). *IEEE Photonics Technol. Lett.*, **20**, 566.

Lichtenstein, N., Manz, Y., Mauron, P., Fily, A., Arlt, S., Thies, A., Schmidt, B., Muller, J., Pawlik, S., Sverdlov, B., and Harder, C. (2004). *Proceedings of the IEEE 19th International Semiconductor Laser Conference, Conf. Dig.*, 45.

Livshits, D. A., Egorov, A. Y., and Riechert, H. (2000). *Electron. Lett.*, **36**, 1381.

Lumics, GmbH (2009). *Diode laser products.* www.lumics.com.

m2k/DILAS, GmbH (2009). *High-brightness diode laser catalogue.* www.m2k-laser.de and www.dilas.com.

Matthews, J. W. and Blakeslee, A. E. (1974). *J. Cryst. Growth*, **27**, 118.

Mawst, L. J., Bhattacharya, A., Lopez, J., Botez, D., Garbuzov, D. Z., DeMarco, L., Connolly, J. C., Jansen, M., Fang, F., and Nabiev, R. F. (1996). *Appl. Phys. Lett.*, **69**, 1532.

McInerney, J., Mooradian, A., Lewis, A., Shchegrov, A., Strzelecka, E., Lee, D., Watson, J., Liebman, M., Carey, G., Cantos, B., Hitchens, W., and Heald, D. (2003). *Electron. Lett.*, **39** (6), 523.

Mehuys, D. G. (1999). Chapter 4 in *Semiconductor Lasers II: Materials and Structures* (ed. Kapon, E.), Academic Press, San Diego, CA.

Mehuys, D. G., Mittelstein, M., Yariv, A., Sarfaty, R., and Ungar, J. E. (1989). *Electron. Lett.*, **25**, 143.

Michel, N., Krakowski, M., Hassiaoui, I., Calligaro, M., Lecomte, M., Parillaud, O., Weinmann, P., Zimmermann, C., Kaiser, W., Kamp, M., Forchel, A., Pavelescu, E., Reithmaier, J., Sumpf, B., Erbert, G., Kelemen, M., Ostendorf, R., Garcia-Tijero, J. M., Odriozola, H., and Esquivias, I. (2008). *Proc. SPIE*, **6909**, 690918.

Mitsubishi Electric Corp., Optical Devices (2009). *Product news and Optical devices general catalogue.* www.mitsubishichips.com and www.mitsubishielectric.com.

Mittelstein, M., Arakawa, Y., Larsson, A., and Yariv, A. (1986). *Appl. Phys. Lett.*, **49**, 1689.

Mittelstein, M., Mehuys, D., Yariv, A., Ungar, J. E., and Sarfaty, R. (1989). *Appl. Phys. Lett.*, **54**, 1092.

Nagarajan, R. and Bowers, J. E. (1999). Chapter 3 in *Semiconductor Lasers I: Fundamentals* (ed. Kapon, E.), Academic Press, San Diego, CA.

Nakamura, S. (1991). *Jpn. J. Appl. Phys.*, **30**, L1705.

Nash, F. R. (1973). *J. Appl. Phys.*, **44**, 4696.

Nichia, Corp. (2009). *Laser diodes: marketed products and engineering samples.* www.nichia.co.jp.

nLight, Corp. (2009). *High-power semiconductor laser products.* www.nlight.net.

O'Gorman, J., Levi, A. F. J., Schmitt-Rink, S., Tanbuk-Ek, T., Coblentz, D. L., and Logan, R. A. (1992). *Appl. Phys. Lett.*, **60**, 157.

Opnext, Inc. (2009). *Product catalogue.* www.opnext.com.

OSRAM GmbH, Opto Semiconductors (2009). *Product catalogue.* www.osram-os.com.

Ostendorf, R., Kaufel, G., Moritz, R., Mikulla, M., Ambacher, O., Kelemen, M. T., and Gilly, J. (2008). *Proc. SPIE*, **6876**, 68760H.

Ostendorf, R., Schilling, C., Kaufel, G., Moritz, R., Wagner, J., Kochem, G., Friedmann, P., Gilly, J., and Kelemen, M. (2009). *Proc. SPIE*, **7198**, 719811.

Paschke, K., Fiebig, C., Feise, D., Fricke, J., Kaspari, C., Blume, G., Wenzel, H., and Erbert, G. (2008). *Proceedings of the IEEE 21st International Semiconductor Laser Conference, Sorrento, Italy, Conf. Dig.*, 131.

People, R. and Bean, J. C. (1985). *Appl. Phys. Lett.*, **47**, 322.

Petrescu-Prahova, I. B., Modak, P., Goutain, E., Bambrick, D., Silan, D., Riordan, J., Moritz, T. and Marsh, J. H. (2008). *Proceedings of the IEEE 21st International Semiconductor Laser Conference, Sorrento, Italy, Conf. Dig.*, 135.

Piprek, J., Abraham, P., and Bowers, J. E. (2000). *IEEE J. Quantum Electron.*, **QE-36**, 366.

QPC, LLC (2009). *Laser products.* www.qpclasers.com.

Riechert, H., Egorov, A. Y., Borchert, B., and Illek, S. (2000). *Compound Semicond.*, **6**, 71.

Ryu, H. Y., Ha, K. H., Lee, S. N., Choi, K. K., Jang, T., Son, J. K., Chae, J. H., Chae, S. H., Paek, H. S., Sung, Y. J., Sakong, T., Kim, H. G., Kim, K. S., Kim, Y. H., Nam, O. H., and Park, Y. J. (2006). *IEEE Photonics Technol. Lett.*, **18**, 1001.

Saito, S., Hattori, Y., Sugai, M., Harada, Y., Jongil, H. and Nunoue, S. (2008). *Proceedings of the IEEE 21st International Semiconductor Laser Conference, Sorrento, Italy, Conf. Dig.*, 185.

SANYO Electric Co., Ltd. (now Panasonic) (2009). Opto Electronics Devices/Laser Diodes. http://panasonic.net/sanyo.

Sebastian, J., Schulze, H., Hülsewede, R., Hennig, P., Meusel, J., Schröder, M, Schröder, D., and Lorenzen, D. (2007). *Proc. SPIE*, **6456**, 64560F.

Sell, D. D., Casey, H. C., and Wecht, K. W. (1974). *J. Appl. Phys.*, **45**, 2650.

Shan, W., Walukiewicz, W., Ager, J. W., and Haller, E. E. (1999). *Phys. Rev. Lett.*, **82**, 1221.

Siegmann, A. (1990). *Proc. SPIE*, **1224**, 2.

Siegmann, A. (1993). *Proc. SPIE*, **1868**, 2.

Singh, R., Heminway, T., Krasnick, R., Grif n, P., Powers, M. (2004). *Proc. SPIE*, **5358**, 29.

Stickley, C. M., Filipkowski, M. E., Parra, E. and Hach III, E. E. (2006). *Proc. SPIE*, **6104**, 42.

Svelto, O. (1998). *Principles of Lasers*, 4th edn, Plenum Press, New York.

Sze, S. M. (1981). *Physics of Semiconductor Devices*, 2nd edn, John Wiley & Sons, Inc., New York.

Tanbun-Ek, T., Olsson, N. A., Logan, R. A., Wecht, K. W., and Seargent, A. M. (1991). *IEEE Photonics Technol. Lett.*, **3**, 103.

Tarasov, I. S., Pikhtin, N. A., Slipchenko, S. O., Sokolova, Z. N., and Vinokurov, D. A. (2004). *Proceedings of the IEEE 19th International Semiconductor Laser Conference, Conf. Dig.*, 37.

Telcordia Technologies, Inc. (2010). *Information Super Store.* https://telecom-info .telcordia.com.

Temkin, H., Tanbuk-Ek, T., Logan, R. A., Cebulla, D. A., and Sergent, A. M. (1991). *IEEE Photonics Technol. Lett.*, **3**, 100.

Tiwari, S., Bates, R. S., and Harder, C. S. (1992). *Appl. Phys. Lett.*, **60**, 413.

Vahala, L. K. and Zah, C. E. (1988). *Appl. Phys. Lett.*, **52**, 1945.

Valster, A., van der Poel, C. J., Finke, M. N., and Boermans, M. J. B. (1992). *Electron. Lett.*, **28**, 144.

Van der Ziel, J. P. (1981). *IEEE J. Quantum Electron.*, **QE-17**, 60.

Vegard, L. (1921). *Z. Phys.*, **5**, 17.

Weisbuch, C. and Vinter, B. (1991). *Quantum Semiconductor Structures: Fundamentals and Applications*, Academic Press, New York.

Wright, P. D., Joyce, W. B., and Craft, D. C. (1982). *J. Appl. Phys.*, **53**, 1364.

Yablonovitch, E. and Kane, E. O. (1988). *IEEE J. Lightwave Technol.*, **LT-4**, 504.

Yariv, A. (1997). *Optical Electronics in Modern Communications*, 5th edn, Oxford University Press, New York.

Yu, P. Y. and Cardona, M. (2001). *Fundamentals of Semiconductors*, 3rd edn, Springer-Verlag, Berlin.

Zah, C. E., Bhat, R., Pathak, B., Caneau, C., Favire, F. J., Andreakdakis, N. C., Hwang, D. M., Koza, M. A., Chen, C. Y., and Lee, T. P. (1991). *Electron. Lett.*, **27**, 1414.

Zah, C. E., Bhat, R., Favire, F. J., Koza, M., and Lee, T. P. (1992). *Proceedings of the IEEE 13th International Semiconductor Laser Conference, Conf. Dig.*, Paper K-5.

Zah, C. E., Li, Y., Bhat, R., Song, K., Visovsky, N., Nguyen, H. K., Liu, X., Hu, M., and Nishiyama, N. (2004). *Proceedings of the IEEE 19th International Semiconductor Laser Conference, Conf. Dig.*, 39.

Zhao, B., and Yariv, A. (1999). Chapter 1 in *Semiconductor Lasers I: Fundamentals* (ed. Kapon, E.), Academic Press, San Diego, CA.

Chapter 2

Design considerations for high-power single spatial mode operation

Main Chapter Topics Page

Semiconductor Laser Engineering, Reliability and Diagnostics: A Practical Approach to High Power and Single Mode Devices, First Edition. Peter W. Epperlein.
© 2013 John Wiley & Sons, Ltd. Published 2013 by John Wiley & Sons, Ltd.

Introduction

The chapter is subdivided into four sections. The first section gives an overview of the different approaches for realizing high-power, edge-emitting diode lasers with a focus on solitary emitters followed by a detailed discussion of these approaches and parameters including the design of effective vertical and lateral waveguide structures, increase of thermal rollover power by cavity length scaling, realization of high internal efficiency and low internal carrier and photon losses, efficient thermal management, suppression of leakage currents, optimization of materials, growth, and processing, and elimination of catastrophic optical damage events at mirrors and in the bulk of the cavity.

Section 2.2 discusses the conditions for single and fundamental transverse vertical and lateral mode behavior of narrow-stripe index-guided diode lasers and gives relevant mathematical expressions for the required index differences and effective vertical and lateral active layer dimensions. Methods are discussed to stabilize the fundamental mode by suppressing the excitation of higher order modes through increasing their threshold gain, for example, by introducing mode-selective losses. Various mode filter schemes such as corrugated waveguides, tilted mirrors, and tapered waveguides are described to enforce fundamental mode operation. Other approaches include the use of low ridge waveguides, thin p-cladding layers, and the extinction of filamentation effects. Methods are also described to suppress so-called shift kinks, which are generated by resonantly coupling power from the lasing fundamental mode to the first-order mode. These include controlling the beat length of the two modes, which can be done by adjusting the cavity length or the difference of the two propagation constants.

Section 2.3 gives a comprehensive account on the design of narrow ridge wave-guide lasers with emphasis on high and kink-free optical output power. It includes numerically modeled and experimental results of the fundamental spatial mode stability regime, mode losses, and slow-axis divergence angle versus relevant ridge dimensions. Further dependencies include: (i) threshold current, kink-free power, and front-facet slope efficiency as a function of the slow-axis far-field angle; (ii) internal optical loss versus front-facet efficiency; (iii) threshold current and slope efficiency versus cladding layer composition; (iv) fast-axis far-field angle versus cladding layer composition and GRIN-SCH layer thickness; and (v) slope efficiency versus threshold current.

Section 2.4 deals with other concepts and techniques to realize diode lasers emitting high power in the fundamental spatial mode or in a diffraction-limited single-lobed far-field radiation pattern. These concepts include unstable resonators, various broad-area laser concepts to realize single-lobe diffraction-limited beams, tapered lasers, monolithic flared amplifier master oscillator power amplifiers, phase-locked coherent diode laser bars, and incoherent standard 1 cm high-power laser bars with high fill factors.

2.1 Basic high-power design approaches

2.1.1 Key aspects

The limiting factors for high-power operation under continuous wave (cw) conditions in narrow-stripe single spatial mode diode lasers can be grouped into two categories. First, in thermal rollover the laser efficiency gradually decreases with increasing drive current due to the increase of temperature in the active layer by Joule heating effects, which eventually leads to a saturation or even decrease of power, and, second, in catastrophic optical damage (COD) mainly the mirrors are damaged by local melting due to the high absorption of laser light at nonradiative recombination centers. However, the damage can also further extend from the surface along the cavity or originate from hot spots in the bulk of the cavity, which are caused by highly nonradiative crystalline phase changes and structural defects such as dislocations.

When operating a single-mode diode laser at high power, single-mode behavior has to be achieved in both transverse vertical and lateral directions. This issue will be dealt with in the next sections. The laser has also to be designed in such a way that high-power operation is obtained with specified transverse vertical and lateral beam divergence angles to maximize the coupling of power into a single-mode fiber, if required. This issue will be discussed in detail further below supported by numerical modeling and experimental results. It should also be noted that highly reliable and long-term operation under high output power has to be realized (see Chapters 3–6).

Some key parameters for realizing high-power operation can be revealed from the equation for the output power $P_{out,f}$ from the front-facet of a single-emitter device

$$P_{out,f} = \frac{h\nu}{q}\eta_i \frac{1}{1 + \dfrac{1}{\dfrac{1 - R_f}{1 - R_r}\sqrt{\dfrac{R_r}{R_f}}}} \frac{\alpha_m}{\alpha_i + \alpha_m} (I - I_{th}) \left(\frac{I_{eff}}{I_{eff} + I_{leak}}\right) \Theta(T) \quad (2.1)$$

where η_i is the internal quantum efficiency, R_f and R_r the reflectivities of the front and rear mirrors, α_m the mirror loss (cf. Equation 1.50), α_i the internal optical loss, I_{th} the threshold current, I_{eff} the effective current contributing to lasing, I_{leak} the leakage current, and $\Theta(T)$ the term representing the thermal power rollover due to heating effects. Equation (2.1) was derived from Equations (1.48) and (1.49); the last two terms were added to consider all major contributing parameters or effects. The reflectivity term is close to unity within ~3% for typical reflectivities $R_f = 0.1$ and $R_r = 0.9$. However, the other factors in Equation (2.1) are essential in achieving high optical power.

These factors and measures such as output power scaling, laser cavity length scaling, vertical and lateral waveguide designs, materials and fabrication optimizations, carrier and photon loss minimizations, and thermal management will be discussed in the following sections.

2.1.2   Output power scaling

It is well established that the laser output power can be dramatically increased by making the laser cavity longer. This is mainly due to a lowering of the thermal resistance leading to an improved cooling of the laser chip with the consequence that the thermal rollover power is increased. However, it has also been observed that the threshold current is increased and the external quantum efficiency is decreased in long-cavity lasers.

To counteract this and to maximize the output power four key figures have to be considered in the length scaling process of a diode laser. According to Harder (2008), these figures are (i) the roundtrip gain g, (ii) the external differential quantum efficiency η_d, (iii) the photon lifetime in the cavity τ_{ph}, and (iv) the asymmetry of the laser cavity defined by the ratio P_r of the power behind the front and rear mirror. Setting the rear-mirror reflectivity equal to unity allows the four key figures to be written as (cf. Equations 1.23, 1.32, 1.50; Agrawal and Dutta, 1993)

$$g = \frac{1}{\Gamma_{tv}}\left[\alpha_i + \frac{1}{2L}\ln\left(\frac{1}{R_f}\right)\right] \tag{2.2}$$

$$\eta_d = \eta_i\frac{\dfrac{1}{2L}\ln\dfrac{1}{R_f}}{\alpha_i + \dfrac{1}{2L}\ln\dfrac{1}{R_f}} = \eta_i\frac{\alpha_m}{\alpha_i + \alpha_m} \tag{2.3}$$

$$\tau_{ph} = \frac{1}{v_{gr}\left[\alpha_i + \dfrac{1}{2L}\ln\left(\dfrac{1}{R_f}\right)\right]} \tag{2.4}$$

$$P_r = \frac{(1 + R_f)}{2\sqrt{R_f}} \tag{2.5}$$

where v_{gr} is the group velocity (cf. Section 1.3.6). It is not possible to adjust the values of Γ_{tv}, α_i, and R_f in such a way to keep all four figures constant while simultaneously increasing the cavity length L to improve the thermal rollover power. Instead, g and η_d are kept fixed and τ_{ph} (*constant photon lifetime scaling* approach) or P_r (*constant power ratio scaling* approach) are adjusted (Harder, 2008). The rules for *constant photon lifetime scaling* are

$$\Gamma_{tv}(L) = \Gamma_{tv}(L_0)\,; \; \alpha_i(L) = \alpha_i(L_0)\,; \; R_f(L) = R_f(L_0)^{L/L_0} \tag{2.6}$$

where $L > L_0$ is the longer cavity length. In this approach, higher power can be obtained by cleaving longer cavities from the same material and reducing R_f according to Equation (2.6) but still maintain the same value for η_d. However, P_r also becomes larger with increasing L, which facilitates the longitudinal spatial hole burning effect. This drawback can be mitigated by using a slightly flared active waveguide (Guermache *et al.*, 2005).

The rules for *constant power ratio scaling* are

$$R_f(L) = R_f(L_0); \; \Gamma_{tv}(L) = \frac{L_0}{L}\Gamma_{tv}(L_0); \; \alpha_i(L) = \frac{L_0}{L}\alpha_i(L_0). \qquad (2.7)$$

In this approach, which is the preferred one for long cavities, also R_f is constant, and hence this scaling type is also called *constant mirror reflectivity scaling*. Both Γ_{tv} and α_i have to be reduced linearly with increasing L, which is considered a demanding task in the design and fabrication of the vertical structure. The constant power ratio scaling approach has the additional advantage that it makes parameters such as the average optical power in the active layer, drive current density, heat generation density, and spectral stability independent of the cavity length (Harder, 2008).

2.1.3   Transverse vertical waveguides

2.1.3.1   Substrate

The effects of substrate orientation on the quality of epitaxial material and laser performance have already been mentioned in Section 1.4.1. These effects are dependent on the materials used and will be discussed in more detail in the following.

Usually the substrates used for devices have a (100)-oriented surface. Some examples will be given on the effects of misorientation from the exact (100) orientation by some degrees. Chand *et al.* (1994) observed that by misorienting (100) GaAs substrates toward $\langle 111 \rangle$A by 3 to 4° the incorporation of impurities like oxygen is reduced in AlGaAs, and the AlGaAs/GaAs heterointerfaces are smoother and sharper. Similar positive results were found in AlGaAs lasers grown on GaAs substrates oriented 2° toward the $\langle 110 \rangle$ direction, which led to a reduction of loop dislocations formed at the interface between the substrate and first epitaxial layer (Epperlein *et al.*, 2000, unpublished).

In general, slight substrate misorientations and lower defect densities result in improved laser performances and lifetimes. These results were confirmed by Chen *et al.* (1987) who demonstrated that similarly tilted substrates led to improved optical quality and lower threshold current densities in (Al)GaAs quantum well (QW) lasers grown on GaAs substrates. More importantly, the surface morphology of the growth on misoriented (100) substrates, and hence threshold current density, is less sensitive to deviations from optimum growth conditions than in (100) substrates, which makes it easier to grow low threshold current material. According to the authors, these enhanced results can be ascribed in part to the fact that misoriented surfaces have steps terminated with Ga. As atoms incident on the surface can then form three bonds, two to the Ga atoms on the (100) surface and one to the Ga atom on the (111) face of the step. This leads to an increased sticking probability of As resulting in smoother AlGaAs layer surfaces. A second mechanism may contribute, because stepped surfaces minimize energetic instabilities at the growth surface (Rode *et al.*, 1977).

The effects of substrate misorientation on material quality and laser performance in other material systems are more diverse. Thus, Mawst *et al.* (1995) report low-temperature photoluminescence measurements on InGaAs/InGaAsP QW structures showing narrow linewidths for growth on exact (100) GaAs substrates. In contrast, growth on (100) substrates misoriented 2 to 10° off toward ⟨110⟩ and 10° toward ⟨111⟩A exhibits broadened luminescence shifted toward longer wavelengths, which can be attributed to interfacial roughness and composition variations due to step bunching growth.

Step bunching refers to the phenomenon where a regular array of monosteps can become unstable and breaks up into regions with high step density or with little or no steps. It has been observed on many surfaces such as vicinal (100) GaAs with surface normals slightly misoriented from specific crystallographic directions resulting in a lowering of the vicinal surface energy by the formation of terraces, steps, or kinks. Several kinetic and thermodynamic mechanisms have been proposed for the formation of step bunching. This includes an asymmetry in the attachment–detachment kinetics of growth units at the step edges, and impurity-induced step bunching where a flux of impurities impinging on the growth surface hampers the motion of a following step, leading to a pinning of steps by impurities (Hata *et al.*, 1998; Cermelli and Jabbour, 2007; Kasu and Kobayashi, 1995).

Corresponding strained-layer InGaAs (active layer)/InGaAsP (confining layer)/InGaP (cladding layer) QW laser structures grown on exact (100) or misoriented substrates show no significant differences in threshold current densities, differential external quantum efficiencies, internal quantum efficiencies, and transparency current densities. However, significant differences are observed in the temperature characteristics with dramatically reduced temperature dependence of I_{th} and η_d for devices grown on exact (100) substrates compared to structures grown on misoriented substrates. This can be explained by the fact that at high temperatures carrier leakage and hence free-carrier absorption in the confinement layers is more pronounced for 2 to 10° off than for 0° off lasers, because of interfacial imperfections caused by the step bunching effect (Mawst *et al.*, 1995).

By contrast, record-high characteristic temperatures of $T_0 = 115$ K and $T_\eta = 285$ K for I_{th} and η_d, respectively, are obtained by growing compressively-strained InGaAsP/InGaAlP/InGaAlP QW laser structures on (100) GaAs substrates misoriented 10° toward ⟨111⟩A (Al-Muhanna *et al.*, 1998a). These high values reflect the strong carrier confinement, which is critical for achieving high cw optical power. One reason for this is that the growth on highly misoriented substrates completely disorders InGa(Al)P, which increases the bandgap by about 70 meV (McKenan *et al.*, 1988). This further increases the high bandgap energy of the InGaAlP confining and cladding layers, which even further reduces carrier leakage from the QW. An additional result of the growth on misoriented substrates is that the p-doping is enhanced, which also improves carrier confinement. Surprisingly, other parameters, such as α_i,

η_i, J_{tr}, and $\Gamma_{tv}g$ are unaffected by the substrate misorientation, which may be due to an increased roughness in misoriented devices, but the effect is overwhelmed by the large decrease in carrier leakage resulting from the disordered materials (Al-Muhanna *et al.*, 1998a).

2.1.3.2 Layer sequence

Figure 2.1 shows a typical, suitable vertical epitaxial structure in its simplest form for high-power operation, with the InGaAs/AlGaAs QW system as a representative example, and which can be considered as generic for all other diode laser layer structures. Here, we do not discuss device-specific details in the QW structure, confinement region, and cladding layers, as well as relevant modifications in the confinement and cladding regions for tailoring the fast-axis beam divergence angle with keeping the threshold current unchanged and maximizing the output power. These specially designed structures will be discussed in the subsections below.

The layer sequence includes a thin GaAs buffer layer on top of the substrate surface, first cladding grading, n-AlGaAs cladding layer, AlGaAs GRIN-SCH layer with embedded InGaAs QW, p-AlGaAs cladding layer, second cladding grading, and p^+-contact layer. The buffer layer is meant to isolate the active device layer from the

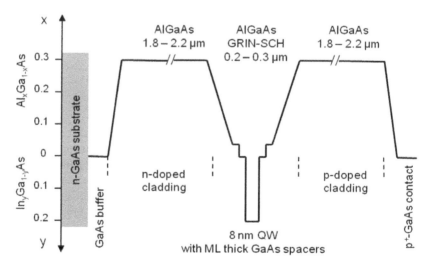

Figure 2.1 Typical schematic vertical laser design of a strained-layer graded-index separate-confinement heterostructure (GRIN-SCH) quantum well (QW) structure for high-power and low threshold current density operation with a weak optical confinement. The triangular GRIN-SCH shape can be replaced by a more efficient carrier-confining parabolic GRIN-SCH. AlGaAs/GaAs superlattices (not shown) are positioned at specific locations throughout the structure to getter segregating surface impurities, trap diffusing ions at heterojunction interfaces, block threading dislocations, or just serve as useful markers to determine layer thicknesses. Monolayer (ML) thick GaAs spacer layers are sufficient to trap detrimental surface-riding oxygen impurities. Not to scale.

substrate, block any threading dislocations and impurity defects from the substrate, and minimize the loop dislocation density at the substrate/epitaxial layer interface by appropriate initial growth conditions. As already discussed in Section 1.4.1.3, several different short AlGaAs/GaAs superlattices (not shown in Figure 2.1) are positioned at specific locations throughout the structure, each with its own purpose, such as to getter segregating surface impurities, to trap diffusing ions at heterojunction interfaces, to block threading dislocations, or just to serve as a useful marker in scanning electron microscopy (SEM) investigations to determine layer thicknesses.

Abrupt interfaces might cause potential spikes that could lead to unintentional conduction band barriers and additional series resistance. Therefore, to avoid this, interfaces between AlGaAs layers with significantly different Al content are graded over a distance of some nanometers. In molecular beam epitaxy (MBE), this could be achieved by ramping the Al cell over some temperature range, though this might be difficult to control due to the large thermal mass of the effusion cell. Alternatively, mechanical shutters (pulsed material supply) or separate Al cells could be used to control the Al flux, or the linear Al grading could be achieved by trading off the effects of source flux and increased Ga desorption as the substrate temperature increases. In organometallic vapour-phase epitaxy (OMVPE), the grading can simply be achieved by ramping the flux of the Al or Ga precursor. In material systems, such as InGaAs/InP or GaAs/GaInP, where the group-V component changes at an interface, the growth is less well controlled and may lead to detrimental effects including the buildup of strain at the graded interface and the formation of parasitic QWs, which can lead to lasing wavelengths different from the designed ones.

2.1.3.3 Materials; layer doping; graded-index layer doping

Materials

Common compounds for semiconductor lasers have been already discussed extensively in Section 1.1.5. In this subsection we concentrate on materials for high-power diode lasers that are nearly exclusively based on material systems grown on GaAs substrates with layer structures using binary, ternary, or quaternary materials from the (AlGaIn)(AsP) compound group. GaAs-based laser devices dominate the market for high-power diode laser products. However, we will also give a brief account of the commercially important InP- and GaN-based materials and laser devices.

Strained InGaAs QWs are usually embedded in AlGaAs/AlGaAs waveguide structures to cover the wavelength range of 880 to 1060 nm and can achieve high optical output powers (Mikulla *et al.*, 1999; Matuschek *et al.*, 2006; Sebastian *et al.*, 2007). Strain-compensating GaAsP spacer layers have been successfully employed for wavelengths > 1000 nm (Bugge *et al.*, 1998). Record-high, single-mode cw powers in the 980 nm wavelength band have been achieved from InGaAs/AlGaAs GRIN-SCH SQW narrow ridge lasers with up to 1.75 W thermal rollover power at 25 °C heat sink temperature and fundamental spatial mode operation in excess of 1.4 W. These devices operate free of catastrophic optical mirror damage (COMD) at maximum power densities above 100 MW/cm^2 (Lichtenstein *et al.*, 2004). Yang *et al.* (2004)

achieved >900 mW cw kink-free power for similar devices with high power and current levels at COMD of 1200 mW cw and 1600 mA, respectively, obtained for a low fast-axis beam divergence angle of 13°. By inserting two low-index AlGaAs layers (high AlAs mole fraction) between the waveguide and cladding layer (see Section 2.1.3.5) the far-field could be tuned and fast-axis beam divergence angles in the range of 13 to 24° have been achieved leading to COMD levels between ∼1200 and 1000 mW cw, respectively.

High-power Al-free diode lasers emitting in the wavelength range >940 nm have been successfully demonstrated by combining InGaAs QWs with InGaP/InGaAsP waveguides (Botez, 1999a; Al-Muhanna et al., 1998b; Zhang et al., 1993); these devices are also commercially available. The use of the Al-free InGaAs/InGaAsP/InGaP material system has apparently several advantages over the Al-containing InGaAs/GaAs/AlGaAs material system for the realization of reliable, high-power diode laser sources, as follows:

- Lower device series resistance.

- Higher electrical and thermal conductivity compared to AlGaAs (Diaz et al., 1994).

- Lower surface oxidation of InGaP compared to AlGaAs greatly facilitates regrowth for the fabrication of single-mode index-guided structures (Groves et al., 1989).

- Lower facet degradation due to the lower surface recombination velocity of InGaP compared to AlGaAs (Wang et al., 1994; Olson et al., 1989).

Further advantages will be discussed as appropriate within the context of relevant topics in the following sections and chapters.

COMD-free output powers of 430 mW cw and 200 mW fundamental transverse mode powers have been obtained from Al-free 980 nm InGaAs/InGaAsP/InGaP triple-QW (TQW)-SCH ridge lasers 4 μm wide (Asonen et al., 1994). Moreover, internal optical power densities at COMD of typically 18 MW/cm² cw have been derived from measurements on aperture devices 100 μm wide (Botez, 1999b; Al-Muhanna et al., 1998b). Apparently, there exists no evidence as yet for any clear improvement in output power and COMD performance of 980 nm Al-free lasers compared to that of 980 nm Al-containing lasers, in spite of the fact that appropriate measures also had been taken to expand the mode in the transverse vertical direction (see Section 2.1.3.5 for a description of diverse established techniques).

By adding Al to the InGaAs active region, high-power, strained AlGaInAs/AlGaAs QW lasers can be realized with emission wavelengths in the range down to 731 nm covering also the important 808 nm wavelength for pumping Nd:YAG (Emanuel et al., 1997; Hanke et al., 1999). The presence of indium in the active layer increases the resistance of the material to dark-line defect (DLD) formation (see Chapter 3) and propagation, and hence improves the optical strength and reliability of these diode laser types (Roberts et al., 1998).

AlGaInAs lasers 100 μm wide with high AlAs mole fraction of 0.24 in the active QW yielded COMD power levels of 2.2 W corresponding to a power density of ~8 MW/cm^2 (Emanuel *et al.*, 1997), which is only lower by a factor of ~2 compared to Al-free lasers (see above). Furthermore, 40 W cw laser bars 1 cm wide comprising 25 groups of AlGaInAs double-QW (DQW) lasers 200 μm wide emitting in the 808 nm band showed low degradation rates of 10^{-6} per hour over 3.3×10^5 accumulated device hours in accelerated life tests (Hanke *et al.*, 1999). The devices showed high slope efficiencies of 1.2 W/A, very low series resistances of 2.2 mΩ, and low internal optical losses of 1.7 cm^{-1}. These results demonstrate that Al-containing lasers can also have the very high reliability usually claimed for Al-free lasers (see also 980 nm high-power lasers above). In conclusion, the fundamental question of whether Al-free lasers are significantly more reliable than Al-containing lasers is still open.

High-power unstrained GaAs/AlGaAs QW lasers are available with emission wavelengths of 800 to 850 nm, although 808 nm devices (Oeda *et al.*, 1998) require very thin QWs with only about 10 to 12 monolayer thicknesses, which is a challenge to control. The wavelength can be further lowered by using AlGaAs QWs and high-power operation of 0.45 W has been demonstrated down to 715 nm for uncoated 12×5 μm stripe AlGaAs QW GRIN-SCH gain-guided lasers (Tihanyi *et al.*, 1994). Maximum cw powers in the 808 nm band of 2.9 and 2.6 W have been obtained for GaAs/AlGaAs SQW-SCH and AlGaAs/AlGaAs SQW-SCH lasers 150 μm wide, respectively. These devices operate without failure at 45 °C, 1 W automatic power controlled (APC) over 2000 h (Shigihara *et al.*, 1991). Al in the active layer leads to strong corrosion and oxidation effects at the laser facets, and hence reduces the power level at the COMD significantly, with the consequence that these lasers possess only a very limited reliability (Yellen *et al.*, 1993).

Quaternary InGaAsP QWs sandwiched in GaInP/InGaAsP waveguide structures offer completely Al-free lasers emitting in the range of 730 to 875 nm with high power and reliability. Uncoated devices 100 μm wide and 700 μm long emitting at 808 nm have achieved maximum COMD-free powers of 5 W cw and long-term reliability without failures at 60 °C, 1 W APC over 3×10^4 h aging tests (Diaz *et al.*, 1997). High-power emission has also been reported for 808 nm, Al-free active layer InGaAsP lasers with GaInP confinement layers and AlGaAs cladding layers (Hayakawa, 1999). In addition, GaInP/AlGaInP waveguides provide higher barriers and hence lasers down to 700 nm emission wavelengths can be realized (Al-Muhanna *et al.*, 1998a; Mawst *et al.*, 1999). The wavelength range of 715 to 810 nm can also be covered by high-power tensile-strained GaAsP/AlGaAs QW lasers showing high efficiencies and good reliability. AR/HR-coated devices 100 μm wide and 4 mm long showed cw powers of 1.8, 3, and 6.4 W at 718, 735, and 785 nm wavelengths, respectively. The output power is COMD limited and power densities of 2.4, 3.7, and 7.1 MW/cm^2, respectively, have been achieved (Erbert *et al.*, 1999).

For the visible wavelength range of 650 to 690 nm, usually compressively-strained GaInP QWs embedded in AlGaInP confinement and AlGaInP cladding layers are used, while tensile-strained GaInP QWs are employed for shorter wavelengths down to 630 nm. High-power operation of lasers emitting in the red wavelength band have been reported (Orsila *et al.*, 1999). Particularly high COMD-free cw powers have been obtained by Watanabe *et al.* (1994) from devices with thin undoped $(Al_{0.7}Ga_{0.3})_{0.5}In_{0.5}P$ window layers grown by low-pressure MOCVD (see Section 4.3 for nonabsorbing mirror approaches) on the cleaved facets of MBE-grown strained SQW AlGaInP ridge lasers 4 μm wide and 1200 μm long emitting with a wavelength of 680 nm. A COMD-free maximum optical output power of 300 mW cw, only limited by thermal rollover, has been obtained, which is about twice as much as that of a conventional laser without the window layer, where the power was limited by COMD. The threshold current of ~100 mA and slope efficiency of ~0.75 W/A were the same as those of the conventional laser, which indicates that the window layer does not affect the laser properties. Slow-axis and fast-axis far-field measurements up to 150 mW confirm the fundamental transverse mode operation. This technique applied to AlGaInP lasers appears to be promising considering that these lasers are very susceptible to facet degradation at high power (Watanabe *et al.*, 1994). Miyashita *et al.* (2000) achieved 121 mW cw for 900 μm long and narrow devices equipped also with a window–mirror structure but emitting in the shorter wavelength band of 659 nm. These authors showed good reliability of lasers 900 μm long emitting at 687 nm at 40 °C and 120 mW APC over 1800 h without failure.

In the short-wavelength regime, diode lasers usually suffer from low efficiencies, which are mainly due to the low conduction band offsets. This can be counteracted by placing superlattices in the upper waveguide structure that act as Bragg reflectors for the electrons and hence reduce carrier escape. The drawback, however, is an increase in the series resistance caused by the additional heterointerfaces. In general, the series resistance is a special issue in (Al)GaInP lasers due to the limitations on p-doping levels (see the next section).

In the blue–UV wavelength regime, great strides have been made in the fabrication of high-power, reliable, GaN-based diode lasers. For a long time effective technological progress was hampered by the mismatch in lattice constants and thermal conductivities between nitride layers and substrates. A recent breakthrough has been achieved by using a triple $In_{0.07}Ga_{0.93}N/In_{0.01}Ga_{0.99}N$ QW structure sandwiched in a GaN SCH and AlGaN claddings, which were deposited on an n-type GaN substrate. Ridge waveguide lasers, 7 μm wide, 600 μm long, and with cleaved uncoated facets, emitted with high efficiency at room temperature record-high cw output powers up to 2 W at wavelengths of 405 nm and operating currents and voltages up to 1 A and 5.7 V, respectively (Saito *et al.*, 2008). Commercial 405 nm blue–violet diode lasers operating in single mode and quasi-continuous wave (q-cw) power of 400 mW are available (Sony Shiroishi Semiconductor Inc., 2010).

Finally, we discuss material systems used for high-power diode lasers emitting with longer wavelengths >1 μm. $In_{1-x}Ga_xAs_yP_{1-y}$ grown on InP substrate is the classical material system delivering wavelengths for long-distance fiber optics communications in the range of 1.1 to 1.65 μm by changing the mole fractions x and y accordingly. Wavelengths at 1.3 and 1.55 μm are of particular interest where the standard silica fiber has minima in total dispersion and loss, respectively. A range of lattice-matched quaternaries extending from InP to the InGaAs ternary line can be grown by complying with the mole fraction condition $x = 0.4y + 0.067y^2$ (see

Figure 1.11). Direct bandgap energies can be achieved ranging from 0.75 eV for the ternary endpoint $In_{0.53}Ga_{0.47}As$ to 1.35 eV for InP. Compared to AlGaAs/GaAs, the band offsets in this system are quite different: only 40% of the band offset is in the conduction band, whereas the band offset in the valence band of 60% is much higher (Piprek *et al.*, 2000). Further drawbacks include high series resistance, inter-valence band absorption, Auger losses, and above all a strong temperature sensitivity of threshold current and efficiency. Internal losses could be reduced to 3 cm^{-1} by using an asymmetric waveguide structure (Nagashima *et al.*, 2004), but, nevertheless, multiple quantum wells (MQWs) are required to provide sufficient modal gain to get control of the internal losses.

Thus, up to 5 W cw output power have been achieved from compressively-strained InGaAsP/InP MQW SCH lasers 100 μm wide emitting at around 1500 nm (Garbuzov *et al.*, 1996). Record-high 1.2 W cw ex-facet power levels have been obtained from junction-side down-mounted, single-mode ridge lasers 3–5 μm wide and 3 mm long with InGaAsP compounds of different compositions used for the compressively strained QWs, barriers, and confinement, and with InP for the claddings. These lasers emitting in the 14xx nm band are suitable for pumping Raman amplifiers. Usually several 14xx nm individual diode lasers with different wavelengths form a Raman gain block to establish a flat gain over a wavelength range of ~70 nm, which is suitable for wavelength time division, multiplexed Raman pumping applications (Garbuzov *et al.*, 2003).

In the 1.3 μm wavelength range, the material system of strained $Al_xGa_yIn_{1-x-y}As$/InP QW has been developed as a potential alternative to the conventional $Ga_xIn_{1-x}As_yP_{1-y}$/InP material system (Zah *et al.*, 1994). The former has a higher electron confinement energy and therefore prevents carrier overflow under high-temperature operation. Strained-layer QWs are chosen to reduce the transparency current and the carrier-dependent loss due to the intervalence band absorption. Both 1.3 μm compressively-strained 5-QW lasers and tensile-strained 3-QW lasers were fabricated using a ridge waveguide structure 3 μm wide. In spite of the Al-containing active layer, no COMD was observed at 25 °C up to 218 mW and 103 mW cw for these lasers, respectively. Specifically, the compressively-strained devices showed excellent temperature characteristics with a threshold current characteristic temperature $T_0 \sim 105$ K, which is about twice as high as that for conventional InGaAsP/InP lasers. For operating the compressively-strained lasers at 85 °C with >5 mW cw, a mean-time-to-failure of 9.4 years has been projected from preliminary life tests. These lasers are interesting for uncooled laser applications, such as fiber-in-the-loop (FITL), where the cost of the laser is an important factor.

There is an approach to extend GaAs-based lasers (the long-wavelength limit for InGaAs QWs is around 1.2 μm) with all their positive characteristics, including high-temperature operation (high T_0), to 1.3 μm and replace InGaAsP lasers with their thermal problems (low thermal conductivity, high electron escape due to low conduction band offset). This has been achieved by InGaAsN/GaAs QW structures with the incorporation of only a very low N content of <2% to avoid a strong degradation of luminescence efficiency at higher concentrations (Kondow *et al.*, 1996, 1997). This material system (see also Section 1.1.5) now holds all the advantages of the GaAs system with access to the commercially important wavelength range of ~1.2 to 1.3 μm, for example, for direct pumping of Raman amplifiers. Excellent laser characteristics have been reported including threshold current densities down to 500 A/cm^2 from narrow-stripe (~4 μm) devices 350 μm long with lasing wavelengths between 1.24 and 1.3 μm and reliable cw operation up to at least 100 °C with high characteristic temperatures T_0 of 110 K (Borchert *et al.*, 2000; Riechert *et al.*, 2000). The latter is caused by the high confinement energy for electrons in these structures, which is due to (i) the high conduction band offset of about 80% and (ii) the increased electron mass compared to InGaAs (Hetterich *et al.*, 2000; Shan *et al.*, 1999). The higher electron mass, however, leads to transparency current densities roughly three times higher than for 980 nm InGaAs QW lasers. The highest known optical powers obtained in the InGaAsN material system are 8 W cw from AR/HR-coated, 1.3 μm InGaAsN SQW diode laser devices 100 μm wide (Livshits *et al.*, 2000).

Layer doping

Some essential laser design parameters depend sensitively on both the doping profile in the vertical structure and doping levels in the individual epitaxial layers, and these include the location of the p–n junction relative to the active layer, the series resistance, in particular in the p-type layers, and the optical losses due to free-carrier absorption. An effective design and control of the doping characteristics is therefore vital to achieve high-power laser operation with low threshold current density.

Layer doping – n-type doping

Silicon is the most common n-type dopant in MBE and MOCVD for all III–V compound semiconductors including (Al)GaAs, (Al)GaInP, and InGaAsP. Doping levels up to 5×10^{18} cm^{-3} in GaAs can be obtained; above this level the amphoteric nature of Si leads to self-compensation. This means the incorporation of Si not only on Ga sites, but also on As sites or as interstitials. InP-based materials and InGaAs can easily be doped electrically active up to 10^{19} cm^{-3} levels.

Layer doping – p-type doping

Beryllium is the most common p-type dopant in MBE for all III–V compounds. Compared to Zn and Mg, Be has some advantages including a smaller diffusion coefficient, which enables better growth control of the required doping profiles and a lower vapor pressure, which decreases the risk of memory effects. Zn supplied by organic precursors, such as dimethyl- or diethyl-Zn, is the preferred p-dopant for MOCVD-grown GaAs and InP-based materials. There are drawbacks with Zn, which include its high diffusion coefficient and re-evaporation tendency. These drawbacks affect the doping profile and incorporation efficiency, respectively, and can be dealt with by optimizing the growth process, usually by reducing the growth temperature and compromising on the material quality. Alternatively, C is a potential p-dopant with a favorably low diffusion coefficient and high incorporation efficiency up to very high levels $>10^{19}$ cm^{-3}. It can be introduced intrinsically by controlling the growth temperature and V/III ratio or by additional external sources such as trimethyl-As, CCl_4, or CBr_4. The second approach offers more independent control of doping and intrinsic defects compared to the first approach. By growing at low temperatures <600 °C and V/III ratios $\cong 1$, doping levels of 10^{19} cm^{-3} can be achieved, a value that can be lowered by growing at higher temperatures reducing also the uptake of oxygen, an impurity, which can form nonradiative recombination centers, particularly in AlGaAs.

Graded-index layer doping

This issue is discussed by means of a GaAs/AlGaAs GRIN-SCH QW laser structure. During growth at ≥ 700 °C Be diffuses from the p-type AlGaAs cladding through the entire p-side GRIN into the QW and beyond into the n-side GRIN where it piles up near the Si dopants close to the edge of the n-AlGaAs cladding with the consequence that the p–n junction is displaced from the QW active region. This is the case when the GRIN-SCH region is nominally undoped. The displacement of the p–n junction is significantly reduced when the p- and n-sides of the GRIN regions are doped with Be and Si, respectively (Chand *et al.*, 1994). Detailed investigations showed that Be diffusion is retarded in the n-GRIN and enhanced in the p-GRIN layer (Swaminathan *et al.*, 1992). Thus, the best option is to dope the n-side of the GRIN with Si and leave the p-side of the GRIN region nominally undoped. The Be atoms will diffuse anyway from the cladding into the p-side of the GRIN region. The Si doping starts high in the n-cladding to keep the series resistance low and is gradually decreased in the GRIN toward the active layer to minimize optical absorption losses. On the p-side the grading is steeper and absolute doping levels are higher to compensate for the lower hole mobilities. The thickness of the inner undoped region is optimized with respect to trading off Ohmic and optical losses. The overall doping profile has to be adjusted to achieve the required mode size.

2.1.3.4 Active layer

Integrity – spacer layers

In an InGaAs/AlGaAs system, there are actually two reasons for growing thin GaAs spacer layers between the active InGaAs layer and the AlGaAs GRIN-SCH regions. First, model experiments on AlGaAs/GaAs structures have shown that Be-doped AlGaAs layers tend to have a higher oxygen content than Si-doped ones, which occurs without any contribution from the Be source itself (Chand *et al.*, 1994). Detailed secondary ion mass spectrometry (SIMS) measurements showed a surface segregation of oxygen atoms in AlGaAs and their subsequent trapping at the AlGaAs/GaAs inverted heterointerfaces. Monolayer-thick GaAs is sufficient to trap the surface-riding oxygen impurities. A similar oxygen accumulation effect occurs in InGaAs/AlGaAs structures and without GaAs spacers the laser performance would be degraded strongly (Choi and Wang, 1990; Chand *et al.*, 1991). GaAs spacers physically separate oxygen atoms from the active InGaAs QW. Second, AlGaAs is grown at high temperatures $\cong 720$ °C, which leads to Ga losses due to desorption from the surface during growth interruption, usually initiated prior to the growth of the InGaAs QW, which occurs at a much lower temperature $\cong 510\ °C$ (cf. Section 1.4.1). Deposition of thin GaAs layers prior to the growth interruption compensates for the anticipated Ga desorption losses.

Integrity – prelayers

Strong extrinsic impurity-related luminescent emissions have been observed in low-temperature (2 K) photoluminescence (PL) spectroscopy measurements on nominally undoped GaAs/AlGaAs MQW structures with different well thicknesses (Epperlein and Meier, 1990). The intensity of these extrinsic PL emissions is strongest in the first GaAs QW grown in the MQW structure and depends sensitively on the thickness of the preceding AlGaAs barrier layer. Measurement of the binding energies of the involved impurities and excitation power dependence of the spectra revealed that the extrinsic PL is due to the radiative recombination of free electrons in the $n = 1$ quantized state of the well with neutral carbon acceptors. It is known that the solubility of impurities in AlGaAs is lower than in GaAs and hence impurities remain afloat on the AlGaAs growth surface and are progressively trapped in a thin layer at the inverted interface (GaAs on AlGaAs) upon deposition of the GaAs. This may lead to interface roughness and the formation of defects due to the growth-inhibiting nature of carbon, for example, by preventing the lateral propagation of the atomic layers because of pinning steps on the surface. These extended and point defects have a negative impact on the performance and reliability of a diode laser. Quantitative measurements showed that carbon-related PL can be suppressed to a low level <0.5% of the intrinsic (impurity-free) PL by a GaAs SQW 5 nm thick grown before the actual, measuring QW. By applying this scheme to a QW diode laser an improved device performance can be expected. In Section 7.2, we will resume the impurity gettering issue and discuss it in more detail.

Integrity – deep levels

Deep-level-transient spectroscopy (DLTS) measurements on MBE-grown GaAs/AlGaAs SQW structures have yielded a series of known electron traps located in the upper AlGaAs layer close to the QW interface within a region of about 15 nm (As *et al.*, 1988). The formation of these traps is due to a growth temperature which is far below the optimum temperature of $\cong 720\ °C$ when ramping up the temperature from the ideal low temperature of $\cong 580\ °C$ for the GaAs growth. It is known that the concentration of each of these traps doubles every 50 °C below 720 °C. The time to stabilize to the optimum temperature of 720 °C is equal to the time to grow by about 25 nm, which agrees well with the observed full width at half maximum (FWHM) of the trap distributions. The growth temperature for the lower AlGaAs cladding layer was at its optimum value throughout the growth and hence no traps have been observed. These traps may impact the performance of AlGaAs diode lasers due to higher internal losses and may also be responsible for enhanced degradation processes. An interrupted growth after the QW may allow the resumption of optimum AlGaAs growth conditions, therefore preventing the formation of these performance-deteriorating defect states. In Section 7.3, we will discuss the experimental details further including the DLTS technique.

Quantum wells versus quantum dots

Today's semiconductor diode laser products are exclusively based on QW active layer concepts and technologies. In principle, this includes all diode laser types, such as narrow-stripe and wide-aperture emitters, and one- and two-dimensional arrays, practically for all commercial applications. This dominance is determined by several factors:

- Impressive performance based on the unique QW strengths, particularly those of strained QW structures, as discussed in Chapter 1.

- Successful exploitation of the fundamental physical parameters into effective, reliable, and reproducible technological concepts, and focused optimizations of the QW/waveguide structures targeted at achieving high gain characteristics, low carrier and photon losses, temperature-stable characteristics, and fit-for-purpose beam divergence angles.

- Utmost reliability figures and extremely long diode laser product lifetimes of greater than 30 years.

In the ultimate case of size quantization realized in a quantum dot (QD), which includes a narrow energy-gap material embedded in a wide-gap matrix, carriers are confined in all three dimensions for QD sizes in the order of the exciton Bohr radius. Energy levels of the carriers are then discrete and separated, and the density of states narrows to delta function-like distributions. QD lasers with these atomic-like density of states are expected to have major advantages over QW lasers, including:

- lower threshold current densities;

- higher temperature stability of characteristics with infinitely high characteristic temperatures T_0 and T_η;

- higher differential gain and tunability of gain spectrum; and

- lower chirp, that is, lower shift of the lasing wavelength with injection current.

Despite the promising potential of QD systems, the best results, obtained to date after nearly three decades from the theoretical inception of QD lasers (Arakawa and Sakaki, 1982) and two decades since the first report on self-assembled QDs (Ledentsov *et al.*, 1994), do not match the high-power performance of commercial QW lasers.

The fabrication of QDs is based on a well-accepted approach by using a Stranski–Krastanow (1937) growth mode, where highly strained semiconductors are epitaxially grown on lattice-mismatched substrates with the formation of coherently strained islands after a few monolayers of growth. Elastic strain relaxation and renormalization of the surface energy are the driving forces for this self-assembled growth process (Bhattacharya et al., 2004). The islands can be subsequently buried to form the QD. The main drawbacks of this in-situ fabrication technique are its high cost and the low control over shape, size, material composition, and positioning of individual dots.

Nevertheless, QD lasers with low threshold and high modulation speed have been realized. However, their temperature stability and narrow linewidth have fallen far short of expectations. This is due to the dispersion of the QD size, shape, composition, and local strain, which leads to an inhomogeneous broadening of the excited state transitions and hence broadening of the ideal gain spectrum. Inhomogeneous line broadening is responsible for the parasitic recombination of carriers residing not in the QDs but primarily in the surrounding optical confinement layer at higher temperatures. These significant hot-carrier effects and associated gain compression are due to a density of states that is far less in the QDs than in the surrounding layers. Additional parasitic recombination currents can be caused by the thermal population of nonlasing QDs and by pumping nonlasing QDs through the inhomogeneous line broadening effect (Asryan and Luryi, 2001, 2002).

In summary, we can conclude that low carrier collection efficiency in the active layer, the various carrier loss processes, as well as thermal broadening of holes in the valence band of QDs, impact the performance of QD lasers by deteriorating the maximum gain, threshold current density, characteristic temperature coefficients T_0 and T_η, internal efficiency, and optical output power.

Several approaches have been proposed to mitigate these negative effects and which include the following:

- Generation of high-density and uniform QDs by optimizing growth parameters, using seeding layers (Mi and Bhattacharya, 2005) or patterned substrates (Kiravittaya et al., 2006).

- Use of a QW tunnel injection structure, first demonstrated in QW lasers and then in QD lasers for reducing hot-carrier effects (Bhattacharya et al., 1998; Bhattacharya, 2000). Here, the carrier collection efficiency is improved by the QW and subsequent phonon-assisted resonant tunnel injection of cooled electrons into the QD. Theoretical considerations have predicted that parasitic recombination of carriers outside the QD can be reduced considerably and large values for $T_0 > 1500$ K could be obtained by using tunnel injection. Further enhancement of T_0 results from the resonant nature of tunneling injection by selectively cutting off the nonlasing QDs (Asryan and Luryi, 2001, 2002).

- Modulation p-doping of the QD barrier. Holes of the p-doped barrier are then transferred into the hole ground state of lower energy in the adjacent QD layer. Therefore, fewer electron–hole pairs are required to be injected from the contacts to compensate for the thermal broadening of the hole distribution. This increase of the hole ground state occupancy increases the gain (Fathpour et al., 2005; Liu et al., 2007).

- Use of short-period $(GaAs)_6(AlAs)_6$ superlattices of indirect bandgap as barrier material for direct bandgap semiconductor QDs (Sun et al., 2004).

- Formation of QDs in the center of a QW, for example, by depositing an InGaAs layer 1.1 nm thick into a GaAs QW layer 10 nm thick surrounded by AlGaAs barriers. This quantum-dots-in-a-well (DWELL) structure minimizes carrier thermal escape from the dot states to barrier levels (Patanè *et al.*, 2000).

Despite the large body of work dealing with QD technology-baselining activities including growth, fabrication, material, and device characterization, relatively very little work has been carried out in diode laser product baselining aimed at achieving high optical output powers, which could be competitive to the state-of-the-art high-power performance of QW lasers. This mismatch is also reflected in the type and number of published data on fundamental and applied QD technology issues. A rigorous search of the published literature in 2009, which includes review articles (not many), textbooks (actually only one), and companies (about two) committed to the development of QD laser products, clearly reveals a disparity between achievements gained on QD technology basic topics and QD diode laser output powers, in stark contrast to the market-leading QW diode laser technology. Certainly, good progress has been made in QD diode laser technology, which includes the following:

- Record-low cw threshold current densities of 19 A/cm^2 in 1.3 μm InAs/GaAs lasers (Park *et al.*, 2000).

- High characteristic temperature coefficients $T_0 \geq 650$ K and $T_n = \infty$ up to 80 °C in 1.3 μm InAs/InGaAs/GaAs devices 100 μm wide (Mikhrin *et al.*, 2005).

- Record-high small-signal modulation bandwidths of $v_{-3\,dB} \cong 25$ GHz measured in 1.1 μm In(Ga)As lasers (Fathpour *et al.*, 2005).

- Low dynamic chirp of 0.1 Å and zero *linewidth enhancement factor* $\alpha \cong 0$ (Mi *et al.*, 2005).

  Note: The linewidth enhancement factor describes the spectral behavior associated with carrier density variations in diode lasers, and $(1 + \alpha^2)$ is a measure for the spectral linewidth broadening. It is defined as $\alpha = -2k_0(\partial n_r/\partial N)/(\partial g/\partial N)$ where n_r is the refractive index, g the material gain, N the carrier density, and k_0 the vacuum wavenumber (Henry, 1982).

- High differential gain dg/d$N \cong 8.5 \times 10^{-14}$ cm^2 at 283 K in InGaAs/GaAs QD ridge lasers (Bhattacharya and Ghosh, 2002).

However, on the optical output power side, the reported achievements are less numerous than in QW laser technologies and only a few can be considered as competitive with established and commercial QW diode laser products. The following data may illustrate the situation:

- 16 W cw total optical powers measured at both uncoated facets of junction side-down on microchannel heat sink mounted, 1.25 μm InAs/GaAs QD lasers 200 μm wide and 4 mm long (Crump *et al.*, 2007). The laser is fabricated in the most common material system used for QD diode lasers. To the author's knowledge, this power is the highest achieved for any QD diode laser device type. The figure has to be compared to the 8 W cw obtained from AR/HR-coated, 1.29 μm InGaAsN SQW lasers 100 μm wide (Livshits *et al.*, 2000). Both QD and QW lasers deliver about the same output power by taking into account the different laser cavity widths and lengths.

- 250 mW single-mode ex-fiber from InAs/GaAs QD diode lasers emitting at available wavelengths 1064, 1210, 1320 nm (Innolume GmbH, 2008). For comparison, a single-mode fiber-coupled InGaAs/AlGaAs SQW laser delivers 400 mW cw ex-fiber and kink-free at 1070 nm and room temperature (Oclaro Inc., 2010). (Note: ex-fiber power is the power emitted from a fiber-coupled laser).

- 3 W from a tapered InGaAs/GaAs QD laser and 3 W from a broad-area InGaAs/GaAs QD laser 100 μm wide both emitting at $\cong$ 920 nm (Kaiser *et al.*, 2007; Deubert *et al.*, 2005). Comparable QW lasers yield optical powers higher by a factor of $\cong$ 2–3 (cf. Table 1.4) and $\cong$ 7 (cf. Table 1.3), respectively.

- 400 mW cw from a single-mode QD laser emitting in the range of 915 nm and consisting of a single layer of self-assembled InGaAs QDs embedded in a GaAs QW 6 nm wide and processed to a ridge waveguide section 1 mm long followed by a taper 2 mm long with a full angle of 4° (Kaiser *et al.*, 2006). Comparable single transverse lateral mode InGaAs/AlGaAs SQW narrow-stripe single ridge waveguide pump lasers emit up to 1800 mW cw COMD free in the wavelength range 9xx nm and at room temperature (Lichtenstein *et al.*, 2004).

Number of quantum wells

The effects of number of active QWs n_{qw} on laser performance and reliability are manifold and have been thoroughly investigated in many publications. We will summarize here the major effects and results.

In general, lasers with MQWs in the active region have higher differential gain and transverse vertical optical confinement factor, which leads to lower threshold current density, weaker temperature dependence of laser characteristics, and higher characteristic temperature coefficient T_0 (Namegaya *et al.*, 1994). However, it may also lead to the generation of misfit dislocations due to the increased total active layer thickness (Namegaya *et al.*, 1994) and stronger facet heating, with the consequence of lower COMD levels (Chapters 7–9; Epperlein, 1997; Epperlein and Bona, 1993). The higher differential gain of MQW lasers leads also to narrower linewidths and higher modulation frequencies due to the greater relaxation oscillation frequency compared to SQW lasers. The last two positive features are more important for high-speed telecommunications than high-power (pump) applications.

The decision to select a SQW or MQW depends on the loss level. In the case of high losses, the MQW is always better because the gain comes from the high-slope part of the modal gain versus current characteristic instead of the saturated part of the

SQW gain curve. The MQW has a higher differential gain in the gain versus current dependence. (Note: In a simple approximation, the MQW gain curve is the SQW gain curve multiplied in both axes by the number of wells.) In this case, the saturated gain of the SQW may not always be large enough to reach threshold gain. However, at low loss, the SQW is always better due to both its lower transparency current J_{tr} (only states of one QW have to be inverted) and its lower internal loss ($\Gamma_{tv,qw}\alpha_i$ scales with number of wells) (cf. Sections 1.3.4.3 and 1.3.5.1). The optimum number of QWs depends on the required gain at threshold (Weisbuch and Vinter, 1991; McIlroy *et al.*, 1985).

In Sections 1.3.4.2 and 1.3.4.3, we discussed the influence of the number of wells on the threshold current of a QW laser and its length dependence. For details, we refer to these sections, including Figure 1.22.

Namegaya *et al.* (1994) have carried out an extensive investigation of the effects of well number n_{qw} on the material and device properties of compressively-strained 1.3 μm $Ga_{0.11}In_{0.89}As_{0.63}P_{0.37}$/InP GRIN-SCH MQW lasers. The wells are 4 nm thick and the number of wells in the MQW devices is in the range of 4 to 12. In the following, we discuss the major calculated and experimental results.

Low-temperature PL spectroscopy showed an increase in both the PL peak wavelength and FWHM for devices with $n_{qw} \geq 10$. In addition, the material with 12 wells showed a strongly reduced PL peak intensity. As has been demonstrated by critical thickness calculations, these PL effects can be ascribed to (i) the total thickness of the wells exceeding the critical thickness for the $n_{qw} \geq 10$ samples and (ii) the consequent degradation of the crystal quality due to strain relaxation. From calculations on the influence of device parameters on T_0, it is found that small values for α_i and α_m and large values for n_{qw}, $\Gamma_{tv,qw}$, and g_0 are effective for high-temperature operation of up to 170 °C. T_0 increases with increasing well number and facet reflectivities and high values of $T_0 \cong 80$ K have been obtained for $n_{qw} \geq 6$ and operating temperatures up to 100 °C. At room temperature, the minimum threshold current is obtained for $n_{qw} = 4$, whereas at higher temperatures >150 °C the minimum is at $n_{qw} = 8$. This is consistent with α_i increasing and g_0 decreasing with increasing temperature and by using the expression $n_{qw} = (\alpha_i + \alpha_m)/(\Gamma_{tv,qw}g_0)$ for achieving a minimum threshold current derived from the condition that at threshold the modal gain has to balance the total losses (see Section 1.3.4.2; McIlroy *et al.*, 1985).

The experiments showed also that the internal quantum efficiency η_i is independent of n_{qw}, and the internal losses α_i increase with n_{qw}, which can be attributed to the increased volume of the absorptive material and total optical confinement factor Γ_{tv} (cf. Section 1.3.4.3). High-power cw emission of >300 mW at 1.3 μm and 730 mA could be achieved at room temperature from optimized single-mode, narrow-stripe, six-QW lasers 1 mm long with low threshold currents of $I_{th} < 13$ mA.

Further useful results include the dependence of the FWHM fast-axis beam divergence angle $\theta_\perp$ on n_{qw}. $\theta_\perp$ increases continuously from ~22° to 32° for $n_{qw} = 4$ to 10. This gives the opportunity to adjust the fast-axis divergence angle to the slow-axis angle $\theta_\parallel$ in order to achieve a circular output beam, which could make it easier to couple power from narrow-stripe lasers with larger $\theta_\parallel$ angles into single-mode fibers. Moreover, accelerated life tests showed that elastically strained-layer MQW lasers with 8 and 10 wells exhibit a significantly higher degradation rate $>10\%$

of the threshold current than corresponding lattice-matched MQW devices. Critical thickness calculations clearly showed that this degradation is caused by a degradation of crystal quality suffering from critical thickness. Devices with 4 and 6 wells are free from inelastic, plastic strain relaxation and dislocation formation effects, whereas devices with more than 8 wells are unstable. Finally, it should be noted that the latter effect can be compensated by using a strain-compensating scheme.

In strained-layer MQW structures the net strain is accumulated, which means that the allowable strain in a SQW is reduced. By using strain-compensating barrier layers with a strain, which is opposite to that in the well layer, each well is then exposed to a similar accumulated strain originating from the underlying layer. This approach can also be applied to generate high strain levels in a SQW with the goal to extend the wavelength range, for example, by using GaAsP barriers in highly strained InGaAs/GaAs QW lasers the lasing wavelength could be extended to 1060 nm (Bugge *et al.*, 1998). *Strain-compensated lasers* also show a higher reliability because of the reduced driving force for strain-activated defect generation. Another positive effect of strained barriers could include the adjustment of the band structure to enhance the performance of certain diode lasers. In this way, the use of tensile-strained GaAlInP barriers in red-emitting lasers could improve the carrier confinement, reduce the absorption of laser light at the mirror facet, and thus enhance the optical strength of the laser (Valster *et al.*, 1997).

2.1.3.5   Fast-axis beam divergence engineering

The beam divergence property is of great importance whenever laser power is required to be coupled efficiently into another device enabling high-power, high-brightness applications including pumping fiber amplifiers, optical storage, and direct material processing. The requirement is not only for high output power but also for narrow beam divergence. Conventional GRIN-SCH QW structures with their tight optical confinement in the transverse vertical (fast-axis) direction usually yield large divergence angles $\theta_\perp > 30°$. However, this results in highly asymmetric elliptical far-field patterns with high aspect ratios >3.5, which require sophisticated optical systems to achieve acceptable coupling efficiencies (cf. Section 1.4.3.3). Tuning the composition of the cladding and GRIN-SCH layers or reducing their thicknesses can lower $\theta_\perp$ to $\approx 25°$ in InGaAs/AlGaAs lasers, but at the expense of lower kink-free powers and efficiencies and higher threshold currents; $\theta_\perp$ decreases at a rate of $\cong 1°$ per 1% AlAs mole fraction reduction in these lasers.

Expanding the optical mode in the transverse vertical direction is now a proven and powerful concept to reduce strongly the divergence angle with the additional advantages of:

- maintaining the low threshold current;

- lower risk of COMD failures at high-power operation;

- single-mode operation and suppression of higher order mode lasing; and

- suppression of beam filamentation effects.

There are many different approaches to realize this concept, each with its own pros and cons. In the following, we try to categorize the different approaches and discuss the major technologies.

Thin waveguides

By thinning the active waveguide layer much below the thickness used in conventional designs, the fast-axis divergence angle can be strongly reduced and the maximum output power limited by COD increased. Narrow-stripe (Al)GaAs lasers with thin active waveguides of only $\cong 0.04$ μm thickness, which is typically about five times less than in conventional lasers, yield $\theta_{\perp} \cong 16°$ and $P \cong 200$ mW (Hamada *et al.*, 1985). However, the strong spreading of the mode far into the cladding layers leads to significant free-carrier absorption resulting in an increase of threshold current and decrease in differential external quantum efficiency. This drawback can be mitigated by the so-called thin tapered-thickness approach where the active waveguide is thicker in the bulk of the laser cavity than near the mirrors. In this way, the fast-axis divergence angle and threshold current can be controlled independently and values of 10° and 60 mA, respectively, from (Al)GaAs devices 3.5 μm wide have been obtained (Murakami *et al.*, 1987). Another approach includes an asymmetrically expanded optical mode toward the substrate side by increasing the refractive index of the n-cladding layer relative to that of the p-cladding. This design can furnish low divergence angles $\theta_{\perp} \cong 23°$, $\theta_{\parallel} \cong 9°$, low threshold currents $I_{th} \cong 40$ mA, and high kink-free powers of 600 mW from 980 nm strained InGaAs/AlGaAs DQW lasers 3.5 μm wide and 1500 μm long (Shigihara *et al.*, 2002).

The above experimental values for $\theta_{\perp}$ can be confirmed to a good approximation with values calculated on the basis of the formula (cf. Equation 1.43) discussed in Section 1.3.5.3. In general, lasers based on the thin-waveguide structure approach may be sensitive to instabilities, which could be caused by the weak localization of the mode, refractive index changes due to current injection, and variations in the fabrication processes.

Broad waveguides and decoupled confinement heterostructures

Broad-waveguide (BW) SCH lasers have been developed primarily to achieve high cw power levels by providing concomitantly both a large equivalent transverse vertical mode spot size d/Γ_{tv} as well as low internal cavity losses $\alpha_i \leq 1$ cm^{-1} with no sacrifice in wall-plug efficiency at high drive current levels; here d is the active layer thickness and Γ_{tv} the transverse vertical confinement factor. From the definition of the internal optical power density at COMD, $\overline{P_{COMD}}$, Botez (1999b) derived an expression for the maximum cw power

$$P_{max,cw} = \left(\frac{d}{\Gamma_{tv}}\right) W \left(\frac{1 - R_f}{1 + R_f}\right) \overline{P_{COMD}} \qquad (2.8)$$

where W is the stripe width and R_f the front-facet reflectivity. One way to increase d/Γ_{tv} is to use a BW-SCH structure by expanding the fundamental mode through increasing the SCH guiding layer thickness t_c. The SCH layer with index $n_{r,w}$ is sandwiched between cladding layers which have an index $n_{r,cl} < n_{r,w}$. Accurate analytical approximations for d/Γ_{tv} and $\theta_\perp$ have been given by Botez (1999b) as

$$\frac{d}{\Gamma_{tv}} \cong t_c \left(0.31 + 2.1/D^{3/2}\right) \sqrt{\pi/2} \quad \text{for } \pi < D < 3\pi \tag{2.9}$$

$$\theta \cong 1.18 \tan^{-1} \left(\lambda/\pi w_0\right) \tag{2.10}$$

where w_0 is the equivalent near-field Gaussian waist

$$w_0 = t_c \left(0.31 + 2.2/D^{3/2} + 30/D^6\right) \tag{2.11}$$

and D is the normalized waveguide thickness defined as

$$D = \left(2\pi/\lambda\right) t_c \left(n_{r,w}^2 - n_{r,cl}^2\right)^{1/2} \tag{2.12}$$

and where $n_{r,w}$ and $n_{r,cl}$ are the refractive indices of the waveguide layer and cladding layer, respectively, λ is the vacuum wavelength, and $n_{r,w} > n_{r,cl}$.

For large $d/\Gamma_{tv} \gtrsim 0.66$ µm values, 0.97 µm emitting InGaAs/InGaAs(P)/GaAs BW-SCH QW lasers 100 µm wide and 2 mm long with high values for $T_\eta \cong 1800$ K and $\eta_d > 85\%$, high cw power levels of 11 W and low fast-axis divergence angles $\theta_\perp \cong 22°$ could be obtained. These devices are designed for $t_c \cong 1$ µm, which is lower than the cutoff thickness for the second-order mode. The experimental $\theta_\perp$ data are in excellent agreement with calculations based on Equation (2.10). Further decrease of $\theta_\perp$ can be obtained by decreasing the index step, $\Delta n_r = n_{r,w} - n_{r,cl}$, which is consistent with the thin-waveguide concept discussed in the previous subsection.

A major problem with the BW concept is that low divergence angles and high powers with low-risk COD failures can be obtained, but at the expense of the excitation of higher order transverse vertical modes at high injection currents.

In the context of the BW-SCH approach, we want to discuss the decoupled confinement heterostructure (DCH) concept (Hausser et al., 1993). In the DCH design, the electronic and optical confinements are decoupled by an internal barrier, and hence both can be optimized independently. It is characterized by a broadened waveguide and thin carrier block layers sandwiching the active region. These barrier layers have to be thick enough to prevent carrier leakage, while being as thin as possible (<40 nm) so as not to appreciably affect the optical waveguiding. Crucial for a proper operation of the barriers is that they are highly n(p)-type doped on the n(p)-side of the junction with typically 3×10^{18} cm^{-3}. Thus, these highly doped thin barrier layers pose no obstacle to majority carrier injection into the active layer, and they act as efficient barriers for the minority carriers in order to prevent carrier leakage. Undoped barrier layers not only lead to carrier leakage of minority carriers, but also inhibit an efficient injection of majority carriers.

Numerical simulations have shown that the leakage currents depend sensitively on barrier width, barrier doping, and barrier material. The simulations showed that hole (electron) leakage in an InGaAsP/InP DCH laser system can be suppressed to

~1% (14%), which compares to ~24% (27%) in a symmetric SCH structure (Hausser *et al.*, 1993). The smaller improvement in electron leakage may be due to the fact that the thin barrier layers are less efficient for electrons due to their much lower mass compared to that of holes. The suppression of carrier leakage in DCH lasers leads to lower internal optical losses and an increase in the characteristic temperature T_0.

By lowering the confinement factor Γ_{tv}, and in combination with the reduced optical losses, the hole burning effect and hence filamentation can be suppressed (see Section 2.2.1.6). This leads to more stable single-mode lasers with higher single-mode power operation. In addition, the DCH structure allows lowering of the Al content in the waveguide and cladding layers of Al-based lasers compared to SCH lasers. The effect results in less laser heating, improved power conversion efficiency, and higher reliability due to a lower electrical and thermal resistivity. Thus, optimized InGaAs/AlGaAs SQW DCH lasers have delivered 9.5 W cw for devices 100 μm wide (thermal rollover limited). Narrow devices with buried ridge waveguides 4–6 μm wide emitted up to 1.3 W cw thermal rollover power and 0.7 W cw single transverse lateral mode, kink-free power in FWHM beam divergence angles of 20° and 8° in the fast-axis and slow-axis directions, respectively (Yamada *et al.*, 1999).

Low refractive index mode puller layers

The intensity profile of the optical mode is engineered by manipulating the spatial variation of the refractive index of the cladding layers in such a way as to achieve both small beam divergence and low threshold current. This is realized by implementing a lower refractive index layer between the GRIN-SCH confinement and cladding layer on both sides of the waveguide.

The design principle is to maximize the mode intensity in the center of the active layer to achieve low threshold currents and to expand the optical field outside into the claddings to achieve small beam divergence angles (Yen and Lee, 1996a). The optical mode in the GRIN-SCH QW region is tightly confined, whereas outside of the low-index layers the mode spreads because the lasing mode index is reduced by the two low-index layers to be close to the index of the claddings.

This effect is illustrated in Figure 2.2 by comparing the calculated near-field profile of the new structure to that of the conventional one in InGaAs/AlGaAs GRIN-SCH QW devices; the calculations have been carried out by using the commercial simulation package LASTIP (Crosslight Software Inc., 2009). The corresponding experimental transverse vertical far-field profiles (Figure 2.3) show a reduction of $\theta_\perp$ from 32° to 19° for these nonoptimized InGaAs/AlGaAs lasers, which also showed no change in the threshold current (Epperlein *et al.*, 2000, unpublished).

In general, $\theta_\perp$ decreases with increasing Al content (corresponds to decreasing index) and thickness of the AlGaAs mode puller layers, and with decreasing Al content in the AlGaAs claddings. The reason is clear that as the cladding index increases to be closer to the lasing mode index, the lasing mode becomes more expanded leading to a smaller $\theta_\perp$ and increased I_{th}. Further decreasing the Al content of the claddings is critical since the cladding index can exceed the fundamental mode index. As a result, there is no guided mode in the waveguide (Lin *et al.*, 1996).

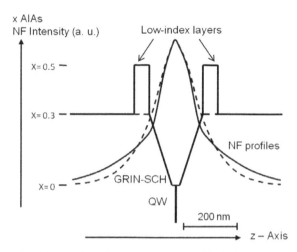

Figure 2.2   Simplified illustration of a transverse vertical InGaAs/AlGaAs GRIN-SCH QW waveguide structure without (dashed line) and with (solid line) low refractive index (high AlAs mole fraction) layers in the cladding layers on both sides of a conventional optical confinement GRIN-SCH layer. Calculated near-field (NF) profiles for the structure without (dashed line) and with (solid line) low-index layers.

Moreover, simulations show that the requirements on the growth conditions are very tight with an Al content to be controlled better than 2% in order to hit $\theta_\perp$ to within 10% of the target value. Anyway, record-low far-field angles of 13° have been reported on optimized 980 nm InGaAs/AlGaAs ridge waveguide lasers 4 μm wide and 3 mm long with low threshold currents of 66 mA, high slope efficiencies of 0.88 W/A, and single-mode operation up to very high powers of 1200 mW cw (Lin *et al.*, 1996; Yang *et al.*, 2004).

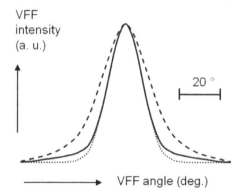

Figure 2.3   Experimental transverse vertical far-field (VFF) profiles for a conventional In-GaAs/AlGaAs GRIN-SCH QW diode laser without (dashed line) and with (solid line) low-index layers (high AlAs mole fraction 0.5) in the cladding layers. Gaussian fit (dotted line) to the VFF profile with low-index layers.

Optical traps and asymmetric waveguide structures

The purpose of the optical trap and asymmetric waveguide is to expand the mode toward the n-side of the structure and restrict its spread in the p-doped region, which minimizes the series resistance and free-carrier losses and leads to low fast-axis divergence angles and threshold currents, high differential quantum efficiencies and high kink-free powers, and suppression of higher order modes at high drive currents. Optical traps can be realized in different ways:

- Make the index profile of the confinement layer (such as LOC, which is the large optical cavity version of a SCH) asymmetric by increasing the index on the n-side (Peters *et al.*, 2005; Li *et al.*, 2008; Shigihara *et al.*, 2002).

- Place a large optical superlattice between the confinement layer and n-cladding (Lichtenstein *et al.*, 2006).

- Insert a graded higher index layer with a ∧-shaped profile (∨-shaped dip of Al content for AlGaAs) and well-defined thickness in the n-cladding at an optimized position far away from the active layer (Qiu *et al.*, 2005).

Experimental results from the latter approach include low fast-axis angles of 18°, low threshold currents of 30 mA, single-mode output powers of 400 mW, and strong suppression or even elimination of higher order modes achieved in narrow ridge waveguide InGaAs/AlGaAs QW lasers. Additional positive effects of this approach are that the incorporation of the optical trap has no adverse impact on threshold current and slope efficiency. The optical trap layer constitutes essentially a much weaker waveguide compared to the waveguide in the active region. To retain single-lobed near-field distribution and hence avoid side lobes in the far field, its thickness has to be optimized. There is also an optimum range of separation between the active layer and the optical trap layer for low fast-axis beam divergence. Moreover, this optical trap approach offers the freedom to design independently the vertical far-field and optical overlap with the active layer and in addition leads to a significantly improved growth tolerance.

An improved suppression of higher order lateral modes can also be achieved within the first approach in the list above by using a layer structure that supports an optical mode asymmetrically expanded toward the substrate into the n-cladding layer of a ridge structure (Shigihara *et al.*, 2002). We will discuss this approach in more detail in Section 2.2.1.2. Furthermore, an asymmetric waveguide, which expands the mode toward the n-side of the waveguide structure, has proven to be very efficient in improving free-carrier losses, optical losses, differential efficiency, internal efficiency, and series resistance. Thus, the highest reported power conversion efficiencies (see Equation 1.53) of 75% have been achieved to date on 808 nm InGaAlAs/AlGaAs broad-area lasers 100 μm wide and 1 mm long (Li *et al.*, 2008). These lasers with ultralow series resistances of 0.07 Ω, high slope efficiencies of 1.4 W/A, and low threshold current densities of 180 A/cm^2 emit >5 W cw in smooth single-lobed far-field patterns with 32° and 8° for the fast-axis and slow-axis beam divergence angles, respectively.

Spread index or passive waveguides

The insertion of passive waveguides with increased refractive index in the cladding layers offers another opportunity to stretch the optical mode in the transverse vertical direction and thus decrease the beam divergence. This coupled waveguide structure, also called spread index (SPIN) structure (Lopata *et al.*, 1996), is composed of an active waveguide, which provides gain and two passive waveguides to modify the far-field pattern.

Figure 2.4 shows the calculated transverse vertical near-field and far-field profiles of a standard InGaAs/AlGaAs structure with and without GaAs layers placed into both the cladding layers for comparison, and demonstrates an effective reduction of the fast-axis divergence angle by 9° down to 16° (Epperlein *et al.*, 2000, unpublished).

The parameters, which affect the mode profile, are the thickness and distance of the high-index GaAs layers from the active waveguide. Since the field distribution

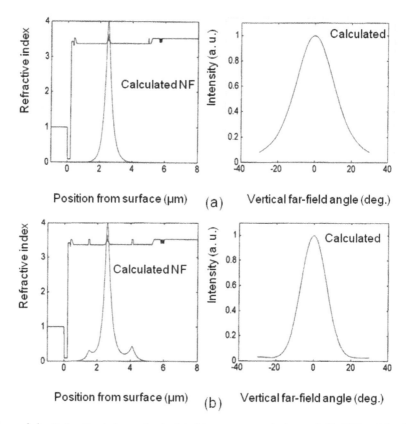

Figure 2.4   Refractive index and calculated transverse vertical near-field (NF) and far-field (FF) profiles of an InGaAs/AlGaAs GRIN-SCH QW laser structure: (a) conventional structure; (b) conventional active waveguide structure with two passive waveguides (with increased index) to expand the near-field and shrink the far-field pattern.

is at maximum in the active QW region, relatively low threshold currents of 24 mA and high slope efficiencies of 0.9 W/A can be expected from SPIN InGaAs/AlGaAs QW lasers exhibiting low fast-axis divergence angles of 17° (Lopata *et al.*, 1996). Further experimental fast-axis divergence angles include even lower values of 15° (Ziari *et al.*, 1995) and 11° (Chen *et al.*, 1990) for similar lasers.

Coupled mode theory and exact numerical calculations (Yen and Lee, 1996b) require that the thicknesses of the passive waveguides are thinner than a critical thickness for the fundamental mode operation. In addition, the separation between waveguides needs to be large enough and the coupling strong enough to effectively modify the far-field distribution. Side lobes can appear in the far-field pattern if the separation between waveguides is too large, which should be avoided. Optimization of the compositional and geometrical parameters of the individual building blocks of the coupled waveguide structure is complex in order to achieve single-mode operation at low thresholds and low fast-axis divergence angles over a wider acceptable parameter range.

An alternative approach utilizes periodic index separate confinement heterostructure (PINSCH) layers as optical confinement to simultaneously reduce the transverse vertical beam divergence and increase the maximum output power (Wu *et al.*, 1991). InGaAs/AlGaAs PINSCH QW lasers with ridge waveguides 5 μm wide and 750 μm long show good performance, which includes $\theta_\perp \cong 20°$, $I_{th} \cong 45$ mA, $\eta_d \cong 90\%$, and single-mode cw powers >620 mW.

Leaky waveguides

Although the leaky waveguide approach (Streifer *et al.*, 1976) has no relevance in commercial laser products, we briefly describe here its basic functionality and essential features by means of the GaAs/AlGaAs system. A thin AlGaAs layer with lower refractive index is placed between the higher index, active GaAs gain region and the GaAs substrate. Thus, optical power can flow through the thin AlGaAs layer and leak into the substrate, which occurs preferentially for higher order modes due to their higher penetration depth. The optical mode experiences only minor absorption losses in the substrate and escapes upon refraction from the cleaved facet into a well-collimated beam with a low divergence angle of only a couple of degrees. There are some drawbacks linked to this approach, which include

- high threshold current densities for high leaking losses;

- significant contribution of lasing in the active waveguide for a weak leaking process; and

- strong laser power losses in the contact layer/electrode layer in case the angle of the leaking modes is too high.

Spot-size converters

Spot-size converter integrated laser diodes are key components for reducing the fabrication cost of optical transmitter modules. They can provide low-loss direct coupling to the optical fiber or silica waveguide without the need for a lens. In addition, they offer expanded tolerance for optical alignment, mode stability, and narrow beam divergence by expanding the near-field fundamental mode. Several types of spot-size converters have been developed to expand the smaller, asymmetrical mode shape of a laser diode to match the larger circular mode of a single-mode fiber (cf. Section 1.4.3.3). These can be categorized into two groups: laterally tapered (Vawter *et al.*, 1997; Bissessur *et al.*, 1998) and vertically tapered waveguides (Kobayashi *et al.*, 1997; Aoki *et al.* 1997) integrated with the active waveguide of the laser. In these lasers, the optical mode is highly confined in the gain region to generate sufficient optical gain, whereas it is expanded in the converter region in both the lateral and vertical directions.

The schematic structure of a Fabry–Pérot laser monolithically integrated with a tapered waveguide thickness spot-size converter is shown in Figure 2.5a. An example of a laterally tapered active waveguide configuration is exhibited in Figure 2.5b. In this so-called double-core taper structure power is transferred from the tapered active ridge waveguide to the underlying passive coupling waveguide.

An effective lateral taper requires tapering the active waveguide width typically from 2 to 0.5 μm over a length of 1 mm. The shape of the taper can be linear over the entire length or in subsections with different taper angles. Adiabatic transformation supplies the best results, but requires a nonlinear taper, which can be approximated by an exponential shape. Fabrication usually implies a single epitaxial growth step, conventional projection lithography, and wet or dry etching. More advanced processing includes electron beam direct write and highly anisotropic chlorine reactive ion beam etching (RIBE) to produce the taper, including the narrow tip with high

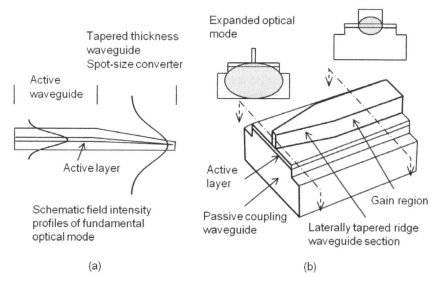

Figure 2.5   Schematic representation of two spot-size converter structures to expand the optical mode size: (a) monolithically integrated (vertically) tapered thickness of active waveguide; (b) laterally tapered active waveguide with underlying coupling passive waveguide (double-core structure).

surface quality and shape fidelity. Considerably expanded near-field spots have been achieved, which translate into very narrow far-field patterns with very low divergence angles of $\theta_\perp \cong 7°$ and $\theta_\| \cong 6°$ forming a nearly circular laser beam (Vawter *et al.*, 1997).

In the vertical taper approach, calculations show that a thickness ratio of the taper of at least 3 must be achieved over a length of typically 200 μm to have efficient fiber coupling and low radiation loss (Kobayashi *et al.*, 1997). In this case, the near-field spot is expanded to about 2.5 μm. There are several techniques for fabricating in-plane thickness modulation, which include selective area growth (SAG) on dielectric masked substrates (Kitamura *et al.*, 1999; Kasukawa *et al.*, 1997; Hirose *et al.*, 1999), selective area etching (Brenner *et al.*, 1995) and shadow-masked growth (Demeester *et al.*, 1990; Aoki *et al.*, 1997). In SAG, the growth rate in a narrow-stripe spacing depends on the width of the patterned SiO_2 masks surrounding the stripe on the substrate. The growth rate decreases with decreasing mask width resulting in thinner layers. Thus, a gradual change of layer thicknesses in the taper and a sufficient coupling between the tapered region and gain region can be achieved. In addition, this technique enables the growth of a smooth core layer that is free from any harmful scattering sites in the laser resonator.

An important advantage of the thickness tapering approach is that the converter region is transparent to the laser light because the thinner active QW layers shift the absorption edge to shorter wavelengths. Experimentally it is found that for achieving the required thickness ratio of 3 requires a mask width of 30 μm for the active layer, which narrows gradually down to 4 μm toward the end of the taper. In this way, a virtually flat area 300 μm long for the gain region and a thickness reducing area 200 μm long with a thickness ratio of 3 for the tapered waveguide can be grown simultaneously (Kobayashi *et al.*, 1997). SAG has the additional advantages of precise dimension control, high uniformity and reproducibility, simultaneous growth of layers with different thicknesses and compositions, control of the bandgap energy of a MQW structure and etching-free process for waveguide formation. The lowest transverse vertical divergence angle achieved with this approach is 9° (Kobayashi *et al.*, 1997). Due to the strong expansion of the near-field mode pattern, COMD-related degradation problems at the laser facets can be practically excluded.

Photonic bandgap crystal

Photonic crystals (Yablonovitch, 1987; Joannopoulos *et al.*, 1995) are composed of periodic dielectric nanostructures affecting the propagation of photons in the same way as the periodic potential in a semiconductor crystal determines the motion of electrons and holes by generating allowed and forbidden electronic energy bands.

Essentially, photonic crystals contain multidimensional structures with a periodic modulation of the refractive index. Photons will pass through regions of high index interspersed with regions of low index. To a photon, the contrast of refractive index looks just like the periodic potential that a charge carrier experiences when moving through the semiconductor crystal. For a large contrast in refractive index, there is a formation of allowed bands in photon energy separated by forbidden regions, the so-called photonic bandgaps. Since the wavelength of the photons is inversely proportional to their energy, the photonic crystal will block light with wavelengths in the photonic bandgap, while allowing other wavelengths to propagate freely throughout the crystal. This gives rise to distinct optical phenomena such as inhibition of spontaneous emission, highly selective optical filters, high-reflecting omnidirectional mirrors, and low-loss waveguides.

Since the basic physical phenomenon is based on diffraction, the periodicity of the photonic crystal structure has to be of the same length scale as half the photon wavelength, which poses high demands on the fabrication of high-quality photonic bandgap crystals (PBCs) operating in the visible part of the spectrum.

For edge-emitting diode lasers a waveguide based on a longitudinal PBC (LPBC) has been developed to realize stable high-power single-mode lasing with a very large modal spot size leading to narrow beam divergence, increased COMD threshold, single-mode operation in broader devices, and suppressed beam filamentation (Maximov *et al.*, 2008). In a LPBC, light propagates in a medium with a refractive index regularly modulated in the direction perpendicular to the propagation axis, which is qualitatively analogous to the operation of a photonic crystal fiber. The basic structure of a LPBC consists of a periodic sequence of layers with alternating high and low refractive index and a localizing optical defect violating the periodic index profile.

Basic types of optical defects are, for example, to increase the index or thickness of one high-index layer compared to the other high-index layers in the periodic sequence. The strength of the defect, which determines the number of optical modes localized by the defect, is then in these cases proportional to the difference in the index and thickness. To achieve single fundamental mode lasing the strength of the optical defect in the LPBC has to be designed in such a way that only the fundamental optical mode is localized by the defect and decays away from the defect, whereas the higher order modes are extended over the entire LPBC.

The general LPBC concept for laser application employs a LPBC on the n-side where the gain region containing the active QWs forms a localizing optical defect due to the high refractive index of these wells. An increase in the number of periods results both in a stronger discrimination of higher order modes by a decrease in the confinement factors of these modes and in a narrowing of the fast-axis divergence angle. Single-mode operation requires strong discrimination between the fundamental mode and the higher order modes in either modal gain and/or loss, which can be realized in different ways:

- Design the confinement factor of the fundamental mode to be much larger than that of the higher order modes.

- Realize a leaky design where all extended higher order modes penetrate into the (absorbing) substrate and contact layer (dependent on laser wavelength), whereas the localized fundamental mode has a very low leaky loss.

- Introduce additional absorbing layers into the structure affecting only all extended modes but not the localized fundamental mode (Maximov *et al.*, 2008).

Even in the case of large leakage losses for all higher order modes, but with confinement factors smaller for the fundamental mode than for the higher order modes, single-mode operation with narrow fast-axis beam divergence can be obtained. By modifying some layers close to the substrate, the preferential leakage of higher order modes can be effectively controlled.

For example, by making the high-index layer closest to the substrate thicker and the low-index layer closest to the substrate thinner by a factor of 2 leads to a substantial leakage of higher order modes into the substrate with the fundamental mode practically unaffected, which finally results in a relative increase of higher order mode losses by about a factor of 10. Together with a larger confinement factor for the fundamental mode, single-mode operation with narrow fast-axis beam emission can be achieved (Maximov *et al.*, 2008).

Figure 2.6 shows the vertical waveguide structure of a typical LPBC laser, considering only the LPBC specific layers (Maximov *et al.*, 2005). High-performance LPBC edge-emitting FP lasers in different material systems have been achieved and include the following:

- GaInP/AlGaInP lasers, 4 μm wide and 1.5 mm long, emit at 658 nm in single-mode with 115 mW cw power into a narrow fast-axis beam divergence angle $\theta_\perp \cong 8°$ and with a characteristic temperature $T_0 \cong 155$ K.

- GaAs/AlGaAs lasers, 4 μm wide and 1 mm long, emit at 850 nm in single-mode 270 mW cw into a far-field pattern of $\theta_\perp \cong 9°$ and $\theta_\parallel \cong 5°$ with an external differential quantum efficiency $\eta_d \cong 87\%$.

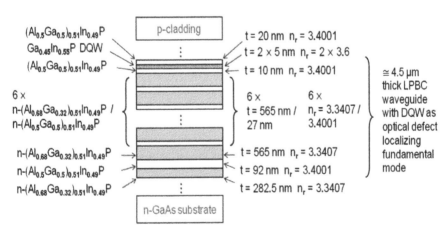

Figure 2.6   Schematic cross-section of the vertical structure of an edge-emitting Fabry–Pérot (FP) diode laser employing a longitudinal photonic bandgap crystal (LPBC) waveguide configuration together with a double quantum well (DQW) as optical defect for single-mode and narrow-beam operation. (Data adapted in amended form from Table I in Maximov *et al.*, 2005.)

- InGaAs/AlGaAs lasers, 10 μm wide, 1.5 mm long, and with a threshold current of 200 mA, emit at 980 nm in single-mode 1.2 W cw into a narrow far-field pattern with angles $\theta_\perp \cong 4°$ and $\theta_\| \cong 3.5°$. The 10 μm wide devices showed single transverse lateral mode operation up to high pump currents, which demonstrates the potential of the LPBC structure for single lateral mode operation in wider stripe lasers (Maximov *et al.*, 2008).

2.1.3.6   Stability of the fundamental transverse vertical mode

A major problem with some of the approaches discussed in the preceding section for expanding the optical mode is that low beam divergence angles and high COMD power levels may be obtained but at the expense of the excitation of higher order transverse vertical modes at high injection currents. This is particularly the case for approaches based on conventional waveguides such as broad waveguides or mode puller schemes employing the insertion of low refractive index layers or passive waveguide layers into the claddings on both sides of the gain-generating active region.

If the modal spot size or more precisely the equivalent transverse spot size d/Γ_{tv} exceeds a critical value, the effective waveguide width (thickness) may exceed the cutoff thicknesses for higher order mode excitations and the differences in the optical confinement factor between fundamental and higher order modes become very small. According to Botez (1999b), this can occur for 970 nm InGaAs/InGaAsP/InGaP BW structures with $d/\Gamma_{tv} \cong$ 0.42, 0.58, and 0.75 μm, leading to effective waveguide widths (thicknesses) of $\cong$ 0.53, 1.1, and 1.58 μm, which correspond to the cutoff thickness for first-, second-, and third-order mode excitation, respectively.

In addition, the power reflectivity of higher order modes is larger than that of the fundamental mode (Casey and Panish, 1978). Consequently, at relatively thin waveguide thicknesses $\gtrsim$0.5 μm multiple transverse vertical mode operation may degrade the far-field pattern and cause kinks in the *P/I* characteristics due to mode switching effects.

Another drawback with these approaches may originate from the high sensitivity of the field confinement at the large spot sizes to minor changes in the refractive index caused by current injection as well as compositional and temperature instabilities.

Altogether, it appears that these approaches enable only in a very limited design and operational space a reliable and robust laser operation with a stable optical mode. Similar limitations may also be true for the thin waveguide structure approach. Spot-size converters, however, may be very effective in achieving the lowest far-field divergence patterns of $7° \times 6°$ with nearly circular beam shapes, but impose great requirements on the design and fabrication of tapered waveguides to achieve the highest possible transformation of the optical mode and power in a reliable and reproducible way.

In the LPBC approach, the mode localization strength depends sensitively on the thickness and refractive index of the core layer comprising the active QW structure and forming the actual optical defect feature. Any variation of the defect thickness may change the localization of the fundamental mode and hence the confinement strength relative to that for higher order modes. However, even for up to 30% variations in the defect thickness, the leakage of higher order modes remains much larger and therefore continues to provide single-mode lasing and low fast-axis beam divergence of typically 8° for the fundamental mode. In principle, the LPBC approach enables the design of ultrabroad waveguides in a very robust way, which includes the insensitivity to variations over a wider range of active region thicknesses and refractive indices.

Finally, it has been demonstrated experimentally that the LPBC laser design is capable of delivering single lateral mode cw powers of up to 3 W from stripe lasers 10–20 μm wide emitted into a low transverse lateral far-field angle of $\lesssim 2°$ (Maximov *et al.*, 2005).

2.1.4 Narrow-stripe weakly index-guided transverse lateral waveguides

2.1.4.1 Ridge waveguide

In Chapter 1, we discussed a weakly index-guiding approach realized in rib and ridge waveguide types of lasers. Common to these designs is that the thickness of at least one layer is laterally nonuniform. In both types of scheme, the lateral laser structure can be modified such that an effective refractive index step of $<10^{-2}$ is generated under the rib or ridge zone. This index step is larger than the carrier-induced index suppression leading then to a relatively stable index-guiding of the lateral mode. In rib lasers, the thickness, for example, of the waveguide layer or the active layer, can be varied laterally. However, current spreading in the p-cladding layer can affect the threshold current density.

In ridge lasers, where the ridge is formed in the upper p-cladding by etching and embedded in a dielectric layer, the loss of effective current by current spreading is less pronounced. However, carrier diffusion in the active layer, which extends beyond the ridge, affects the threshold current, but also produces a continuous lateral variation of gain and index. The partial overlap of the mode with the dielectric layer forms an effective index step with a size determined by the height of the ridge and the residual thickness to the active layer, which is the thickness of the remaining p-cladding layer outside the ridge. A sensitive adjustment of the etch depth is required to provide enough effective lateral index step for single lateral mode operation. The beam quality and fundamental mode operation of ridge waveguide lasers are sensitively dependent on the design of critical ridge dimensions and their control during device fabrication. These topics will be discussed in more detail in Section 2.3 below.

Narrow ridge waveguide, single-emitter lasers have been extensively investigated and are widely employed in many key application areas because of their:

- simple fabrication technology requiring only one single epitaxial growth step;

- low optical losses;

- low threshold currents;

- low parasitic capacitances;

- high reliability;

- high control of the lateral index step providing single-mode operation for devices with a wide active region;

- easy integration; and

- record-high optical powers emitted in a single transverse vertical and lateral mode.

2.1.4.2   Quantum well intermixing

Quantum well intermixing (QWI) has proven to be an extremely useful and important technology for patterning in a post-growth process the refractive index and bandgap in the plane of the QW layers, with two principal objectives. First, to obtain high-performance device applications including low-loss lateral waveguides and nonabsorbing mirror structures in diode lasers. Second, to integrate monolithically optical devices such as diode lasers, detectors, modulators, and optical switches in a photonic integrated circuit.

The QWI process can be localized to selected regions of the QW structure so that the optical properties of only the selected areas are modified in this bandgap engineering process. It involves the interdiffusion of constituent atoms across the well/barrier interface resulting in a controlled modification of the material composition, which leads to a change in the bandgap and shift of the absorption edge to higher energies. Figure 2.7a shows a schematic representation of the QWI process.

A number of intermixing techniques have been successfully developed and can be categorized as follows:

- Impurity-induced disordering (Laidig et al., 1981) through ion implantation with subsequent high-temperature annealing (Thornton et al., 1985; Welch et al., 1987; Epperlein et al., 1987, unpublished) and without subsequent thermal treatment (Kuttler et al., 1998) or Zn diffusion (Itaya et al., 1996) or ion beam intermixing at elevated temperatures.

- Laser-induced disordering (Epler et al., 1988).

- Impurity-free vacancy disordering (Kowalski et al., 1998), which is based on the generation of group-III vacancies during the deposition of SiO_2 capping layers; these vacancies then diffuse through the structure during thermal treatment in a rapid thermal annealer leading to enhanced intermixing and an increase in bandgap energy.

In Chapter 4, we discuss these techniques in detail, in the context of enhancing the optical strength and robustness of diode lasers by developing efficient nonabsorbing mirror concepts.

Here, in this section, we restrict our discussion to the development of lateral waveguide structures by QWI processes. Decisive for this application is that the

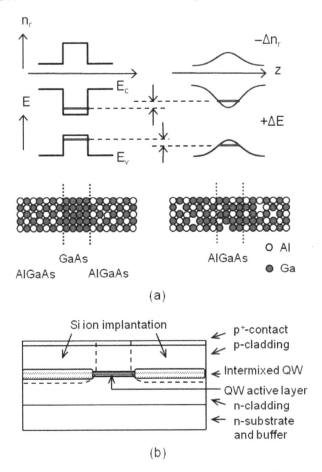

Figure 2.7 Schematic diagrams of the quantum well intermixing (QWI) process and its application in transverse lateral waveguiding. (a) Simplified conduction and valence band structure of a GaAs/AlGaAs QW before and after QWI. Resulting increase in bandgap energy and decrease in refractive index of the intermixed QW are also indicated. Only group-III atoms Ga and Al are shown. (b) Simplified illustration of a transverse lateral laser waveguiding structure based on QWI by an ion implantation process.

shift of the bandgap energy and the linked refractive index change are sufficiently high. Blue shifts of the bandgap wavelength can be as high as 60 and 50 nm obtained by impurity-free vacancy disordering of 1.55 μm InGaAs/AlGaInAs MQW structures (Bubke *et al.*, 2002) and ion implantation-induced disordering of 850 nm GaAs/AlGaAs SQW structures (Epperlein *et al.*, 1987, unpublished), respectively. Refractive index reductions of 0.18 (5.2%) have been measured on Zn-disordered 850 nm GaAs/AlGaAs QW lasers (Gray and Marsh, 1996). Such figures have enabled the fabrication of low-threshold, single-mode, real refractive index waveguided, planar buried heterostructure diode lasers using a silicon impurity-induced disordering process (Thornton *et al.*, 1985; Welch *et al.*, 1987). Figure 2.7b shows the schematic cross-section of a QW laser with a transverse lateral waveguide realized by QWI through an ion implantation process. One of the leading high-power, single-mode, single-emitter pump laser products is based on this technology.

2.1.4.3   Weakly index-guided buried stripe

This type of laser structure has the potential to deliver a stable transverse lateral mode, high kink-free output power, high reliability at high power operation, and a reasonable aspect ratio of the far-field pattern. Useful features of this structure are that the thickness of the p-type cladding layer, which significantly affects the transverse lateral mode can be controlled more precisely than that in a ridge waveguide and that the fabrication process is simpler and more reproducible due to the self-alignment involved. Precise design and control of the composition and thickness of the cladding and current blocking layers are essential to realize a low effective refractive index step in the horizontal direction of about $2\text{--}5 \times 10^{-3}$ for a stable transverse lateral mode operation up to high power. Typical buried-stripe widths are between 2.5 and 5 μm (Figure 2.8).

The fabrication consists of a two-step growth method and standard photolithography, dry and wet selective etching processes. The first epitaxial step includes the following layers: n-buffer, n-cladding, active region with optical confinement, first p-cladding, etch stop, n-current blocking, and possibly a suitable cap layer. The latter is recommended, in case the current blocking layer is of AlGaAs to avoid rapid oxidation of its exposed surface after etching the buried stripe. A thin GaAs cap layer is appropriate, but it must be damage-free to ensure optimum regrowth and also a high reliability of the finished laser device. Equally crucial is that all etched surfaces are free of defects so that a defect-free regrowth of the second p-cladding layer and the top contact layer can be obtained. Figure 2.8 shows a schematic generic cross-section of the structure and a cross-sectional view of a real device (Epperlein *et al.*, 2000, unpublished).

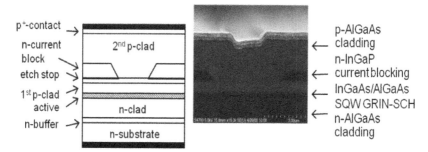

Figure 2.8   Schematic cross-section of a weakly index-guided, self-aligned buried-stripe diode laser depicting the essential components for single transverse lateral mode operation. Cross-sectional scanning electron microscopy image of an InGaAs/AlGaAs QW laser with n-type InGaP current blocking layer.

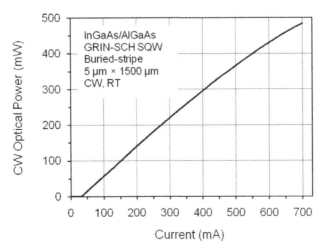

Figure 2.9   Optical output power versus drive current characteristic of a strained-layer, single-mode 980 nm InGaAs/AlGaAs GRIN-SCH QW diode laser with a self-aligned buried-stripe structure and n-type InGaP current blocking layers.

High kink-free powers of 980 nm, strained-layer InGaAs/AlGaAs GRIN-SCH QW lasers have been obtained from buried-stripe types of structures (Figure 2.9): 475 mW cw for 5 μm × 1500 μm large devices with InGaP current blocking layers and typical threshold currents of 32 mA (Epperlein *et al.*, 2000, unpublished) and 545 mW for devices 2.2 μm wide and 1000 μm long with AlGaAs current blocking layers and threshold currents around 22 mA (Horie *et al.*, 2000). In both cases, the maximum slope efficiency is ≅ 0.88 W/A. The lower kink-free power level in the former case may be due to the larger buried-stripe width (opening in the current blocking layer) which is more than twice that in the latter case.

2.1.4.4   Slab-coupled waveguide

The slab-coupled optical rib waveguide laser is a high-power, high-brightness diode laser that emits light in a single spatial, fundamental mode with a nearly circular profile and large modal diameter of several micrometers. The concept of this laser is based on results from a coupled-mode analysis between a rib and a slab region (Marcatili, 1974), and states that by appropriately selecting the slab thickness *t*, rib height *h*, and rib width *w*, the slab region acts as a mode filter and removes higher order modes from the rib region (i.e., these modes are coupled to the continuum of slab modes, which then radiate energy laterally). The criteria for a single-mode rib waveguide are determined by the ratios of effective slab thickness/effective rib height and effective slab thickness/effective rib width where the effective waveguide dimensions are the actual waveguide dimensions increased by the field decay lengths in the adjacent layers (Marcatili, 1974).

This mode filtering scheme allows a much larger mode area and a lower fundamental mode loss than in conventional ridge lasers. The large circular mode area strongly reduces the power density at the facets and therefore reduces significantly the risk for COMD-related laser failures. It also allows butt coupling to single-mode fibers with high coupling efficiency and without the use of lenses. The low modal loss makes possible longer devices, which would have reduced heat dissipation at high-power operation and therefore reduced thermal waveguiding. Gain is added to the rib region by a MQW structure so that the lowest order mode will lase without causing sufficient gain guiding in a higher order mode. Low modal gain implies, however, that the loss in the waveguide must also be kept very small (Donnelly *et al.*, 2003).

The structure is grown in a single epitaxial growth step and the device, schematically shown in Figure 2.10, looks similar to a ridge laser except that the ridge is etched through the active region into the waveguide layer forming the rib region (Donnelly *et al.*, 2003). Defects generated at the etched surfaces could lead to degradation in long-term reliability. Deposition of SiO_2 or Al_2O_3 passivation layers or regrowth of appropriate semiconductor layers in the etched grooves could mitigate these detrimental effects. The thickness of the waveguide layer of typically 4 μm is much larger than that used in standard single-mode lasers, and the height and width of the rib region are larger and nearly the same. The slab thickness is typically in the range of 3 to 3.5 μm. The grooves are etched 30–100 μm wide, and have to be wide enough so that the unpumped regions outside the grooves do not affect the lowest order mode, which is confined in the rib region but narrow enough so that the optical absorption in the unpumped regions can contribute to the loss of the slab-coupled higher order modes, possibly enhancing mode stability. The MQW gain region is placed on top of the waveguide in order to avoid having a waveguide inside a waveguide and having the lowest order mode localized around the MQW region (Donnelly *et al.*, 2003).

These lasers operating in a large, low aspect ratio, lowest order, single-lobed spatial mode have a large nearly circular near-field spot of typically 4.2 and 3.8 μm,

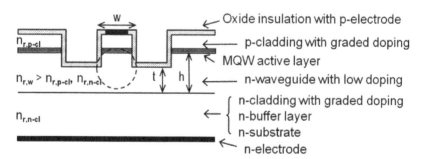

Figure 2.10   Schematic cross-section of a generic slab-coupled optical waveguide diode laser structure with a multiple quantum well (MQW) active region placed on top of the thick passive waveguide. The fundamental, bound spatial mode in the rib waveguide is also shown (dashed line). (Adapted in amended form from Donnelly *et al.*, 2003.)

and far-field divergence angles of 11 and 12° in the slow-axis and fast-axis directions, respectively, enabling butt coupling of power with high efficiency up to 88% into a single-mode fiber of 4.2 μm mode diameter. Single-mode 980 nm InGaAs/AlGaAs QW lasers with an optimum length of 1 cm have low internal losses of ∼0.8 cm^{-1}, high internal efficiencies close to unity, and high cw output powers > 1 W (Donnelly *et al.*, 2003).

2.1.4.5   Anti-resonant reflecting optical waveguide

The lateral structure of the anti-resonant reflecting optical waveguide (ARROW) consists of a low refractive index core region, which defines the lateral spot size of the device, surrounded by highly reflecting, high index cladding layers (Mawst *et al.*, 1992a; Yang *et al.*, 1998). The reflecting cladding layers are designed with a thickness and refractive index that correspond to an odd number of quarter lateral wavelengths $\lambda_1/4$ of the radiation leakage from the fundamental ARROW mode where λ_1 is the lateral wavelength in the high-index, anti-resonant reflecting layers. In this way, the anti-resonant fundamental mode suffers low radiation loss. However, higher order modes, which are not anti-resonant, suffer much higher losses, preventing them from reaching the threshold for lasing.

Typical ARROW structures have a core width between 4 and 6 μm and a built-in lateral index step $\Delta n_r > 0.03$, which provides mode stability to high output power, strong discrimination against higher order lateral modes, strong stability against gain–spatial hole burning, and which makes the device insensitive to index variations caused by temperature and carrier injection. An additional positive feature of the device is its buried-type structure with a planar top configuration for efficient heat sinking.

Figure 2.11 shows a simplified generic cross-section of a single-core ARROW laser structure along with the lateral index profile. The structure is usually grown by

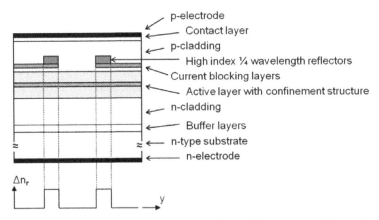

Figure 2.11   Schematic cross-section of the generic vertical structure of the anti-resonant reflecting optical waveguide (ARROW) semiconductor diode laser and the transverse lateral effective refractive index profile. (Adapted in amended form from Yang *et al.*, 1998.)

a two-step MOCVD process. The first step is up to the high-index guide layer (to be patterned to the lateral anti-resonant reflectors) and includes the vertical n-cladding, active waveguide structure and a current blocking layer used in the finished so-called self-aligned stripe (SAS) device geometry to restrict current injection to the low-index core region. The ARROW pattern is defined by conventional photolithography and wet chemical etching prior to the second growth step, which includes the thick vertical p-type cladding layer and the p^+-doped contact layer. The lateral effective index step is designed such that the reflector regions with typical widths of 0.9 µm are anti-resonant for the fundamental ARROW mode. The injected current is self-aligned to the low-index core region by the current blocking layers (reverse-biased junctions). The thicknesses of these layers are chosen such that the effective refractive index in the regions outside the high-index reflectors is identical to that in the core region (Yang et al., 1998).

Mode calculations on a 2D ARROW waveguide structure (Yang et al., 1998) show that the dominant mechanism for higher order mode discrimination is lateral edge radiation loss resulting in a low loss of ~ 2 cm^{-1} for the fundamental mode and high losses of > 16 cm^{-1} for the higher order modes calculated for a 970 nm InGaAs/InGaAsP laser structure. Such a large difference in edge losses and the immunity to gain–spatial hole burning are essential prerequisites for operating the laser in single spatial mode to high drive currents in excess of 10 times threshold. Calculations also show that the mode discrimination is large over a wide range of effective index step size, which leads to more relaxed fabrication tolerances and hence reproducible device parameters.

Measured far-field patterns show transverse lateral divergence angles as low as 4.5° from core devices 6.5 µm wide in single-mode operation (Yang et al., 1998). The best single-mode output powers of 300 mW cw are from single-core structures 4 µm wide emitting into a narrow transverse lateral far-field pattern with half-width angle $\theta_\parallel = 9°$ (Mawst et al., 1992a).

2.1.4.6 Stability of the fundamental transverse lateral mode

In the preceding sections, we have briefly described the strengths and weaknesses of the various approaches with respect to stable and reproducible operation of diode lasers in the fundamental transverse lateral mode. In this section, we discuss and summarize the major features of each technique relating to this stability issue.

In general, a major challenge for high-power diode lasers is the lateral mode instabilities that arise from the conflicting design requirements for high output power and single transverse lateral mode operation. In the sections above, we have shown that a large optical mode size is required for various reasons to achieve a narrow and preferably circular far-field pattern and to avoid damage to the facets. In the fast-axis direction, the optical confinement is determined by the layer structure, and the waveguide in this direction has to be thin enough to support only the fundamental mode even at high output powers. In the transverse lateral direction, a small refractive index step is required to suppress the higher order modes in the waveguide with a large lateral mode size. In various studies, it has been shown that the coherent superposition of the various transverse lateral modes is responsible for the formation of kinks in the P/I characteristics and far-field beam steering effects (Guthrie *et al.*, 1994; Schemmann *et al.*, 1995). Therefore, to achieve high-power single-mode output, two approaches can be taken to design waveguides either that support only the fundamental lateral mode or that allow higher order lateral modes but with a gain insufficient to reach lasing threshold.

The mode stability of ridge lasers is primarily determined by the waveguide geometry and layer composition where the dominating parameters are the ridge width and residual waveguide thickness in the etched regions outside the ridge. In Section 2.3, we discuss the results of a sensitivity analysis performed on the dependence of the fundamental mode and far-field pattern in the transverse lateral direction on these parameters. These results show a very limited window for fundamental mode operation determined by relatively narrow ranges of ridge widths, residual thicknesses, and slow-axis beam divergence angles for a given vertical laser structure. Proven practical approaches to obtain in the etching process the optimum residual thickness leading to single-mode devices with low threshold current and slow-axis beam divergence angle include applying the ridge etch, first at specific locations of a companion wafer, before applying a beveled etch across the actual laser wafer. This procedure can enhance the yield of fit-for-purpose laser devices.

For a given ridge width, the number of modes supported by a ridge waveguide, and their lasing conditions, depend on both the difference in effective refractive index between the regions within and outside the ridge and the transverse lateral gain distribution which couples to the field. However, this built-in index step can be very different to the actual index profile caused by detrimental effects such as index changes by carrier injection (Xu *et al.*, 1996), local thermal heating, and mechanical stress, resulting in the emergence of higher order lateral modes. For instance, time-dependent measurements on the laser beam quality degradation show that the temperature profile in the cavity plays a significant role in the transverse lateral guiding of the lasing modes (Hunziker and Harder, 1995).

Other approaches affecting the discrimination between the fundamental and higher order modes as well as the threshold current are to reduce the ridge height and p-cladding layer thickness (Wu *et al.*, 1995). In Section 2.2, we describe how, from the point of view of a lateral index step, a low-ridge and thin p-cladding device can be considered as equivalent to a high-ridge, thick p-cladding device and can deliver the same low-index step of $\sim 3 \times 10^{-3}$ required for stable single-mode operation. In addition, we describe results from numerical studies (Chen *et al.*, 2009) on the lateral mode behavior impacted by effects such as self-heating, spatial hole burning, lateral carrier distribution, and gain profile variation with increasing input current.

Most of the topics discussed above as potential parameters determining the stable fundamental lateral mode operation of ridge waveguide lasers are also valid for planar BH lasers realized by a QWI process, which leads to real refractive index lateral waveguides. This includes parameters such as effective waveguide width and height, p-cladding thickness, and all perturbing contributions with the potential to cause the generation of higher order optical modes as discussed above. The compositional modification in the QW structure, especially in strained material systems, raises the question of QW material quality after the intermixing process, which could give rise to mode instabilities. In general, it has been demonstrated that no degradation in the quality of the QW takes place: for instance, PL spectra show that strained QWs are still coherently strained, and InGaAs/GaAs SQWs disordered by shallow As implantation and thermal annealing show good structural integrity. However, on the other hand, compositional interdiffusion in lattice-matched InGaAs/InP systems can give rise to a strained structure after the QWI process. High-resolution transmission electron microscopy lattice images show no misfit dislocations in these disordered structures and the lasers show no degradation in threshold current.

The weakly index-guided buried-stripe laser has the built-in advantage of maintaining stable fundamental transverse lateral mode operation up to high power even with the occurrence of spatial hole burning. However, the laser device has to be carefully designed and fabricated, requiring a low transverse lateral index step of $\sim 3.5 \times 10^{-3}$, a narrow buried-stripe width of 2.2 µm, and a defect-free etching of the stripe and optimum regrowth of the upper layers. A novel double etch-stop structure and associated selective etching processes can realize these requirements to maintain the stability of the mode. The geometrical and compositional tolerances acceptable for stable transverse lateral mode operation are not known. One reason for achieving stable mode operation is that in this structure the thickness of the p-type cladding layer outside the buried-stripe region, which significantly affects the transverse lateral mode, can be controlled more precisely than, for example, in a ridge waveguide. In addition, the fabrication process involves a self-alignment process, which supports the geometrical specifications to be met for transverse lateral mode operation. Furthermore, long cavity structures provide better thermal conditions and reduce the effective injected carrier density in the active region, which, in turn, leads to a very stable transverse lateral mode. In addition, the thickness of the cladding layers is 2.2 µm in order to suppress resonant mode coupling between the ordinary mode in the laser waveguide (laser mode) and an unusual mode (substrate mode) propagating in the transparent substrate, which has a larger refractive index than the cladding layers. The thick cladding layers lead to an improved linearity of the *P/I* characteristic due to the elimination of substrate mode-induced mode hopping. Finally, stable transverse lateral mode operation is also demonstrated in long-term laser stress tests performed at high optical output power and high temperature (Horie *et al.*, 2000).

Slab-coupled waveguide lasers with a large, properly designed passive rib waveguide support in principle only one bound spatial mode due to coupling of the higher order modes into the slab modes. However, only slight changes in waveguide geometry can lead to mode changes, such as devices with wider ribs (e.g., increase from 4.6 to 5.4 µm) and shallower etch depths have lower lateral index confinements, which can contribute to the emergence of mode instabilities at higher power levels. These instabilities are thermal in origin as demonstrated in pulsed measurements with increasing pulse length at constant current. This means that the emergence of higher order lateral modes is due to a thermally induced increase in refractive index in the rib region. The low modal loss of these lasers permits very long devices, making it easier to handle power dissipation at high-power operation without thermal waveguiding, and therefore can lead to more stable modes. It has been found that the cw power at which mode instabilities occur increases with cavity length and correlates with the electrical power dissipation per unit length, which indicates again the role of thermal gradient effects in the mode instability issue by enabling the formation of higher order modes. It has also been shown that there is a maximum cw output power, which depends on the modal optical loss and the normalized series resistance, which is the product of the series resistance and device length. Devices with calculated optimum lengths of typically 1 cm show stable mode behavior over the entire drive current range up to the COMD power level with no signs of beam steering, but with only a slight widening of the beam divergence angles at higher currents, most likely due to thermal effects (Donnelly et al., 2003).

The single-mode operation of the ARROW diode laser relies on the large built-in lateral refractive index step formed by the central low-index core region and the surrounding high-index, quarter-wave anti-resonant reflecting regions. The fundamental spatial ARROW mode exhibits low loss over a relatively large range in index step while the first-order mode suffers a large loss. This ensures stable single-mode operation to high output power levels and strong discrimination against higher order lateral modes. The large mode discrimination over a wide range of effective index steps demonstrates a relatively large tolerance window of fabrication parameters over which single-mode operation can be obtained. However, unequal widths of the two quarter-wave reflectors due, for instance, to a slight deviation of the photolithography alignment, can cause the emergence of an asymmetrical shoulder in the transverse lateral far-field beam pattern (Yang et al., 1998).

Regarding transverse vertical mode stability, it has been shown that ARROW devices with a high transverse vertical optical confinement factor $\Gamma_{tv} = 3\%$ reach the threshold for higher order modes much earlier than devices with low $\Gamma_{tv} = 1\%$, due both to gain profile distortion and to distortion of the effective index profile in the device core with increasing drive current. Devices with cores 8.5 μm wide and $\Gamma_{tv} = 1\%$ can stay single-mode to more than 40 times threshold, which permits the projection of stable single-mode operation up to >1 W power levels. In contrast, core lasers 10 μm wide become multimode at about 10 times threshold with experimental stable single-mode powers to 300 mW for $\Gamma_{tv} = 1.5\%$ diode lasers (Chang et al., 2002).

2.1.5 Thermal management

The previous discussions in this chapter have dealt intensively with design approaches aimed at achieving high optical output power of narrow-stripe diode lasers. These approaches were elaborated in Sections 2.1.1–2.1.4. Whenever relevant in the course of the previous chapter and this chapter, we have pointed to the significance of having an effective thermal management in place. Equation (2.1) summarizes the direct and indirect involvement of relevant temperature-related parameters for realizing high optical output power. Thermal management is an important factor not only to achieve high output powers, but also, as we will see in later chapters, to obtain long operational lifetimes and high reliability by minimizing temperature-dependent laser degradation effects.

Good thermal management includes the design and fabrication of laser structures with:

- high characteristic temperature coefficients T_0 and T_η;

- low carrier and internal optical losses;

- high differential quantum efficiency;

- high electrical-to-optical power conversion efficiency;

- low laser series resistance; and

- low thermal resistance and high heat removal efficiency of the heat sink.

Restrictions in output power are primarily governed by the heat generated and carriers lost in the device. Carrier escape from the gain region leads to lower characteristic temperatures and efficiency, and a lower thermal rollover power (cf. Section 1.3.7.3). The heat emanates from the losses of the energy supplied to the laser device that are caused by nonradiative recombination events, absorption processes, and Joule heating of the drive current (series resistance), and these losses are reflected in low values of the external differential quantum efficiency and electro-optical power conversion efficiency (cf. Section 1.3.7.1). The waveguide losses can be reduced, for example, by reducing the carrier losses due to scattering, leakage, and absorption through an optimization of material composition, doping, growth, and processing (cf. Sections 1.4, 2.1.3.3, and 2.1.3.4).

The electrical-to-optical power conversion efficiency η_c, also called wall-plug efficiency (see Equation 1.53), is determined among others by material parameters including the mobility and free-carrier absorption of holes in the p-type layer of the waveguide as well as the thermal resistance of the waveguide and claddings. Promising material systems include InGaAs/InGaAsP/InGaP on GaAs substrates used for Al-free lasers. These systems, which are less reactive to oxygen, have very low series resistances in the order of 30 mΩ, high thermal conductivity of the InGaP cladding layers, and high characteristic temperatures $T_0 \cong 210$ K and $T_\eta \cong 1800$ K can deliver conversion efficiencies $\eta_c > 60\%$ (Al-Muhanna et al., 1998b). Also for these Al-free lasers, η_c decreases much more slowly with increasing drive current than it does for Al-containing devices with AlGaAs claddings. It seems, however, that AlGaAs material systems have the leading edge concerning carrier mobility and thermal conductivity of the waveguide and cladding layers (Peters et al., 2005). Low internal losses and low temperature sensitivity with high characteristic temperatures can be obtained from QW lasers in particular strained-layer systems (cf. Sections 1.1.4.1, 1.3.5.1, 1.3.7.3, and 2.1.3.4).

By making the laser longer, the thermal resistance can be reduced and the output power maximized; however, the latter requires scaling the internal losses in the waveguide correspondingly (cf. Section 2.1.2). As we have seen in Section 2.1.2, the internal optical losses and the transverse vertical optical confinement factor both have to be reduced linearly with increasing laser length (cf. Equation 2.7). This has to be carried out by adjusting the series resistance, which is responsible for the Joule heating (cf. Section 2.1.3.3), the thermal resistance, which governs the heat removal (cf. Section 1.4.3), and the characteristic temperature coefficients, which determine the temperature sensitivity of the laser (cf. Sections 1.1.4.1 and 1.3.7.3).

2.1.6 Catastrophic optical damage elimination

As discussed in Section 1.2.3.3, another very important temperature-related effect is catastrophic optical damage (COD). COD is an irreversible process which can occur at laser mirror facets and in the bulk of the laser cavity and is caused by strong heating due to high optical power densities, nonradiative carrier recombinations, or thermally accelerated decomposition processes triggered by small amounts of oxygen breaking the atomic bonds at the laser facets.

Catastrophic optical mirror damage (COMD) is a major problem limiting the maximum output power and arises when the power density at the mirror facet exceeds a critical level, which is a characteristic for the given material system. COMD is the result of strong surface recombination via traps, which causes a depletion of charge carriers at the crystal surface. The depleted bands of the active region then become absorbing at the lasing wavelength. The heat generated in this process raises the local temperature very strongly. At a critically high optical flux density, the raised temperature causes a sufficient shrinkage of the local bandgap energy with the consequence that the optical absorption and hence the temperature become even higher. This positive feedback can cause a thermal runaway with the ultimate melting of the end facet of the laser diminishing irreversibly any useful laser output power. Chapters 3 to 9 will give more details on the physics of laser degradation processes.

One way to maximize the optical output power is to increase the modal spot size in the transverse vertical direction, which increases the power level at which COMD occurs, or to eliminate COD processes. In Section 2.1.3.5, we discussed several mode expansion concepts within the context of fast-axis beam divergence engineering. Other approaches to suppress or even eliminate COMD processes include different facet passivation techniques and nonabsorbing mirror schemes, as well as appropriate reflectivity coating configurations, which will be discussed in Chapter 4.

2.2 Single spatial mode and kink control

2.2.1 Key aspects

This section discusses first the conditions for single and fundamental transverse vertical and lateral mode behavior of narrow-stripe index-guided diode lasers and gives relevant mathematical expressions for the required index differences and effective vertical and lateral active layer dimensions. Then the main design principles are described for achieving strong single-mode operation and high kink-free output powers up to high drive currents.

The effect of different ridge waveguide structures on the lateral mode behavior is investigated. The focus here is on ridge width and residual thickness, which is the total thickness of the p-type cladding layer outside the ridge region after etching. The internal physical mechanisms such as spatial hole burning, lateral carrier distribution, gain profile variation, and temperature-induced changes in the built-in refractive index profile are also discussed in this context. It is shown that low-ridge, thin p-cladding lasers can operate in a single transverse lateral mode with cw performance characteristics. Furthermore, asymmetrical expansion of the optical mode in the vertical direction toward the substrate, mirror reflectivity, and laser length are parameters used to enhance kink-free power operation in ridge waveguide lasers. All of them will be addressed in the following sections. The use of longitudinal photonic bandgap crystals, already discussed in the context of single-mode and beam divergence engineering in the transverse vertical direction (cf. Section 2.1.3.5 above), will also be discussed for achieving single-mode operation in the transverse lateral direction. Finally, a quantitative figure of merit is derived for evaluating the transverse mode operation over a wide range of ridge waveguide geometries.

Furthermore, techniques are discussed to stabilize the fundamental mode by suppressing the excitation of higher order modes through increasing their threshold gain. This can be accomplished, for example, by introducing mode-selective losses such as forming highly resistive regions at both sides of the ridge waveguide stripe or coupling the optical higher order mode to the absorptive metal contact layers outside the ridge through a sufficiently thin insulator dielectric layer, which embeds and defines the ridge structure. Various mode filter schemes such as corrugated waveguides, curved waveguides, tilted mirrors, and tapered waveguides are described to enforce fundamental mode operation by discriminating against higher order modes.

Methods are also described to suppress so-called beam-steering kinks, which are generated by resonantly coupling power from the lasing fundamental mode to the first-order mode. These include controlling the beat length of the two modes, which can be done by adjusting the cavity length or the difference of the two propagation constants, and thus maximizing the kink-free output power.

The filamentation effect, which is formed through gain saturation and self-focusing, leads to beam quality degradation preferentially in broader devices through lateral mode break-up overriding the built-in lateral mode control. We discuss briefly the various methods, which have been developed for controlling and suppressing the filament formation mechanisms, and hence promoting single transverse lateral mode performance.

Further details on single transverse lateral mode design issues and operating parameters of ridge waveguide diode lasers will be given in Section 2.3.

2.2.1.1 Single spatial mode conditions

As discussed in Section 1.3.3, the vertical structure of a diode laser can be well approximated by a three-layer slab waveguide comprising the active layer of thickness d, which is sandwiched between cladding layers (see Figure 1.19), and also the thickness of the optical confinement layers in case the active layer consists of a QW structure. The remaining layers in the structure can be ignored if the cladding layers are sufficiently thick so that the optical mode is confined largely in the three-layer slab. The mode analysis of this slab waveguide structure has been extensively studied in various publications (Adams, 1981; Marcuse, 1991; Agrawal and Dutta, 1993). The slab waveguide supports TE and TM modes with the electric and magnetic fields polarized along the junction plane, respectively. However, we consider only TE modes, because these are generally favored over TM modes in heterostructure semiconductor diode lasers due to their higher modal mirror reflectivity (Ikegami, 1972; Kardontchik, 1982) and lower threshold gain (Coleman, 1993). Because of the periodic nature of the trigonometric functions in the eigenvalue equations, multiple solutions do exist for the TE mode eigenvalues. The number of allowed, confined waveguide modes, however, is limited and is determined by the cutoff condition (Agrawal and Dutta, 1993), which is, in its final form,

$$k_0 d \left(n_{r,a}^2 - n_{r,cl}^2 \right)^{1/2} = p\pi \tag{2.13}$$

where $k_0 = 2\pi/\lambda$, p is an integer with even and odd values corresponding to even and odd TE modes, respectively, $n_{r,a}$ and $n_{r,cl}$ are the refractive indices of the active and cladding layer, respectively.

In Section 1.3.3, we introduced the expression $D = k_0 d \left(n_{r,a}^2 - n_{r,cl}^2 \right)^{1/2}$ (see Equation 1.28), the normalized waveguide thickness, which is a crucial parameter in the determination of the mode characteristics of the three-layer slab waveguide. The waveguide can only support the lowest order ($p = 0$), that is, the fundamental TE mode in case $D < \pi$. This, in combination with Equation (1.28), results in the single transverse vertical mode condition for the active layer thickness

$$d < \frac{\lambda}{2} \left(n_{r,a}^2 - n_{r,cl}^2 \right)^{-1/2}. \tag{2.14}$$

For example, for an InGaAsP/InP laser emitting in the wavelength range 1.1–1.65 μm, Botez (1981) obtained to a good approximation $\lambda \left(n_{r,a}^2 - n_{r,cl}^2 \right)^{-1/2} \cong 0.95$ μm, which then leads to the condition $d < 0.48$ μm for single transverse vertical mode emission.

To describe the transverse lateral mode behavior, we have to distinguish between gain-guiding and index-guiding. In contrast to gain-guided devices, where the effective modal index $n_{r,eff}(y)$ as given by Equation (1.29) is constant in the slow-axis y direction, index-guided devices are laterally structured with a higher index central region of width w surrounded by areas with lower effective index:

$$n_{r,eff}(y) = \begin{cases} n_{r,eff}^{in}, & |y| \le w/2 \\ n_{r,eff}^{out}, & |y| > w/2 \end{cases}; \qquad \Delta n_{r,tl} = n_{r,eff}^{in} - n_{r,eff}^{out} \tag{2.15}$$

where $n_{r,eff}^{in}$ and $n_{r,eff}^{out}$ are the effective refractive indices in these two regions, and $\Delta n_{r,tl}$ is the transverse lateral index step between them. This index step determines the strength of index guiding.

The transverse lateral modes are obtained by solving the wave equation for the three-layer slab waveguide problem in the two regions given by Equation (2.15). We can apply a similar procedure as for the transverse vertical modes and find in analogy to Equations (1.28) and (2.13) for the normalized waveguide width W and its cutoff condition the following expressions (Agrawal and Dutta, 1993):

$$W = k_0 w \left\{ \left(n_{r,eff}^{in} \right)^2 - \left(n_{r,eff}^{out} \right)^2 \right\}^{1/2} \tag{2.16}$$

$$W = q\pi \tag{2.17}$$

where q is an integer with even and odd values corresponding to even and odd transverse lateral modes, respectively. The lowest order ($q = 0$) mode is supported by a waveguide with $W < \pi$ and results in an effective active layer width w for fundamental mode operation

$$w < \frac{\lambda}{2} \left\{ \left(n_{r,eff}^{in} \right)^2 - \left(n_{r,eff}^{out} \right)^2 \right\}^{-1/2} = \frac{\lambda}{2} \left(2 \overline{n_{r,eff}} \Delta n_{r,tl} \right)^{-1/2} \tag{2.18}$$

where $\overline{n_{r,eff}} = (n_{r,eff}^{in} + n_{r,eff}^{out})/2$ is the average effective modal index. An example may illustrate the value of Equation (2.18). We take an InGaAs/AlGaAs ridge laser emitting with a wavelength $\lambda = 0.98$ μm. Using ~ 3.4 for the average effective modal index and $\sim 10^{-3}$ for a typical transverse lateral index step, Equation (2.18) yields $w \lesssim 6 \times \lambda \lesssim 6$ μm for the upper bound of the effective ridge width to achieve single transverse lateral mode operation, in good agreement with experiments.

We can also obtain in analogy to Equations (1.27) and (1.29) the following useful expressions for the transverse lateral confinement factor Γ_{tl} and effective refractive index for the fundamental transverse lateral mode:

$$\Gamma_{tl} \cong \frac{W^2}{(2 + W^2)} \tag{2.19}$$

and

$$n_{r,eff}^2 \cong (n_{r,eff}^{out})^2 + \Gamma_{tl} \left\{ \left(n_{r,eff}^{in} \right)^2 - \left(n_{r,eff}^{out} \right)^2 \right\}. \tag{2.20}$$

Γ_{tl} is defined in analogy to Equation (1.26) as the degree of overlap of the electric field intensity profile in the transverse lateral direction with the central high-index region of width w.

Single transverse vertical and lateral mode operation can be achieved for active layer dimensions complying with Equations (2.14) and (2.18). In this case, the overall confinement factor is given by $\Gamma = \Gamma_{tv} \Gamma_{tl}$ and represents the fraction of the mode energy contained within the active region in both the transverse vertical and lateral directions.

Finally, we determine the condition to match the slow-axis numerical aperture (NA) of a diode laser to the NA of a fiber. In analogy to the NA of a fiber (see Equation 1.66b) we can write

$$(NA)^2 = (n_{r,eff}^{in})^2 - (n_{r,eff}^{out})^2$$

$$= \left(n_{r,eff}^{out} + \Delta n_{r,tl}\right)^2 - (n_{r,eff}^{out})^2$$

$$\cong 2n_{r,eff}^{out} \Delta n_{r,tl} \qquad (2.21)$$

from which the index step as a function of the NA is obtained as

$$\Delta n_{r,tl} = \frac{(NA)^2}{2n_{r,eff}^{out}}. \qquad (2.22)$$

Typical NAs of single-mode and multimode fibers are 0.1 and 0.2, which lead to small index steps of 1.5×10^{-3} and 6×10^{-3}, respectively, by using a typical value of 3.3 for the refractive index. In Section 2.1.4, we described various technologies that can be used to fabricate transverse lateral waveguides with small-index steps.

2.2.1.2 Fundamental mode waveguide optimizations

Waveguide geometry; internal physical mechanisms

The ridge waveguide structure has proven to be the simplest and most straightforward way of achieving high-power single spatial mode diode laser operation. Far-field patterns and single-mode operation can be controlled easily. However, stringent dimensional tolerances are required to achieve good performance in these weakly index-guided laser devices. This includes particularly ridge width and height as well as the residual thickness of the p-type cladding layer, which can be linked to the ridge height. Detrimental effects include changes in refractive index profile by local thermal heating and carrier injection, spatial hole burning, lateral current spreading, and gain profile variations.

Numerical studies performed on the lateral mode behavior of (Al)GaInP QW lasers with ridge widths of 2.4–3.6 μm, residual thicknesses of 0.1–0.2 μm, and a constant p-cladding thickness of 2 μm have produced the following major results (Chen *et al.*, 2009). The cutoff condition for single fundamental lateral mode operation is mainly dependent on the effective lateral index step and ridge width. The emergence of the first-order mode is sensitive to the ridge height as the ridge width is increased, and can be effectively suppressed by narrow and shallow ridge geometries. The lateral carrier distribution in the active region is strongly affected by the ridge height, which changes the lateral gain profile and influences the lateral modes. Devices with wider ridges have sufficient modal gain to meet the threshold condition of the first-order lateral mode. The threshold current of the first-order lateral mode increases with decreasing ridge width and height.

This can be understood by the fact that narrower ridge widths provide better lateral confinement of electrons and holes. Consequently, carriers confined in the ridge center support the fundamental mode at low drive currents. At higher currents above the threshold current of the fundamental mode, however, the lateral carrier distributions, which overlap with the optical mode profile of the fundamental mode, are used up by the increased stimulated emission. This means a more pronounced lateral spatial hole burning effect at higher currents with the consequence that the higher order mode is then favored due to the improved match between the optical mode profile and the lateral gain distribution.

In addition to carrier spatial hole burning, temperature-induced refractive index change between the inside and outside ridge regions is another mechanism responsible for the emergence of higher order modes. Although narrower ridge widths may cause anti-guiding effects due to strong increases in carrier density at high drive currents, the temperature-induced index difference becomes larger with increasing current and may push the laser device beyond the cutoff condition. This effect is also stronger for higher ridge structures due to their poor heat dissipation. Lasers with higher ridge heights are also more susceptible to spatial hole burning and therefore more prone to the emergence of higher order lateral modes.

Different ridge heights have also a different effect on current spreading in the lateral direction. Simulations show (Chen et al., 2009) that the electron and hole concentration within the active region is larger for a high-ridge than for a low-ridge structure, which leads to a higher lateral interband gain profile, in particular at the ridge boundary of the high-ridge structure. This higher gain profile, however, supports the emergence of higher order lateral modes in high-ridge laser devices.

These results are confirmed by independent self-consistent 2D modeling (Xu et al., 1996), in particular that low-ridge devices have a higher first-order lateral mode threshold current than high-ridge devices, which is also consistent with the considerable increase in kink power measured in laser devices just by decreasing the ridge height. Furthermore, the modal gain of the first-order lateral mode of low-ridge lasers increases with injected current at a slower rate ($\sim$6%) than that of high-ridge devices. Depending on the value of the mirror loss, the gain/current curves of low-ridge and high-ridge devices may cross each other before the first-order mode lasing occurs, which would reverse the order in the occurrence of the kink in the P/I characteristics of the two laser types. The simulations show that for a ridge waveguide laser, which supports only one lateral mode in the "unbiased" state, it is still possible for the first-order mode to emerge at higher injection currents, caused by a carrier-induced strong change of the built-in lateral refractive index profile and spatial hole burning (self-focusing) effects. They also support the experimental results where the use of low-ridge and thin p-claddings results in ridge lasers with low thresholds and high single-mode output power (Wu et al., 1995).

The latter study shows that, from an index-step point of view, a low-ridge ($\cong$ 130 nm), thin p-cladding ($\cong$ 250 nm) InGaAs/AlGaAs QW laser device is equivalent to a high-ridge ($\cong$ 1200 nm), thick p-cladding ($\cong$ 1300 nm) device leading to an index step of $\cong 3 \times 10^{-3}$. The fact that low-ridge, thin p-cladding devices do not show the strong threshold effect as the ridge width is narrowed, demonstrates that the index step is the key parameter that determines device performance. These devices operate in a single spatial mode up to high cw power levels if the ridge width is sufficiently narrow (Wu *et al.*, 1995).

Figures of merit

In another approach, systematic simulations have been carried out (Laakso *et al.*, 2008) to investigate the dimensional range that ensures stable single transverse mode operation of InGaAs/AlGaAs ridge waveguide edge-emitting lasers over the whole bias range by employing both a fast 2D mode solver and the commercial software package LASTIP (Crosslight Software Inc., 2010). A quantitative figure of merit, based on the "under-the-ridge" active layer total optical confinement factor ($\Gamma = \Gamma_{tv}\Gamma_{tl}$), indicates the likelihood of single transverse modal behavior over a broad range of ridge widths and residual thicknesses. The definition of a figure of merit for stable single transverse mode operation is associated with the maximization of the following expression (Laakso *et al.*, 2008):

$$\Gamma_{1m} = (g\Gamma_1 - g\Gamma_m)/g\Gamma_1 = (\Gamma_1 - \Gamma_m)/\Gamma_1 \qquad (2.23)$$

where Γ_1 and Γ_m are the confinement factors of the $m = 1$ (fundamental) and $m > 1$ transverse modes, respectively, and g is the local gain.

The evaluation of the single transverse mode operation space can be simplified by studying just Γ_{12} and Γ_{13}, because $\Gamma_2 > \Gamma_4$, Γ_6, etc., and $\Gamma_3 > \Gamma_5$, Γ_7, etc. Therefore, Γ_{12} and Γ_{13} as well as the product $\Gamma_{12} \times \Gamma_{13}$ can be considered as useful *figures of merit*. For large Γ_{12} (resp. Γ_{13}) values close to one, the second (third) and all higher even (odd) modes are suppressed, whereas for lower values the ridge waveguide is likely to operate in a transverse multimode regime. Stable single mode operation is achieved when both Γ_{12} and Γ_{13}, that is, the product $\Gamma_{12} \times \Gamma_{13}$, have large values close to unity.

Figure 2.12 summarizes the major trend of results obtained from simulations carried out for ridge widths in the range of 2 to 8 μm and residual p-cladding layer thicknesses between 0 and 600 nm of InGaAs QW lasers with GaAs waveguide layers 140 nm thick, $Al_{0.6}Ga_{0.4}As$ cladding layers 1500 nm thick, and p^+-GaAs contact layers 200 nm thick (Laakso *et al.*, 2008).

Stable single-mode operation can be achieved over a relatively wide range of residual thicknesses, in particular for each ridge width up to about 4 μm. However, the lowest possible thickness ensuring a large value for $\Gamma_{12} \times \Gamma_{13}$ should be targeted, because high residual thicknesses cause a reduction in confinement and gain of the fundamental transverse mode resulting in an increased threshold current. Stable single-mode operation is harder to achieve for ridge widths around 5 μm and above, because of the very precise control required to get the high $\Gamma_{12} \times \Gamma_{13}$ value. The size of a high $\Gamma_{12} \times \Gamma_{13}$ area depends on the overlap of high Γ_{12} and high Γ_{13} areas, which can be tuned by the transverse vertical optical mode profile by changing the waveguide thickness and/or the index contrast between the waveguide and cladding layers.

A sensitivity analysis performed by changing the waveguide thickness by ±60 nm around the original value of 140 nm and AlAs mole fraction by ±0.1

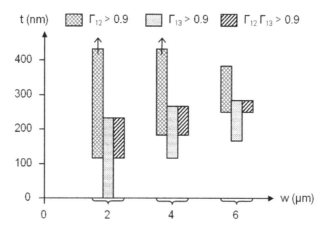

Figure 2.12   Relative and normalized confinement factors Γ_{12} and Γ_{13} and their product $\Gamma_{12}\Gamma_{13}$ calculated for a series of InGaAs/AlGaAs QW ridge waveguide lasers with different widths w and residual thicknesses t. Large values $\Gamma_{12} > 0.9$ ($\Gamma_{13} > 0.9$) mean that the second (third) and all even (odd) modes are suppressed. Single transverse mode operation is achieved when $\Gamma_{12} \times \Gamma_{13} > 0.9$. (Selected data adapted from contour plots of Figure 3 in Laakso *et al.*, 2008.)

around the original value of 0.6 shows that, for example, by reducing the waveguide thickness to 80 nm or decreasing the AlAs mole fraction to 0.5, the area of high $\Gamma_{12} \times \Gamma_{13}$ values can be increased (Laakso *et al.*, 2008).

To demonstrate the size of the effect, Table 2.1 gives as an example for the low and high residual thickness values found for $\Gamma_{12} \times \Gamma_{13} > 0.9$, a ridge width of 2 μm, and four waveguide thickness/AlAs mole fraction combinations. Finally, the investigations showed that, by expanding the transverse vertical near-field pattern into the p-cladding region, single-mode operation could be achieved even with relatively wide and shallow ridge structures having a lower voltage and series resistance. However, the advantages of a wider ridge might be cancelled by an increased threshold current due to a lower confinement factor and higher free carrier absorption of the transverse vertical mode with a higher portion of intensity now in the p-side cladding layer.

Transverse vertical mode expansion; mirror reflectivity; laser length

In Section 2.1.3.5, we discussed concepts to maintain fundamental transverse mode operation at high output power emission and prevent among other things COMD by enlarging the near-field size of the fundamental mode. An improved suppression of higher order lateral modes can be achieved, in particular, by using a layer structure that supports an optical mode asymmetrically expanded toward the substrate into the n-cladding layer of a ridge structure (Shigihara *et al.*, 2002) or by employing a longitudinal photonic bandgap crystal approach for the mode expansion (Maximov *et al.*, 2008). This single-mode improvement can be explained by the facts that, first, the field expansion reduces the influence of the refractive index caused by the ridge profile and, second, the interaction is weaker between the optical field and the ridge

Table 2.1   Low and high residual thickness t_{low} and t_{high} limits at ridge width $w = 2\ \mu m$, respectively, for $\Gamma_{12}\Gamma_{13} > 0.9$ ranges calculated for InGaAs/AlGaAs QW ridge lasers with 140 nm GaAs waveguide (WG) and 1500 nm $Al_{0.6}Ga_{0.4}As$ cladding (original), 80 nm WG and $Al_{0.6}Ga_{0.4}As$ cladding (a), 200 nm WG and $Al_{0.6}Ga_{0.4}As$ cladding (b), 140 nm WG and $Al_{0.7}Ga_{0.3}As$ cladding (c), and 140 nm WG and $Al_{0.5}Ga_{0.5}As$ cladding (d). The difference ($t_{high} - t_{low}$) gives the range for single transverse mode operation and the shift of this range is indicated in the bottom row. Both quantities are listed as a function of the WG thickness and AlAs mole fraction of the cladding ((a)–(d)). (Selected data adapted from the contour plots of Figure 9 in Laakso *et al.*, 2008.)

	Original	(a)	(b)	(c)	(d)
	140 nm WG $Al_{0.6}Ga_{0.4}As$	80 nm WG $Al_{0.6}Ga_{0.4}As$	200 nm WG $Al_{0.6}Ga_{0.4}As$	140 nm WG $Al_{0.7}Ga_{0.3}As$	140 nm WG $Al_{0.5}Ga_{0.5}As$
t_{low} [nm] at $w = 2\ \mu m$	$\cong 110$	$\cong 170$	$\cong 60$	$\cong 90$	$\cong 110$
t_{high} [nm] at $w = 2\ \mu m$	230	330	180	210	260
$t_{high} - t_{low}$ [nm]	120	160	120	120	150
t_{low} shift [nm]		60	−50	−20	0

edges. This enables the ridge stripe width to be increased by maintaining single transverse lateral mode operation.

The maximum kink-free output power is influenced by the facet reflectivity, which affects the refractive index changes of the ridge region via the total optical power and temperature rise in the laser cavity. Calculations show that the kink-free output power can be increased by decreasing the front-facet reflectivity and an increase by a factor of about 3 could be achieved experimentally by using a 4% reflectivity (Shigihara *et al.*, 2002). Moreover, the suppression of higher order modes and an increase of kink-free output power can also be achieved by making the cavity length longer. This positive effect can be ascribed to the inverse dependence of the thermal resistance on cavity length, which leads to lower refractive index increases and less impact on the built-in refractive index profile.

2.2.1.3   Higher order lateral mode suppression by selective losses

Absorptive metal layers

A technologically very simple method to introduce additional losses for the first-order transverse lateral mode in ridge waveguide diode lasers is to decrease the thickness of the dielectric layer, which defines and embeds the ridge waveguide. Thus, the

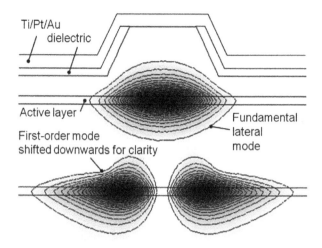

Figure 2.13   Calculated equidistant intensity contour lines 5–100% for fundamental and first-order (shifted downward for clarity) transverse lateral modes in a simplified ridge waveguide structure with a centerline width of 4 μm and residual thickness of the p-cladding layer outside the ridge region of 0.2 μm (not to scale). The p-metallization and ridge-embedding insulating dielectric layer are shown schematically.

optical field can penetrate into the p-type metallization layer resulting in a strong absorption outside the ridge region, which is significantly larger for the first-order than the fundamental mode.

Figure 2.13 shows the intensity plots of the fundamental and first-order lateral optical mode of a ridge waveguide device. The lateral field distributions were calculated by solving the wave equation in the semiconductor waveguide structure. This basic eigenvalue problem was handled by using the MATLAB® Partial Differential Equation Toolbox simulation package (The MathWorks, Inc., 1997). The only material parameter required for the simulations is the square of the propagation constant $\beta^2 = (2\pi/\lambda)^2 n_r^2$ with values of 448.4 μm^{-2} in the claddings under the ridge and the active waveguide, 461.8 μm^{-2} in the active waveguide, and 39.48 μm^{-2} outside the ridge in the ambient. These were calculated for a wavelength $\lambda = 1$ μm and refractive indices $n_r = 3.37$ and 3.42 of the cladding layers (AlGaAs) and active waveguide (InGaAs QW/AlGaAs GRIN-SCH), respectively. By using the software package LASTIP, we obtained the same mode shapes.

If the thickness of the insulator layer is less than 200 nm, the penetration of the optical field into the absorptive Ti/Pt/Au layer increases, leading to a selective loss of the first-order lateral mode. The presence of Ti and Pt is crucial, because the real part of the refractive index of these metals is higher at a lasing wavelength of about 1 μm than the effective index of the lasing mode, which is, for example, 3.3 compared to $n_{r,Ti} = 3.315 - 3.275i$ and $n_{r,Pt} = 3.42 - 5.765i$. Thus, the optical field is distorted and leaks into the metallization where it is absorbed. Kink-free operation can be improved by up to 50% in 980 nm InGaAs/AlGaAs ridge lasers for reduced

SiO_2 insulator thicknesses in the range of 50 to 75 nm. There is no influence on the kink-free power for a Au-only metallization layer, because Au has only a negligible effect on the field distribution due to its low real-part index $n_{r,Au} = 0.095 - 6.2i$ (Buda *et al.*, 2003).

The thinning of the insulator layer to thicknesses below 200 nm generates an additional absorption loss for the fundamental mode of ~1.3 and ~2.5 cm^{-1} for ridge devices 4 and 3 µm wide, respectively. The higher loss in the narrower device is because, in this case, the relative extension of the fundamental lateral mode outside the ridge region is larger. Finally, a thinner insulator layer would also have the benefit of reducing the stress level on the ridge structure, which could have a positive effect on laser reliability (Buda *et al.*, 2003).

Highly resistive regions

Another approach for improving single-mode operation and hence the kink-free power is to suppress the lateral expansion of drive current in a ridge laser by forming highly resistive regions at both sides of the ridge, which reduces the gain and therefore increases the threshold current for higher order lateral mode emission. The highly resistive regions are formed when the etched p-type layer outside the ridge stripe is exposed to a plasma using a mixture of methane and hydrogen in a reactive-ion etching chamber. The hydrogen passivation generates carrier compensation down to a depth of about 700 nm or, in other words, the thickness of the resistive layer between the active QW and the etched surface is 700 nm (Yuda *et al.*, 2004) (Figure 2.14).

LASTIP simulations predict a decrease of local gain in the highly resistive regions for the emergence of higher order lateral modes and an improvement in kink-free

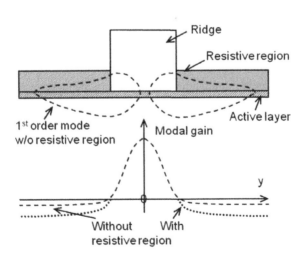

Figure 2.14   Schematic cross-section of a simple rectangular ridge waveguide with lateral resistive regions. For clarity, only the first-order lateral mode TE$_1$ is depicted and the qualitative modal gain diagram including the additional losses by the resistive regions is shown in the lower part of the figure.

output power. Experimental data confirm these predictions and demonstrate a power improvement by ∼20% of laser devices emitting kink-free at around 530 mW cw with slope efficiencies improved by ∼10%. The higher slope efficiencies can be explained by a more efficient use of the drive current to activate the fundamental mode. However, life tests carried out at high power and temperature show a degradation of the kink-free power level, which can be attributed to a redistribution of hydrogen in the resistive layers. Other methods such as proton implantation may be effective in preventing the change of the kink-free power in high-power and high-temperature stress tests (Yuda *et al.*, 2004).

In a similar approach, by introducing lateral absorbing regions on both sides of the ridge waveguide, a 25% increase of kink-free power up to 900 mW could be achieved, but with a 10% decrease of the slope efficiency indicating a slight negative effect of the absorbing layers on the fundamental lateral mode (Pawlik *et al.*, 2002). In contrast to the preceding devices, these InGaAs/AlGaAs QW ridge lasers with lateral absorber regions show stable and reliable operation in accelerated life tests.

2.2.1.4   Higher order lateral mode filtering schemes

It is common practice to design single transverse lateral mode index-guided diode lasers with a narrow waveguide and a small refractive index step in order to cut off the emergence of higher order modes. However, these design features have a negative impact on the laser performance including a lower output power due to a smaller gain volume, higher thermal resistance, higher slow-axis beam diffraction angle, and higher risk of COMD. In addition, the small "cold" index step profile is more susceptible to perturbations such as local heating, mechanical stress, and carrier injection, which can lead to dramatic changes of the index profile in the "hot" state and hence to the propagation of higher order lateral modes or even to a collapse of the waveguide in the worst case. Other designs and mechanisms, beyond the narrow and small-index step waveguide approach, are required to discriminate against higher order lateral modes. These include some that have already been described in the sections above, adding optical loss structures outside the active waveguide, confining the gain to the waveguide core, enabling strong radiation losses for higher-order lateral modes, designing waveguides laterally corrugated, or laterally flaring and vertically tapering. Further techniques are curved waveguides and tilted mirrors, which will be described in the following.

Curved waveguides

A refined curved waveguide is used as a lateral spatial mode filter to increase the propagation loss and therefore threshold for higher order modes in high-power narrow-stripe index-guided diode lasers (Swint *et al.*, 2004). Optimization of the curved waveguide structure is performed with a beam propagation method (BPM). The final structure consists of an S-shaped curve stretching over the entire cavity length with no straight sections, which eliminates the potential mode mismatch problem in the guide and distributes the radiant loss over a larger region. The optimized design also includes two cosine-shaped curvatures of opposite sign for the S-bend, which creates a smoother transition for the mode. Calculations showed an increase of the bend loss for higher order lateral modes with decreasing index step and decreasing radius of curvature. Typical values are between about 0.5 and 3.5 cm^{-1} for an index step 3×10^{-3} and radii of curvature R in the range $\cong 21$–12 mm, respectively (Swint et al., 2004). These bend losses are large and produce a significant increase in the threshold of higher order modes. Compared to straight waveguide devices, bend devices with $R = 12$ mm produce a higher kink-free power by $\sim 160\%$ up to $\cong 600$ mW for InGaAs/AlGaAs QW ridge diode lasers, 2 μm × 2000 μm in active layer size, with only a slight increase in threshold by 5–10% and no decline in differential quantum efficiency. The performance of curved waveguide lasers can be improved further by designing waveguides that distribute the bend loss over an even larger region and avoid any mode mismatch between waveguide sections (Swint *et al.*, 2004).

Tilted mirrors

A theoretical analysis based on a 3D modal reflectivity model has been carried out to investigate the feasibility of using a tilted mirror to discriminate against higher order lateral modes and to increase the kink-free output power (Tan *et al.*, 1998). The reflectivity seen by different modes has been calculated as a function of the tilt angle, thickness, and refractive index of the mirror coating.

A key result of the calculations is that the first-order mode exhibits a reflectivity minimum at a smaller angle than that of the fundamental mode, and that with the selection of a suitable coating and mirror tilt the undesired first-order lateral mode can be strongly suppressed. A typical result for the first-order mode reflectivity of a 980 nm InGaAs/AlGaAs QW ridge laser 3 μm wide is that it reaches its first minimum at $\cong 1.4°$ independent of the coating thickness. The largest reflectivity ratio between the fundamental and first-order mode of $\cong 10^3$ is achieved for an optimum tilt angle of 1.4° and a mirror coating with a thickness $\cong 0.143$ μm and refractive index $\cong 1.9$ (Tan *et al.*, 1998). This ratio is equivalent to an increase in mirror loss by a factor of 3 compared to that of the fundamental mode. An uncertainty in angle of $\pm 0.3°$ still results in a high reflectivity and mirror loss ratio of more than 10^2 and 2, respectively.

The optimum tilt angle is more determined by the waveguide structure than by the thickness and index of the coating. For example, a strongly index-guiding waveguide such as a buried heterostructure with the same lateral dimensions has a much higher optimum mirror tilt angle of $\approx 3°$ (Tan *et al.*, 1998). At these small tilt angles the reflectivity of the fundamental mode is still sufficiently high, leading to lasers with only a slight increase in threshold current and decrease in differential quantum efficiency, but with a strongly improved kink-free output power and far-field stability and divergence.

2.2.1.5   Beam steering and cavity length dependence of kinks

Beam-steering kinks

Beam instabilities such as bilateral steering of the beam in the order of $\pm(1$–$3)°$ compromise the coupling efficiency of laser emission into a single-mode fiber. This can lead to kinks (beam-steering kinks) in the coupled power versus injected current characteristics, even when there are no nonlinearities or kinks in the emitted power versus current dependence.

Simulations based on a 2D self-consistent two-mode model (Guthrie et al., 1994) show at higher currents a significant spatial hole burning of the gain distribution due to carrier transport limitations and a continuous increase in the first-order lateral mode gain, as the overlap between the gain profile and the fundamental mode is compromised by this hole burning effect. This means that high-power fundamental mode operation is limited by the coherent coupling of the fundamental mode into the first-order lateral mode, which is mediated by any kind of slight asymmetry or imperfection in the waveguide structure (Guthrie et al., 1994; Tan et al., 1997; Herzog et al., 2000).

The simulations demonstrate that the dynamic evolution of the effective waveguide and the coherent lasing of emergent multiple lateral modes of the waveguide under high current injection can lead to the beam-steering effects. They also confirm the decline in the fundamental mode differential efficiency beyond the threshold of the first-order mode.

Kink versus cavity length dependence

A periodic dependence of the kink power on cavity length has been observed in weakly index-guided diode lasers of different waveguide geometries and material systems (Schemmann et al., 1995). Periods vary between 100 and 350 μm depending on laser type. Relative kink power differences exceeding a factor of 4 have been observed. Facet coatings lead to differences in amplitude but not in the period of the length of the oscillations. These results indicate that the kink mechanism is of the same origin for all laser types.

A new model has been proposed, which assumes that phase-locked fundamental and first-order lateral modes exist at certain preferred laser lengths and propagate in a laser cavity above the kink power level (Schemmann et al., 1995).

A necessary condition for phase locking is that both fundamental and first-order modes fit into the laser resonator at the same vacuum wavelength while propagating with different propagation constants β_0 and β_1, respectively:

$$\beta_0 = \left(2\pi/\lambda_0\right) n_{r,0}; \ \beta_1 = \left(2\pi/\lambda_1\right) n_{r,1}; \ \lambda_0 = \lambda_1. \tag{2.24}$$

Both modes have to fulfill simultaneously the mirror boundary conditions; that is, they have to be in phase after each roundtrip (cavity roundtrip phase condition). The period of variation is the modal beat length

$$L_b = 2\pi/\beta_0 - \beta_1. \tag{2.25}$$

The maximum kink-free power can be achieved by choosing the proper laser length L. A kink in the P/I characteristic will occur whenever the laser length meets the condition

$$L = mL_b/2; \ m = \text{integer}. \tag{2.26}$$

For other lengths, a kink will only occur when $\beta_0 - \beta_1$ has been changed accordingly due to carrier injection. In this way, the periodic change of kink power versus laser length can be explained. The beam-steering kinks described above cannot be suppressed by one of the techniques discussed in the preceding subsections. They can only be controlled by adjusting the beat length of the phase-locked fundamental and first-order modes either through the laser length or through the propagation constant difference. The proposed model for phase-locked lateral modes is in full agreement with the experimental kink power versus cavity length characteristics (Van der Poel *et al.*, 1994; Schemmann *et al.*, 1995).

2.2.1.6   Suppression of the filamentation effect

Typically, at relatively high power levels, lateral mode break-up through filamentation can override the built-in lateral mode control with the consequence that an array of parasitic optical waveguides can form inside the cavity. This filamentary lasing process is particularly pronounced in diode lasers with wide apertures.

Filamentation is caused by spatial hole burning, which leads to a decline in local carrier density and gain and consequently to an increase of the refractive index and a strong self-focusing of the beam in the active region. In combination with thermal and lateral carrier diffusion effects, filamentation can develop into a highly dynamic process producing a nonuniform and spatially incoherent near-field intensity pattern, which breaks up into filaments unstable in time.

This instability of the individual filaments produces unpredictable kinks in the *P/I* characteristic and reduces the coherence of the laser light. The onset of filamentation drastically increases the slow-axis beam divergence angle, which is many times the diffraction limit, and therefore significantly reduces the brightness of the laser and coupling efficiency of light into a single-mode fiber. For further physical details on the filamentation effect, see Section 2.4.1, below.

Many techniques have been proposed to control the filament formation mechanisms and to overcome this beam-quality issue and achieve single lateral mode and high-power performance in particular with wide-aperture laser structures. Such single-emitter laser types include ARROW lasers (for single-core ARROW and principle, see Section 2.1.4.5; for three-core, wide-aperture ARROW, see Zmudzinski *et al.*, 1995), α-distributed feedback (DFB) lasers, integrated master oscillator power amplifiers (MOPAs), or tapered devices; the last three structures will be discussed in Section 2.4, below. The most promising laser structure in achieving single lateral mode and filamentation-free performance up to high power is the tapered diode laser.

Simulated and experimental results show that low modal gain tapered devices with low confinement factors (1.35%) produce a tenfold improvement in the beam quality and insensitivity against filamentation compared to high modal gain devices with 2.7% confinement factors (Mikulla *et al.*, 1998).

Independent simulations of the carrier-induced filament formation process lead to expressions for the growth rate of sinusoidal perturbations or filaments superimposed on the steady-state field in a diode laser (Dente, 2001). The highly useful equations for the maximum filament gain $g_{f,max}$ and the period P_f of the filaments are given by

$$g_{f,max} = \frac{[(\alpha^2 + 1)^{1/2} - 1]}{2} \times \frac{(I_0/I_s)}{1 + (I_0/I_s)} \times (\alpha_i + \alpha_m) \qquad (2.27)$$

and

$$P_f = 2\pi \left(\frac{\alpha (I_0/I_s) \alpha_i k}{1 + (I_0/I_s)} \right)^{-1/2} \qquad (2.28)$$

where α is the anti-guiding parameter or linewidth enhancement factor, which describes the carrier-induced coupling of the gain change to the refractive index change in the active layer and is given by $\alpha = -2k_0(\partial n_r/\partial N)/(\partial g/\partial N)$ (see the topic on QWs versus QDs in Section 2.1.3.4). Furthermore, α_i is the internal waveguide loss, α_m the distributed mirror losses, k the wavenumber, I_0 the local lateral intensity, and I_s the saturation intensity. The equation for the saturation intensity is

$$I_s = \frac{(h\nu) \dfrac{\partial R}{\partial N}}{\Gamma_{tv} \dfrac{\partial g}{\partial N}} = \frac{\eta_i V_g}{\Gamma_{tv} \dfrac{\partial g}{\partial J}} \qquad (2.29)$$

where we used the common relations for the recombination rate R and bandgap voltage V_g; η_i is here the carrier injection efficiency, N the carrier density expressed in carriers/cm^2, and J the current density (Dente, 2001).

There are two requirements for achieving lateral coherence and suppressing filament formation. First, the growth rate of filaments has to be controlled by keeping the filament gain as small as possible. Second, the filament period has to be larger than the device width, in which case filament formation cannot develop.

These requirements can be realized according to Equations (2.27) and (2.28) by minimizing the anti-guiding factor α, minimizing the optical losses α_i and α_m, and maximizing the saturation intensity I_s. The latter is a very efficient approach for achieving a significant control of filaments (Dente, 2001).

Equation (2.29) shows that a low confinement factor has a large effect on the saturation intensity through the inverse dependence on both the confinement factor and the differential gain. A plausible explanation of the latter is: as the confinement factor is reduced, the threshold gain is increased, which leads to a smaller differential gain. These effects will lead to a significant increase in the saturation intensity.

Equations (2.27) and (2.28) predict the carrier-induced filament gain and period, respectively, while Equation (2.29) relates the saturation intensity to measurable quantities. These quantitative results extend the qualitative results of Mikulla et al. (1998), which suggest that the active region nonlinear index of refraction is proportional to the square of the confinement factor. Moreover, these quantitative predictions are also in agreement with the experimental improvement of the beam quality of tapered lasers achieved for low modal gain with low confinement factors (Mikulla *et al.*, 1998).

There is a tradeoff between low threshold current and low filamentation when changing the confinement factor. However, a laser design optimized for good lateral mode performance and strongly reduced filamentation tendencies requires a low confinement factor with the simultaneous positive effect of a lower differential gain but at the expense of a slightly increased threshold current. There is direct evidence of low confinement factors increasing the COD level in high-power 980 nm InGaAs/AlGaAs lasers. Simultaneously, the slow-axis far-field data of these lasers suggest that significant filament suppression has also been realized (Yamada *et al.*, 1999).

2.3 High-power, single spatial mode, narrow ridge waveguide lasers

2.3.1 Introduction

In addition to the many design considerations for high-power and single-mode operation described in detail in Sections 2.1 and 2.2, the purpose of this section is to give more insight into specific details of quantitative dependencies between crucial parameters of ridge waveguide diode lasers.

This laser type is particularly suitable for this type of demonstration, because its structure in both the transverse vertical and lateral directions is clear and relatively simple to build and has proven to deliver highly reliable, high-power, and single spatial mode laser products suitable for a broad range of applications. Far-field patterns and single-mode operation can be controlled easily. However, stringent dimensional tolerances are required to achieve good performance in these weakly index-guided laser devices. This includes, in particular, ridge width w and height, as well as the residual thickness t of the p-type cladding layer outside the ridge, which can be linked to the ridge height. Moreover, it is known that, as the ridge narrows, the sidewall integrity becomes an important factor in the laser performance (Legge *et al.*, 2000). Nonradiative recombination and optical losses at etched sidewalls, in particular with narrow ridge widths, degrade the laser performance. In addition, the small lasing volume in narrow ridge lasers may increase the optical losses due to process-related scattering leading to increased threshold current densities and limited high-temperature operation.

Experimental and calculated data will be presented below (Epperlein *et al.*, 2000, unpublished), the latter achieved by employing various kinds of calculation codes. These include a 2D mode solver (Litchinitser and Iakhnine, 2010), an effective index method (Agrawal and Dutta, 1993; Buus, 1982), and LASTIP (Crosslight Software Inc., 2010), which among others self-consistently combine QW band structure calculations, carrier recombination effects, carrier drift and diffusion, the heat flux equation and self-heating effects, and optical mode calculations. It would be beyond the scope of this text to elaborate on the details of these different computation codes; however, the given references should give the reader sufficient information to become familiar with these different modeling techniques. Details of the ridge structure used in the investigations are shown in Figure 2.15a.

2.3.2   Selected calculated parameter dependencies

Here we focus on various dependencies, as follows:

- Cutoff conditions of the first-order mode as a function of w and t.

- Mode losses versus t depending on the thickness d of the ridge embedding dielectric layer.

- Slow-axis near-field (NF) spot size versus slow-axis beam divergence angle $\theta_\parallel$.

- Slow-axis far-field (FF) angle $\theta_\parallel$ versus transverse lateral index step $\Delta n_{r,tl}$ and t depending of the parameters w and d.

- Fast-axis NF spot size as a function of the AlAs mole fraction x in the $Al_xGa_{1-x}As$ cladding layers and the fast-axis beam divergence angle $\theta_\perp$.

- Fast-axis FF angle $\theta_\perp$ as a function of x and the GRIN-SCH layer thickness.

- Internal loss coefficient α_i versus the differential external quantum efficiency η_d as a function of the parameter d.

- Transverse lateral index step $\Delta n_{r,tl}$ as a function of t.

2.3.2.1   Fundamental spatial mode stability regime

The mode stability regime for a 980 nm $In_{0.18}Ga_{0.82}As/Al_{0.28}Ga_{0.72}As$ SQW GRIN-SCH ridge diode laser has been calculated by changing w and t and using the data in Figure 2.15a. The fundamental mode region in the w/t parameter space including the cutoff boundary of the first-order mode is displayed in Figure 2.15b.

These data are less extensive compared to those reported by Laakso et al. (2008) and described in Section 2.2.1.2 above, but are in remarkably good agreement with those data when compared to similar waveguide and cladding layer parameters. In the lower w/t regime of 2 µm/150 nm, the lateral beam divergence angle of $\theta_\parallel \cong 17°$ represents an upper design limit for $\theta_\parallel$, whereas the upper 5 µm/550 nm regime with $\theta_\parallel \cong 7°$ is a lower design limit. Considering the fabrication tolerances of ± 0.5 µm and ± 50 nm for w and t, respectively, and the requirement for a low slow-axis divergence angle of around 8°, possible target values could be $w = 4$ µm and $t = 450$ nm.

2.3.2.2   Slow-axis mode losses

The waveguide losses $g(TE_0)$ and $g(TE_1)$ of the fundamental and first-order transverse lateral mode, respectively, calculated (Harder and Achtenhagen, private communication) for a ridge waveguide $In_{0.18}Ga_{0.82}As/Al_{0.28}Ga_{0.72}As$ SQW GRIN-SCH laser

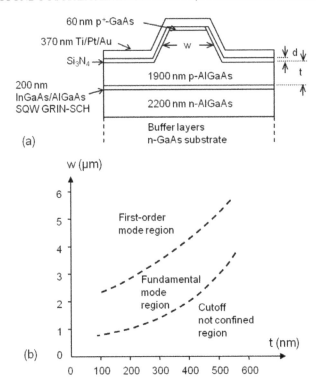

Figure 2.15   (a) Schematic cross-section of an index-guided ridge waveguide diode laser with typical geometrical dimensions, definitions, and compositions of layers used in the calculations. (b) Calculated single-mode regime in the ridge width w/residual thickness t space with adjacent first-order mode region and unconfined mode region.

3.5 μm wide as a function of t for various Si_3N_4 ridge embedding layer thicknesses d, are plotted in Figure 2.16a.

As expected from the discussion in Section 2.2.1.3, the losses increase with decreasing d and are higher for the first-order mode. A loss margin of $\sim$5 cm^{-1} at $t \cong$ 400 nm between the two modes can be derived from Figure 2.16b for d values in the range of 50 to 100 nm, allowing the laser to operate in a single transverse horizontal mode.

2.3.2.3   Slow-axis near-field spot size

Figure 2.17 shows the sensitivity of the fundamental transverse horizontal NF spot size at FWHM on the transverse horizontal divergence angle for a ridge waveguide InGaAs/AlGaAs SQW laser 3 μm wide with a GRIN-SCH layer 300 nm thick.

The figure clearly shows two distinct regimes: first, for divergence angles $\theta_\parallel \lesssim$ 6° the NF spot size changes rapidly with small decreases of $\theta_\parallel$; and, second, for NF

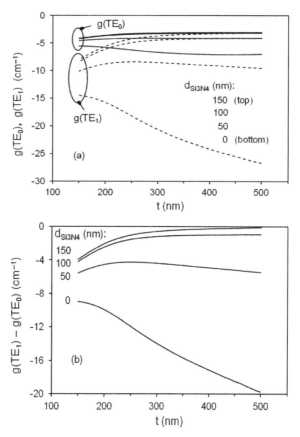

Figure 2.16 (a) Waveguide losses $g(TE_0)$ and $g(TE_1)$ and (b) losses margin $g(TE_1) - g(TE_0)$ of the fundamental TE_0 and first-order TE_1 transverse lateral mode, respectively, calculated for a ridge waveguide $In_{0.18}Ga_{0.82}As/Al_{0.28}Ga_{0.72}As$ GRIN-SCH SQW laser 3.5 μm wide as a function of the residual thickness t for various Si_3N_4 ridge-embedding layer thicknesses d.

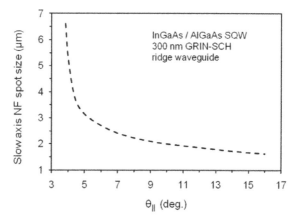

Figure 2.17 Calculated fundamental transverse lateral near-field (NF) spot size at full width half maximum (FWHM) as a function of the transverse lateral beam divergence angle $\theta_\parallel$ for a ridge waveguide InGaAs/AlGaAs SQW laser 3 μm wide with a GRIN-SCH layer 300 nm thick.

spot sizes $\lesssim 2.5$ µm the divergence angle $\theta_\parallel$ increases rapidly with small reductions in the NF spot size.

2.3.2.4 Slow-axis far-field angle

Here, we demonstrate the dependence of $\theta_\parallel$ on $\Delta n_{r,tl}$ and t with w and d as parameters. Figure 2.18a shows a crossover point at $\cong 9°$ and 420 nm in the $\theta_\parallel/t$ plot with a much stronger dependence on w below this point. There is only a slight dependence in the $\theta_\parallel/t$ plot on d with a maximum decrease by $\sim 1°$ when increasing d from 0 to 150 nm (Figure 2.18b). The higher $\theta_\parallel$ values for $d = 0$ nm can be understood in terms of a

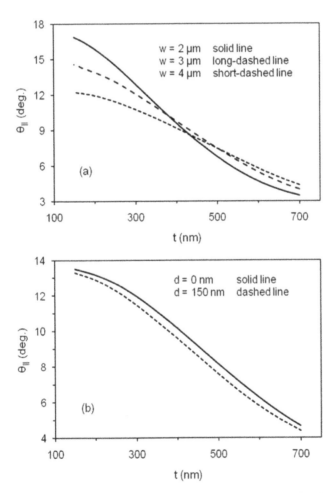

Figure 2.18 Transverse lateral far-field beam divergence angle $\theta_\parallel$ calculated as a function of the residual p-waveguide thickness t for various (a) widths w and (b) dielectric embedding layer thicknesses d of an InGaAs/AlGaAs GRIN-SCH SQW ridge diode laser.

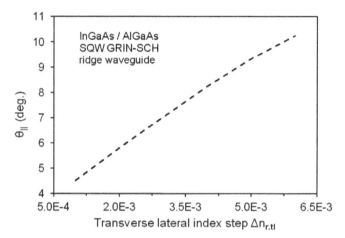

Figure 2.19   Slow-axis beam divergence angle $\theta_\parallel$ calculated as a function of the transverse lateral refractive index step $\Delta n_{r,tl}$ of a ridge waveguide InGaAs/AlGaAs SQW GRIN-SCH diode laser 4 μm wide.

stronger confinement under the ridge and negligible expansion of the fundamental mode into the highly absorbing Ti/Pt/Au contact layer.

The slow-axis beam divergence angle $\theta_\parallel$ is plotted versus the transverse lateral refractive index step $\Delta n_{r,tl}$ of a ridge waveguide InGaAs/AlGaAs SQW GRIN-SCH diode laser 4 μm wide in Figure 2.19. Typical index steps are around 3×10^{-3} corresponding to divergence angles of about 7°. These figures are consistent with the requirements for a single-mode diode laser emitting into a single-mode fiber (see Section 2.2.1.1).

2.3.2.5   Transverse lateral index step

Figure 2.20 shows the transverse lateral refractive index step $\Delta n_{r,tl}$ calculated as a function of the residual p-cladding thickness t outside the ridge region of an InGaAs/AlGaAs laser device 4 μm wide.

An index step of, for example, 3×10^{-3} is given by $t = 550$ nm, a value which leads to $\theta_\parallel \cong 7°$ according to Figure 2.19, which is in agreement with Figures 2.15b, 2.18a and 2.18b.

2.3.2.6   Fast-axis near-field spot size

The dependence of the fast-axis NF spot size on the AlAs mole fraction x in the cladding layers and fast-axis beam divergence angle $\theta_\perp$ of a ridge $In_yGa_{1-y}As/Al_xGa_{1-x}As$ SQW laser 4 μm wide is shown in Figure 2.21 and Figure 2.22, respectively.

The figures show that by lowering x the mode can be expanded in the transverse vertical direction (due to less confinement caused by increased index), which results

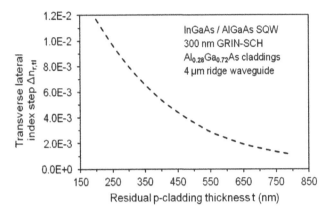

Figure 2.20 Transverse lateral refractive index step $\Delta n_{r,tl}$ calculated as function of the residual p-cladding thickness t outside the ridge region of an InGaAs/AlGaAs laser device 4 μm wide.

in a lowering of $\theta_\perp$ at a rate of roughly 1° per 1% AlAs mole fraction x. However, as pointed out in the introductory remarks of Section 2.1.3.5, this occurs at the expense of lower kink-free powers and efficiencies, and higher threshold currents. Alternatively, a series of different, more suitable approaches for fast-axis beam divergence engineering has been discussed.

2.3.2.7 Fast-axis far-field angle

The dependence of $\theta_\perp$ on x inherently given in Figures 2.21 and 2.22 is explicitly displayed in Figure 2.23.

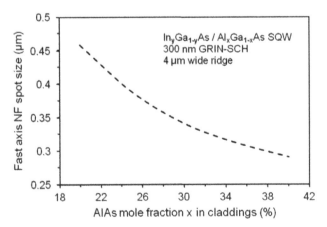

Figure 2.21 Calculated dependence of the transverse vertical near-field (NF) mode spot size on AlAs mole fraction x in the cladding layers of ridge $In_{0.18}Ga_{0.82}As/Al_xGa_{1-x}As$ SQW lasers 4 μm wide with GRIN-SCH layers 300 nm thick.

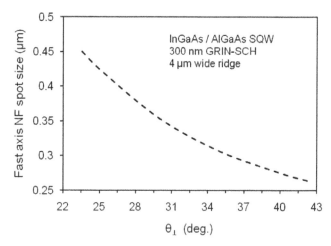

Figure 2.22   Transverse vertical near-field (NF) mode spot size calculated versus fast-axis beam divergence angle $\theta_\perp$ of ridge $In_yGa_{1-y}As/Al_xGa_{1-x}As$ SQW lasers 4 μm wide with GRIN-SCH layers 300 nm thick.

The dependence of $\theta_\perp$ on the GRIN-SCH layer thickness is shown in Figure 2.24. A decrease of the latter leads to a reduction of $\theta_\perp$, but at the expense of an increase in the threshold current and decrease in efficiency due to significant free-carrier absorption in the claddings caused by the strong spreading of the mode into these layers. This approach has been discussed in detail in Section 2.1.3.5.

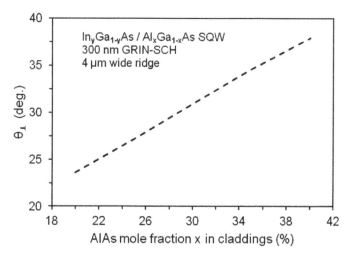

Figure 2.23   Calculated transverse vertical beam divergence angle $\theta_\perp$ as function of AlAs mole fraction x in the cladding layers of ridge $In_{0.18}Ga_{0.82}As/Al_xGa_{1-x}As$ SQW lasers 4 μm wide with GRIN-SCH layers 300 nm thick.

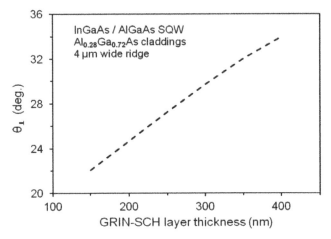

Figure 2.24   Calculated transverse vertical beam divergence angle $\theta_\perp$ as a function of GRIN-SCH layer thickness of ridge $In_{0.18}Ga_{0.82}As/Al_{0.28}Ga_{0.72}As$ SQW lasers 4 μm wide.

2.3.2.8   Internal optical loss

By changing the thickness d of the ridge-embedding Si_3N_4 layer, the internal loss coefficient α_i and hence the external differential quantum efficiency η_d are altered.

Figure 2.25 shows α_i as a function of η_d for a ridge $In_{0.18}Ga_{0.82}As/Al_{0.28}Ga_{0.72}As$ SQW laser 4 μm wide with a GRIN-SCH waveguide layer 300 nm thick. Low values of $d < 50$ nm lead to high losses >6 cm^{-1} and low efficiencies $<80\%$, whereas low losses <3 cm^{-1} and high efficiencies $>90\%$ can be achieved for $d > 150$ nm values.

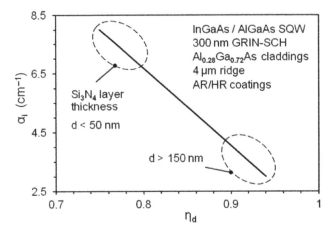

Figure 2.25   Internal optical loss coefficient α_i versus external differential quantum efficiency η_d of ridge $In_{0.18}Ga_{0.82}As/Al_{0.28}Ga_{0.72}As$ SQW lasers 4 μm wide with GRIN-SCH waveguide layers 300 nm thick. The calculated data have been achieved by changing the thickness d of the dielectric ridge-embedding layer.

2.3.3   Selected experimental parameter dependencies

This section consists of two parts. First, the experimental parameter dependencies of uncoated $In_yGa_{1-y}As/Al_xGa_{1-x}As$ SQW GRIN-SCH lasers with apertures 80 μm wide, including:

- threshold current density J_{th} versus AlAs mole fraction x; and
- slope efficiency η_{slope} versus x and J_{th}.

Second, experimental parameter dependencies of AR/HR-coated, ridge $In_{0.18}Ga_{0.82}As/Al_{0.28}Ga_{0.72}As$ SQW devices 3.5 μm wide with GRIN-SCH layers 300 nm thick, including:

- threshold current I_{th} depending on slow-axis divergence angle $\theta_\parallel$ at various thicknesses d of the ridge-embedding Si_3N_4 dielectric layer;
- slope efficiency η_{slope} as a function of $\theta_\parallel$ at different values of d; and
- kink-free power P_{kink} versus residual p-cladding layer thickness t.

2.3.3.1   Threshold current density versus cladding layer composition

The dependence of the threshold current density J_{th} of a broad-area InGaAs/AlGaAs SQW diode laser on the AlAs mole fraction x of the $Al_xGa_{1-x}As$ cladding layer is shown in Figure 2.26.

The decrease of J_{th} with increasing x is caused by an increase of the confinement factor and modal gain resulting from less mode expansion due to the refractive index, which decreases with x of the cladding layers.

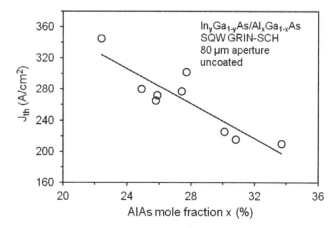

Figure 2.26   Experimental dependence of the threshold current density J_{th} of broad-area $In_{0.18}Ga_{0.82}As/Al_xGa_{1-x}As$ SQW diode lasers on AlAs mole fraction x of the $Al_xGa_{1-x}As$ cladding layers. Solid line is trendline.

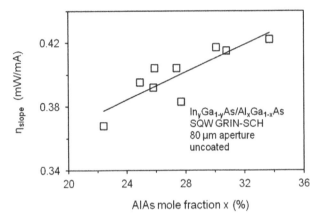

Figure 2.27   Slope efficiency η_{slope} measured on broad-area $In_{0.18}Ga_{0.82}As/Al_xGa_{1-x}As$ SQW lasers as a function of the Al content x in the $Al_xGa_{1-x}As$ cladding layers. Solid line is trendline.

2.3.3.2   Slope efficiency versus cladding layer composition

Figure 2.27 shows the slope efficiency η_{slope} measured on broad-area InGaAs/AlGaAs SQW lasers as a function of the Al content x in the $Al_xGa_{1-x}As$ cladding layers.

The increase of η_{slope} with x can be explained by the fact that the refractive index of the cladding layers decreases with increasing x leading to a stronger confinement of the mode and less spreading into the absorptive cladding layers and therefore to less optical losses.

2.3.3.3   Slope efficiency versus threshold current density

Figure 2.28 shows the experimental dependence of η_{slope} on J_{th}. It can be understood by combining Figures 2.26 and 2.27 and considering both the various explanations given in the relevant two sections and the fact that decreasing J_{th} values in Figure 2.28 correspond to increasing x values.

However, there is a tradeoff between low J_{th}/high η_{slope} and large $\theta_\perp$ values found with increasing x values. This issue has been repeatedly discussed at various locations including Sections 2.1.3.5, 2.3.2.6, and 2.3.2.7. In Section 2.1.3.5, we discussed different, more powerful approaches for engineering the fast-axis beam divergence angle with much less impact on other diode laser parameters.

2.3.3.4   Threshold current versus slow-axis far-field angle

The effect of the ridge-embedding Si_3N_4 layer thickness d on the threshold current I_{th} versus slow-axis divergence angle $\theta_\parallel$ dependence of AR/HR-coated, ridge In-GaAs/AlGaAs SQW lasers 3.5 μm wide is illustrated in Figure 2.29. The devices are from the same epitaxial wafer material (Harder and Achtenhagen, private communication), the Si_3N_4 layers with different thicknesses have been deposited on different

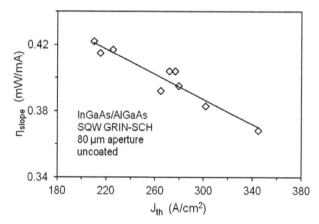

Figure 2.28 Experimental dependence of slope efficiency η_{slope} on threshold current density J_{th} of broad-area InGaAs/AlGaAs diode lasers. Solid line is trendline.

parts of the same wafer, and the individual laser devices with changing residual thicknesses t, that is, divergence angles $\theta_\parallel$, have been obtained by bevel-etching laser bars with a specific technique.

The decrease of I_{th} with $\theta_\parallel$ is mainly due to a diminished carrier loss caused by less current spreading in the etched, residual p-cladding layer decreasing in thickness t with increasing $\theta_\parallel$. The increase of I_{th} at a certain $\theta_\parallel$ value with decreasing d is caused by an increased optical loss of the fundamental mode into the highly absorbing Ti/Pt/Au contact layer.

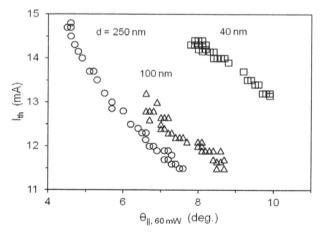

Figure 2.29 Threshold current I_{th} measured as a function of slow-axis beam divergence angle $\theta_\parallel$ (at 60 mW cw) of AR/HR-coated narrow ridge InGaAs/AlGaAs SQW lasers 3.5 μm wide at different locations of the same wafer material, but with different ridge-embedding Si_3N_4 layer thicknesses d.

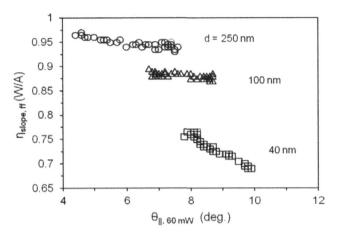

Figure 2.30 Slope efficiency η_{slope} measured as a function of slow-axis beam divergence angle $\theta_\parallel$ of AR/HR-coated narrow ridge $In_{0.18}Ga_{0.82}As/Al_{0.28}Ga_{0.72}As$ SQW lasers 3.5 μm wide at different locations of the same wafer material but with different ridge-embedding Si_3N_4 layer thicknesses d.

2.3.3.5 Slope efficiency versus slow-axis far-field angle

The experimental procedure here is similar to that in the previous section but with the slope efficiency η_{slope} plotted versus the transverse lateral far-field angle $\theta_\parallel$ at different d parameter values. The devices used in Figure 2.30 are the same as used in Figure 2.29.

In general, η_{slope} decreases with increasing $\theta_\parallel$ (decreasing t), which is a consequence of the strong overlap of the mode with the dielectric and mode leaking out into the absorbing Ti/Pt/Au contact layer resulting in increasing optical losses. This effect is more pronounced in thinner dielectric layers and is reflected by the steeper degradation rate of η_{slope} with $\theta_\parallel$. The increase of η_{slope} at a certain $\theta_\parallel$ value with increasing d is caused by a decreased optical loss of the fundamental mode into the highly absorbing Ti/Pt/Au contact layer.

2.3.3.6 Kink-free power versus residual thickness

Figure 2.31 shows the kink-free power P_{kink} as a function of the residual p-cladding layer thickness t of 7 nm strained-layer $In_{0.18}Ga_{0.82}As/Al_{0.28}Ga_{0.72}As$ SQW devices with GRIN-SCH layers 300 nm thick, ridge waveguides 3.5 μm wide, and AR/HR-coated mirrors. By changing the residual thickness, the ridge height is varied, which is a crucial parameter in investigating the lateral mode behavior where the total thickness of the p-cladding layer is maintained at 1.9 μm.

The kink power variation of the type exhibited in the figure consists of two distinct features: first, an increase of P_{kink} with t; and, second, a drop of P_{kink} at a certain t value and then rising again. We will discuss the most likely root causes for both effects briefly by referring to the discussion in Section 2.2.1.2 on approaches to optimize the

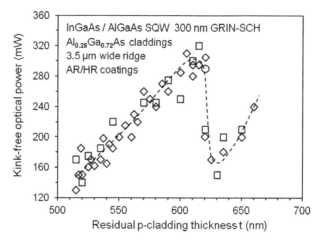

Figure 2.31   Experimental kink-free optical power P_{kink} as a function of residual p-cladding layer thickness t of 7 nm thick strained-layer $In_{0.18}Ga_{0.82}As/Al_{0.28}Ga_{0.72}As$ SQW devices with GRIN-SCH layers 300 nm thick, ridge waveguides 3.5 μm wide, and AR/HR-coated mirrors. Data points are from measurements on different bevel-etched laser bars with linearly varying t of the same laser wafer material.

waveguide for single-mode operation by considering waveguide geometry aspects. In addition, we will consider detrimental internal physical mechanisms such as changes in refractive index profile by local thermal heating and carrier injection, spatial hole burning, lateral current spreading, and gain profile variations.

First, different ridge heights (residual thicknesses) result in different effective refractive indices: a higher ridge height (lower t) leads to a larger effective index, with the consequence that the lateral optical mode is more strongly confined than in a structure with a lower ridge height (higher t). In the case of a high ridge height, the mode intensity profile of the first-order lateral mode matches the lateral gain profile closely and the threshold condition of the first-order lateral mode is decreased. Also, there is a strong carrier lateral spatial hole burning of the local gain profile by the fundamental mode and severe heat accumulation in higher ridge structures at high intracavity optical intensities leading to a temperature-induced increase of the refractive index and hence to an actual lateral index profile very different from that of the "cold" (unbiased) state. This "hot" (biased) index profile features increased index amplitudes in the center, but also near both slopes of the ridge, which thus promotes the emergence of higher order lateral modes (Figure 2.32).

In addition, lateral current spreading is also an important issue for structures with different ridge heights. For lower ridge heights (high t) the current density spreads widely in the active region, whereas for higher ridges (low t) the current density is larger within the active region. This means that the electron and hole concentrations within the active region of high-ridge structures are larger than those of devices with low-ridge heights leading to a lateral interband gain profile, which is higher in high-ridge than in low-ridge structures, in particular in the ridge boundary (ridge slope)

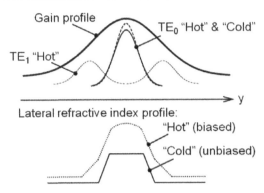

Figure 2.32 Schematic representation of the fundamental and first-order lateral modes in a laser ridge waveguide structure at "cold" (unbiased; solid line) and "hot" (biased; dotted lines) conditions. Qualitative illustration of the emergence of the first-order transverse lateral mode with increasing temperature based on the change of the built-in (cold) refractive index profile and available gain distribution.

regions (Chen *et al.*, 2009; Xu *et al.*, 1996; Wu *et al.*, 1995). The higher gain profile in high-ridge (low-t) structures will then support the emergence of the first-order lateral mode.

Second, and finally, the periodic shape of the kink power versus residual thickness plot in Figure 2.31 (only one period or cluster shown) can be explained with the hybrid mode kink model, which phase-locks the fundamental and first-order lateral mode in coherent conditions. This model has been described in Section 2.2.1.5 and has been developed to explain the periodic kink power versus cavity length dependence (Schemmann *et al.*, 1995). The condition for the emergence of a kink is that Equation (2.26) has to be met. In the case of a fixed cavity length L (as in Figure 2.31), this can be achieved by adjusting the beat length L_b (cf. Equation 2.25) via a change in the propagation constants (refractive indices) of both the lateral modes (cf. Equation 2.24) caused by changing the residual thickness (ridge height). Phase locking occurs only if the difference between the refractive indices of the two lateral modes is not too big, which for a 980 nm laser is around 5×10^{-4}.

2.4 Selected large-area laser concepts and techniques

2.4.1 Introduction

In the previous sections of this chapter, in particular, but also in the previous chapter, we have discussed various common approaches and techniques used in design, materials, and process technologies to achieve single-mode diode laser waveguides capable of delivering the highest kink-free optical powers of 1 W and above from a few micrometer-wide apertures. High-power, narrow-stripe, single-mode lasers can easily be coupled into single-mode fibers and have a dominant application in pumping Er-doped fiber amplifiers for optical communication network systems. To overcome power limitations generally encountered with conventional narrow-stripe single-mode lasers, such as ridge waveguide or buried-heterostructure devices, numerous novel structures have been developed with an increased transverse horizontal mode size for output powers in excess of 1 W, but still emitting in a single mode or diffraction-limited beam to deliver high-brightness, broad-area semiconductor lasers. High-power coherent light sources with diffraction-limited beam profiles are needed to allow beam propagation over large distances and good focusing of high optical power into small spots or single-mode fibers in applications such as solid-state laser end pumping, harmonic generation (frequency doubling), free-space optical communication systems, optical memories, printing, direct diode material processing, and military applications.

The simplest configuration is the single-element broad-area laser with a wide output facet in the transverse horizontal direction and a large optical cavity in the transverse vertical direction to achieve a large optical mode volume in order to reduce intrinsic heating effects, optical power density, and therefore facet degradation for high-power operation. However, these gain-guided devices suffer from a strong loss of beam quality, mainly caused by self-focusing and filamentation processes (see Section 2.2.1.6) and multiple lateral mode oscillations, effects that usually occur as the lateral waveguide width exceeds $\sim$10 µm. As discussed in Section 2.2.1.6, the formation of filaments is caused by self-focusing of the intracavity laser field in the gain medium. The overlap of the fundamental mode with the gain region is strongly reduced, because the mode's lateral dimension is constricted when the filamentation has emerged.

The consequence is that the inversion, which is not depleted by the fundamental mode, then becomes available to the higher order lateral modes, dramatically increasing the likelihood of multiple mode operation. It has been demonstrated that broad-area devices, in which two counter-propagating beams are present, are susceptible to filamentation at low optical power densities (Lang *et al.*, 1994; Tamburrini *et al.*, 1992). This is in contrast to devices, which avoid counter-propagating beams, such as the unstable resonator laser (see Section 2.4.3), and which are more resistant to filamentation formation up to much higher powers (Goldberg and Mehuys, 1992; Walpole *et al.*, 1992).

Monolithic and multi-chip solutions have been proposed and demonstrated aimed at achieving high-power lasers with single-lobe diffraction-limited beams that can be focused with relatively simple optics. Multi-chip laser devices include external cavity geometries and external injection of broad-area amplifiers, which, however, will not be discussed in this text.

On the other hand, monolithic structures include different conceptual techniques but are not limited to (i) broad-area lasers in different configurations including tailored gain profiles, Gaussian reflectivity facets, and lateral grating-confined tilted waveguides, (ii) unstable resonator lasers, (iii) tapered devices, (iv) master oscillator flared power amplifiers, and (v) laser bar arrays. We will discuss these approaches and techniques in detail in the following sections of this chapter. However, we will not discuss the LPBC concept for application in the 2D structure of a broad-area laser device. We discussed this concept in Section 2.1.3.5 in the context of the 1D structure of a narrow single-mode laser and showed the potential of the LPBC structure for the development of high-power single-mode broad-area lasers (Maximov *et al.*, 2008).

A number of issues had to be resolved in the development of some of these techniques. These include an efficient coupling of the output power into a single-lobed far-field pattern, high operating stability over a wide range of powers and temperatures, and a high discrimination of the modes up to high power at sufficient uniformities of gain and refractive index for achieving single-mode operation. In particular, the high nonlinearity and strong coupling between gain and refractive index in semiconductor diode lasers pose a great challenge to the realization of single-mode broad-area lasers.

In the following, we describe the principle of operation, key characteristics, and key performance data of the various monolithic solutions listed above. We should mention that a series of state-of-the-art optical output power data for standard broad-area lasers 100 μm wide, tapered amplifier lasers, and 1 cm laser bar arrays can be found in Tables 1.3, 1.4, and 1.5, respectively.

2.4.2   Broad-area (BA) lasers

2.4.2.1   Introduction

Conventional broad-area (BA) lasers already show at low powers degraded and un-desirable slow-axis far-field patterns with non-Gaussian, so-called top-hat shapes superimposed by multiple lobes. In addition to the causes mentioned above, reasons for this degradation can be seen in three aspects (Harder, 2008).

First, higher order lateral modes are extracting the lateral gain profile more effectively than the fundamental mode in particular due to the increased spatial hole burning at high power. In Section 2.4.2.2, we show how this detrimental effect can be mitigated by tailoring the spatial gain profile via nonuniform current injection into the laser device.

Second, at high-power operation the temperature in the waveguide rises and the increase is higher in the center than at the edges due to a poorer heat extraction efficiency, which causes a temperature difference ΔT leading to a lateral difference in the refractive index Δn_r and hence to the formation of a thermal waveguide. Analogous to the formula derived for the NA of a fiber (see Equation 1.66b), the NA of this thermal waveguide can be evaluated according to

$$\text{NA}^2 = (n_r + \Delta n_r)^2 - n_r^2 \cong 2n_r(\Delta n_r/\Delta T)\Delta T \qquad (2.30)$$

where n_r is the index of the waveguide at the edges. Taking only a small $\Delta T = 5$ K (see Chapter 9), a typical value of 4×10^{-4} K^{-1} for the temperature coefficient of the index (cf. Section 1.3.6) and $n_r = 3.3$ for AlGaAs results in NA $\cong 0.11$. Thermal waveguiding is certainly an issue for the slow-axis NA degradation and can be reduced by efficient heat extraction and heat sinking of the laser chip. However, an ideal solution would be an athermal waveguide structure showing no dependence of its modal index on temperature.

Third, gain guiding can be caused by changes in the lateral gain profile and refractive index due to carrier injection. In Sections 2.2 and 2.3, we discussed various approaches and techniques to improve the slow-axis beam divergence pattern of narrow-stripe diode lasers. Similarly, mode filter schemes for BA diode lasers have been developed. We will present two of them below in Sections 2.4.2.3 and 2.4.2.4.

For completeness, we should mention a technique that is outside the mainstream of common techniques for engineering the intrinsic materials and design parameters of a BA laser to achieve a single-lobed diffraction-limited spot. A spectrally resolved phase manipulation technique, external to the laser cavity, is used to reshape the multiple lateral mode emission of a conventional BA diode laser into a diffraction-limited single-lobe spot. A mode conversion efficiency of 60% with a total insertion loss of 25% has been demonstrated with conventional bulk optical elements and a binary phase mask for a commercially available 980 nm BA laser with a width of 100 μm and length of 1 mm. Using custom designs of the optical elements will allow a laser-to-fiber coupling efficiency of >50%. Advanced designs can reduce the device to a single optical element inserted between the diode laser and optical fiber (Stelmakh, 2007).

2.4.2.2   BA lasers with tailored gain profiles

Conventional BA semiconductor lasers have a nearly uniform lateral gain profile. This results in devices with very wide and unstable slow-axis far-field patterns, which are caused by filamentation and lateral multimode effects. Poor mode discrimination between the fundamental and higher order modes leads to the multi-lobed far-field pattern. Without making the device narrower, this issue can only be solved by creating a nonuniform spatial gain distribution within the device, which favors the fundamental mode and suppresses the higher order modes, then resulting in a single-lobed, diffraction-limited far-field pattern.

This led to the proposal and demonstration of a new type of semiconductor laser: the tailored-gain BA semiconductor diode laser (Lindsey and Yariv, 1988). The 2D gain tailoring is achieved by a predetermined pattern of current injecting and no-current injecting contacts over the surface of the laser device by varying the fractional coverage per unit area of injecting to no-injecting contact. This can be achieved by using a halftone pattern of dots (or any other shapes) which has been formed on a photoresist mask for structuring the contact layer. The technique is reminiscent of the halftone process used in the graphics industry to reproduce a photograph on the printed page. The lateral current injection profile can vary, for example, from a minimum at one side to a maximum at the other, or from a minimum to a maximum to a minimum, resulting in the formation of an asymmetric or symmetric, linear, gain-tailored BA laser, respectively.

The following example may illustrate the effect: in the asymmetric version, the peak gain at one side is fixed by the requirement that at threshold the modal gain of the lasing mode must equal the optical losses, whereas the gain at the opposite side is set at transparency. Calculations for a 0–60 cm^{-1} linear gain profile across 60 μm width produced theoretical modal gains for the higher order lateral modes, which are between 5 and 8 cm^{-1} less than that of the fundamental mode, which makes it difficult for these modes to reach threshold. This mode discrimination is much higher than with conventional uniform gain BA lasers 60 µm wide, which have modal gains only <1 cm^{-1} lower for higher order modes than for the fundamental mode.

Another important difference between uniform and nonuniform gain profiles is the shape of the far-field (FF) patterns. Unlike uniform gain profile lasers, the FF patterns of higher order modes in nonuniform gain profile devices are single-lobed and only slightly displaced from that of the fundamental mode. This means that when gain saturation at high power causes the emergence of higher order modes, the slow-axis FF pattern will stay single lobed albeit somewhat wider. This theoretical prediction is in agreement with experiments and can be considered as a useful feature of BA lasers with asymmetric linear gain profiles (Lindsey and Yariv, 1988). In contrast, higher order modes of symmetric gain structures do not have single-lobed FF patterns.

The experimental slow-axis FF angles of BA lasers 60 µm wide with asymmetrically tailored gain profiles are $\cong$ 2.5°, that is, very close to the diffraction limit of $\cong$ 2°, and this is also the same for the symmetric gain structure. This value has to be compared to the $\cong$ 15° of the multi-lobed FF pattern obtained for a conventional, uniform gain BA laser with the same width. The gradient in the gain profile plays a crucial role in device performance at high power. Above threshold, gain saturation reduces the gain preferentially in the high-gain areas and thus the nonuniform gain profile tends to become more like that of a conventional BA laser with a uniform gain. This negative effect can be minimized by increasing the gradient in the gain profile. Doubling the value of 0.25%/µm used in the last example produces a single-lobed FF pattern up to much higher power levels and with practically unchanged widths of $\cong$ 2.5°. The threshold current of this new class of diode laser is only about 5% higher than that of a conventional, uniform gain BA laser.

Finally, filamentation does not occur in asymmetric gain profile BA lasers in contrast to conventional BA lasers with uniform gain. This result is unexpected and not yet fully understood (Lindsey and Yariv, 1988).

2.4.2.3   BA lasers with Gaussian reflectivity facets

The basic idea of this approach is to achieve lateral mode discrimination in BA diode lasers by applying a smooth spatial mode filtering at the output mirror. This can be accomplished by laterally varying the reflectivity profile of the output facet. The spatial filtering takes place only in the transverse lateral direction parallel to the active layer. Numerical simulations using a beam propagation scheme based on a fast Fourier transform procedure have been used to calculate the optical field and lateral modes in gain-guided AlGaAs double-heterostructure BA lasers (McCarthy and Champagne, 1989). The lasers used in the simulations were 40 µm wide and had a length L between 100 and 400 µm. The rear facet was uncoated with a uniform reflectivity and the front-facet intensity reflectivity profile is a Gaussian function in transverse lateral position y as follows:

$$R_f(y) = R_{f,0} \exp\left\{-(y/w_f)^2\right\} \tag{2.31}$$

Table 2.2 Modal discrimination ratio between fundamental transverse lateral mode and higher order lateral modes calculated as a function of the reflectivity waist w_f of a broad-area laser with a Gaussian reflectivity profile of the front facet for different laser lengths L. (Selected data adapted from Figure 1 in McCarthy *et al.*, 1990.)

L [μm]	W_f [μm]						
	2	3	4	5	10	15	20
400	6.0	4.0	3.0	2.2	1.5	1.4	1.3
100	2.7	2.0	1.6	1.5	1.2	1.1	1.0

where $R_{f,0}$ is the maximum reflectivity at the center of the current stripe and falls to $R_{f,0}/e$ at a distance w_f from it. A facet with an infinite waist w_f corresponds to a conventional facet with uniform reflectivity. The modal discrimination defined as the ratio between the field amplitudes of the fundamental mode and higher order lateral modes has been calculated as a function of w_f and L. Some typical data adapted from McCarthy and Champagne (1989) are listed in Table 2.2.

The modal discrimination for a certain length L increases slowly as w_f becomes smaller from $w_f = \infty$ (conventional facet) to $w_f \approx 7$ μm and increases significantly for $w_f \lesssim 7$ μm. The highest calculated discrimination is about 6 for $w_f \cong 3$ μm and $L = 400$ μm. The modal discrimination with a Gaussian facet of any reflectivity waist w_f is better than the discrimination with a conventional facet. The discrimination increases linearly with L with a slope for $w_f = 7$ μm, twice that obtained for a conventional facet. Therefore, single lateral mode is easier to obtain for lasers with longer cavities and output facets with narrower reflectivity profiles.

Moreover, the FF pattern of the fundamental lateral mode has been calculated as a function of w_f for different L values. Table 2.3 lists some typical data selected from McCarthy and Champagne (1989).

Table 2.3 Transverse lateral beam divergence angle $\theta_\parallel$ in degrees calculated as a function of the reflectivity waist w_f of a broad-area laser with a Gaussian reflectivity profile of the front facet for different laser lengths L. (Selected data adapted from Figure 3 in McCarthy *et al.*, 1990.)

L [μm]	W_f [μm]						
	2	3	4	5	10	15	20
400	4.2	2.5	2.0	1.7	1.2	1.1	1.1
100	4.0	3.3	2.8	2.4	1.8	1.5	1.3

The slow-axis beam divergence angle $\theta_\parallel$ is almost independent of L but is more sensitive to w_f for shorter lasers and increases strongly for small values $w_f < 7$ μm. However, broadening of the fundamental mode FF profile for a Gaussian reflectivity facet is much lower than broadening in a conventional BA laser, which is very susceptible to the emergence of multiple lateral modes. The slow-axis FF profiles remain Gaussian down to $w_f \approx 5$ μm. The Gaussian reflectivity facet increases the threshold current slightly by $\approx 10\%$ as w_f is lowered from infinity to 5 μm, and decreases the efficiency. Both are caused by a roundtrip loss increasing with decreasing w_f, because of a reduced optical feedback from the front mirror. These smaller drawbacks, however, are alleviated by the fact that the Gaussian reflectivity facet inhibits multimode oscillation and preserves the beam coherence over a wide range of optical output powers.

2.4.2.4 BA lasers with lateral grating-confined angled waveguides

BA low-loss diode lasers with high transverse lateral mode discrimination can be realized using embedded gratings as the confining layers. The highest modal discrimination is obtained when the grating extends uniformly through the gain stripe and on either side, as has been demonstrated with the angled-grating distributed feedback laser (also known as α-DFB), the best-known device in the class of grating-confined BA lasers with impressive performance (Lang et al., 1998). Single-lobed, near-Gaussian output power levels of about 1 W cw diffraction-limited with $<0.2°$ lateral divergence have been achieved. The new device operates in both single spectral mode and single spatial mode and has a low threshold current density ~430 A/cm^2 and high slope efficiency >0.5 W/A (DeMars et al., 1996).

The angled-grating laser consists of a BA gain stripe and embedded grating parallel to the gain stripe. The gain stripe, which has the shape of a parallelogram, is typically a few hundred micrometers wide and a few millimeters long. Both the stripe and the grating are oriented at an angle (typically $\sim15°$) to the cleaved AR/HR-coated facets. Despite the angled geometry, the output radiation is emitted perpendicular to the AR-coated output facet.

The operation of the device is based on the narrow angular and wavelength selectivity of distributed reflection from a tilted-angle embedded grating. The gratings provide lateral optical confinement by multiple diffraction events that occur as light propagates through the cavity. Waveguides whose sidewalls are Bragg gratings can be very selective filters of incident angle. For example, at 90° normal incidence the grating has an acceptance bandwidth $\sim2.5°$, whereas at 15° incidence the angular acceptance is 0.029°, almost two orders of magnitude lower (Lang et al., 1998). The angled grating supplies both feedback and highly selective spatial filtering over a large-area active region, enforcing single spatial mode oscillation and single longitudinal oscillation.

The device is a hybrid combination of a DFB laser and a FP laser. It exhibits the characteristics of a DFB cavity due to the evenly distributed grating, but it also shows characteristics of a FP cavity, because the facets are essential for resonance; that is, only those waves that are normal to the facets will be efficiently coupled back into the cavity. Other waves are reflected entirely by total internal reflection and are deflected out of the resonator. The angled facet forces the laser to select the mode with the highest energy and lowest modal loss, which then results in the best-quality output beam.

The optical waves can be viewed as continuously bouncing from side to side due to Bragg reflection as they propagate along the cavity. A ray of light traveling in a medium with refractive index n_r will have a maximum reflectivity when the wavelength λ complies with the Bragg condition

$$\lambda = (2n_r \Lambda \sin\theta)/m \qquad (2.32)$$

where Λ is the grating period, θ is the angle of the ray with respect to the grooves of the grating (waveguide axis), and m is the diffraction order of the grating. The facets reflect part of these waves back into the cavity to form the resonant mode. This resonant mode is distributed along a serpentine path within the gain region and between the two facets.

The field inside the cavity is caused in fact by the beat pattern between a set of resonant modes, and the field distribution varies periodically over a beat length L_b with typical values in the range of 300 to 400 µm (Lang et al., 1998). The lateral modes are phase locked to form a single resonator mode, which oscillates from side to side within the laser and produces the unique characteristic of a single-lobed output.

If the angle of the facets is not chosen precisely, the laser will automatically operate at the sole wavelength satisfying Equation (2.32). Then the laser cavity has to be an integral number of beat lengths. The beat length (see Equations 2.24 to 2.26) is set by the path (propagation constant) difference between the two lowest loss modes, which depends on the coupling coefficient, the width of the pumped region, and strongly on pump gain, that is, the difference in saturated gain between the pumped region and the surrounding region (Lang et al., 1998). To fit the laser length as-cleaved, the beat length can be adjusted by operating the device at a slightly higher or lower saturated gain. In addition, according to Lang et al. (1998), if the device happens to be cleaved to a nonintegral number of beat lengths, then the feedback into the mode from the facet will be lowered, the threshold will increase, and the beat length will self-adjust to fit the cavity length.

2.4.3   Unstable resonator (UR) lasers

2.4.3.1   Introduction

In general, an unstable resonator (UR) is described as when a ray bouncing back and forth between the two mirrors diverges indefinitely away from the resonator axis. This is in contrast to a stable resonator whose rays remain bounded. For small apertures the oscillation is confined to the fundamental mode; however, the power available in the output beam is necessarily limited. For an unstable resonator a large mode volume in a single transverse mode is available, because the field does not tend to be confined to the resonator axis and rays tend to escape from the cavity, consequently experiencing much greater losses than rays in a stable resonator. This fact can be exploited to advantage if the escape losses are converted into useful output coupling, while retaining single-mode for the resonator and thereby overcoming the limitations linked to narrow-aperture lasers. If the active region is sufficiently large, the central core region can act as a narrow-stripe master oscillator feeding an angled extraction beam into the off-axis active regions, which results in both an amplification to high power and suppression of filamentation. In principle, the UR design provides mode selectivity in BA devices by suppressing higher order lateral modes.

A broad variety of techniques has been developed and employed to provide means of diverging the lateral fields within the active region aimed at counterbalancing and ameliorating the self-focusing effect of lasing filaments and providing excellent mode control for large-gain volume lasers producing nearly diffraction-limited, high-power laser beams.

In general, on-the-chip UR techniques can be grouped into three categories:

1. Curved-mirror URs where one or both facets are curved to cause a diverging mode.

2. Continuous URs, which have a continuous variation of the refractive index in the transverse lateral direction along the cavity and have the end facets simply cleaved forming planar mirrors.

3. Quasi-continuous URs, which have a train of weak negative cylindrical, diverging lenses grown into the resonator structure to cause the mode to expand laterally as it propagates along the cavity.

In the following we will first focus the discussion on principles and results of the methods in category 1 before dealing briefly with the approaches in categories 2 and 3.

2.4.3.2   Curved-mirror UR lasers

Figure 2.33a shows the schematics of a typical half-symmetric design, where the diverging wavefront is generated by a diverging mirror at one facet, which can have, for example, a convex cylindrical or parabolic shape. Dry-etching techniques such as chemically assisted ion beam etching (CAIBE) or focused ion beam (FIB) micromachining are widely used mirror fabrication techniques. The reactive-ion etching technique CAIBE with typical etch conditions of 15 sccm $SiCl_4$ gas flow, 20 mTorr pressure, 0.15 W/cm^2 power density, and -180 V dc self-bias can deliver high-quality smooth mirror surfaces with very low rms roughnesses, typically in the range of 3 to 5 nm (Biellak *et al.*, 1997). In the FIB process, a focused high-energy beam of Ga ions sputters away material in an extremely controllable way. Scanning electron microscopy (SEM) photographs show smooth and featureless mirror surfaces at a resolution of <100 nm (Tilton *et al.*, 1991) and interferometer measurements show only minor surface abrasion with scratches and digs around 38 nm (Ongstad

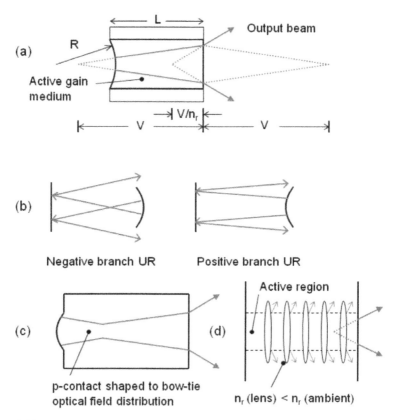

Figure 2.33   Schematic diagrams of unstable resonator (UR) diode laser types. (a) Curved-mirror UR with virtual source points V. (Selected data adapted in amended form from Figure 1 in Ongstad *et al.*, 2010.) (b) Negative branch and positive branch UR comprising a concave and convex mirror, respectively, and a cleaved flat output mirror in each case. (c) Negative branch UR with bow-tie-shaped current injection contact. (d) Regrown-lens-train UR with flat cleaved facets.

et al., 2010). The FIB technique has the advantage of being maskless and sample orientation independent; however, in contrast to etching FIB is not suited to high-volume batch fabrication.

The UR geometry with a length L and convex mirror radius R forms two virtual source points (Ongstad *et al.*, 2010) at locations

$$V = \pm \left(L^2 + LR\right)^{1/2} \tag{2.33}$$

from the flat mirror. The V_+ virtual point is at an object distance $V+L$ from the diverging convex mirror with a focal length $-R/2$. Upon reflection from the curved facet, the radiation forms a virtual image at the other virtual source point V_-, which is at a distance $V - L$ away from the convex mirror. According to the known geometrical optics rules and sign conventions, we obtain for this setup the imaging forming condition (Ongstad *et al.*, 2010)

$$(V + L)^{-1} - (V - L)^{-1} = -2R^{-1}. \tag{2.34}$$

An important characteristic parameter of an UR is the magnification M defined as the roundtrip geometric mode expansion factor, and which determines in large part the threshold gain. M is given by (Tilton *et al.*, 1991)

$$M = \frac{\left(L^2 - LR\right)^{1/2} + L}{\left(L^2 - LR\right)^{1/2} - L} \tag{2.35}$$

where R, the mirror radius of curvature, is by convention negative for the convex diverging case. A typical magnification is 3.7 for a UR with $L = 5$ mm and with a radius $R = 10$ mm for a convex, diverging mirror (negative R). For this device, the virtual source points V_+ and V_- are located at $+8.66$ mm and -8.66 mm from the flat mirror, respectively. If the radiation is coupled out from the flat mirror facet, the virtual waist of the transverse lateral mode is located behind the output mirror at a reduced distance V/n_r inside the device (Ongstad *et al.*, 2010). It amounts to 2.62 mm for our example and by using $n_r \cong 3.3$ for AlGaAs.

Substantial quasi-cw powers (27% duty cycle) of >10 W have been achieved by optically pumping UR lasers of the type illustrated in Figure 2.33a. Far fields, realized by reimaging the high-brightness virtual source point at V/n_r inside the cavity, show nearly diffraction-limited beam qualities up to 25 times threshold. In comparison, standard FP lasers produce more than 10 times diffraction-limited beams under similar conditions (Ongstad *et al.*, 2010).

Impressive results, however, have been achieved from GaAs/AlGaAs, In-GaAs/GaAs, and GaInP/AlGaInP UR laser devices with optimized mirror curvatures realized by a computer-controlled FIB milling process. Devices with $M \cong 2.5$ generate up to 1.5 W single-facet, nearly diffraction-limited cw output power with record-high brightness values of 400 MW cm^{-2} sr^{-1} (DeFreez *et al.*, 1993).

Other curved-mirror UR laser approaches include (i) curved-grating UR lasers and (ii) negative branch UR diode lasers with real internal focus.

In a grating UR diode laser, one or both mirrors are replaced by a curved grating (Pepper and Craig, 1989). This configuration can then be referred to as a distributed Bragg reflector (DBR) unstable resonator (DBR-UR; Eriksson *et al.*, 1998), or if the grating is extended throughout the active region, as a distributed feedback (DFB) unstable resonator (DFB-UR). The curvature changes throughout the grating to match the changing wavefront of the beam as it propagates into and out of the grating region and each groove of the curved grating has essentially a unique shape (Lang, 1994).

A grating UR offers the high spatial mode selectivity of a grating resonator, which enhances the mode selectivity of a common curved-mirror UR. High powers of 460 mW of single spatial and single longitudinal mode have been demonstrated (Dzurko *et al.*, 1993) from BA grating ring oscillator devices with 368 µm × 1000 µm active dimensions, 200 A/cm^2 threshold current density, and >60% external differential quantum efficiency. Half-symmetric DFB-UR diode lasers produced >400 mW cw single-frequency output power at 765 nm lasing wavelength with a brightness >100 MW cm^{-2} sr^{-1} (Baird *et al.*, 1997). These devices used a core DFB grating of period 240 nm, a FIB fabricated HR convex cylindrical mirror with a radius of 12.77 mm and a low-reflectivity flat output facet, and were based on an AlGaAs SQW GRINSCH structure with a resonator magnification of 2.5.

Negative branch UR diode lasers with real internal focus contain a concave mirror and usually a plane cleaved output facet. Figure 2.33b shows for illustration the positive and negative branch UR configuration with straight diverging wavefront and asymmetric bow-tie-shaped diverging wavefront, respectively. Devices have been designed (Tanguy *et al.*, 2003; Corbett *et al.*, 2004) with an asymmetric bow-tie injection contact in order to improve the overlap between the carrier distribution and the optical field within the active region (Figure 2.33c). The radius of the concave mirror ranges between about 300 and 600 µm, and device lengths are between 1 and 1.5 mm, which results in geometric magnification factors for the cavity between −7.8 and −11.2 (R is positive for concave curvature in Equation 2.35). These GaAs-based devices operate up to >1 W cw with a diffraction-limited central lobe, when corrected for spherical phase, and achieve a brightness >400 MW cm^{-2} sr^{-1} for uncoated curved facets and AR-10% coated front mirrors, which leads to a split in power emitted from the front mirror (76%) and rear mirror (24%).

The technology can also be used to etch both facets of the resonator to obtain a collimated output beam, which can be realized by a confocal UR geometry, either in a negative branch configuration with two concave end mirrors, or in a positive branch configuration with a convex and concave mirror.

2.4.3.3   UR lasers with continuous lateral index variation

Transverse lateral variations in refractive index play a decisive role in controlling the mode behavior of diode lasers. A continuous UR has a continuous variation of the index in the lateral direction and has its end facets simply cleaved, forming conventional planar end mirrors. The index increases with distance from the center of the resonator and consequently the rays of the resonator mode are bent away from the center of the laser cavity.

A quadratic lateral refractive index profile with weak gain guiding and strong index anti-guiding gives high calculated gain difference >20 cm^{-1} between the fundamental mode and higher order lateral modes and hence strongly supports high-power single-mode operation of wide-aperture diode lasers. In this case, the lateral field distribution for the fundamental mode has a FWHM power density of about 100 µm (Chan *et al.*, 1988).

A realization of the required parabolic gain and index profile can be achieved by designing the p-guiding layers such that the thickness of the lower half with higher index increases laterally from the center of the wide stripe, which finally results in the required anti-guiding effective index profile. The weakly guiding gain profile can be produced by a halftone electrode design, similar to that briefly described in Section 2.4.2.2. This is accomplished by continuously decreasing the widths of the individual electrode stripes from the center to both sides in the lateral direction. A smooth variation in the lateral gain profile will be caused by the carrier diffusion for an electrode structure with a spatial resolution of better than 1 μm.

Alternatively, a suitable temperature variation in the lateral direction can create a continuous thermal negative lens inside the laser cavity via the positive temperature coefficient of the index and consequently a UR with cleaved flat end mirrors. This can be achieved either by heating the edges of the stripe or by cooling the center of the cavity. In first experiments, thermally induced lateral index tailoring by an external focused laser heat source was used to modify the lateral index profile in a BA laser and thus control the lasing mode. An absorbed heating power of ∼50 mW was sufficient to counteract the effects of internal junction heating and restore fundamental cw mode operation up to about twice the threshold current (Hohimer *et al.*, 1988).

A simpler and more controllable method is to use on-chip resistive microheaters adjacent to the stripe edges. Sun *et al.* (1993) showed that both NF and FF patterns exhibit a significant improvement in beam quality when a suitable amount of heating power was applied. Single-lobed far fields with near diffraction-limited widths are obtained by applying typical heating powers of around 130 mW to temperature-induced UR diode lasers 60 μm wide and 320 μm long.

2.4.3.4 Quasi-continuous unstable regrown-lens-train resonator lasers

This so-called regrown-lens-train UR approach (Paxton *et al.*, 1993) uses a train of discrete, negative cylindrical diverging lenses embedded in the laser cavity along the cavity direction to cause the necessary anti-guiding effect by expanding the mode laterally parallel to the junction as it propagates (Figure 2.33d). As a result, fundamental lateral mode operation can be attained from a BA laser.

The index variation in the lateral direction can be controlled by changing the geometric parameters and the density of lenses. Roundtrip magnifications between two and six can be achieved. The diverging lens elements are incorporated above the GRIN-SCH layer in a GaAs layer. Their fabrication includes a wet chemical etch process through the cladding to an etch-stop layer close to the active region and a subsequent MOCVD regrowth of the etched pits with a lower index material. The field distribution outside the active region is affected by these lower index regions causing the effective index to be lowered relative to the non-regrown regions of the heterostructure, with the consequence that negative lenses, which have biconvex shapes, are formed and which then diverge and spread the light laterally as it propagates.

Up to 500 mW single-facet power has been obtained from lasers 100 μm wide with near diffraction-limited performance and with good lateral mode discrimination up to several times the threshold current with magnifications of around six (Srinivasan *et al.*, 1992).

2.4.4   Tapered amplifier lasers

2.4.4.1   Introduction

The use of a tapered gain region offers a variety of advantages including the provision of high-power and low-divergence laser sources, which can be fabricated cost effectively by using standard ridge waveguide processing techniques. Tapered waveguide amplifier laser sources can be categorized into: (i) master oscillator power amplifier (MOPA) lasers, which monolithically integrate a conventional single spatial and spectral mode DBR laser with a flared-contact amplifier; and (ii) tapered waveguide lasers operating as URs. The former configuration will be discussed in Section 2.4.4.3 whereas the discussion of the latter is in Section 2.4.4.2 next.

2.4.4.2   Tapered lasers

Figure 2.34a shows the schematics of the tapered laser geometry. Mode control is achieved by using a narrow index-guided ridge waveguide section at the back mirror, which effectively operates as a mode-selective filter allowing only the fundamental

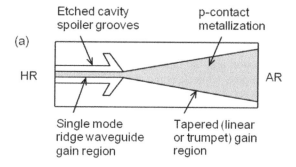

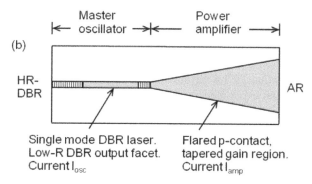

Figure 2.34   Schematic diagrams of (a) a tapered diode laser with cavity spoiling structures to suppress FP modes outside the ridge waveguide, and (b) monolithic flared-amplifier master oscillator power amplifier with separate oscillator and amplifier contacts.

transverse lateral mode. The gain-guided tapered section is defined in the p-contact layer typically with a length of 2 mm and an opening angle of $\sim 6°$, which leads to an emitting aperture ~ 200 μm wide. The standard shape of the taper has a linearly increasing width, but other shapes have also been investigated, such as a trumpet-shaped taper (Kristjánsson et al., 2000) or a Gaussian patterned p-contact layer to modify the lateral injection profile with the $1/e^2$ width varying with length (Walpole et al., 2000).

A single, nonsegmented metallization is usually applied over the patterned p-contact layer to provide electrical contact to the ridge and tapered regions. However, tapered lasers with electrically isolated ridge waveguide and tapered sections with separate control of the injected current to each section have also been developed (Odriozola et al., 2009). This adds a new degree of freedom to increase the output power and beam quality (cf. Section 2.4.4.3 next). Device performance can be improved, for example, by stronger pumping of the ridge waveguide section. In addition, this device offers the opportunity that, at constant current in the tapered section, the optical output power can be modulated effectively by just varying the ridge waveguide laser current. This is an easier realization than modulating the much larger amplifier current (Paschke et al., 2005).

To maximize the available power, the back mirror is HR ($>90\%$) coated and the front output facet AR ($<1\%$) coated. The lengths of the ridge and tapered sections are crucial parameters in achieving efficient laser devices with good output beam performance. Design strategies for tapered lasers show that a careful balance between the ridge waveguide section and the tapered section is necessary to maximize the output power and brightness while keeping the beam quality near the diffraction limit (Borruell et al., 2008). The angle of the taper is designed to accommodate the diffraction of a well-behaved Gaussian beam propagating from the narrow end of the taper.

Only a small portion of the small amount of power reflected from the low-reflectivity output mirror enters the narrow aperture of the ridge waveguide. If the curvature of the wavefront is negligible across the narrow aperture, it will efficiently fill the tapered gain region after both, the reflection from the back mirror and free diffraction from the aperture along the cavity length before filamentation sets in. This will result in the formation of a self-consistent lowest order transverse lateral mode.

As tapered lasers are intrinsically multipass devices, they are susceptible to filamentation, which leads to a broadening of the far field from the diffraction limit. We discussed in Section 2.2.1.6 various approaches for suppressing the filamentation effect. Simulated and experimental results show that low modal gain tapered laser devices with low confinement factors (1.35%) produce a tenfold improvement in the beam quality and insensitivity against filamentation compared to high modal gain devices with 2.7% confinement factors (Mikulla et al., 1998). The astigmatism of the output beam of a tapered laser will be discussed in the next section.

Cavity spoiling grooves are etched through the active region on either side of the narrow end of the taper (see Figure 2.34a). The grooves, which are tilted deflect and scatter unwanted radiation away from the tapered region and suppress a direct feedback path between the front and back mirror outside the narrow ridge region and hence the backward-traveling part of the optical field. There are alternative approaches for the so-called beam spoilers to suppress the propagation of unwanted modes outside the cavity with only a minor loss in power. These include passive absorber sections formed by ion-implanted impurities on both sides of the ridge waveguide (Delépine et al., 2001), active reverse-biased absorber sections next to the ridge waveguide (Fiebig et al., 2009) and an optimized rear facet with a lower reflectivity (Fiebig et al., 2010).

Of the numerous results obtained from the many tapered laser approaches, only some characteristic findings will be reported here.

To explore the limitation on power available from 970 nm InGaAs/AlGaAs QW GRIN-SCH structures, linearly tapered devices 3 mm long were biased to the point of failure. The lasers had threshold currents of ∼0.5 A and linear P/I characteristics, which thermally rolled over at higher powers and had relatively low slope efficiencies of ∼0.54 W/A due to the long device lengths. A maximum power of 4.2 W cw was achieved at 10 A, where the FF profile was predominantly single-lobed with about half of the power contained in the central lobe with a width of 0.26° FWHM, which is only ∼1.7 times the diffraction limit for the 325 μm width of the output mirror (Kintzer et al., 1993).

Wavelength-stabilized InGaAsN-based tapered lasers with emission wavelengths around 1230 nm showed maximum output powers of 1 W limited by thermal rollover. The vertical structure included an InGaAsN SQW 6.5 nm thick, a large optical cavity 1100 nm thick, and 1200 nm AlGaAs cladding layers. The 2 mm long devices with a first-order DBR grating at the back facet had threshold currents ∼0.9 A and slope efficiencies ∼0.3 W/A. Bisping et al. (2009) observed single-mode emission with (i) a divergence angle of 49° FWHM in the fast-axis, (ii) a slow-axis angle of 9° FWHM, and (iii) a record-high brightness of 23 MW cm^{-2} sr^{-1} at 1230 nm.

The maximum cw power observed for Gaussian patterned contact 1.55 μm In-GaAsP/InP tapered lasers was 1.3 W at 4.5 A with differential slope efficiencies of 0.37 W/A and threshold currents of 0.8 A. The fraction of power in the central FF lobe remained large at >80% throughout the current range up to 4 A, and, above all, the central lobe width was typically 1.2 times the diffraction limit up to 3 A. The purpose of the Gaussian-like distribution of the lateral current is to match the lateral optical profile and hence reduce self-focusing and beam instability effects. Its use provides improved performance of tapered lasers operating near 1.55 μm wavelength. For devices having the Gaussian lateral injection profiles with improved stable single lateral modes, a record-high power of 0.52 W at 1.55 μm has been coupled into a single-mode optical fiber with an overall coupling efficiency of 52% (Walpole et al., 2000).

High-power, 980 nm tapered InGaAs/AlGaAs DQW lasers with electrically separated ridge waveguide and tapered sections have achieved >12 W cw optical output powers at currents of 16 A to the tapered amplifier and 400 mA to the single-mode ridge waveguide laser in a single longitudinal mode and a nearly diffraction-limited high-quality beam with a conversion efficiency of 44%. The laser resonator has a total length of 6 mm and consists of a single-mode ridge laser 2 mm long and 4 μm wide and a tapered amplifier 4 mm long with a full taper angle of 6°. The ridge waveguide section contains a sixth-order surface Bragg grating 1 mm long defined by projection lithography and fabricated by a reactive-ion etching process. The laser chip is soldered junction-side up onto a conductively cooled package. The threshold (turn-on) current of the amplifier is 1.8 A, and the *P/I* characteristic is linear up to 15 A with an average slope efficiency of 0.85 W/A (Paschke *et al.*, 2008).

2.4.4.3   Monolithic master oscillator power amplifiers

The design of the monolithic, flared-amplifier master oscillator power amplifier (MFA-MOPA) is schematically illustrated in Figure 2.34b. In the MFA-MOPA architecture, a DBR master oscillator is monolithically integrated onto the same chip as a high-power flared-contact power amplifier (Welch and Mehuys, 1994).

The master oscillator is a single-stripe, single-mode, index-guided laser, typically 4 μm wide, and whose reflectors are not cleaved crystal facets but integrated DBR gratings. These gratings provide frequency-selective feedback at a wavelength proportional to the grating period, which causes the device to lase in a single spectral mode. The front grating is significantly shorter than the back grating in order to lower its reflectivity aimed at transmitting a large amount of the single spatial mode output power of the DBR laser into the flared, gain-guided power amplifier where it spreads laterally by diffraction and is amplified as it propagates through the tapered gain region.

Current to the amplifier is restricted to the tapered contact region, whose width varies linearly up to typically 300 μm at the end of its ∼2.5 mm length. The optical output power scales to a first order linearly with the width of the amplifier output mirror at a set current. The taper angle of ∼5° is usually wider than the lateral beam diffraction angle of $\theta_\parallel = \lambda/Dn_{r,eff} = 4.25°$ FWHM (cf. Section 1.2.2) obtained for an example with $\lambda = 0.98$ μm, $D = 4$ μm, and $n_{r,eff} = 3.3$, which is the effective index of the DBR lateral mode propagating in an AlGaAs waveguide.

The output mirror is AR ($R < 0.1\%$) coated to minimize feedback from the amplifier to the master oscillator and to suppress parasitic lasing of the amplifier alone. The standard vertical QW structure is grown by a two-step MOCVD process with the fabrication of the second-order gratings intermediate to the first and second growth. Usually, the gratings are defined holographically and chemically etched into the surface of the p-cladding layer.

An important advantage of the MFA-MOPA design is that master oscillator powers as low as ∼10 mW are sufficient to saturate the gain in the tapered amplifier. However, powers ∼100 mW cw have been applied to achieve a more complete saturation leading to lower turn-on currents and higher slope efficiencies (O'Brien *et al.*, 1997). When the oscillator is turned off, the amplifier operates in a superluminescent mode and only low-power amplified spontaneous emission (ASE) noise up to high amplifier currents is emitted from the output mirror. As the ASE is both spectrally and spatially broad, it does not significantly overlap with the actual diffraction-limited amplifier output beam when the oscillator is on.

The NF and FF patterns determine the beam quality of the MFA-MOPA emission. The NF has been numerically investigated by applying a beam propagation method (Lang *et al.*, 1993). Initially, at low amplifier currents, the NF power intensity versus lateral dimension pattern is nearly Gaussian. This is due to both the output of the fundamental lateral mode with near-Gaussian intensity profile from the master oscillator and a negligible effect of gain saturation at low currents. However, at higher currents, gain saturation is effective, in particular in the center of the amplifier, leading to a top-hat NF power shape with flattened intensity and intensity peaks near both the edges of the output aperture. This distinct calculated distribution becomes more pronounced with increasing current and has been experimentally confirmed (Ostendorf *et al.*, 2008). The calculated FFs remain near-diffraction limited showing only a small increase in the side lobes at higher amplifier currents that is caused by the fact that the NF is top-hat rather than Gaussian (Lang *et al.*, 1993). The wavefront of the amplified beam is near-spherical, consistent with the diffraction from the narrow aperture 4 μm wide at the amplifier input.

The radiation pattern from the tapered output aperture is a broadly diverging beam emitting from the master oscillator source with typical dimensions $d \cong 1$ μm in the transverse vertical and $D \cong 4$ μm in the transverse lateral directions. The output beam is highly astigmatic caused by the difference of diffraction in the two directions. The fast-axis beam diverges at the output mirror from the narrow waveguide with thickness d, whereas the slow-axis beam appears to originate from a virtual source about $L/n_{r,eff}$ inside the amplifier with length L. Lensing effects caused by thermal gradients and carrier density gradients from nonuniform gain saturation across the amplifier width may have an effect on the longitudinal position of the virtual source (O'Brien *et al.*, 1997; Mehuys, 1999). The same arguments on astigmatism are also valid for tapered lasers discussed in Section 2.4.4.2 above.

Measurement of the slow-axis FF profile has been performed with a spherical objective lens placed one focal length away from the output mirror. This collimates the fast-axis beam and forms a waist of the slow-axis beam component after the lens. The waist profile represents the slow-axis FF after the dominant quadratic phase curvature across the beam in the transverse lateral direction has been removed. This FF, scaled by the focal length of the lens, represents the actual slow-axis FF. The quadratic phase curvature is caused by the natural divergence of the beam and to a lesser extent by thermal effects and carrier density variations in the amplifier (Welch and Mehuys, 1994).

Since the early 1990s the MFA-MOPA has advanced by more than an order of magnitude in cw power emitted in a coherent, diffraction-limited, and single longitudinal beam, which can be small-signal modulated at high speeds with power gain. First devices (Welch *et al.*, 1992) with amplifiers 2 mm long and 250 μm output aperture width delivered ≥1.1 W cw in a single spatial and spectral mode throughout the operating range. The FF maintained a single-lobe radiation pattern with a diffraction-limited FWHM of 0.18° and showed no beam steering as a result of changes in injection current.

Record-high, nearly diffraction-limited, single-frequency powers of >11 W cw at >13 A amplifier currents have been achieved from a single-chip diode laser at a wavelength of 980 nm and with 300 mA input current to the master oscillator, which has holographically defined and wet chemically etched second-order Bragg gratings into the cavity. The total length of the MFA-MOPA is 8 mm and the length of the power amplifier with a total flare angle of 6° is between 4 and 6 mm (Wenzel *et al.*, 2007).

In general, power limitations at higher power levels may be caused by slight inhomogeneities of gain and refractive index within the amplifier. These can lead to small intensity ripples and phase modulation on the diffracting beam, in particular across the central part of the NF pattern, and can result in a filamentary emission. In addition to the central intensity modulation the NF profile displays two large intensity spikes near the stripe edges. The corresponding FF pattern is dominated by a central diffraction-limited lobe with a peak height increasing sublinearly with output power. However, at higher powers the intensity modulation of the output increases, with the result that an increasing fraction of the total output power contains nondiffraction-limited components (Mehuys, 1999).

2.4.5 Linear laser array structures

2.4.5.1 Introduction

Diode laser arrays can be classified into two groups.

First, phase-locked arrays of diode lasers have been developed to deliver high coherent powers in narrow, diffraction-limited beams in the Watt-level range for applications such as free-space optical communications, frequency doubling, and optical interconnects.

Second, spatially incoherent arrays with the standard 1 cm long laser bar as the basic building block originally attracted greatest interest in the development of diode-pumped solid state lasers, such as Nd:YAG or Er:YAG. These laser bars can be stacked to create a 2D aperture emitting optical powers in the kW-range, which are heavily employed in many industrial applications such as cutting and welding processes. Since spatial and temporal coherence is usually not required for pumping solid state lasers, the pumps are primarily configured to supply the highest possible output powers in a reliable way. This means that the focus of the design is on epitaxial structure, material, thermal management, packaging, and reliability issues to achieve high-power conversion efficiency and ultimately high-power capability.

Although the topic of this section is beyond the scope of this text – namely, high-power, single spatial mode, single-edge-emitting diode lasers – we will briefly discuss in the following sections the principles of operation and relevant performance data of edge-emitting, phase-locked coherent laser bars and high-power incoherent standard laser bars 1 cm wide, since these linear arrays can deliver diffraction-limited or narrow slow-axis divergence beams, respectively.

2.4.5.2 Phase-locked coherent linear laser arrays

It has been demonstrated that monolithic phase-locked arrays can deliver high co-herent output powers in the >1 W range emitted in single-mode, diffraction-limited beams (Botez, 1994; Botez and Mawst, 1996). There are four basic types of phase-locked arrays: namely, leaky-wave coupled, evanescent-wave coupled, Y-junction coupled, and diffraction-coupled arrays. The arrays make use of a periodic variation of the real part of the effective refractive index formed in the junction plane through closely spaced individual emitters on the same wafer along the 1D bar array, and operate as a coherent array by an adequate coupling between oscillating lateral modes of adjacent emitters.

Decisive for the positive development of high-power arrays was the revelation that strong overall coupling between individual emitters occurs only when each element couples equally to all others, which is called *parallel coupling*, in contrast to *series coupling*, where only nearest neighbors are coupled (Botez, 1994). The latter scheme is plagued by weak overall coherence and poor intermodal discrimination, whereas parallel-coupled systems have high intermodal discrimination, full coherence, uniform near-field intensity profiles, and therefore are insensitive to the emergence of higher order modes at high drive currents.

Evanescent-wave coupled devices can produce parallel coupling, but only by lowering the optical confinement, which then makes the devices vulnerable to thermally and injected carrier induced refractive index changes. This is due to the low built-in index step, typically $\leq 5 \times 10^{-3}$, which has to be below the cutoff for high-order modes. Only strongly index-guided (index steps ≥ 0.01), leaky-wave coupled systems can deliver parallel coupling realized in strong optical confinement structures and as a consequence then meet the requirement for both full coherence and stability (Botez, 1994).

In leaky-wave coupled arrays, the lasing modes are peaked in the low-index array regions. Evanescent-wave coupled devices have field intensity peaks that reside in the high-index array regions. For both types, the locking condition can be in-phase (leading to single central lobe far fields), when the fields in each element are cophasal, or out-of-phase (leading to twin-lobe far-fields), when the fields in adjacent elements are π shifted apart. In Y-junction coupled and diffraction-coupled arrays, the in-phase operation is selected via interference and diffraction, respectively (Botez, 1994).

Positive-index guided arrays with evanescent-type coupling exhibit only weak overall coupling, resulting in poor intermodal discrimination. Devices operated in the out-of-phase condition could produce diffraction-limited powers as high as 200 mW (Botez et al., 1988). These devices are vulnerable to thermal gradients and gain spatial hole burning. By contrast, leaky-mode coupling allows parallel coupling among array elements exhibiting strong intermodal discrimination, which leads to high-power, single-lobe, diffraction-limited operation with coherent powers up to 2 W pulsed (Mawst et al., 1992b) and 10 W pulsed in a beam twice the diffraction limit (Yang et al., 1996).

In the leaky-wave coupling approach, light is coupled between anti-guides and the resulting array lases on a set of leaky modes characterized by their radiation loss. The index of the anti-guide core is lower than that of the surrounding cladding layer, and consequently light is partially reflected at the anti-guide core boundaries and refracted into the cladding layers (Botez, 1994). This causes a radiation loss, which is roughly proportional to the square of the lateral leaky mode number. In this way, the anti-guide acts as a lateral mode discriminator.

In closely spaced anti-guides of linear arrays, the device losses are significantly reduced, because radiation leakage from individual elements serves primarily the purpose of coupling the array elements. The intermodal discrimination is maximized when adjacent anti-guides resonantly couple. These resonant arrays of anti-guides, also called resonant optical waveguide (ROW) arrays, are the best candidates for achieving high coherent output powers from phase-locked parallel-coupled arrays. Unlike evanescent-type devices, in anti-guided arrays there is no limitation on the effective index step. This allows the fabrication of single-lobe emitting structures of high-index steps in the range of 0.05 to 0.20 that are stable against temperature- and/or carrier-induced index variations (Botez, 1994).

There are many ways to design the lateral effective index step structure for a phase-locked, anti-guided array (Botez, 1994). Figure 2.35 shows a schematic representation of a typical practical device structure of the self-aligned stripe array type (cf. Figure 2.8 for strong similarities). The fabrication process consists of two epitaxial growth steps and is very similar to that described in Section 2.1.4.3. The first step involves the growth of the interelement regions. After etching down to the high-index guide layer, regrowth is performed in the second step in the element regions. The lateral anti-guiding property results from periodically placing the guide layer in close proximity of <0.2 μm to the active layer. In these locations, the fundamental transverse mode is mainly confined to the passive guide layer, that is, the modal gain is low in these areas. An alternative approach, which does not involve regrowth, is the formation of the necessary index step by using Zn diffusion-induced intermixing of a superlattice upper cladding layer (Gray *et al.*, 1998).

Continuous wave output powers are 1 W emitted in a beam 1.7 times the diffraction limit and 0.5 W diffraction limited from an element bar 20×120 μm wide with an AlGaAs/GaAs structure emitting at around 850 nm wavelength (Zmudzinski *et al.*, 1993). The highest pulsed powers are 10 W at twice the diffraction limit from a linear array with elements 200 μm wide (Yang *et al.*, 1996). Anti-guided array lasers

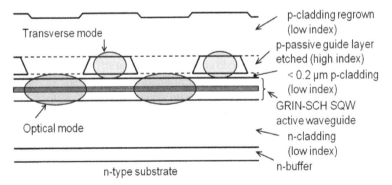

Figure 2.35   Schematic representation of a self-aligned-stripe structure for realizing a phase-locked, anti-guided coherent laser array (cf. Figure 2.8 for similarities).

fabricated by the superlattice disordering technique produce 1.6 W quasi-cw (100 µs pulses) powers at 1.2 times the diffraction limit (1° FWHM) (Gray *et al.*, 1998). These AlGaAs/GaAs DQW laser bars emitting at a wavelength of 860 nm consist of only five elements with 10 µm anti-guide–interelement center spacings.

2.4.5.3   High-power incoherent standard 1 cm laser bars

The standard 1 cm long laser bar is the basic building block for high-power incoherent laser pumps. The size and type of the individual emitters and their configuration in the bar depend on the specified optical power and application. A bar is subdivided into an array of current-pumped regions, which in turn may be further subdivided into many parallel narrow emitters (see Figure 1.12). The individual emitters can be BA, single-mode, or tapered-type devices with cavity lengths up to 5 mm. The periodically spaced active elements (pumped area) constitute a percentage of the total bar area, which is called the fill factor. Fill factors can range from about 10% up to close to 100% depending on application and operation condition. A typical high-power bar with about 90% fill factor consists of emitters 110 µm wide on 120 µm centers.

Laser array bars have a geometry that spreads the total optical output power over a large width, which finally leads to a very low optical power density at the output facet. An example may illustrate the effect: a typical "lasing" width of 7000 µm leads to an optical power density of ~14 mW/µm for a total power of 100 W cw. This density, however, is far from the point of facet failure. The bars can be operated under cw conditions to low peak power levels. Due to thermal limitations under cw operation, the bars are usually operated in quasi-cw mode to pump solid state lasers, where high-energy pulses are required. In quasi-cw operation, pulse widths > 1 µs are applied, which allows effective heat transfer out of the laser chip mounted on a heat sink when the laser is operated with a low duty cycle. For example, AlGaAs/GaAs laser bars with 90% fill factors can be reliably operated in quasi-cw mode with pulses 200 µs long at a repetition rate of 50 Hz (1% duty cycle) to power levels greater than 200 W. However, this quasi-cw pulsed mode operation involves a thermal cycling of the laser chip, which may lead to a gradual degradation of the output power over the bar's operational time. Typical operating lives of quasi-cw bars with 75% fill factors can be expected to exceed several billion 200 µs pulses (>1 Gshot) with >200 W peak powers at duty cycles up to 10% (Deichsel *et al.*, 2008). This projected lifetime corresponds to >5000 h of "on-time."

Commercial laser array bars are usually soldered p-side down onto a heat spreader and contacted on the n-side with an array of wire bonds that are bridged to an insulated n-contact. Then 2D arrays are formed by stacking such tested, burned-in, and yielded modular laser bar submounts into appropriate packages equipped with efficient cooling systems. As already mentioned above, the discussion of these high-power laser systems is beyond the scope of this text.

In the following, we discuss some key approaches for increasing the optical output power of incoherent laser bars, which are, however, valid for any semiconductor diode laser device. There are many mechanisms contributing to the loss of optical power, including Joule heating, scattering and absorption losses, carrier leakage, below-threshold losses, and band alignment. One of the primary issues limiting the maximum optical power of any semiconductor diode laser including laser arrays is self-heating, which leads to thermal rollover at high drive currents. As mentioned in the introduction above, effective thermal management is a crucial requisite for achieving high optical powers. There are two key measures to counteract self-heating effects, which are to minimize the thermal resistance R_{th} and to maximize the electrical-to-optical power conversion efficiency η_c of the laser (cf. Section 1.3.7.1).

The thermal resistance results from the finite thermal conductivity of the epitaxial layers, solders, submounts, and heat sinks that are part of the diode laser assembly. In the case of a laser bar, the bar geometry is an additional factor in the thermal resistance budget and includes the fill factor and the cavity length (valid also for a single laser). A larger thermal footprint enables a larger 2D parallel heat flow, and consequently the thermal resistance is lowered almost in direct proportion to the active length, resulting in a lower operating temperature at a given heat load. However, the laser length also affects the threshold current I_{th} and the external differential quantum efficiency η_d and therefore has to be chosen correspondingly.

The most direct way to increase the maximum sustainable optical power is to maximize η_c, which primarily means to lower the internal operating temperatures. Positive side effects of this measure are among others that also the impact of heating on the built-in index profile, differential efficiency, defect mobility, and reliability is reduced. The dissipated electrical power of a diode laser is given by

$$P_{diss} = I(V_d + R_s I) \text{ for } I < I_{th} \tag{2.36a}$$

$$P_{diss} = I(V_d + R_s I) - \eta_d \frac{h\nu}{q}(I - I_{th}) \text{ for } I > I_{th} \tag{2.36b}$$

where V_d is the diode voltage (cf. Section 1.3.7), which is actually the overall built-in voltage (Bour and Rosen, 1989), R_s is the series resistance, $(V_d + R_s I)$ is the terminal voltage across the diode laser, and the last term in Equation (2.36b) is the total optical output power P_{out} (cf. Equation 1.51). The dissipated power has to be minimized in order to maximize the power conversion efficiency η_c, which is defined as the ratio of the optical power to electrical power and is given by Equation (1.53). The temperature rise ΔT_j of the active region in cw operation is

$$\Delta T_j = R_{th} P_{diss} \tag{2.37a}$$

which can be approximated at high drive currents $I \gg I_{th}$ by

$$\Delta T_j \cong I R_{th} \left[V_d + I R_s - \frac{h\nu}{q} \eta_d(T) \right] \tag{2.37b}$$

where the temperature rise in the junction ΔT_j is usually defined with respect to a controlled temperature reference T_h of a heat sink.

The optical output power P_{out} is a function of temperature, since both I_{th} and η_d are temperature sensitive. Both parameters are exponentially dependent on temperature and a characteristic temperature coefficient T_0 and T_η, respectively (cf. Equations 1.55 and 1.56)

$$P_{out}(T) = \eta_d(T) \frac{h\nu}{q} [I - I_{th}(T)] \tag{2.38a}$$

which can be approximated for high drive currents $I \gg I_{th}$ by

$$P_{out}(T) \cong \eta_d(T) \frac{h\nu}{q} I \cong \eta_d(T_h) \exp\left\{-\Delta T_j / T_\eta\right\} \frac{h\nu}{q} I \tag{2.38b}$$

where Equation (1.56) has been rewritten to reflect the temperature dependence of the efficiency directly.

It is clear from Equation (1.53) that a maximization of η_c requires a minimization of V_d, R_s, and I_{th} and a maximization of η_d. Maximizing η_d includes minimizing the internal optical losses α_i and maximizing the internal differential quantum efficiency η_i, for example, by decreasing the free-carrier absorption of the optical field in the doped regions. An asymmetric waveguide, which expands the mode toward the n-side of the waveguide structure, has proven to be very efficient in improving free-carrier losses, optical losses, differential efficiency, internal efficiency, and series resistance (cf. Section 2.1.3.5).

Thus, the highest reported power conversion efficiencies of 75% have been achieved to date on 808 nm InGaAlAs/AlGaAs BA lasers 100 μm wide and 1 mm long (Li *et al.*, 2008). These lasers with ultralow series resistances of 0.07 Ω, high slope efficiencies of 1.4 W/A, and low threshold current densities of 180 A/cm^2 emit >5 W cw in smooth single-lobed far-field patterns with 32° and 8° for the fast-axis and slow-axis beam divergence angles, respectively.

It is also of paramount importance for ensuring high η_d and low I_{th} values at high temperatures that T_η and T_0 are as high as possible, so that the onset of thermal rolling with drive current is strongly increased. The conditions on these parameters are required to maximize η_c and constitute an overall set of requirements, which contains subsets of conditions to maximize P_{out}, minimize P_{diss} as well as ΔT_j, with the additional requirement to minimize R_{th} according to Equations (2.36a)–(2.38b), above.

To maximize T_0 and T_η (see Section 1.3.7.3 for details), the carrier leakage needs to be minimized, among others, for example, by providing carriers with sufficiently high-energy barriers or additional barriers near the active region; in addition, the cavity length should be shortened (Mawst *et al.*, 1995) for high T_η values. However, these are conflicting requirements for T_η, since for short-cavity lengths, band filling (see Section 1.3.4.3) causes carrier leakage. Short cavities are also required for high η_d values. Carrier leakage from short-cavity effects can be prevented among others by using MQW active regions (cf. Section 1.3.4.3) as demonstrated by short-cavity lasers with very high η_c values (Botez *et al.*, 1996). The use of strained barriers in

QW active layer structures can also reduce carrier leakage from the active region (Mawst *et al.*, 1999). By introducing strain in the barriers the heavy- and light-hole bands split, which results in a higher effective barrier height for holes and hence a suppressed hole leakage. Furthermore, simulations and experiments show that V_d can be reduced by grading the p-side SCH doping profile. Reduced values up to 60 mV have been obtained in Al-free InGaAs(P)/InGaP/GaAs active diode lasers, which corresponds to a 4% improvement in η_c (Kanskar *et al.*, 2005). These lasers also have superior η_c values (Botez *et al.*, 1996) compared to conventional Al-containing devices due to their low differential series resistance and higher thermal conductivity (Al-Muhanna *et al.*, 1998b).

In optimizing R_s, the doping profile has to be selected such that it balances the series resistance, which favors high doping levels, with the bulk optical absorption of the lasing mode, which favors low doping levels (cf. Section 2.1.3.3). The ideal doping profile minimizes the sum of the resistance loss and the bulk optical loss, thus maximizing η_c. A theoretical treatment shows that the expected operating current must be known at the time the doping is specified, because a laser device operating at 100 mA will have a different optimal doping profile than a device operating at 1 A (Kanskar *et al.*, 2005).

The diverse measures discussed in this section to optimize power conversion efficiency and optical output power are not restricted to laser bars, but are in principle general and hence applicable to any semiconductor diode laser.

Using the above improvement steps of η_d, V_d, and R_s in bars 1 cm long with 1 mm cavity length individual emitters, record-high η_c values of 73% and 71% at T_h of 10 °C and 25 °C, respectively, could be obtained. The diode laser bar with a lasing wavelength near 970 nm operated with a high slope efficiency of 1.14 W/A, a threshold current of 4 A, and a turn-on voltage of 1.3 V (Kanskar *et al.*, 2005).

Within the last decade tremendous progress has been achieved in increasing dramatically the cw output power of standard 1 cm laser array bars from 100 W per bar to heroic powers of 1000 W. We select some characteristic data from the vast amount of published material available to demonstrate the advances.

High-brightness, single-mode array laser bars consisting of 200 single-mode stripe emitters 5 μm wide on 50 μm centers (fill factor ∼10%) produce >115 W cw at 140 A with a power conversion efficiency of ≅ 52%. The bars are mounted junction-side down using AuSn solder onto expansion-matched submounts, which are in turn mounted on water-cooled microchannel coolers. The devices emitting in the 9xx nm wavelength band have threshold currents ≲5 A and show excellent beam quality as demonstrated by the smooth Gaussian far-field patterns with divergence angles of 8° FWHM (same for single emitter) and 24° FWHM in the slow- and fast-axis, respectively (Lichtenstein *et al.*, 2004). High optical powers of 325 W cw at 420 A and 16 °C have been achieved in the 9xx nm wavelength band from InGaAs/AlGaAs SQW GRIN-SCH 1 cm laser bars with 50% fill factors and cavities 3.6 mm long by leveraging the enhanced thermal properties of the long cavities on standard microchannel coolers. High-power conversion efficiencies of 62% can be ascribed to the low internal losses of ∼1 cm^{-1} and the low thermal and electrical resistance values of the laser chip (Lichtenstein *et al.*, 2005).

New chip, packaging, and cooling technologies have led to first output powers beyond the 500 W barrier with 509 W cw at 540 A from 1 cm InGaAs/AlGaAs bars with a 50% filling factor and 3 mm cavity length. The high power is due to a low-loss waveguide structure with a low fast-axis far-field angle of 27°, high-power conversion efficiency >68%, optimized facet coatings, and efficient active cooling. Quasi-cw operation with 100 µs pulses and 0.1% duty cycle show the first record-high 1000 W output from the 980 nm bar material (Jenoptik Diode Lab, 2006; Sebastian *et al.*, 2007; Schröder *et al.*, 2007).

Optimized 980 nm laser bar structures with low internal loss $\alpha_i < 0.6$ cm^{-1}, high internal quantum efficiency $\eta_i > 98\%$, high $T_0 \cong 200$ K and $T_\eta \cong 700$ K enable first record-high cw output powers of 950 W at 1120 A (supply limit) from a microchannel cooled 1 cm bar (Li *et al.*, 2007). The bar consists of 65 emitters, each 115 µm wide and 5 mm long, has a 77% filling factor, and performs with a conversion efficiency η_c of 70% and 67% for 4 mm and 5 mm cavity lengths, respectively.

Far-field beam divergence angles for laser bars in the transverse lateral (slow-axis) and transverse vertical (fast-axis) direction are typically in the range of 5 to 10° FWHM and 20 to 35° FWHM, respectively, depending on the technology and layout of the bar used.

References

Adams, M. J. (1981). *An Introduction to Optical Waveguides*, John Wiley & Sons, Inc., New York.

Agrawal, G. P. and Dutta, N. K. (1993). *Semiconductor Lasers*, 2nd edn, Van Nostrand Reinhold, New York.

Al-Muhanna, A., Wade, J. K., Earles, T., Lopez, J., and Mawst, L. J. (1998a). *Appl. Phys. Lett.*, **73**, 2869.

Al-Muhanna, A., Mawst, L. J., Botez, D., Garbuzov, D. Z., Martinelli, R. V., and Connolly, J. C. (1998b). *Appl. Phys. Lett.*, **73**, 1182.

Aoki, M., Komori, M., Sato, H., Tsuchiya, T., Taike, A., Takahashi, M., Uomi, K., and Tsuji, S. (1997). *IEEE J. Sel. Top. Quantum Electron.*, **3**, 1405.

Arakawa, Y. and Sakaki, H. (1982). *Appl. Phys. Lett.*, **40**, 939.

As, D. J., Epperlein, P. W., and Mooney, P. M. (1988). *J. Appl. Phys.*, **64**, 2408.

Asonen, H., Ovtchinnikov, A., Zhang, G., Näppi, J., Savolainen, P., and Pessa, M. (1994). *IEEE J. Quantum Electron.*, **QE-30**, 415.

Asryan, L. V. and Luryi, S. (2001). *IEEE J. Quantum Electron.*, **QE-37**, 905.

Asryan, L. V. and Luryi, S. (2002). *Proc. SPIE*, **4656**, 59.

Baird, B. W., DeFreez, R. K., Evans, G. A., and Kirk, J. B. (1997). *Proceedings of the Conference on Lasers and Electro-Optics, CLEO'97, OSA Tech. Dig.*, 230.

Bhattacharya, P. (2000). *Opt. Quantum Electron.*, **32**, 211.

Bhattacharya, P. and Ghosh, S. (2002). *Appl. Phys. Lett.*, **80**, 3482.

Bhattacharya, P., Zhang, X., Yuan, Y., Kamath, K., Klotzkin, D., Caneau, C., and Bhat, R. (1998). *Proc. SPIE*, **3283**, 702.

Bhattacharya, P., Ghosh, S., and Stiff-Roberts, A. D. (2004). *Annu. Rev. Mater. Res.*, **34**, 1.

Biellak, S. A., Fanning, C. G., Sun, Y., Wong, S. S., and Siegman, A. E. (1997). *IEEE J. Quantum Electron.*, **QE-33**, 219.

Bisping, D., Pucicki, D., Fischer, M., Koeth, J., Zimmermann, C., Weinmann, P., Hoefling, S., Kamp, M., and Forchel, A. (2009). *IEEE J. Sel. Top. Quantum Electron.*, **15**, 968.

Bissessur, H., Graver, C., Le Gouezigou, O., Michaud, G., and Gaborit, F. (1998). *IEEE Photonics Technol. Lett.*, **10**, 1235.

Borchert, B., Egorov, A. Y., Illek, S., and Riechert, H. (2000). *IEEE Photonics Technol. Lett.*, **12**, 597.

Borruell, L., Odriozola, H., Tijero, J., Esquivias, I., Sujecki, S., and Larkins, E. (2008). *Opt. Quantum Electron.*, **40**, 175.

Botez, D. (1981). *IEEE J. Quantum Electron.*, **QE-17**, 178.

Botez, D. (1994). Chapter 1 in *Diode Laser Arrays* (eds. Botez, D. and Scifres, D.), Cambridge University Press, Cambridge.

Botez, D. (1999a). *Proc. SPIE*, **3628**, 2.

Botez, D. (1999b). *Appl. Phys. Lett.*, **74**, 3102.

Botez, D. and Mawst, L. J. (1996). *IEEE Circuits Dev. Mag.*, **12**, 25.

Botez, D., Hayashida, P., Mawst, L. J., and Roth, T. J. (1988). *Appl. Phys. Lett.*, **53**, 1366.

Botez, D., Mawst, L. J., Bhattacharya, A., Lopez, J., Li, J., Kuech, T. F., Iakovlev, V. P., Suruceanu, G. I., Caliman, A., and Syrbu, A. V. (1996). *Electron. Lett.*, **32**, 2012.

Bour, D. P. and Rosen, A. (1989). *J. Appl. Phys.*, **66**, 2813.

Brenner, T., Hess, R., and Melchior, H. (1995). *Electron. Lett.*, **31**, 1443.

Bubke, K., Sorel, M., Robert, F., Bryce, C., Arnold, J., and Marsh, J. (2002). *Proceedings of the IEEE/LEOS 2002 Workshop on Fiber and Optical Passive Components*, 153.

Buda, M., Tan, H. H., Fu, L., Josyula, L., and Jagadish, C. (2003). *IEEE Photonics Technol. Lett.*, **15**, 1686.

Bugge, F., Zeimer, U., Sato, M., Weyers, M., and Tränkle, G. (1998). *J. Cryst. Growth*, **183**, 511.

Buus, J. (1982). *IEEE J. Quantum Electron.*, **QE-18**, 1083.

Casey, Jr., H. C. and Panish, M. B. (1978). *Heterostructure Lasers, Part A: Fundamental Principles and Part B: Materials and Operating Characteristics*, Academic Press, New York.

Cermelli, P. and Jabbour, M. E. (2007). *Phys. Rev.*, **B75**, 165409.

Chan, A. K., Lai, C. P., and Taylor, H. F. (1988). *IEEE J. Quantum Electron.*, **QE-24**, 489.

Chand, N., Becker, E. E., van der Ziel, J. P., Chu, S. N., and Dutta, N. K. (1991). *Appl. Phys. Lett.*, **58**, 1704.

Chand, N., Chu, S. N. G., Dutta, N. K., Lopata, J., Geva, M., Syrbu, A. V., Mereutza, A. Z., and Yakovlev, V. P. (1994). *IEEE J. Quantum Electron.*, **QE-30**, 424.

Chang, J. C., Lee, J. J., Al-Muhanna, A., Mawst, L. J., and Botez, D. (2002). *Appl. Phys. Lett.*, **81**, 4901.

Chen, H. Z., Ghaffari, A., Morkoc, H., and Yariv, A. (1987). *Appl. Phys. Lett.*, **51**, 2094.

Chen, Y. C., Waters, R. G., and Dalby, R. J. (1990). *Electron. Lett.*, **26**, 1348.

Chen, J. R., Wu, Y. C., Lu, T. C., Kuo, H. C., Kuo, Y. K., and Wang, S. C. (2009). *Opt. Rev.*, **16**, 375.

Choi, H. K. and Wang, C. A. (1990). *Appl. Phys. Lett.*, **57**, 321.

Coleman, J. J. (1993). Chapter 8 in *Quantum Well Lasers* (ed. Zory, P. S.), Academic Press, San Diego, CA.

Corbett, B., Justice, J., and Lambkin, P. (2004). *Proc. SPIE*, **5365**, 164.

Crosslight Software Inc. (2010). *Products: LASTIP.* www.crosslight.com.

Crump, P., Patterson, S., Elim, S., Zhang, S., Bougher, M., Patterson, J., Das, S., Dong, W., Grimshaw, M., Wang, J., Wise, D., DeFranza, M., Bell, J., Farmer, J., DeVito, M., Martinsen, R., Kovsh, A., and Toor, F. (2007). *Proc. SPIE*, **6456**, 64560E.

DeFreez, R. K., Bao, Z., Carleson, P. D., and Felisky, M. K. (1993). *Proc. SPIE*, **1850**, 75.

Deichsel, E., Schröder, D., Meusel, J., Hennig, P., Hülsewede, R., Sebastian, J., and Ludwig, S. (2008). *Proc. SPIE*, **6876**, 68760K.

Delépine, S., Gérard, F., Pinquier, A., Fillion, T., Pasquier, J., Locatelli, D., Chardon, J., Bissessur, H., Bouché, N., Boubal, F., and Fapet, P. (2001). *IEEE J. Sel. Top. Quantum Electron.*, **7**, 111.

DeMars, S. D., Dzurko, K. M., Lang, R. J., Welch, D. F., Scifres, D. R., and Hardy, A. (1996). *Proceedings of the Conference on Lasers and Electro-Optics, CLEO'96, OSA Tech. Dig.*, 77.

Demeester, P., Buydens, L., and Van Daele, P. (1990). *Appl. Phys. Lett.*, **57**, 168.

Dente, G. C. (2001). *IEEE J. Quantum Electron.*, **QE-37**, 1650.

Deubert, S., Debusmann, R., Reithmaier, J., and Forchel, A. (2005). *Electron. Lett.*, **41**, 1125.

Diaz, J., Eliashevich, I., Mobarhan, K., Kolev, E., Wang, L., Garbuzov, D., and Razeghi, M. (1994). *IEEE Photonics Technol. Lett.*, **6**, 132.

Diaz, J., Yi, H. J., Razeghi, M., and Burnham, G. T. (1997). *Appl. Phys. Lett.*, **71**, 3042.

Donnelly, J. P., Huang, R. K., Walpole, J. N., Missaggia, L. J., Harris, C. T., Plant, J. J., Bailey, R. J., Mull, D. E., Goodhue, W. D., and Turner, G. W. (2003). *IEEE J. Quantum Electron.*, **QE-39**, 289.

Dzurko, K. M., Hardy, A., Scifres, D. R., Welch, D. F., Waarts, R. G., and Lang, R. J. (1993). *IEEE J. Quantum Electron.*, **QE-29**, 1895.

Emanuel, M. A., Skidmore, J. A., Jansen, M., and Nabiev, R. (1997). *IEEE Photonics Technol. Lett.*, **9**, 1451.

Epler, J. E., Ponce, F. A., Endicott, F. K., and Paoli, T. L. (1988). *J. Appl. Phys.*, **55**, 3439.

Epperlein, P. W. (1997). *Proc. SPIE*, **3001**, 13.

Epperlein, P. W. and Bona, G. L. (1993). *Appl. Phys. Lett.*, **62**, 3074.

Epperlein, P. W. and Meier, H. P. (1990). *Impurity trapping in nominally undoped GaAs/AlGaAs quantum wells*. In *Defect Control in Semiconductors* (ed. Sumino, K.), *Proceedings of the International Conference on Science and Technology of Defect Control in Semiconductors, The Yokohama 21st Century Forum*, Elsevier Science (North-Holland), Amsterdam, **2**, 1223.

Epperlein, P. W., Meier, H. P., Heuberger, W., and Graf, V. (1987). For further information see www.pwe-photonicselectronics-issueresolution.com.

Epperlein, P. W., Parry, M., Helmy, A., Drouot, V., Harrell, R., Moseley, R., Stacey, S., Clark, D., Harker, A., and Shaw, J. (2000). For further information see www.pwe-photonicselectronics-issueresolution.com.

Erbert, G., Bugge, F., Knauer, A., Sebastian, J., Thies, A., Wenzel, H., Weyers, M., and Tränkle, G. (1999). *IEEE J. Sel. Top. Quantum Electron.*, **5**, 780.

Eriksson, N., Larsson, A., Uemukai, M., and Suhara, T. (1998). *IEEE J. Quantum Electron.*, **QE-34**, 858.

Fathpour, S., Mi, Z., and Bhattacharya, P. (2005). *J. Phys. D: Appl. Phys.*, **38**, 2103.

Fiebig, C., Feise, D., Eppich, B., Paschke, K., and Erbert, G. (2009). *IEEE Photonics Technol. Lett.*, **21**, 1755.

Fiebig, C., Eppich, B., Paschke, K., and Erbert, G. (2010). *IEEE Photonics Technol. Lett.*, **22**, 341.

Garbuzov, D. Z., Menna, R. J., Martinelli, R. U., DiMarco, L., Harvey, M. G., and Connolly, J. C. (1996). *Proceedings of the Conference on Lasers and Electro-Optics, CLEO'96, OSA Tech. Dig.*, postdeadline paper CPD10.

Garbuzov, D., Komissarov, A., Kudryashov, I., Maiorov, M., Roff, R., and Connolly, J. (2003). *Proceedings of the Optical Fiber Communication Conference, OFC 2003, OSA Tech. Dig.*, **1**, 394.

Goldberg, L. and Mehuys, D. (1992). *Appl. Phys. Lett.*, **61**, 633.

Gray, J. and Marsh, J. (1996). *Proceedings of the Conference on Lasers and Electro-Optics, CLEO'9, OSA Tech. Dig.*, paper CTuC3, 78.

Gray, J., Marsh, J., and Roberts, J. (1998). *IEEE Photonics Technol. Lett.*, **10**, 328.

Groves, S., Liau, Z., Palmateer, S., and Walpole, J. (1989). *Appl. Phys. Lett.*, **56**, 312.

Guermache, A., Voiriot, V., Locatelli, D., Legrand, F., Capella, R.-M., Gallion, P., and Jacquet, J. (2005). *IEEE Photonics Technol. Lett.*, **17**, 2023.

Guthrie, J., Tan, G. L., Ohkubo, M., Fukushima, T., Ikegami, Y., Ijichi, T., Irikawa, M., Mand, R. S., and Xu, J. M. (1994). *IEEE Photonics Technol. Lett.*, **6**, 1409.

Hamada, K., Wada, M., Shimizu, H., Kume, M., Susa, F., Shibutani, T., Yoshkawa, N., Itoh, K., Kano, G., and Teramoto, I. (1985). *IEEE J. Quantum Electron.*, **QE-21**, 623.

Hanke, C., Korte, L., Acklin, B. D., Luft, J., Grötsch, S., Herrmann, G., Spika, Z., Marciano, M., deOdorico, B., and Wilhelmi, J. (1999). *Proc. SPIE*, **3628**, 64.

Harder, C. S. (2008). Chapter 5 in *Optical Fiber Telecommunications V A: Components and Subsystems* (eds. Kaminow, I. P. et al.), Academic Press, San Diego, CA.

Hata, K., Shigekawa, H., Ueda, T., Akiyama, M., and Okano, T. (1998). *Phys. Rev.*, **B57**, 4500.

Hausser, S., Meier, H., Germann, R., and Harder, C. (1993). *IEEE J. Quantum Electron.*, **QE-29**, 1596.

Hayakawa, T. (1999). *Proc. SPIE*, **3628**, 29.

Henry, C. H. (1982). *IEEE J. Quantum Electron.*, **QE-18**, 259.

Herzog, W. D., Goldberg, B. B., and Ünlü, M. S. (2000). *IEEE Photonics Technol. Lett.*, **12**, 1604.

Hetterich, M., Dawson, M. D., Egorov, A. Y., Bernklau, D., and Riechert, H. (2000). *Appl. Phys. Lett.*, **76**, 1030.

Hirose, S., Yoshida, A., Yamaura, M., Hara, K., and Munekata, H. (1999). *Jpn. J. Appl. Phys.*, **38**, 1516.

Hohimer, J. P., Hadley, G. R., and Owyoung, A. (1988). *Appl. Phys. Lett.*, **52**, 260.

Horie, H., Arai, N., Mitsuishi, Y., Komuro, N., Kaneda, H., Gotoh, H., Usami, M., and Matsushima, Y. (2000). *IEEE Photonics Technol. Lett.*, **12**, 1304.

Hunziker, G. and Harder, C. (1995). *Appl. Opt.*, **34**, 6118.

Ikegami, T. (1972). *IEEE J. Quantum Electron.*, **QE-8**, 470.

Innolume GmbH (2008). *High-Power Laser Products*. www.innolume.com.

Itaya, K., Mondry, M., Floyd, P., Coldren, L., and Merz, J. (1996). *J. Electron. Mater.*, **25**, 565.

Jenoptik Diode Lab, GmbH (2006). *Press release Jan. 2006*. www.jenoptik.com.

Joannopoulos, J. D., Meade, R. D., and Winn, J. N. (1995). *Photonic Crystals: Molding the Flow of Light*. Princeton University Press, Princeton, NJ.

Kaiser, W., Deubert, S., Reithmaier, J., and Forchel, A. (2006). *Proceedings of the Conference on Lasers and Electro-Optics, CLEO 2006, OSA Tech. Dig.*, paper CFG3, 1.

Kaiser, W., Reithmaier, J., Forchel, A., Odriozola, H., and Esquivias, I. (2007). *Appl. Phys. Lett.*, **91**, 051126.

Kanskar, M., Earles, T., Goodnough, T., Stiers, E., Botez, D., and Mawst, L. J. (2005). *Proc. SPIE*, **5738**, 47.

Kardontchik, J. E. (1982). *IEEE J. Quantum Electron.*, **QE-18**, 1279.

Kasu, M. and Kobayashi, N. (1995). *J. Appl. Phys.*, **78**, 3026.

Kasukawa, A., Nishikata, K., Yamanaka, N., Arakawa, S., Iwai, N., Mukaihara, T., and Matsuda, T. (1997). *IEEE J. Sel. Top. Quantum Electron.*, **3**, 1413.

Kintzer, E. S., Walpole, J. N., Chinn, S. R., Wang, C. A., and Missaggia, L. J. (1993). *IEEE Photonics Technol. Lett.*, **5**, 605.

Kiravittaya, S., Rastelli, A., and Schmidt, O. (2006). *Appl. Phys. Lett.*, **88**, 043112.

Kitamura, S., Hatakeyama, H., Hamamoto, K., Sasaki, T., Komatsu, K., and Yamaguchi, M. (1999). *IEEE J. Quantum Electron.*, **QE-35**, 1067.

Kobayashi, H., Yamamoto, T., Ekawa, M., Watanabe, T., Ishikawa, T., Fujii, T., Soda, H., Ogita, S., and Kobayashi, M. (1997). *IEEE J. Sel. Top. Quantum Electron.*, **3**, 1384.

Kondow, M., Uomi, K., Niwa, A., Kitatani, T., Watahiki, S. and Yazawa, Y. (1996). *Jpn. J. Appl. Phys.*, **35**, 1273.

Kondow, M., Kitatani, T., Nakatsuka, S., Larson, M. C., Nakahara, K., Yazawa, Y., Okai, M., and Uomi, K. (1997). *IEEE J. Sel. Top. Quantum Electron.*, **3**, 719.

Kowalski, O., Hamilton, C., McDougall, S., Marsh, J., Bryce, C., De La Rue, R., Vogele, B., Stanley, C., Button, C., and Roberts, J. (1998). *Appl. Phys. Lett.*, **72**, 581.

Kristjánsson, S., Eriksson, N., Modh, P., and Larsson, A. (2000). *IEEE Photonics Technol. Lett.*, **12**, 1319.

Kuttler, M., Strassburg, M., Pohl, U., Bimberg, D., Behringer, M., and Hommel, D. (1998). *Appl. Phys. Lett.*, **73**, 1865.

Laakso, A., Dumitrescu, M., Pietilä, P., Suominen, M., and Pessa, M. (2008). *Opt. Quantum Electron.*, **40**, 853.

Laidig, W., Holonyak, Jr., N., Camras, M., Hess, K., Coleman, J., Dapkus, P., and Bardeen, J. (1981). *Appl. Phys. Lett.*, **38**, 776.

Lang, R. J. (1994). *IEEE J. Quantum Electron.*, **QE-30**, 31.

Lang, R. J., Hardy, A., Parke, R., Mehuys, D., O'Brien, S., Major, J., and Welch, D. (1993). *IEEE J. Quantum Electron.*, **QE-29**, 2044.

Lang, R. J., Mehuys, D., Welch, D. F., and Goldberg, L. (1994). *IEEE J. Quantum Electron.*, **QE-30**, 685.

Lang, R. J., Dzurko, K., Hardy, A. A., DeMars, S., Schoenfelder, A., and Welch, D. F. (1998). *IEEE J. Quantum Electron.*, **QE-34**, 2196.

Ledentsov, N. N., Ustinov, V. M., Egorov, A. Y., Zhukov, A. E., Maximov, M. V., Tabatadze, I. G., and Kop'ev, P. S. (1994). *Semiconductors*, **28**, 832.

Legge, M., Bacher, G., Bader, S., Forchel, A., Lugauer, H. J., Waag, A., and Landwehr, G. (2000). *IEEE Photonics Technol. Lett.*, **12**, 236.

Li, H., Towe, T., Chyr, I., Brown, D., Nguyen, T., Reinhardt, F., Jin, X., Srinivasan, R., Berube, M., Truchan, T., Bullock, R., and Harrison, J. (2007). *IEEE Photonics Technol. Lett.*, **19**, 960.

Li, L., Liu, G., Li, Z., Li, M., Li, H., Wang, X., and Wan, C. (2008). *IEEE Photonics Technol. Lett.*, **20**, 566.

Lichtenstein, N., Manz, Y., Mauron, P., Fily, A., Arlt, S., Thies, A., Schmidt, B., Muller, J., Pawlik, S., Sverdlov, B., and Harder, C. (2004). *Proceedings of the IEEE 19th International Semiconductor Laser Conference, Conf. Dig.*, 45.

Lichtenstein, N., Manz, Y., Mauron, P., Fily, A., Schmidt, B., Mueller, J., Arlt, S., Weiss, S., Thies, A., Troger, J., and Harder, C. (2005). *Proc. SPIE*, **5711**, 1.

Lichtenstein, N., Fily, A. C., and Reid, B. (2006). *US Patent* No. 7085299.

Lin, G., Yen, S. T., Lee, C. P., and Liu, D. C. (1996). *IEEE Photonics Technol. Lett.*, **8**, 1588.

Lindsey, C. P. and Yariv, A. (1988). *US Patent* No. 4791646.

Litchinitser, N. and Iakhnine, V. (2010). *Optical Waveguides: Numerical Modeling.* http://optical-waveguides-modeling.net/index.jsp.

Liu, C. Y., Yoon, S. F., Cao, Q., Tong, C. Z., and Li, H. F. (2007). *Appl. Phys. Lett.*, **90**, 041103.

Livshits, D. A., Egorov, A. Y., and Riechert, H. (2000). *Electron. Lett.*, **36**, 1381.

Lopata, J., Vakhshoori, D., Hobson, W. S., Han, H., Henein, G. E., Wynn, J. D., deJong, J., Schnoes, M. L., and Zydzik, G. J. (1996). *IEEE Lasers and Electro-Optics Conference, Proceedings of LEOS'96*, **1**, 346, and *Electron. Lett.*, **32**, 1007.

Marcatili, E. A. J. (1974). *Bell. Syst. Tech. J.*, **53**, 645.

Marcuse, D. (1991). *Theory of Dielectric Optical Waveguides*, Academic Press, New York.

Matuschek, N., Pliska, T., Troger, J., Mohrdiek, S., and Schmidt, B. (2006). *Proc. SPIE*, **6184**, 618402.

Mawst, L. J., Botez, D., Zmudzinski, C., and Tu, C. (1992a). *IEEE Photonics Technol. Lett.*, **4**, 1204.

Mawst, L. J., Botez, D., Zmudsinski, C., Jansen, M., Tu, C., Roth, T. J., and Yun, J. (1992b). *Appl. Phys. Lett.*, **60**, 668.

Mawst, L. J., Bhattacharaya, A., Nesnidal, M., Lopez, J., Botez, D., Morris, J. A., and Zory, P. (1995). *Appl. Phys. Lett.*, **67**, 2901.

Mawst, L. J., Rusli, S., Al-Muhanna, A., and Wade, J. K. (1999). *IEEE J. Sel. Top. Quantum Electron.*, **5**, 780.

Maximov, M. V., Shernyakov, Y. M., Novikov, I. I., Kuznetsov, S. M., Karachinsky, L. Y., Gordeev, N. Y., Kalosha, V. P., Shchukin, V. A., and Ledentsov, N. N. (2005). *IEEE J. Quantum Electron.*, **41**, 1341.

Maximov, M. V., Shernyakov, Y. M., Novikov, I. I., Karachinsky, L. Y., Gordeev, N. Y., Ben-Ami, U., Bortman-Arbiv, D., Sharon, A., Shchukin, V. A., Ledentsov, N. N., Kettler, T., Posilovic, K., and Bimberg, D. (2008). *IEEE J. Sel. Top. Quantum Electron.*, **14**, 1113.

McCarthy, N. and Champagne, Y. (1989). *J. Appl. Phys.*, **67**, 3192.

McIlroy, P., Kurobe, A., and Uematsu, Y. (1985). *IEEE J. Quantum Electron.*, **QE-21**, 1985.

McKenan, S., Carter, C. B., Bour, D. P., and Shealy, J. R. (1988). *J. Mater. Res.*, **3**, 406.

Mehuys, D. G. (1999). Chapter 4 in *Semiconductor Lasers II: Materials and Structures* (ed. Kapon, E.), Academic Press, San Diego, CA.

Mi, Z. and Bhattacharya, P. (2005). *J. Appl. Phys.*, **98**, 023510.

Mi, Z., Bhattacharya, P., and Fathpour, S. (2005). *Appl. Phys. Lett.*, **86**, 153109.

Mikhrin, S. S., Kovsh, A. R., Krestnikov, I. L., Kozhukhov, A. V., Livshits, D. A., Ledentsov, N. N., Shernyakov, Y. M., Novikov, I. I., Maximov, M. V., Ustinov, V. M., and Alferov, Z. I. (2005). *Semicond. Sci. Technol.*, **20**, 340.

Mikulla, M., Chazan, P., Schmitt, A., Morgott, S., Wetzel, A., Walther, M., Kiefer, R., Pletschen, W., Braunstein, J., and Weimann, G. (1998). *IEEE Photonics Technol. Lett.*, **10**, 654.

Mikulla, M., Schmitt, A., Walther, M., Kiefer, R., Moritz, R., Mueller, S., Sah, R. E., Braunstein, J., and Weimann, G. (1999). *Proc. SPIE*, **3628**, 80.

Miyashita, M., Shima, A., Katoh, M., Sakamoto, Y., Ono, K., and Yagi, T. (2000). *Proc. SPIE*, **3947**, 72.

Murakami, T., Ohtaki, K., Matsubara, H., Yamawaki, T., Saito, H., Isshiki, K., Kokubo, Y., Shima, A., Kumabe, H., and Susaki, W. (1987). *IEEE J. Quantum Electron.*, **QE-23**, 712.

Nagashima, Y., Onuki, S., Shimose, Y., Yamada, A., and Kikugawa, T. (2004). *Proceedings of the IEEE 19th International Semiconductor Laser Conference, Conf. Dig.*, 47.

Namegaya, T., Matsumoto, N., Yamanaka, N., Iwai, N., Nakayama, H., and Kasukawa, A. (1994). *IEEE J. Quantum Electron.*, **QE-30**, 578.

O'Brien, S., Lang, R., Parke, R., Major, J., Welch, D. F., and Mehuys, D. (1997). *IEEE Photonics Technol. Lett.*, **9**, 440.

Oclaro Inc. (2010). *High-Power Diode Laser Products. LC96A1070-20R.* www.oclaro.com.

Odriozola, H., Tijero, J., Borruel, L., Esquivias, I., Wenzel, H., Dittmar, F., Paschke, K., Sumpf, B., and Erbert, G. (2009). *IEEE J. Quantum Electron.*, **QE-45**, 42.

Oeda, Y., Fujimoto, T., Yamada, Y., Shibuya, H., and Muro, Y. (1998). *CLEO'98, OSA Tech. Dig. Ser.*, 10.

Olson, J., Ahrenkiel, R., Dunlavy, D., Keyes, B., and Kibbler, A. (1989). *Appl. Phys. Lett.*, **55**, 1208.

Ongstad, A. P., Dente, G. C., Tilton, M. L., Chavez, J. C., Kaspi, R., and Gianardi, D. M. (2010). *CLEO/QELS 2010, OSA Tech. Dig. Ser.*, paper CTUE6.

Orsila, S. L., Toivonen, M., Savolainen, P., Vilokkinen, V., Melanen, P., Pessa, M., Saarinen, M. J., Uusimaa, P., Fang, F., Jansen, M., and Nabiev, R. (1999). *Proc. SPIE*, **3628**, 203.

Ostendorf, R., Kaufel, G., Moritz, R., Mikulla, M., Ambacher, O., Kelemen, M., and Gilly, J. (2008). *Proc. SPIE*, **6876**, 68760H.

Park, G., Shchekin, O. B., Huffaker, D. L., and Deppe, D. G. (2000). *IEEE Photonics Technol. Lett.*, **12**, 230.

Paschke, K., Sumpf, B., Dittmar, F., Erbert, G., Staske, R., Wenzel, H., and Tränkle, G. (2005). *IEEE J. Sel. Top. Quantum Electron.*, **11**, 1223.

Paschke, K., Fiebig, C., Feise, D., Fricke, J., Kaspari, C., Blume, G., Wenzel, H., and Erbert, G. (2008). *Proceedings of the IEEE 21st International Semiconductor Laser Conference, Conf. Dig.*, 131.

Patanè, A., Polimeni, A., Eaves, L., Henini, M., Main, P. C., Smowton, P. M., Johnston, E. J., Hulyer, P. J., Herrmann, E., Lewis, G. M., and Hill, G. (2000). *J. Appl. Phys.*, **97**, 1943.

Paxton, A. H., Schaus, C. F., and Srinivasan, S. T. (1993). *IEEE J. Quantum Electron.*, **QE-29**, 2784.

Pawlik, S., Traut, S., Thies, A., Sverdlov, B., and Schmidt, B. (2002). *Proceedings of the IEEE 18th International Semiconductor Laser Conference, Conf. Dig.*, 163.

Pepper, D. M. and Craig, R. R. (1989). *US Patent* No. 4803696.

Peters, M., Rossin, V., and Acklin, B. (2005). *Proc. SPIE*, **5711**, 142.

Piprek, J., Abraham, P., and Bowers, J. E. (2000). *IEEE J. Quantum Electron.*, **QE-36**, 366.

Qiu, B., McDougall, S. D., Liu, X., Bacchin, G., and Marsh, J. (2005). *IEEE J. Quantum Electron.*, **41**, 1124.

Riechert, H., Egorov, A. Y., Borchert, B., and Illek, S. (2000). *Compound Semicond.*, **6**, 71.

Roberts, J. S., David, J. P., Smith, L., and Tihanyi, P. L. (1998). *J. Cryst. Growth*, **195**, 668.

Rode, D. L., Wagner, W. R., and Schumaker, N. E. (1977). *Appl. Phys. Lett.*, **30**, 75.

Saito, S., Hattori, Y., Sugai, M., Harada, Y., Jongil, H. and Nunoue, S. (2008). Proceedings of the *IEEE 21st International Semiconductor Laser Conference, Conf. Dig.*, 185.

Schemmann, M. F., van der Poel, C. J., van Bakel, B. A., Ambrosius, H. P., Valster, A., van den Heijkant, J. A., and Acket, G. A. (1995). *Appl. Phys. Lett.*, **66**, 920.

Schröder, D., Meusel, J., Hennig, P., Lorenzen, D., Schröder, M., Hülsewede, R., and Sebastian, J. (2007). *Proc. SPIE*, **6456**, 64560N.

Sebastian, J., Schulze, H., Hülsewede, R., Hennig, P., Meusel, J., Schröder, M., Schröder, D., and Lorenzen, D. (2007). *Proc. SPIE*, **6456**, 64560F.

Shan, W., Walukiewicz, W., Ager, J. W., and Haller, E. E. (1999). *Phys. Rev. Lett.*, **82**, 1221.

Shigihara, K., Nagai, Y., Karadida, S., Takami, A., Kokubo, Y., Matsubara, H., and Kakimoto, S. (1991). *IEEE J. Quantum Electron.*, **27**, 1537.

Shigihara, K., Kawasaki, K., Yoshida, Y., Yamamura, S., Yagi, T., and Omura, E. (2002). *IEEE J. Quantum Electron.*, **38**, 1081.

Sony Shiroishi Semiconductor Inc. (2010). *optics.org News & Analysis.* http://optics.org/news/1/4/26.

Srinivasan, S. T., Schaus, C. F., Sun, S. Z., Armour, E. A., Hersee, S. D., McInerney, J. G., Paxton, A. H., and Gallant, D. J. (1992). *Appl. Phys. Lett.*, **61**, 1272.

Stelmakh, N., (2007). *IEEE Photonics Technol. Lett.*, **19**, 1392.

Stranski, I. N. and Krastanow, L. (1937). *Sitzungsber. Akad. Wiss. Wien*, **146**, 797.

Streifer, W., Burnham, R. D., and Scifres, D. R. (1976). *IEEE J. Quantum Electron.*, **QE-12**, 177.

Sun, G., Khurgin, J. B., and Soref, R. A. (2004). *IEEE Photonics Technol. Lett.*, **16**, 2203.

Sun, Y., Biellak, S. A., Fanning, G., and Siegman, A. E. (1993). *IEEE Lasers and Electro-Optics Conference, Proceedings of LEOS'93*, **1**, 600.

Swaminathan, V., Chand, N., Geva, M., Anthony, P. J., and Jordan, A. S. (1992). *J. Appl. Phys.*, **72**, 4648.

Swint, R. B., Yeoh, T. S., Elarde, V. C., Coleman, J. J., and Zediker, M. S. (2004). *IEEE Photonics Technol. Lett.*, **16**, 12.

Tamburrini, M., Goldberg, L., and Mehuys, D. (1992). *Appl. Phys. Lett.*, **60**, 1292.

Tan, G. L., Mand, R. S., and Xu, J. M. (1997). *IEEE J. Quantum Electron.*, **QE-33**, 1384.

Tan, G. L., Sargent, E. H., and Xu, J. M. (1998). *IEEE J. Quantum Electron.*, **QE-34**, 353.

Tanguy, Y., Voignier, V., O'Neill, E., McInerney, J. G., Huyet, G., and Corbett, B. (2003). *IEEE Photonics Technol. Lett.*, **15**, 637.

The MathWorks, Inc. (1997). *Products: MATLAB*. www.mathworks.com.

Thornton, R. L., Burnham, R. D., Paoli, T. L., Holonyak, N., and Deppe, D. G. (1985). *Appl. Phys. Lett.*, **47**, 1239.

Tihanyi, P. L., Jain, F. C., Robinson, M. J., Dixon, J. E., Williams, J. E., Meehan, K., O'Neill, M. S., Heath, L. S., and Beyea, D. M. (1994). *IEEE Photonics Technol. Lett.*, **6**, 775.

Tilton, M. L., Dente, G. C., Paxton, A. H., Cser, J., DeFreez, R. K., Moeller, C. E., and Depatie, D. (1991). *IEEE J. Quantum Electron.*, **QE-27**, 2098.

Valster, A., Meney, A. T., Downes, J. R., Faux, D. A., Adams, A. R., Brouwer, A. A., and Corbijn, A. J. (1997). *IEEE J. Sel. Top. Quantum Electron.*, **3**, 180.

Van der Poel, C. J., Schemmann, M. F., and Acket, G. A. (1994). *Proceedings of the IEEE 14th International Semiconductor Laser Conference, Conf. Dig.*, 241.

Vawter, G. A., Sullivan, C. T., Wendt, J. R., Smith, R. E., Hou, H. Q., and Klem, J. F. (1997). *IEEE J. Sel. Top. Quantum Electron.*, **3**, 1361.

Walpole, J. N., Kintzer, E. S., Chinn, S. R., Wang, C. A., and Missaggia, L. J. (1992). *Appl. Phys. Lett.*, **61**, 740.

Walpole, J. N., Donnelly, J. P., Missaggia, L. J., Liau, Z. L., Chinn, S. R., Groves, S. H., Taylor, P. J., and Wright., M. W. (2000). *IEEE Photonics Technol. Lett.*, **12**, 257.

Wang, M., Hwang, D., Lin, P., Dechiaro, L., Zah, C., Ovadia, S., Lee, T., and Darby, D. (1994). *Appl. Phys. Lett.*, **64**, 3145.

Watanabe, M., Tani, K., Sasaki, K., Nakatsu, H., Hosada, M., Matsui, S., Yamamoto, O., and Yamamoto, S. (1994). *Proceedings of the IEEE 14th International Semiconductor Laser Conference, Conf. Dig.*, 251.

Weisbuch, C., and Vinter, B. (1991). *Quantum Semiconductor Structures: Fundamentals and Applications*, Academic Press, New York.

Welch, D. F., Scifres, D. R., Cross, P. S., and Streifer, W. (1987). *Appl. Phys. Lett.*, **51**, 1401.

Welch, D. F., Parke, R., Mehuys, D., Hardy, A., Lang, R., O'Brien, S., and Scifres, D. (1992). *Electron. Lett.*, **28**, 2011.

Welch, D. F. and Mehuys, D. G. (1994). Chapter 2 in *Diode Laser Arrays* (eds. Botez, D. and Scifres, D. R.), Cambridge University Press, Cambridge.

Wenzel, H., Paschke, K., Brox, O., Bugge, F., Fricke, J., Ginolas, A., Knauer, A., Ressel, P., and Erbert, G. (2007). *IEEE European Conference on Lasers and Electro-Optics and International Quantum Electronics Conference*. doi: CLEOE-IQEC.2007.4385978.

Wu, C. H., Zory, P. S., and Emanuel, M. A. (1995). *IEEE Photonics Technol. Lett.*, **7**, 718.

Wu, M. C., Chen, Y. K., Hong, M., Mannaerts, J. P., Chin, M. A., and Sergant, M. A. (1991). *Appl. Phys. Lett.*, **59**, 1046.

Xu, M. L., Tan, G. L., Clayton, R., and Xu, J. M. (1996). *IEEE Photonics Technol. Lett.*, **8**, 1444.

Yablonovitch, E. (1987). *Phys. Rev. Lett.*, **58**, 2059.

Yamada, Y., Okubo, A., Oeda, Y., Yamada, Y., Fujimoto, T., and Muro, K. (1999). *Proc. SPIE*, **3626**, 231.

Yang, G., Smith, G. M., Davis, M. K., Kussmaul, A., Loeber, D. A., Hu, M. H., Nguyen, H. K., Zah, C. E., and Bhat, R. (2004). *IEEE Photonics Technol. Lett.*, **16**, 981.

Yang, H., Mawst, L. J., Nesnidal, M., Lopez, J., Bhattacharya, A., and Botez, D. (1996). *IEEE Lasers and Electro-Optics Conference, Proceedings of. LEOS'96*, paper MM3.

Yang, H., Nesnidal, M., Al-Muhanna, A., Mawst, L. J., Botez, D., Vang, T. A., Alvarez, F. D., and Johnson, R. (1998). *IEEE Photonics Technol. Lett.*, **10**, 1079.

Yellen, S., Shepard, A., Dalby, R., Baumann, A., Serreze, H., Guido, T., Soltz, R., Bystrom, K., Harding, C., and Waters, R. (1993). *IEEE J. Quantum Electron.*, **29**, 2058.

Yen, S. T. and Lee, C. P. (1996a). *IEEE J. Quantum Electron.*, **32**, 1588.

Yen, S. T. and Lee, C. P. (1996b). *IEEE J. Quantum Electron.*, **32**, 4.

Yuda, M., Hirono, T., Kozen, A., and Amano, C. (2004). *IEEE J. Quantum Electron.*, **40**, 1203.

Zah, C., Bhat, R., Pathak, B., Favire, F., Lin, W., Wang, M., Andreadakis, N., Hwang, D., Koza, A., Lee, T., Wang, Z., Darby, D., Flanders, D., and Hsieh, J. (1994). *IEEE J. Quantum Electron.*, **30**, 511.

Zhang, G., Ovtchinnikov, A., Näppi, J., Asonen, H., and Pessa, M. (1993). *IEEE J. Quantum Electron.*, **29**, 1943.

Ziari, M., Verdiell, J. M., and Welch, D. F. (1995). *IEEE Lasers and Electro-Optics Conference, Proceeedings of LEOS'95*, **1**, 5.

Zmudzinski, C., Botez, D., and Mawst, L. J. (1993). *Appl. Phys. Lett.*, **62**, 2914.

Zmudzinski, C., Botez, D., Mawst, L. J., Bhattacharya, A., Nesnidal, M., and Nabiev, R. F. (1995). *IEEE J. Sel. Top. Quantum Electron.*, **1**, 129.

Part II

DIODE LASER RELIABILITY

Overview

Responsible for the significant improvement of the high-power performance of diode lasers in the past years has also been a remarkable progress in the optical strength of semiconductor materials and device structures. Optical strength is the capability to resist optical damage during laser operation. Optical damage phenomena can happen in the bulk of the laser cavity and above all at the susceptible laser mirror facets. They belong to the group of sudden laser degradation modes and depend strongly on the optical emission density, laser material system, structure, type, and concentration of defects in the laser cavity, at interfaces, and at mirror surfaces. The power limits at catastrophic optical damage (COD) events could be increased significantly by various technological concepts and approaches, which will be discussed in detail in this part.

To illustrate the progress, record-high, continuous wave (cw) critical power densities to COD at mirror facets of >100 MW/cm^2 have been reported for the demanding single-mode, edge-emitting 980 nm GaAs-based pump diode lasers (Lichtenstein *et al.*, 2004). These devices delivered COMD-free output powers >1.7 W within a near-field area of $\sim 3~\mu\text{m} \times 0.6~\mu\text{m}$ at FWHM $\cong 2 \times 10^{-8}$ cm^2 assuming a uniform optical filling of the near-field spot.

Excellent laser reliability has been usually associated with InP-based materials mainly because of their relatively high insensitivity to rapid degradation processes due to nonradiative recombination enhanced defect motion effects, such as dislocation climb and/or glide, which, on the other hand, can occur more easily in GaAs-based materials. However, recent advances in laser wafer growth and processing have resulted in dramatically improved reliability of GaAs-based compounds.

The most striking lifetimes have been reported for commercial 980 nm single-mode InGaAs/AlGaAs pump laser sources, key elements for Er-doped fiber amplifiers, imposing the most stringent requirements on utmost kink-free optical output powers and reliability for their demanding deployment in optical communications networks, in particular in submarine environments. Accelerated long-term life tests have resulted in random failure rates of <80 FITs at 200 mW and, 25 °C to a 95% upper confidence limit, which is equivalent to $<2\%$ cumulative failures in 27 years, whereas wear-out lifetimes can be larger than 10 million hours (Epperlein *et al.*, 2001, unpublished). FIT is one of the failure rate unit definitions (see Section 5.1.7) and stands for failure in time defined by failures per hour times 10^9). These figures are very much in line with the fact that no intrinsic failures were reported by leading pump laser manufacturers for field-deployed components over an accumulated operating time of more than 10 billion device hours of actual use.

Before a laser product can be put on the market, it has to be demonstrated that operation of the laser will meet the agreed specifications under certain conditions for a specified period of time without failure. Reliability is expressed as the probability the laser has this ability, and failure is defined at the time when it loses this ability.

Fundamental reliability engineering terms, concepts, techniques, and a laser reliability test plan will be discussed in detail in the last two chapters, Chapters 5 and 6, of Part II, which consists of four chapters. This will include also the definition of reliability specifications, product failures, and a risk assessment program. Critical laser parameters for long-term stability and reliability will be determined as well as suitable test equipment, sample sizes, and test durations defined. At the core of Chapter 6 is the discussion of specifications, conditions, and results of the well-established test procedures for laser chip, subcomponent, and module qualifications. In the final section of Chapter 6, the benefits of a reliability engineering and growth program will be discussed, which includes a reliability cost model. Chapter 3 discusses the various basic diode laser degradation phenomena and modes. It also gives an account of stability conditions for critical laser parameters and discusses factors relevant for achieving strong laser robustness. Chapter 4 deals with optical strength engineering and discusses the various approaches and techniques to enhance the optical strength of diode lasers.

Chapter 3

Basic diode laser degradation modes

Main Chapter Topics   Page

Introduction

This chapter starts with a discussion of possible causes leading to a degradation of critical diode laser parameters. We then define criteria that are necessary to ensure a high and long-term stability of these parameters. The bulk of this chapter deals with the classification of diode laser degradation modes and with a description of the typical features and physical mechanisms of these degradation modes. We discuss rapid, gradual, and sudden degradation mechanisms along with potential procedures for eliminating laser failures in these individual categories. Degradation phenomena will be classified also by the location and relevant degradation modes in each category discussed. Finally, key factors that determine laser robustness will be elaborated.

Semiconductor Laser Engineering, Reliability and Diagnostics: A Practical Approach to High Power and Single Mode Devices, First Edition. Peter W. Epperlein.
© 2013 John Wiley & Sons, Ltd. Published 2013 by John Wiley & Sons, Ltd.

3.1 Degradation and stability criteria of critical diode laser characteristics

In the following, we discuss the conditions of some crucial electrical and optical parameters to facilitate stable operation of diode lasers under constant current or constant optical output power conditions. Usually the optical output power decreases for a diode laser degrading during constant current operation, in contrast to an increase in operating current in case the laser degrades during constant power operation. In general, these changes are primarily caused by a decrease in injected carrier lifetimes and an increase in internal optical losses. These root-cause parameters directly affect the optical output power P versus current I characteristics by increasing the threshold current and decreasing the slope efficiency or external differential quantum efficiency.

3.1.1 Optical power; threshold; efficiency; and transverse modes

The main mechanisms, which can affect these performance parameters in different strengths include degradation of the:

- active region;
- mirror facets;
- lateral confinement; and
- Ohmic contacts.

Therefore all provisions have to be taken to prevent them from occurring.

3.1.1.1 Active region degradation

Operation of diode lasers at high injection current densities creates high-energy carriers and thermal gradients, which have the potential to generate strain fields and high nonradiative recombination rates inside the active gain block. These factors can promote the precipitation of host atoms as well as the motion, multiplication of isolated defects into clusters, and the growth of dislocations, which can significantly impact the performance of lasers by increasing the threshold current and lowering the efficiency and output power level. Thermally assisted degradation can be due to defect migration into the active layer, which is manifested by the generation of so-called dark-line defects (DLDs). The activation energies E_a of these degradation processes are in the range of ~ 0.2 to 1.0 eV depending on the type of active nonradiative defect center involved. The time to laser failure t_f due to thermally activated degradation is given by

$$t_f = t_0 \exp\{E_a / k_B T\} \tag{3.1}$$

within the Arrhenius model (see Chapter 5 for details) where t_0 is a constant depending on the laser structure and k_B is the Boltzmann constant. The DLD structures are regions of greatly reduced radiative efficiency linked to a dislocation network, which grows by the nonradiative recombination-enhanced defect motion mechanism. We will describe this defect type in more detail in one of the following sections. Additional nonradiative traps can migrate into the active region by a current density effect. This is due to the electric field created by the current density, which lowers the effective binding energy of many impurities. The process is not temperature dependent and activation energies cannot be assigned.

To minimize temperature-driven degradation effects, it is crucial to lower the laser operating temperature by lowering factors such as internal absorption losses, series resistance, and thermal resistance. This will contribute to achieving not only a high stability in kink-free and single-mode operation up to high power, but also a high reliability and long lifetime of the laser chip.

3.1.1.2 Mirror facet degradation

Mirror facet degradation is a serious issue, particularly in Al-containing lasers and lasers emitting visible light. At the high optical densities at the facets in excess of some tens of megawatts per square centimeter of single-mode high-power lasers, facet degradation can be triggered by optical absorption of the laser light at the facets. There are two modes of degradation: first, the oxidation at the active region on the mirror facet through a photo-assisted reaction; and, second, the catastrophic optical mirror damage (COMD), which shows a sudden degradation due to meltdown of the facet. The reason for this is that the free surface of the facet is an imperfect crystal lattice with many dangling bonds, which attract impurities. These sites form various types of defects that cause excess optical absorption. At a sufficiently high optical density, the localized heating can be large enough to destroy the facet.

In Section 3.2, we will discuss in detail the key processes leading to facet oxidation and especially to COMD. Section 4.1 deals with facet surface properties acting as microscopic origins for COMD. In Part III, we present measured local facet temperatures and in particular the critical temperature at the onset of COMD.

In addition, we will discuss the multiple experimental relationships between COMD threshold level, facet temperature increase, and atomic defect type and concentration in the laser facet surface responsible for the COMD process.

3.1.1.3 Lateral confinement degradation

In Sections 1.3.5.2 and 2.1.4, we discussed various forms of lateral confinement structures. Current-restrictive layers are used to achieve an effective injection of current to the active region, which is required to obtain low threshold currents and high output powers. For example, index-guided buried heterostructures use reverse-biased junctions and stripe lasers utilize ridge-type structures or quantum well intermixed (QWI) regions for confining the current in the transverse lateral direction to the active layer.

During laser operation or accelerated aging tests, defects may be created in the current-confining junctions resulting in the degradation of current confinement and increase in leakage current flowing outside the active region, which leads to an increased threshold current and reduced external differential quantum efficiency. A soft turn-on in the I/V characteristic is usually an indicator for the presence of a resistive shunt path leading to the leakage current.

Similarly, in ridge lasers, any defect formation in the etched p-waveguide layer outside the ridge structure has to be suppressed to prevent any change in composition and conductivity of the waveguide, which could lead to an increased leakage current. The lasing mode in a ridge laser partially overlaps with the dielectric ridge embedding layer and thus introduces the necessary effective lateral index step. Therefore, great care has to be taken to prevent any compositional change and delaminating effect of this dielectric layer, which would otherwise lead to a significant degradation of threshold current, efficiency, and output power.

In contrast, diode lasers employing QWI regions for transverse lateral waveguiding (see Section 2.1.4.2 and Figure 2.7) (Welch *et al.*, 1987; Kuttler *et al.*, 1998; Andrew *et al.*, 1992) and bandgap tuned active regions (Nagai *et al.*, 1995; Noël *et al.*, 1996) have proven to show high reliability and no degradation in threshold current and output power compared to the as-grown nonintermixed structures during normal operation or accelerated aging conditions. In these devices, QWI has been achieved (i) for tuning the bandgap of the active layer by a Zn diffusion process in GaAs/AlGaAs QW lasers (Nagai *et al.*, 1995) and phosphorus ion implantation with subsequent SiO_2 cap annealing of InGaAsP/InP QW lasers (Noël *et al.*, 1996); and (ii) for lateral waveguiding by silicon implantation-induced disordering in InGaAs/GaAs GRIN-SCH QW structures (Welch *et al.*, 1987).

In particular, the latter technology has found a dramatic industrial application in high-power, highly reliable, single-mode 980 nm pump laser products (JDSU Inc., 2010). Parameters controlling the ion-induced QWI process, such as ion doses, fluxes, energies, and post-implantation annealing time and temperature, have to be optimized to achieve high integrity of the QWI regions and dislocation-free crystal quality of strained and unstrained material systems required for long-term stability, robustness, and compatibility with high-temperature processes.

3.1.1.4 Ohmic contact degradation

For establishing Ohmic contacts, usually a Schottky-type electrode with its inert interface between the metal and semiconductor is used at the device side, which is closer to the active layer, and an alloy-type electrode on the substrate side farther away from the active region. The corresponding contact resistances R_c are closely related to the potential barrier height ϕ_b between the metal and semiconductor (difference between work functions of metal and semiconductor) by

$$R_{c,Sch} \propto \exp\left\{\phi_b/\sqrt{N}\right\} \tag{3.2}$$

and

$$R_{c,alloy} \propto \exp\{\phi_b/k_B T\} \tag{3.3}$$

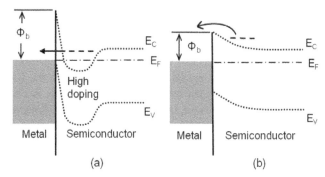

Figure 3.1   Schematic band structure representation of basic concept for Ohmic contacts in n-type semiconductors: (a) Schottky-type: high doping concentration, tunneling; (b) alloy-type: low barrier height, thermionic emission.

for Schottky-type and alloy-type contacts, respectively (Sze, 1981), where N is the dopant concentration in the semiconductor and T the temperature.

Equation (3.2) describes the metal/semiconductor contact for high dopings ($>10^{19}$ cm^{-3}) and dominating tunneling processes. It shows that, in the tunneling range, the contact resistance depends strongly on doping concentration and varies exponentially with the factor $\phi_b/N^{1/2}$. Figure 3.1a illustrates the basic concept of the energy band diagram for a Schottky-type Ohmic contact in n-type semiconductors where the depletion layer width (tunneling barrier thickness)

$$d_{m/s} \propto \sqrt{\phi_b/N} \tag{3.4}$$

decreases with increasing dopings. Typical Schottky-type contacts are made of Ti/Pt/Au in which Ti is ~20 nm thick and serves as an adhesion layer, Pt ~150 nm thick is used as a diffusion barrier for Au, and Au ~200 nm thick is required for low electrode resistance and easy bonding with Au wire. The diffusion of Au into the semiconductor has to be prevented, because Au can form nonradiative recombination centers and carrier traps in the active layer under operation. This process is facilitated by the high treatment temperature during bonding of the laser chip onto a heat sink with solder. The Au-related defects decrease laser efficiency and reliability. The strength of the diffusion barrier depends not only on a sufficiently thick layer, but also on the crystalline fine structure of Pt, which is sensitively dependent on its deposition technique and process parameters. For example, Pt is one of the major contributors to stress in the overall laser structure and the stress level in the Pt layers doubles for a deposition carried out at a 90° angle compared to one at 45° (see Part III). Stress in the electrode layer stack has to be minimized to prevent wafer warping and stress-induced laser degradation. Ti/Pt/Au deposited on GaAs produces a tensile stress at the interface of the metallization of typically $+1.3 \times 10^{10}$ dyn/cm^2, whereas GaAs is under compressive stress, leaving the electrode layer/GaAs with a concave/convex curvature of the outer surfaces (conversion: 1 dyn/cm^2 = 0.1 Pa = 0.1 N/m^2).

Under normal lasing operating conditions, Schottky-type Ohmic contacts are quite stable due to the inert interface. However, at extreme operating conditions of >200 °C and >30 kA/cm^2, degradation can develop due to the formation of a depletion layer of the group-III elements at the interface. This is caused by out-diffusion of group-III elements leading to a high electrical resistance of the depletion layer. Due to the higher resistance of the metal/group-V element alloy layer, a preferential formation of metal/group-III element alloy spikes with lower resistance can occur, which can finally lead to damage of the active layer (Fukuda, 1999). For a standard Schottky barrier, the barrier height is determined primarily by the character of the metal and the metal/semiconductor interface properties, and to a lesser extent by the doping. This means that the semiconductor surface has to be free of any contaminants such as oxides and residues from processing steps such as etching.

Equation (3.3) is for a metal/semiconductor contact with lower doping concentration and thermionic emission dominating the current transport for a low barrier height ϕ_b structure (Figure 3.1b). The task is to develop a low-energy barrier height at the interface, which can be achieved by using a noble metal such as Cu, Ag, or Au, which is very electronegative. These metals easily react with the electropositive group-III elements (such as Ga and In) of compound semiconductors having bandgap energies $\lesssim 2.5$ eV. They form alloy-type electrodes by sintering or alloying, resulting finally in both a reduced depletion layer width and a reduced barrier height ϕ_b. Typical alloy-type electrodes are AuGeNi and AuZnNi for n-type and p-type semiconductors, respectively.

Alloy-type electrodes are susceptible to degradation under practical laser operating conditions. During operation, alloy layers of the metal/group-V elements and metal/group-III elements are formed where the former layers have a higher electrical resistance than the latter ones. This forces the current to flow in the metal/group-III element alloy layer with the consequence that alloy spikes can grow under the current flow, which can finally lead to degradation of the active region. For example, in the case of a Au electrode on InP, Au$_3$In and Au$_2$P$_3$ clusters are formed at the interface with increasing temperature and time. These clusters finally grow inhomogeneously where Au$_2$P$_3$ has a larger electrical resistance than Au$_3$In (Fukuda, 1999). Therefore, appropriate conditions need to be developed to achieve alloy-type electrodes with low resistance and high stability. The presence of Zn favors the formation of an alloy layer with low electrical resistance.

The series resistance of a diode laser is a crucial parameter, because it determines the power dissipation via Joule heating (see Equation 2.36). Minimizing the series resistance is mandatory for maximizing the power conversion efficiency and minimizing the junction temperature (see Equation 2.37). This leads to high efficiency, low threshold current, and long operating life. The contact resistance can be a significant contributor to the total series resistance. In this context, we measured the specific contact resistivity of p-GaAs with Ti/Pt/Au metallurgy as a function of the p-doping

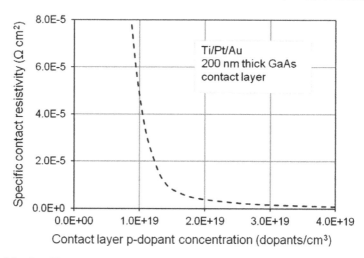

Figure 3.2 Specific contact resistivity of a Ti/Pt/Au metallization layer on a p-doped GaAs layer 200 nm thick measured as a function of the p-doping concentration.

concentration in a p-GaAs contact layer 200 nm thick (Figure 3.2). Typically, Zn and C were used as p-dopants, and the films were e-beam deposited and furnace-annealed at 360 °C for 15 min (Epperlein *et al.*, 2000, unpublished). The contact resistance depends sensitively not only on the doping level (Figure 3.2), but also on the contact layer thickness. It decreases typically by one order of magnitude for an increase in thickness from 50 to 500 nm.

The figure clearly shows a dramatically rapid increase in the specific contact resistivity for a p-doping concentration $<2 \times 10^{19}$ cm^{-3}, whereas above this value it is negligibly small. The specific contact resistivity is defined as $\varrho_{co} = (\partial V / \partial J)_{V=0}$ and has the units Ω cm^2, J is the current density, and V the voltage drop across the contact area. Contact resistance occurs as the current flows normal to and through the interface between the two layers. Taking a typical effective contact area on top of a ridge waveguide of $A_{co} = 2$ μm × 750 μm $= 1.5 \times 10^{-5}$ cm^2, the contact resistance $R_{co} = \varrho_{co}/A_{co}$ can be derived from the measured ϱ_{co} values. For example, at a doping level of 1×10^{19} cm^{-3}, the specific contact resistivity is $\sim 5 \times 10^{-5}$ Ω cm^2, which gives a contact resistance of 3.3 Ω; similarly for 3×10^{19} cm^{-3}, the resistivity is already very low with $\sim 1.5 \times 10^{-6}$ Ω cm^2 leading to a negligible contact resistance of ~ 0.1 Ω. In this high-doping regime, the tunneling process will dominate. A too low doping density by design or processing, or any small removal of a layer at the surface of the highly doped semiconductor contact layer, can have a devastating effect on the contact resistance. It would be desirable to obtain an accurate measure of the effective doping profile near the top side of the contact layer, for example, by performing a secondary ion mass spectrometry (SIMS) depth profile (Runyan and Shaffner, 1998) of the dopant concentration in the contact layer before depositing the p-metallization.

3.1.2 Lasing wavelength and longitudinal modes

Changes in lasing wavelength can be caused by several factors with the most relevant ones including:

- changes in the transition energy or bandgap energy due to band filling;

- changes in the refractive index caused by the injected carrier density or the so-called plasma effect;

- changes in the refractive index caused by an increased temperature due to Joule heating at increased injected currents; and

- changes in the effective bandgap energy caused, for example, by a degrading strained active QW layer structure or by increased temperature due to Joule heating at higher injected currents.

The first two effects in the above list are nearly constant after the start of lasing because the carrier density injected into the active layer is clamped and becomes nearly constant above threshold. Depending on the operating conditions, the wavelength can change. During normal operation or life tests under constant optical output power, the lasing wavelength can increase because the drive current increases to keep the output power constant. Thus, the Joule heating increases, which causes the bandgap energy and refractive index to change with increasing temperature. However, the wavelength tends to decrease during stress tests under constant current operation due to the band filling and plasma effects that are intensified if the threshold carrier density increases during degradation. High optical cavity losses can cause a decrease in the wavelength primarily due to a stronger band filling effect as a consequence of the correspondingly higher injected threshold carrier density.

In addition, defects can act as *saturable absorbers* with the potential to induce *self-sustained pulsations* (see Section 1.3.5.2) that strongly impact the transmission quality of optical communication systems. Such oscillations can be observed at frequencies around 1 GHz in lasers under cw operation. In simple terms, these pulsations may be caused by the absorption of photons at defect sites in the active region. This will create more carriers, more recombinations, and more gain up to the point where the enhanced stimulated emission reduces the gain, leading again to an increase in absorption and a repetition of the whole cycle.

Furthermore, spectral broadening of longitudinal modes caused by a reduced quality factor of the laser cavity due to any laser device degradation process can lead to issues as well, particularly in wavelength division multiplexing communication systems, because the transmission speed in a fiber is different for the different wavelengths involved.

Spectral broadening is enhanced when the laser is modulated, which causes *chirping*, a phenomenon that leads to a rapid change in the center wavelength of the emitted laser light during modulation. Briefly, this is caused by the abrupt change in the carrier flux density and temperature in the laser cavity causing changes in the refractive index. In general, chirp in Fabry–Pérot (FP) lasers is dominated by changes in the energy gap, whereas in distributed feedback (DFB) lasers changes in refractive index are the major source. Chirp in semiconductor diode lasers increases the lasing wavelength (so-called "downward" chirp means a decrease of instantaneous frequency with time). For further details on the chirp effect and its reduction, the reader is referred to textbooks such as those by Agrawal and Dutta (1993), Fukuda (1999), and Carroll et al. (1998).

To keep the wavelength changes to a minimum, it is mandatory to develop low threshold current density laser devices and to establish the best possible thermal management, which includes low internal optical losses, low electrical series resistances, low thermal resistances, and very effective heat sinks.

An optical element can be added to control the temperature dependence of the laser's spectral properties. For example, fiber Bragg gratings (FBGs) (Ventrudo and Rogers, 1996) or bulk gratings (Volodin et al., 2004) forcing the laser to emit at a single wavelength can lead to a significant improvement in both wavelength temperature stability and spectral width.

A FBG can be created by inscribing periodic or aperiodic variations of refractive index into the photosensitized core of an optical fiber using intense illumination with an ultraviolet laser, usually in a holographic arrangement. This generates structural changes and thus a permanent modification of the index where a weak index modulation of only 10^{-4} can already be very effective if the grating is sufficiently long in the order of some millimeters. The FBG is usually placed in the pigtail of a fiber-coupled laser chip.

A volume grating is similar in that it is implemented externally to the laser chip, but instead of the fiber, bulk collimating optical elements are used between the laser output and grating. To circumvent the complexity and delicacy in aligning this external cavity grating mirror structure, the grating can be added to the standard diode laser structure. It is evident that the grating acts only on the longitudinal modes, and the laser can still oscillate in multiple transverse modes. The grating can be defined holographically and integrated at wafer level well outside the active region but close enough to interact with the tail of the beam propagating in the cavity (Williamson, 2007). This location of the grating is different to that in a DFB laser where the grating is usually placed at the edge of the waveguide layer.

These techniques reduce the wavelength drift with temperature from typically 0.3 nm/K for a normal diode laser to ∼0.01 nm/K for a grating-stabilized laser depending on the material system and grating type and configuration. Up to ten-fold improvements in spectral width have been observed in stabilized diode lasers (Williamson, 2007; Volodin et al., 2004).

A very important mode has to be considered in the laser–FBG (external grating) configuration. The laser can be a regular, unmodified FP diode laser. If the FBG is positioned far enough away from the output mirror, the laser light loses its coherence outside its coherence length. This is not strictly an external cavity laser, but rather a regular FP laser operated with strong external feedback. Reflection from the FBG injects a very large signal at the FBG resonant wavelength into the laser cavity and competes with the cavity modes. High-power lasers with their usually low front-facet reflectivities are particularly susceptible to external feedback effects. And, indeed, in these cases this reflection dominates the cavity modes. This has the effect of broadening the laser linewidth, but it locks the wavelength to the FBG so that multiline operation and *mode hopping* are suppressed. These external feedback effects can drive the laser in the high-power regime into a bistable state between single-mode and *coherence collapse* states (Tkach and Chraplyvy, 1986; Davis *et al.*, 2005). While the laser device experiences an increase in high-frequency noise, low-speed operation is very significantly enhanced, because instabilities associated with mode hopping are removed.

3.2 Classification of degradation modes

This section describes the different degradation modes of a diode laser that can be observed during normal operation or accelerated aging. There are several types of degradation modes in relation to the initial period of degradation, degradation rate, and degree of degradation and which can be grouped by the location of degradation. In the following, we give a classification of these degradation modes by type and location, and describe their specific characteristics and techniques to reduce or eliminate them.

3.2.1 Classification of degradation phenomena by location

Degradation phenomena can be classified by the location of degradation: external degradations occur outside the laser crystal, whereas internal degradations occur only inside the laser crystal. In the first category are degradations of the mirror, contact electrode, and chip solder. The second category includes degradations of the active layer including the p–n junction.

The description of degradation modes in the following two subsections is only meant to be brief within this overview. Full details on every degradation mode can be found in Section 3.1 above and following Section 3.2.2.

3.2.1.1 External degradation

Mirror degradation

This degradation comprises mirror oxidation and COMD.

Mirror oxidation occurs with a much higher rate in 0.85 μm high-Al content AlGaAs/GaAs systems than in 1.3 μm and 1.5 μm band InGaAsP/InP lasers where the oxidation phenomenon is not conspicuously observed due to a very slow oxidation rate. The long-term stability and lifetime of the laser are affected by facet oxidation. The thickness of the oxide layer is proportional to the square root of operating time, and depends on the output power energy, power density, moisture, and composition of the active layer. The oxidation process occurs even in an atmosphere with a low partial pressure of oxygen of around 10^{-6} Torr and its rate increases with increasing Al content. Oxygen in the mirror surface has been clearly detected by Auger electron spectroscopy (Kajimura, 1980). An alternative spectrochemical analysis technique is energy-dispersive x-ray (EDX). Both techniques are performed by exciting the sample with the electron beam using scanning electron microscopy (SEM). X-ray production and Auger electron production are, however, competing processes; that is, the excited atom will relax either by emission of an x-ray or an Auger electron, with the x-ray emission more probable for elements of high atomic number.

With respect to the facet oxidation mechanism, it is enhanced effectively by the formation of photoinduced carriers resulting in breakage of the bonds at the facet. Defects can form at the interface between oxide and semiconductor with the potential to enable nonradiative carrier recombination and thus generate facet heating. This in combination with other processes, including strong optical absorption of emitted laser light at the facet and large surface currents due to the large concentrations of native traps at the cleaved facet surface, can trigger a positive feedback loop. The loop comprises in the sequence the processes of optical absorption, carrier generation, nonradiative recombination, facet heating, bandgap reduction, and back to the starting point with an even stronger optical absorption. This process can ultimately lead to thermal runaway and thus COMD and meltdown of the mirror facet. (See Sections 3.2.2.2 and 3.2.2.3 for details on degradation mode features and degradation elimination. See Section 4.1 for facet surface properties acting as microscopic origins for mirror degradation.)

Contact degradation

Contact degradation is caused by diffusion of the electrode metal and by the alloy between the electrode metal and the semiconductor during laser operation at high currents associated with Joule heating of the contact region. The degraded region is outside the active region where an alloy reaction occurred, but high currents and strong heating can drive the alloy reaction toward the active region. There are two ways of establishing an Ohmic contact: by Schottky-type and alloy-type electrodes. In Section 3.1.1.4, we discussed in detail the degradation of both Ohmic contact types in the context of investigating the stability criteria of critical diode laser parameters.

According to these discussions, alloy-type electrodes are susceptible to degradation under practical laser operating conditions. During operation, alloy layers of the metal/group-V elements and metal/group-III elements are formed where the former layers have a higher electrical resistance than the latter ones (Piotrowska *et al.*, 1981). This forces the current to flow in the metal/group-III element alloy layer with the consequence that alloy spikes can form under the current flow, which can finally lead to degradation of the active region. In contrast, Schottky-type contacts are quite stable under practical operating conditions because of the inert metal/semiconductor interface. However, degradation can occur under severe operating conditions. A depletion layer of the group-III elements can form at the interface caused by out-diffusion of group-III elements. The electrical resistance of the depletion layer is high, and therefore the current tends to concentrate in that part of the layer consisting of metal/group-III element alloy with lower electrical resistance. Thus, alloy spikes can grow and ultimately damage the active layer. Moreover, in another reaction at a Schottky-type electrode, grain boundary diffusion of the group-III element into the electrode metal can occur and thereby cause an increase in thermal and electrical resistance during long-term operation (Fukuda *et al.*, 1988).

Solder degradation

At the interface between metal and solder, effects such as metal diffusion, intermetallics formation, and thermal fatigue can occur. These effects depend on the materials used for the electrode and solder. They can influence the laser performance and reliability because of the resulting changes in electrical and thermal characteristics. Thermal resistance can increase due to thermal fatigue, which results from crack formation at the bonding solder metal under thermal cycle stresses or power cycling. This is more likely to happen with soft solders having a low melting point such as Sn, In, and Sn-rich AuSn than with hard solders including Au-rich AuSn or AuSi. Although soft solders act initially as absorbers of mechanical stress, they degrade gradually during long-term aging, however.

In cases where the current flow path is not homogeneous, voids, hillocks, and whiskers can form at the bonding site. Voids form in regions where the current density and the mass flux of the metal ions change from low to high, whereas hillocks and ultimately whiskers grow in regions where the current density and mass flux change from high to low. Apart from the current density, variations in temperature, diffusion coefficient of metal ions, and metal layer resistivity are further driving forces for the mass transport due to *electromigration*. Whiskers and reactions at the semiconductor/solder interface can cause local short circuits in the active p–n junction with sudden damage to the laser. Examples are the growth of In whiskers up to 2 μm long at high current densities or Sn whiskers generated from soft Sn-rich AuSn (Mizuishi, 1984) and penetration of Sn from the AuSn solder into the semiconductor (Mizuishi *et al.*, 1982).

Another type of void formation can occur at the bonding site where interdiffusion between solder material such as Sn and electrode material such as Au can take place, ultimately forming SnAu intermetallics. Voids can form at the interface in the Sn layer due to a Sn grain boundary diffusion rate that is higher than that of bulk diffusion of Au into Sn (Nakahara and McCoy, 1980). Voids of this type are called Kirkendall voids (Smigelskas and Kirkendall, 1947). Grain boundary diffusion is a surface-activated process and takes place at the surfaces of the grains of the Au layer, whereas bulk diffusion is a direct bulk-to-bulk process. Intermetallic and void formations increase the thermal and electrical resistance.

The use of hard solders having a high melting point reduces or eliminates the instability and reliability issues induced by the use of soft solder metals, which, however, are easier to use in the fabrication process.

3.2.1.2 Internal degradation

Active region degradation and junction degradation

The majority of degradation phenomena occur in the active region. Detailed accounts on the various relevant failure mechanisms are given in Section 3.1.1.1 above and in the Sections 3.2.2.1 and 3.2.2.2 below dealing with rapid degradation and gradual degradation of the active region. We want to refer to these sections for insightful information on the specific features and root causes of these individual degradation modes as well as approaches and techniques for suppressing or eliminating them.

The structure and fabrication process of buried-heterostructure (BH) lasers (see Section 1.3.5.2 and Figure 1.27) introduce new degradation modes that cannot be observed in planar-type, non-BH lasers. The usual BH structure is formed by a two-step crystal growth process involving mesa etching after the first growth step down to the n-cladding including the active layer, and then burying the active region in a regrowth process in the second growth step. The BH degradation is induced by the increase in defects at the defective BH interface under current injection. The increased defect density leads to an increase in nonradiative recombination current. The injected carrier lifetime decreases and thus increases the threshold current. Cross-sectional electron beam-induced current (EBIC) (see details of the technique in Part III) images can show local breakdown of the p–n junction accompanied by leakage current (Mizuishi *et al.*, 1983). The reason for this degradation is the formation of deep-level defects at the p–n junction leading to nonradiative recombination. This degradation mode can be suppressed by achieving a defect-free BH interface, for example, by an appropriate treatment of the etched sidewalls of the active layer before regrowth.

3.2.2 Basic degradation mechanisms

Figure 3.3 shows a schematic diagram of the major failure modes at constant current operation of the laser device. Based on the rate of change in device characteristics, in

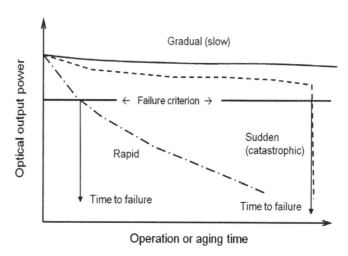

Figure 3.3 Schematic representation of typical degradation modes and time-to-failure condition of diode lasers operated under constant current.

this case in optical output power, we can roughly categorize them into rapid, gradual, and sudden modes. At constant current operation a decrease in optical output is observed during degradation. In contrast, if a laser is operated at a constant optical output power the drive current is increased during laser degradation in order to maintain constant output power. If rapid degradation is eliminated, there is still the gradual degradation that occurs over a long period and which determines ultimately the lifetime of the laser. A device can also degrade suddenly after an initial rapid or gradual course of degradation, which usually occurs in a catastrophic way at laser mirror surfaces (see Sections 3.1.1.2, 3.2.1.1, 3.2.2.3, and 4.1) or in the bulk of the cavity at certain defects. These defects can include point defects such as vacancies and interstitials, or line defects such as dislocations, or planar and 3D defects such as dislocation loops, point defect clusters, precipitates, and voids. The formation of these defects can occur during nonoptimized epitaxial growth and device fabrication processes, but some of the defects might also be created during high-power operation of the laser device itself.

The figure also shows schematically the level of degraded power that the laser has to reach to become a failure. The definition of such failure criteria is dependent on the requirements of the laser application and determines the individual time to failure. A typical failure criterion for diode lasers is a 10% drop in initial optical output power. As we will see in Chapters 5 and 6, by life testing a large ensemble of laser devices, we can get statistical data on both the median time to failure at which 50% of the devices tested failed and the failure rate.

3.2.2.1 Rapid degradation

Features and causes of rapid degradation

- The degradation process occurs in the active region, which causes the internal absorption loss to increase and the injected carrier lifetime to be shortened, because of an increase in nonradiative recombination (cf. Section 3.1.1.1 above). The result is a rapid decrease in external differential quantum efficiency and optical output power and an increase in threshold current. Typical lifetimes $\lesssim 100$ h.

- Main causes are the generation and growth of dislocations, and sometimes precipitate-like defects with an excess of host atoms such as In and P for InP-based devices. The latter can occur during epitaxy by thermal decomposition of InP prior to the growth of an InP buffer layer. It can also occur by thermal decomposition at the interface between InP and the InGaAsP active layer. These root causes give rise to the formation of DLDs, which are extended, linear regions of greatly reduced radiative efficiency. They can grow in the $\langle 100 \rangle$ and $\langle 110 \rangle$ directions when observed perpendicular to the (001) substrate, which is usually the active region plane. The $\langle 100 \rangle$ DLD crosses an active stripe oriented along the $\langle 110 \rangle$ direction at 45° (Figure 3.4). These nonradiative regions, acting also as light absorbers, can be observed, for example, in electroluminescence

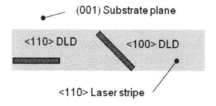

Figure 3.4   Schematic representation of typical dark-line defects (DLDs) oriented in the ⟨100⟩ and ⟨110⟩ directions on a (001) substrate plane with the laser stripe in the ⟨110⟩ direction.

(EL) topographs, EBIC images, and photoluminescence (PL) images. These techniques will be described in Chapters 7 to 9.

- The ⟨100⟩ dislocation network grows by the so-called recombination-enhanced dislocation climbing (REDC) motion from a dislocation that originally existed in the active region. In the *phonon-kick model* (Gold and Weisberg, 1964), the energy required for defect reaction is emitted in a nonradiative recombination process between electronic states in the dislocation or other defects such as point defects. It is then transformed into lattice vibrational states (phonons) via multi-phonon generation at defect sites (Henry and Lang, 1977) giving rise to the climb process. In this model, the defects are mainly interstitial atoms or vacancies. By absorbing interstitial atoms at a dislocation (Petroff and Kimerling, 1976) or emitting vacancies from a dislocation (O'Hara *et al.*, 1977), the dislocation network grows and extends along the ⟨100⟩ direction (Figure 3.5).

- The dominant parameters for REDC include (Ueda, 1996): (i) deep levels associated with point defects or dangling bonds in dislocation cores; (ii) activation energy for point defect generation and migration; and (iii) nonradiative recombination rate at defects. DLD growth is intrinsic to the material. It may be influenced by effects such as dopant type and density and migration of metal atoms from the electrode.

- In materials with wider bandgaps, REDC and therefore dislocation network growth occur more easily. It exists in GaAs, AlGaAs, GaP, GaAsP, Ga-rich InGaAsP lattice matched to GaAs, but not in InP and In-rich InGaAsP lattice matched to InP. The last two experimental results are in agreement with calculations made on the deep trapping levels of anion vacancies in the $In_{1-y}Ga_yAs_{1-x}P_x$ alloy system (Buisson *et al.*, 1982; Dow and Allen, 1982). According to these calculations, vacancy levels are located outside the bandgap in In-rich InGaAsP on InP, whereas vacancy levels are located deep within the bandgap in Ga-rich InGaAsP on GaAs. This confirms the ease of REDC and DLD formation in the latter system and demonstrates that rapid degradation due to DLD formation cannot occur that easily in the former material.

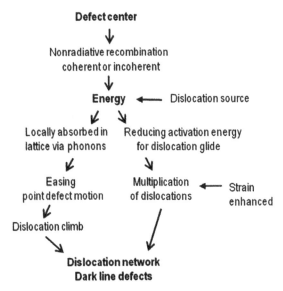

Figure 3.5   Simplified formation process of dislocation networks in a semiconductor diode laser leading to dark-line defects.

- The growth velocity of $\langle 100 \rangle$ DLD in degraded GaAs/AlGaAs lasers is also much higher than that in degraded InGaAsP/InP lasers: at room temperature, it is ~ 100 μm/h (Imai *et al.*, 1979) compared to ~ 0.5 μm/h (Fukuda *et al.*, 1983), respectively.

- The dislocation network for $\langle 110 \rangle$ DLD is caused by mechanical stress. Its growth rate depends on the magnitude of stress and on the bonding strength of the host atoms. In III–V compound semiconductors, the slip plane under mechanical stress is the (111) plane and the projection of this plane onto the (001) plane is the $\langle 110 \rangle$ direction (Fukuda, 1999). If the mechanical stress is $\gtrsim 10^8$ dyn/cm^2, the slip dislocation grows from the surface or interface under current injection and, when it reaches the active layer, $\langle 110 \rangle$ DLDs are formed. The dislocations may be introduced initially by relaxation of the stress via recombination-enhanced dislocation glide (REDG) motion during operation (Figure 3.5).

- Rapid degradation caused by the generation of $\langle 110 \rangle$ DLDs due to REDG occurs in AlGaAs/GaAs devices. However, in InGaAsP/InP lasers no $\langle 110 \rangle$ DLDs are observed in the active region during operation at room temperature.

- Driving forces for REDG include (Ueda, 1996): (i) local temperature rise at the active p–n junction; (ii) background strain field due to factors such as lattice mismatch in the active region, formation of contact electrodes, dielectric layer for current confinement; and (iii) nonradiative recombination energy and rate.

Elimination of rapid degradation

- Use low dislocation density or dislocation-free substrates that are commercially available to avoid propagation of threading dislocations into the active layer (see Section 1.4.1.1).

- Avoid crystal growth-induced defects, particularly dislocation-type defects at the interface between substrate and epitaxial layers. The elimination of $\langle 100 \rangle$ DLDs is related to the quality of the epitaxial growth process of the structure.

- Minimize internal stress in the laser structure, in particular at the interface between the active waveguide and cladding layers to reduce strain-enhanced formation of REDC and REDG leading to DLDs. In particular, the elimination of $\langle 110 \rangle$ DLDs is related to the processing steps after the epitaxial layer growth, which includes minimizing mechanical stress.

- Add Al (AlAs mole fraction $\cong 0.05$) to the active GaAs layer of AlGaAs/GaAs lasers, which reduces the defect density by gettering oxygen and changes the stress type in the active layer from tension to compression (Olsen and Ettenberg, 1977). This drastic change in stress is correlated with improved lifetimes up to two orders of magnitude (Thompson, 1979) through a reduced DLD concentration (Kishino *et al.*, 1976).

- Eliminate defects induced during fabrication processes such as diffusion, dielectric and metal depositions, ion implantation, and annealing by optimizing the conditions for these processes.

- Avoid stress and mechanical damage during laser device fabrication: (i) elastic strain due to difference in thermal expansion between epitaxial layers and contact metallization and dielectrics; (ii) external stress due to bonding and soldering processes and laser wafer cleaving; (iii) scratches and cracks (source of dislocations) induced during laser device handling.

3.2.2.2   Gradual degradation

Features and causes of gradual degradation

- Gradual and slow long-term decrease of optical output power and increase of threshold current with operating or aging time. The gradual degradation mode determines the maximum lifetime, that is, over 30 years for modern high-power single-mode diode lasers at specified operating conditions.

- Degradation rate tends to increase as the Al content is increased in the active layer of AlGaAs/GaAs lasers to much higher AlAs mole concentrations than recommended to reduce the DLD density (see Section 3.2.2.1 above). Increased stress due to mismatch between substrate and active layer may be a root cause, but also as the AlAs mole fraction in the active layer approaches that of an indirect bandgap AlGaAs (~ 0.45).

- Degradation is enhanced strongly by internal stress in AlGaAs/GaAs visible lasers caused by precipitate-like microdefects (clusters) and microdislocation loops (Ueda, 1996).

- Degradation not due to dislocation network growth, formation of dark-spot defects (DSDs) with reduced light emission is possible in InGaAsP/InP devices at high temperatures due to a recombination-enhanced growth process at localized defects. DSDs lack the linear form of DLDs and can be linked to low-quality epitaxial growth. Ga- and As-rich regions in the active layer of InGaAsP/InP lasers are correlated with the location of DSDs that have activation energies of $\sim$0.16–0.2 eV (Fukuda *et al.*, 1983). However, Yamakoshi *et al.* (1979) report much higher activation energies of $\sim$1.2 eV that suggest that, due to details in device fabrication, different mechanisms of DSD generation may be active in the last two device structures.

- Oxidation of facets unprotected from the atmosphere after cleavage and enhanced by emitted light. Formation of point defects at the interface between oxide film and semiconductor (see Section 3.2.1.1). These defects can act as nonradiative recombination centers and hence contribute to the long-term degradation.

- Slow degradation is dominated by an increase in the concentration of nonradiative deep-level defect centers in the active region measured by deep-level transient spectroscopy (details of the DLTS technique will be given in Part III). Trap (with deep energy levels in the bandgap of 0.24–0.88 eV) concentrations ($\lesssim 10^{17}$ cm^{-3}) in AlGaAs/GaAs lasers correlate with threshold current increasing (up to 100%) with accelerated aging time (Uji *et al.*, 1980; Epperlein, 1987, unpublished). Formation of point defect clusters and possibly microloops by defect generation and condensation effects. Deep levels related to single impurities may not be associated with gradual degradation.

Elimination of gradual degradation

- Suppress the introduction of deep-level defects and traps during crystal growth and device fabrication processes, because point defect complexes in as-grown epitaxial layers may enhance gradual degradation.

- Reduce the concentration of residual impurities during epitaxial growth. This suppresses the generation of dislocation loops by condensation of excess point defects such as vacancies or interstitials through nonradiative recombinations of minority carriers at nucleation centers, for example, residual impurities.

- Suppress facet oxidation by cleaving the laser chip in an appropriate environment (ultrahigh vacuum or inert gas) and deposit without delay both suitable passivation layers to saturate dangling bonds and dielectric coatings to achieve the required facet robustness and reflectivity (see Chapter 4 and Section 1.3.8).

- Eliminate internal and external stress around the active region, which itself can be a pseudomorphic strained-layer QW structure without forming misfit dislocations through relaxation of strain as long as the well layer thickness is below a certain critical value. Apart from such strained thin layer systems, good lattice-matching in the laser structure is mandatory during crystal growth.

- Eliminate migration of electrode metals such as Au in Au-based alloy-type Ohmic contacts, which forms deep-level defects and therefore can cause long-term degradation due to the increase of nonradiative recombination in the active region. Use instead Schottky-type electrodes, where Ti/Pt/Au is an excellent electrode material for p-contacts; Pt acts as a diffusion barrier for Au (see Section 3.1.1.4 above).

3.2.2.3 Sudden degradation

Features and causes of sudden degradation

- Catastrophic optical damage (COD) takes places suddenly during operation due to a current surge and is concomitant with high optical power density.

- COD occurs mainly at laser mirror surfaces (COMD) but also at defects in the bulk of the laser resonator. For laser operation below COD threshold, degradation is accelerated by both temperature and operating current. Thermal acceleration follows the usual Arrhenius law, characterized by an activation energy E_a with a lifetime $t_f \propto \exp(E_a/k_B T)$ (see Section 3.1.1.1 above and Chapter 5). Drive current and optical power acceleration follows a power law characterized by an exponent n, and the time to failure can be written as

$$t_f \propto J^{-n} \quad \text{or} \quad t_f \propto P^{-n} \tag{3.5}$$

where J is the injected current density and P the light output power density, and n is typically in the wide range of ~ 1 to 3 depending on the specific laser type and properties (see Chapters 5 and 6).

- COD events are more likely to occur in wider bandgap material systems such as AlGaAs/GaAs and InGaAsP/InGaP on n-GaAs than in lower bandgap materials such as InGaAsP/InP on n-InP substrates with typical COMD power density levels $\gtrsim 10$ MW/cm^2 and $\gtrsim 100$ MW/cm^2, respectively (Ueda, 1996). COD is observed in the bulk of InGaAsP/InP DH lasers with DLDs in random directions but there is no observation of COMD events.

- COMD is more likely to occur in laser materials showing stronger nonradiative recombination-enhanced defect motion (reaction), which is directly linked to increased nonradiative recombination and higher facet temperatures (see Section 3.2.2.1 above). In Part III, we describe measurements of mirror temperatures as a function of optical output power in diode lasers of different material systems and mirror technologies.

- COMD power level decreases in inverse proportion to the square root of pulse width in a pulsed operation (Eliseev, 1973, 1996).

- COMD power level decreases with increasing active layer thickness and number of active QWs in GaInP/AlGaInP ridge lasers 5 μm wide (Epperlein, 1997), which is consistent with the measured facet temperature rise increasing with the QW number (Epperlein and Bona, 1993).

- Degradation of QW lasers is very similar to that of bulk DH lasers made of the same materials. Degradation rate, however, is lower in QW lasers due to the smaller optical confinement (larger mode size in transverse vertical direction; see Section 2.1.3.5), which leads to a higher power threshold level to COMD in QW lasers compared to DH lasers with approximately the same power density at COMD in both laser types.

- Compressively strained-layer QW lasers show enhanced COMD by bandgap shrinkage. Uniform biaxial stress is in the bulk of the cavity with strain components parallel and perpendicular to the facet. At the facet, the latter becomes zero because the crystal lattice is free to relax, leaving a uniaxial stress at the facet. This strain release causes a bandgap reduction, which can be as high as 38 meV (Okayasu *et al.*, 1990), and which contributes to the positive feedback loop triggering the COMD event (see below for more details).

- COMD occurs at that part of the facet where the density of emitted light is highest. This can generate a facet temperature that locally exceeds the melting point of the semiconductor with a sudden decrease of light output. After the cooling-down and recrystallization time of the molten part, dislocations and DLDs extend perpendicular from the damaged facet along the cavity in usually the $\langle 110 \rangle$ direction.

- A COMD event is not reversible, but may be repeated with reduced optical output power and with the COMD event occurring at a different spot within the near-field pattern. COMD power levels tend to decrease with increasing operating time, because additional defects enhancing nonradiative recombination and optical absorption, and hence mirror heating, may be introduced, for example, by facet oxidation (see below).

- COMD is triggered both by optical absorption of emitted laser light at the facet and by large surface currents due to the large concentrations of native traps at the cleaved facet surface (Fermi-level pinning). Photoinduced carriers recombine nonradiatively and together with the large surface currents result in strong heating of the facet. This further decreases the bandgap energy at the facet, which in turn leads to increased light absorption and thus increased heating in a thin surface space-charge region with an absorption rate dominating the stimulated emission rate, which finally leads to a thermal runaway (Henry *et al.*, 1979) with ultimate facet meltdown (Figure 3.6). To initiate the positive feedback loop leading to thermal runaway, a critical facet temperature rise is required, which has been measured quasi-cw at $\Delta T_{cr} \sim 120$ K for GaAs/AlGaAs QW lasers (Epperlein, 1997; see Section 9.5.2.1).

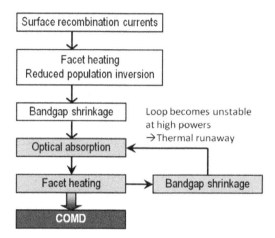

Figure 3.6 Simplified generation mechanisms and positive feedback loop leading to thermal runaway at a laser facet with ultimate catastrophic optical mirror damage (COMD) in edge-emitting semiconductor lasers.

- Introduction of point defects such as vacancies during facet oxidation. Oxidation is enhanced by bond breaking in the photoinduced electron–hole pair generation process. At the interface between oxide and semiconductor defects are left behind, because of the nonuniform removal and oxidation of semiconductor elements. These point defects may become the origins of extended structural defects acting as nonradiative recombination centers increasing the heating of the facet, reducing the bandgap energy, and ultimately increasing the absorption coefficient at the facet. This effect is larger at AlGaAs/GaAs facets than at InGaAsP/InP facets, because of the larger surface recombination velocity of $\sim 4 \times 10^5$ cm/s. Thus, AlGaAs/GaAs lasers are inferior to InGaAsP/InP lasers in both facet-related and inner region-related nonradiative recombination-mediated degradation.

- In general, no facet meltdown due to COMD is observed in InGaAsP/InP lasers, but facet degradation can occur, which may then be caused, however, indirectly by local heating at the top electrode or due to a large current flowing through the laser near the facet (Ueda *et al.*, 1984). Degradation may not be catastrophic, but can increase the threshold current and decrease the external differential quantum efficiency of BH InGaAsP/InP lasers. The *I*/*V* characteristics of these lasers may show a soft turn-on indicating a leakage path in the BH structure, which employs reverse-biased junctions for lateral current confinement (Ueda *et al.*, 1984).

Elimination of sudden degradation

- Cut the positive feedback loop leading to COD by the following approaches and technologies:

- Protect the mirror surface by appropriate passivation layers and dielectric films to minimize the surface state density N_s and thus the surface recombination velocity, and hence strong optical absorption and heating (see Chapter 4 for details). The surface recombination velocity is given by the well known expression

$$v_{s,n} = N_s v_{th} \sigma_{s,n} \tag{3.6}$$

where it is assumed that the material is p-type and electrons are minority carriers, and where v_{th} is the carrier thermal velocity and $\sigma_{s,n}$ the electron capture cross-section at the surface trap level. A similar expression holds for holes as minority carriers in n-type material. The COMD power level P_{COMD} can be increased by using a facet-coating material with relatively high thermal conductivity. For example, Al_2O_3 facet coatings are more efficient than SiO_2 coatings with cw $P_{COMD}(Al_2O_3) \cong 3 \times P_{COMD}(\text{uncoated})$ and cw $P_{COMD}(SiO_2) \cong 1.5 \times P_{COMD}(\text{uncoated})$ (Imai *et al.*, 1978).

- Apply effective nonabsorbing mirror structures (see Chapter 4 for details).

- Control and minimize during crystal growth and device fabrication processes the generation of defects participating in nonradiative recombination.

- Apply appropriate mode spot widening techniques (see Section 2.1.3.5 and Chapter 4) to decrease the output power density at the facet and thus the damage.

3.3   Key laser robustness factors

In summarizing the detailed descriptions in this chapter on diode laser degradation modes and mechanisms, the following parameters can be considered as the key factors determining the robustness of a diode laser:

- *Materials of the active region.*

- *Transverse lateral confinement.*

- *Thermally assisted defect migration into the active layer.*

- *Stress and defects.*

- *Current density-enhanced degradation effects.*

- *Catastrophic optical damage.*

- *Electrical overstress and electrostatic discharge.*

At the end of this section, we will give additional information and information not yet discussed on the last two parameters in the list above, respectively. First, however, we derive a quantitative figure of merit for evaluating the occurrence of COMD. Second, we discuss the basics of electrostatic discharge (ESD), which is a subset of the electrical overstress (EOS) category. We describe the impact of both effects on diode lasers and state some standard precautions to control and eliminate ESD.

EOS and ESD are external factors causing diode lasers to fail due to their extreme sensitivity to excessive current levels and current spikes. The high sensitivity of diode lasers to fast overshoot events such as ESD is due to the lasers' fast response time with rise times in the picosecond regime. In a wider sense, the prevention of these effects can be viewed as a measure to enhance laser robustness, although these detrimental effects actually belong within a laser protection strategy. Such a strategy comprises (ILX Lightwave Corp., 2003):

- test and measurement instrumentation, which may give rise to damage mechanisms like overcurrent and spikes;

- system setup components responsible for radiated electrical transients;

- laboratory environment inducing fast transients;

- human contact, laser handling, and packaging releasing ESD.

Nevertheless, we consider EOS/ESD as relevant factors determining the robustness of a diode laser and hence justify within this context an appropriate description of their relevant features and methods to suppress or eliminate them.

First, we discuss the COMD-related issue. The optical near-field (NF) area is one of the crucial parameters for maximizing the power level at COMD. In general, the COMD levels of lasers with smaller NF areas are obviously lower than COMD levels of lasers with larger NF areas. The spreading of the mode in the transverse vertical and transverse lateral directions determines the size of the NF area. The former is determined by the vertical layer structure and can be as large as ~ 1 µm by using one of the mode expansion methods discussed in Section 2.1.3.5. In contrast, the latter is given by the width of the lateral waveguide, which is typically in the order of 2–4 µm for single-mode lasers.

We can now define an optical intensity factor (OIF) as the inverse value of the NF area (with units of $(\mu m^2)^{-1}$); that is, the higher this factor, the greater the probability for COMD to occur. Further, we define an optical acceleration factor (OAF) as the optical intensity factor to the power n (with units of $(\mu m^2)^{-n}$), where the exponent n is a characteristic number of the specific laser type, roughly between about 1 and 3. This acceleration factor actually determines the time to the COMD event; that is, the larger this factor, the shorter the time to COMD. Its definition corresponds to the general definition of lifetime of a device or system exposed to an applied stress (see Chapters 5 and 6).

NF spot areas, OIFs, and OAFs have been calculated for 980 nm InGaAs/AlGaAs GRIN-SCH SQW ridge waveguide lasers 4 µm wide as a function of the transverse vertical and transverse lateral far-field beam angles that are the more easily accessible parameters. Tables 3.1a–c list OIF data and data of OAFs equal to OIFs to the power of 2 and 3, respectively. The data are grouped in three categories according to their severity to COMD failure. This classification is based on empirical experience with

Table 3.1 Optical intensity factors (OIFs) and optical acceleration factors (OAFs) calculated for 980 nm InGaAs/AlGaAs GRIN-SCH SQW ridge waveguide lasers 4 μm wide as a function of slow-axis and fast-axis far-field beam divergence angles. (a) OIFs defined as the inverse of near-field spot areas have units of $(\mu m^2)^{-1}$. (b) OAFs defined as OIFs to the power 2 have units of $(\mu m^2)^{-2}$. (c) OAFs defined as OIFs to the power 3 have units of $(\mu m^2)^{-3}$. Data in dark-grey areas predict high probability of COMD events or low COMD levels, whereas data in unshaded areas indicate safe operation with respect to COMD. The data in the light-grey areas represent some kind of transitional regime between high and low degrees of COMD robustness. The data highlighted in bold are from different sources of leading state-of-the-art high-power, highly reliable, single-mode, and single-emitter in-plane diode laser products. The OAF can be considered as a useful figure of merit for assessing the likelihood for the occurrence of COMD events over the entire far-field beam divergence angle range.

Fast-axis far field angle [deg] ⟶

Slow-axis far field angle [deg] ↓

(a) Optical Intensity Factor OIF: Inverse of near field area

Slow \ Fast	25	30	35	40	45	50
6	**0.43**	**0.51**	0.59	0.68	0.76	0.83
8	**0.57**	**0.68**	0.79	0.90	1.01	1.11
10	0.71	0.85	0.99	1.12	1.26	1.39
12	**0.85**	1.02	1.19	1.35	1.51	1.67
14	1.00	1.19	1.38	1.57	1.76	1.94
16	1.14	1.36	1.58	1.80	2.01	2.22
18	1.28	1.53	1.78	2.02	2.26	2.49

(b) Optical Acceleration Factor OAF : $OAF = (OIF)^{n=2}$

Slow \ Fast	25	30	35	40	45	50
6	**0.18**	**0.26**	0.35	0.46	0.57	0.70
8	**0.32**	**0.46**	0.63	0.81	1.01	1.24
10	0.51	0.72	0.98	1.27	1.58	1.93
12	**0.73**	1.04	1.41	1.82	2.28	2.78
14	0.99	1.42	1.91	2.47	3.10	3.78
16	1.29	1.85	2.49	3.23	4.04	4.93
18	1.63	2.33	3.15	4.08	5.10	6.22

(c) Optical Acceleration Factor OAF : $OAF = (OIF)^{n=3}$

Slow \ Fast	25	30	35	40	45	50
6	**0.08**	**0.13**	0.21	0.31	0.43	0.58
8	**0.18**	**0.32**	0.50	0.73	1.02	1.38
10	0.36	0.62	0.97	1.42	1.99	2.69
12	**0.62**	1.06	1.67	2.46	3.44	4.63
14	0.99	1.69	2.64	3.89	5.45	7.34
16	1.47	2.51	3.94	5.80	8.12	10.93
18	2.09	3.57	5.59	8.23	11.53	15.53

the numbers highlighted used in leading commercial single-mode, high-power, pump laser products showing practically no COMD failures and ultralong lifetimes. Thus, the OIF and OAF data $\ll 1$ stretching along both far-field angle directions in the upper left corner of the tables can be considered as safe regarding COMD, whereas lasers with data $\gg 1$ around the lower right corner show very low COMD threshold levels and very low degrees of COMD robustness. The intermediate area with OIF and OAF data in the range of ~ 0.7 to 1.3 running diagonally across the tables represents some kind of transitional regime between safe and unsafe laser operation with respect to COMD. The OAF can therefore be used as a figure of merit to assess the likelihood for COMD over the whole far-field angle range.

Second, we discuss the EOS/ESD-related issue. When a laser is operated outside the safe operating regime defined by the specifications for parameters such as voltage, current, and power, an EOS event can cause damage to the weakest part of the device that ruins the device. Similarly, an ESD event can damage the laser. ESD can affect the functioning of the device at any stage during device fabrication, testing, handling, or field use. ESD takes place when charge accumulates on a surface for any reason and is dissipated into the device. However, damage can also be caused by ESD from the device or by field-induced discharges. ESD events need not always lead to immediate failure but they can cause a latent defect, which may be undetected during testing. However, in such a case, the laser may fail later during field operation and the failure may then wrongly be attributed to root causes other than the original ESD event.

The level of robustness and type of failure are dependent on the polarity of the electrical surge applied to the p–n junction of the diode laser. A forward electrical surge causes a current surge, whereas a reverse surge causes a voltage surge. In general, a forward surge causes COD in short-wavelength lasers such as AlGaAs/GaAs devices, whereas in more COD-resistant long-wavelength InGaAsP/InP lasers the interface between the electrode and semiconductor is affected. In the latter event, a large local current pushes the temperature up at the interface within a short time (~ 100 ns) of the current spike, which is much shorter than the time to diffuse away the large amount of heat dissipated into a small volume within the short time of the spike. This will lead to a strong alloy reaction for alloy-type electrodes and melting of the semiconductor for Schottky-type electrodes at the interface, with rapid movement of these alloying or molten zones to the active region and finally destruction of the laser. A reverse surge affects the p–n junction, which after breakdown can cause a large current and strong local heating resulting in the melting of the semiconductor at the p–n junction. The effect of reverse surge on the P/I characteristic of an InGaAsP/InP laser has been demonstrated in controlled ESD stress experiments at various voltages (DeChiaro and Unger, 1991). The test method used the human body model (HBM, see details further down) where a 100 pF capacitor is discharged through a switching component and a 1.5 kΩ series resistor leading to a minimum rise time of the pulse of 10 ns as defined by Military Standards (Defense Logistics Agency, 2011). Up to 1.2 kV pulses, there is not much change in the P/I characteristic. However, there is an abrupt change at 1.4 kV in which the threshold current increased by a factor of two and the slope efficiency decreased by $\sim 60\%$. Further increase of the voltage continuously increases the threshold and decreases the efficiency with a tripling of the threshold current and decrease of the slope efficiency by ~80% compared to the original unstressed device.

Finally, we briefly describe industry-proven models used to simulate ESD events, list standard precautions to control ESD, and provide some relevant ESD-related literature and references.

To cover the different mechanisms leading to ESD damage during the various fabrication processes and from handling by personnel, three primary models of ESD events have been developed in industry: the human body model (HBM), machine model (MM), and charged device model (CDM). Test procedures based on these models have shown that most of the ESD field failure signatures could be reproduced. The ESD stress tests are usually performed using an automatic test system in applying a defined number of "zaps" per voltage stress level and polarity to the device under test.

The HBM test method simulates the direct transfer of electrostatic charge from the charged human body to the device under test (regarding its circuit and components; see above). For example, a person walking on a carpet and wearing shoes with highly insulating soles can build up a voltage in excess of 15 kV. Of course, other "nonhuman" materials that accumulate and transfer charge in a similar manner are also covered by the HBM. The HBM peak current pulse in a device with very low impedance is typically 0.66 A per 1 kV within ~50 ns effective time. Relevant industry standards are ESD STM5.1 (ESD Association, 2011), MIL-STD-883-Method 3015 (Defense Logistics Agency, 2011), and JESD22-A114-B (JEDEC, 2011).

The MM module emulates the rapid direct transfer of electrostatic charge from a charged conductive object such as a metallic tool to the device under test. Its circuit consists of charging up a capacitor of 200 pF to a certain voltage and then discharging it directly into the device through a 0.5 μH inductor and no series resistor. The stored energy is obviously twice that in the HBM model due to the higher capacitance. However, the energy dissipated into the device under test is also dependent on its series resistance. It can be shown that the energy dissipated, for example, in a 5 Ω device in a 1 kV test, is about 10 times higher for the MM test than for the HBM test. The series inductance shapes the form of the oscillating MM wave, which has a period of ~100 ns and a peak current fall to ~30% at ~200 ns compared to its initial value of ~16 A at 1 kV. The MM test is more stringent than the HBM test; however, there are ESD situations in reality that are well represented by MM. Relevant industry standards are ESD STM5.2 (ESD Association, 2011) and JESD22-A115-A (JEDEC, 2011).

The CDM covers ESD events involving the transfer of charge from the charged device to another body which is at a lower potential. This can occur by triboelectrification where charge builds up on the device by frictional movements during any fabrication and packaging process followed by rapid discharge via contact with a conductive object. Such an event can be even more destructive than HBM or MM ESD, because of its high current and despite its much shorter pulse duration. In the so-called socketted CDM test principle, the device is inserted in a socket, charged from a high-voltage source, and then discharged through a 1 Ω resistor to the ground plane. This gives rise to high peak currents of $\sim$40 A per 1 kV and short pulse durations of less than a few nanoseconds. The CDM discharge is about 100 times faster than that of HBM and MM, and the peak current is about 40 times that of a HBM pulse. Relevant industry standards are ESD STM5.3.1 (ESD Association, 2011) and JESD22-C101-A (JEDEC, 2011).

Finally, simple precautionary measures can be taken to prevent ESD problems before they occur rather than fighting fires:

- Wear a protective wrist strap designed to drain built-up electric charge safely to ground. Ground all tools including tweezers and soldering irons.

- Use anti-static worktables with dissipative work surfaces. Connect all parts of the workstation to a common point ground.

- Use anti-static floor coverings and wear ESD control footwear or foot straps. Use ionized air blowers. Wear clothing resistant to charge buildup (cotton).

- Keep unprotected devices away from charge-generating sources.

- Store devices in a closed conductive environment and short the pins of the device by a conductive material or shunts.

In the microelectronics industries, effective ESD protection schemes have been successfully developed, which are directly implemented into the actual integrated circuit chip to protect inputs and outputs from ESD pulses. In principle, such an on-chip scheme could be adjusted to the needs of a diode laser system operated under harsh conditions by providing an input circuit with a low-impedance shunt path to prevent the occurrence of excessive voltages on the laser chip. Such a measure and the standard precautions listed above can therefore be considered as essential factors contributing to the robustness of a diode laser.

We briefly describe a review of common ESD protection approaches. The review originally prepared by LASORB, a division of Pangolin Laser System (2012), shows that existing design solutions for ESD and surge protection do not work well or at all for diode lasers. To solve the ESD protection problem LASORB (from LASer and ESD absORBer) developed a new hybrid component, which will also be discussed below. For the following discussion we assume a 15 kV ESD event and maximum limits for the reverse bias voltage of 2 V and forward bias voltage of 2.2 V for the diode laser. In the common ESD protection methods, a simple device is directly connected across the terminals of the diode laser. Such devices include:

- Resistor: a 44 mΩ resisistor is required to meet the above voltage limits for a 15 kV ESD event. However, this is not practical, because during operation much more power would be expended in the resistor by-pass than in the diode laser path.

- Capacitor: a 1 mF capacitor appears to be sufficient to prevent a 15 kV ESD event from exceeding the voltage limits of the laser. However, a parasitic resistance and parasitic inductance in series with the nominal capacitance have to be taken into account along with the frequency response of the components between around 20 MHz and 1 GHz, which is equivalent to an ESD event speed in the range of typically one nanosecond to some tens of nanoseconds. This leads to an impedance of the capacitor of $\lesssim 44$ mΩ in the above frequency range to be effective in protecting the laser. Standard 1 mF electrolytic capacitors have typically 1 Ω series resistances and 15 nH series inductances resulting in an impedance > 44 mΩ. Best 1 mF tantalum capacitors have series resistances of ~ 50 mΩ and series inductances of ~ 1 nH leading to an impedance > 6 Ω at 1 GHz. Considering also the fact that the capacitance would impact the performance in an application requiring high modulation frequencies the simple capacitive approach has clear drawbacks.

- Schottky diode: a Schottky diode (Sze, 1981) is connected anti-parallel with the diode laser. However, there are several problems with this approach including that, first, Schottky diodes are not designed to withstand nanosecond pulses of up to 50 A caused by an ESD event and, second, this configuration would only protect from negative ESD events that cause to reverse-bias the laser. Considering also that positive ESD events can pass through the diode laser without attenuation makes this approach not fit-for-use.

- Zener diode: a Zener diode (Sze, 1981) is connected anti-parallel with the diode laser. However, there are two major problems making a Zener diode not practicable for protecting a diode laser. First, a Zener is too slow and second, it is very difficult to choose a Zener voltage that is sufficiently close to the forward-bias at lasing threshold of the diode laser.

- Switch, relay or MOSFET: the switch can be implemented as a relay or a depletion-mode metal-oxide semiconductor field-effect transistor (MOSFET) (Sze, 1981). The switch would be normally closed in the quiescent state of the laser. In this way an ESD event is conducted by the switch rather than the diode laser; however, the switch will have no effect and not protect the laser because it has to be open once the laser power is turned on. In case of a relay, further drawbacks include that the impedance of < 44 mΩ (see above) required to be effective in protecting the laser from a 15 kV ESD event, may deteriorate due to wear-out during lifetime. The drawbacks of using a depletion-mode MOSFET include that typical turned-on resistances are too high in the range of some Ohms.

LASORB's approach is to prevent the negative polarity of a diode laser while also prohibiting discharge from exceeding the maximum forward-bias limits. Integrated into the circuit of the hybrid component containing active silicon and passive devices is a fast-acting p-n junction that will not exceed 2 V and hence is able to protect the laser from reverse bias and negative 15 kV ESD events. The component also includes a slew-rate detector that monitors the voltage across the pins of the laser, and can distinguish between normal laser operation and ESD and power surges. If there is a fast change of voltage detected, a special diode is activated, which conducts the voltage and thus current away from the laser by passing it into the LASORB component path. The theoretical reaction time is between 200 and 800 ps, fast enough to react to ns-level ESD events.

References

Agrawal, G. P. and Dutta, N. K. (1993). *Semiconductor Lasers*, 2nd edn, Van Nostrand Reinhold, New York.

Andrew, S., Marsh, J., Holland, M., and Kean, A. (1992). *IEEE Photonics Technol. Lett.*, **4**, 426.

Buisson, J. P., Allen, R. E., and Dow, J. D. (1982). *J. Phys. (Paris)*, **43**, 181.

Carroll, J., Whiteaway, J., and Plumb, D. (1998). *Distributed Feedback Semiconductor Lasers*, IEE, London.

Davis, M. K., Ghislotti, G., Balsamo, S., Loeber, D. A., Smith, G. M., Hu, M. H., and Nguyen, H. K. (2005). *IEEE J. Sel. Top. Quantum Electron.*, **11**, 1197.

DeChiaro, L. F. and Unger, B. A. (1991). ESD Association, *Proceedings of the EOS/ESD Symposium*, **EOS-13**, 91172.

Defense Logistics Agency (2011). *Military Standard MIL-STD-883E, Method 3015*. www.dscc.dla.mil/programs/milspec.

Dow, J. D. and Allen, R. E. (1982). *Appl. Phys. Lett.*, **41**, 672.

Eliseev, P. G. (1973). *J. Lumin.*, **7**, 338.

Eliseev, P. G. (1996). *Prog. Quantum Electron.*, **20**, 1.

Epperlein, P. W. (1987). For further information see www.pwe-photonicselectronics-issueresolution.com.

Epperlein, P. W. (1997). *Proc. SPIE*, **3001**, 13.

Epperlein, P. W. and Bona, G. L. (1993). *Appl. Phys. Lett.*, **62**, 3074.

Epperlein, P. W., Parry, M., Helmy, A., Drouot, V., Harrell, R., Moseley, R., Stacey, S., Clark, D., Harker, A., and Shaw, J. (2000). For further information see www.pwe-photonicselectronics-issueresolution.com.

Epperlein, P. W., Hawkridge, A. R., and Skeats, A. (2001). For further information see www.pwe-photonicselectronics-issueresolution.com.

ESD Association (2011). *Standards ESD STM5.1, ESD STM5.2, ESD STM5.3.1*. www.esda.org.

Fukuda, M. (1999). *Optical Semiconductor Devices*. John Wiley & Sons, Inc., New York.

Fukuda, M., Wakita, K., and Iwane, G. (1983). *J. Appl. Phys.*, **54**, 1246.

Fukuda, M., Fujita, O., and Uehara, S. (1988). *J. Lightwave Technol.*, **6**, 1808.

Gold, R. D. and Weisberg, L. R. (1964). *Solid State Electron.*, **7**, 811.

Henry, C. H. and Lang, D. V. (1977). *Phys. Rev.*, **B15**, 989.

Henry, C. H., Petroff, P. M., Logan, R. A., and Meritt, F. R. (1979). *J. Appl. Phys.*, **50**, 3721.

ILX Lightwave Corp. (2003). *Application Note* #3. www.ilxlightwave.com.

Imai, H., Morimoto, M., Sudo, H., Fujiwara, T., and Takusagawa, M. (1978). *Appl. Phys. Lett.*, **33**, 1011.

Imai, H., Fujiwara, T., Segi, K., Takusagawa, M., and Takanashi, H. (1979). *Jpn. J. Appl. Phys.*, **18**, 589.

JDSU Inc. (2010). The developer SDL Inc. of diode lasers with QWI lateral waveguides merged with JDS Uniphase Corp. in 2001. www.jds.com.

JEDEC (2011). Standards JESD22-A114-B, JESD22-A115-A, JESD22-C101-A. www.jedec.org.

Kajimura, T. (1980). *J. Appl. Phys.*, **51**, 908.

Kishino, S., Chinone, N., Nakashima, H., and Ito, R. (1976). *Appl. Phys. Lett.*, **29**, 488.

Kuttler, M., Strassburg, M., Pohl, U., Bimberg, D., Behringer, M., and Hommel, D. (1998). *Appl. Phys. Lett.*, **73**, 1865.

LASORB (2012). *ESD absorber for laser diodes.* http://lasorb.com/.

Lichtenstein, N., Manz, Y., Mauron, P., Fily, A., Arlt, S., Thies, A., Schmidt, B., Muller, J., Pawlik, S., Sverdlov, B., and Harder, C. (2004). *Proceedings of the IEEE 19th International Semiconductor Laser Conference, Conf. Dig.*, 45.

Mizuishi, K. (1984). *J. Appl. Phys.*, **55**, 289.

Mizuishi, K., Tokuda, M., and Fujita, Y. (1982). *Proceedings of the IEEE 8th International Semiconductor Laser Conference, Conf. Dig.*, 140.

Mizuishi, K., Sawai, M., Todoroki, S., Tsuji, S., Hirao, M., and Nakamura, M. (1983). *IEEE J. Quantum Electron.*, **QE-19**, 1294.

Nagai, Y., Shigihara, K., Karakida, S., Kakimoto, S., Otsubo, M., and Ikeda, K., (1995). *IEEE J. Quantum Electron.*, **QE-31**, 1364.

Nakahara, S. and McCoy, R. J. (1980). *Thin Solid Films*, **72**, 457.

Noël, J. P., Melville, D., Jones, T., Shepherd, F. R., Miner, C. J., Puetz, N., Fox, K., Poole, P. J., Feng, Y., Koteles, E. S., Charbonneau, S., Goldberg, R. D., and Mitchell, I. V. (1996). *Appl. Phys. Lett.*, **69**, 3516.

O'Hara, S., Hutchinson, P. W., and Dobson, P. S. (1977). *Appl. Phys. Lett.*, **30**, 368.

Okayasu, M., Fukuda, M., Takeshita, T., and Uehara, S. (1990). *IEEE Photonics Technol. Lett.*, **2**, 689.

Olsen, G. H. and Ettenberg, M. (1977). *J. Appl. Phys.*, **48**, 2543.

Petroff, P. M. and Kimerling, L. C. (1976). *Appl. Phys. Lett.*, **29**, 461.

Piotrowska, A., Auvray, P., Guivarc'h, A., Pelous, G., and Henoc, P. (1981). *J. Appl. Phys.*, **52**, 5112.

Runyan, W. R. and Shaffner, T. J. (1998). *Semiconductor Measurements & Instrumentation*, 2nd edn, McGraw-Hill, New York.

Smigelskas, A. D. and Kirkendall, E. O. (1947). *Trans. AIME*, **171**, 130.

Sze, S. M. (1981). *Physics of Semiconductor Devices*, 2nd edn, John Wiley & Sons, Inc., New York.

Thompson, A. (1979). *IEEE J. Quantum Electron.*, **QE-15**, 11.

Tkach, R. W. and Chraplyvy, A. R. (1986). *J. Lightwave Technol.*, **4**, 1655.

Ueda, O. (1996). *Reliability and Degradation of III-V Optical Devices*, Artech House, Boston, MA.

Ueda, O., Imai, H., Yamaguchi, A., Komiya, S., Umebu, I., and Kotani, T. (1984). *J. Appl. Phys.*, **55**, 665.

Uji, T., Suzuki, T., and Kamejima, T. (1980). *Appl. Phys. Lett.*, **36**, 655.

Ventrudo, B. F. and Rogers, G. (1996). *US Patent* No. 5485481.

Volodin, B. L., Dolgy, S. V., Melnik, E. D., Downs, E., Shaw, J., and Ban, V. S. (2004). *Opt. Lett.*, **29**, 1891.

Welch, D. F., Scifres, D. R., Cross, P. S., and Streifer, W. (1987). *Appl. Phys. Lett.*, **51**, 1401.

Williamson, III, R. S. (2007). *Integrated wavelength control for diode pumped lasers*: In *Photonics Spectra*, June.

Yamakoshi, S., Abe, M., Komiya, S., and Toyamer, Y. (1979). *Proceedings of the International Electron Devices Meeting*, 122.

Chapter 4

Optical strength engineering

Introduction

This chapter is organized as follows. First, we describe the properties of the laser facet after the cleavage procedure and then give the physical origins of facet degradation and failure. A concise yet comprehensive account follows on approaches and techniques to improve and enhance the optical strength in the bulk of the laser and at its mirror facets. This includes a discussion of most different facet passivation techniques, each with its own strengths and weaknesses, nonabsorbing mirror concepts, noninjecting or current blocking mirror technologies, heat spreader layers and laser chip mounting techniques, low optical con nement structures, and other mode spot widening approaches such as thin tapered active layer and ared lateral waveguide con gurations.

4.1 Mirror facet properties – physical origins of failure

When a semiconductor laser facet is cleaved, a set of unsaturated bonds appears at the exposed surface. This is caused by the interruption to the periodicity of the crystal lattice, which results in dangling bonds at the surface. The dangling bonds interact strongly with atoms and molecules of the ambient. Thus, surface states can form that have energy levels in the bandgap and lead to band bending at the surface (Sze, 1981; Wolfe *et al.*, 1989; Yu and Cardona, 2001). Surface states can give or receive electrons and thus connect the crystal to the ambient.

For a semiconductor with 10^{22} atoms/cm^3, the density of surface states could be $(10^{22}\ \text{cm}^{-3})^{2/3} \cong 5 \times 10^{14}\ \text{cm}^{-2}$ assuming that there is at least one dangling bond per atom due to the discontinuity of the crystal lattice at the surface. However, an actual surface usually has a much lower density of surface states of $\sim 10^{11}$ detectable states per square centimeter depending on the arrangement of the surface atoms. The perturbation introduced by the loss of translational symmetry in the surface lattice may result in a rearrangement of the surface atoms resulting in a stretching and twisting of bonds (surface reconstruction with band structure modi cation). The microscopic details vary with the properties of the material and its crystal orientation, adsorbed or chemisorbed impurities, and preparation conditions. Figure 4.1 shows a schematic representation of all these processes.

The surface lattice formed by the periodic distribution of atoms gives rise to an $E(k)$ dispersion diagram for surface states. Surface states may be grouped deep in the bandgap with a low $E(k)$ dispersion. This results in a high density of surface states clustered in a narrow distribution, which pins the Fermi-level by the overall charge state at the facet surface rather than by charge neutrality in the bulk. Surface states may also have a large $E(k)$ dispersion and partly overlap an intrinsic band; in this case, however, the overlapping states are indistinguishable from intrinsic bulk states. In addition, in ionic crystals, the surface cations and anions function as donors and acceptors, respectively, and hence act as compensated pairs.

The effect of Fermi-level pinning leads to a bending of the energy bands near the surface. In equilibrium, the Fermi-level must remain constant from bulk to surface, meaning that the band edges must then adjust to be compatible with the Fermi-level pinning at the surface and the bulk Fermi-level. The bands bend upward if the surface state acts as an acceptor and traps an electron, and if the surface state is a donor-like trap it can lose an electron to form a positive charge at the surface, which causes the bands to bend down. If an adsorbing atom has a greater electron af nity than the work function of the semiconductor, it may capture an electron from the surface leading to an upward bending of the bands due to the negative charge formed at the surface. Oxygen is such a candidate with a very high electronegativity causing

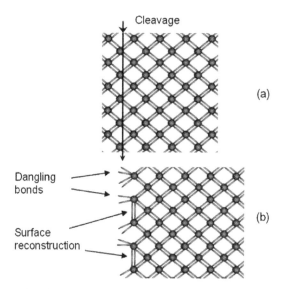

Figure 4.1   Schematics of physical origins of facet degradation mechanisms. (a) Simpli ed representation of a crystal lattice before the facet cleavage process. (b) Perturbed facet surface after cleavage showing dangling bonds and partially reconstructed sites. Dangling bonds give rise to the formation of point defects, structural defects, oxides, and the adsorption of ambient impurity atoms, all of which lead to strong local nonradiative recombination rates and facet heating with the ultimate failing of the laser mirror.

an upward band bending. When oxygen is adsorbed on an n-type laser facet, the depletion layer increases and the surface conductance decreases. In contrast, when oxygen is adsorbed on a p-type surface, the surface becomes more negative. This increases the hole concentration near the surface and builds up an accumulation layer resulting in an increase of the surface conductance.

Due to the large density of surface states, there is an enhanced nonradiative re-combination rate at the surface leading to a depletion of minority carriers in the vicinity of the facet surface. In general, the nonradiative recombination rate at the facet surface is higher than that in the inner region, which leads to a reduction of carrier density within the carrier diffusion length from the surface. As a consequence, minority carriers ow into the surface region from the surrounding, higher concentra-tion regions. The surface recombination rate is limited by the rate at which minority carriers move toward the surface, which is the thermal velocity of carriers with a value near 10^7 cm/s for most semiconductors. Under steady-state conditions, both rates must be equal, and we can write the boundary condition at the facet surface $x = 0$ for the surface recombination current density. For example, for p-type material with electrons as minority carriers

$$J_{s,n}(x = 0) = q D_n \frac{\partial n}{\partial x}(x = 0) = q v_{s,n} \partial n (x = 0) \qquad (4.1)$$

where D_n is the diffusion constant for electrons, $\partial n/\partial x$ the carrier concentration gradient, $v_{s,n}$ the surface recombination velocity for electrons (see Equation 3.6), and ∂n the net reduction of electrons at the facet. Equation (4.1) takes into account that, rst, the direction of the carrier concentration gradient is opposite to the direction of the carrier motion and, second, the diffusion current density is the carrier ux density multiplied by the carrier charge. A similar expression can be written for the surface recombination current density carried by holes as minority carriers, but then there would be a negative sign at the diffusion current density term in Equation (4.1).

The surface recombination velocity for an InP facet surface is about two orders of magnitude smaller than that for a GaAs facet, and hence tends to be much more of a problem in GaAs-based diode lasers emitting in the 800 nm regime than in InP-based lasers with wavelengths > 1.3 µm. Surface recombination velocities have been determined by measuring, for example, the absolute photoluminescence quantum ef ciency (Boroditsky et al., 2000) or the effective minority carrier lifetime using time-resolved photoluminescence (Wang et al., 2005). Typical v_s data for common material systems can be found in Table 4.1. In principle, the recombination rate at a heterostructure interface can be tailored to very low levels, which gives it a big advantage over semiconductor facet surfaces exposed to air. This explains qualitatively the higher COMD power level in window-type lasers (see Section 4.3.2 below) relative to that of conventional lasers, where in the former the active layer is separated from the facet surface by a heterostructure barrier. Furthermore, in general, the interface recombination velocity increases with an increase in the lattice mis t at a semiconductor interface such as in InGaP/GaAs with $\sim 6 \times 10^5$ cm/s for a mis t of $\sim 2.5\%$.

Table 4.1 Experimental surface recombination velocities for some common material systems.

Material	Surface recombination velocity [cm/s]	Reference
n-GaAs	$2–3 \times 10^6$	Wittry and Kyser (1966), Jastrzebski et al. (1975)
GaAs/AlGaAs	$2–4 \times 10^6$	Kappeler et al. (1982)
GaAs, GaAs/AlGaAs QW	$4–6 \times 10^5$	Swaminathan et al. (1990)
InGaAs/GaAs QW	$1–2 \times 10^5$	Hu et al. (1994)
InGaAs	$< 1.5 \times 10^4$	Boroditsky et al. (2000)
GaN	$< 3 \times 10^4$	Boroditsky et al. (2000)
n-InP p-InP	$\sim 1 \times 10^4$ $\sim 1 \times 10^5$	Hoffman et al. (1980)
AlGaAsSb/GaInAsSb DH	30	Wang et al. (2005)

As we will see in Section 4.2.2, the facet surface quality can be controlled effectively by using a suitable passivation technology, hence causing a signi cant reduction in the surface recombination velocity.

As detailed in Section 3.2.2.3, surface nonradiative recombination and surface current effects mediated by surface states result in strong facet heating, and hence can be extremely detrimental to the operating performance and reliability of diode lasers, particularly regarding the optical strength of their facet mirrors. These effects can be minimized by passivating the surface, which reduces the number of surface states and hence the recombination intensity.

In the following sections, we discuss appropriate mirror protection and passivation technologies, as well as structures to minimize the absorption of the emitting laser light at the laser facet by using a nonabsorbing mirror structure. Moreover, we include further optical strength enhancement factors and methods such as material optimizations, current blocking mirrors, heat spreader techniques, and mode spot widening techniques (see Section 2.1.3.5).

4.2 Mirror facet passivation and protection

The effort to enhance the optical robustness of a laser facet usually consists of two processing steps. In the rst step, the integrity and stabilization of the freshly prepared facet is re-established by minimizing the surface state density. After this pre-treatment, a dielectric overcoat is applied to strengthen the mechanical and chemical strength of the sensitive laser mirror facet and to modify its re ectivity, as required. Dielectric coatings only, such as SiO_2, Al_2O_3, and Ga_2O_3 layers, have also been employed successfully to meet the electrical, optical, thermal, and mechanical re-quirements at the facet (see Sections 1.3.8 and 4.2.3 for details).

4.2.1 Scope and effects

The preparation of a stable mirror facet with a reduced degradation potential includes cleaning, chemical treatment, and coating of the facet surface and has to meet the following speci c objectives:

- Establish a defect-free and conductive-layer-free surface by tying up dangling bonds and removing impurities to minimize facet heating.

- Establish an oxide-free surface and passivate states that may trigger oxidation.

- Establish a scatter-free stable surface crystal structure to improve ef ciency and beam quality.

- Establish a diffusion barrier to prevent both the incorporation of oxygen into the semiconductor from the dielectric overcoat and the ambient and the movement of semiconductor atoms into the coating.

- Establish an absorption-free facet surface including passivation and coating layers to prevent facet heating at the lasing wavelength.

- Ensure the coatings meet the optical, electrical, thermal, and mechanical conditions described in Section 1.3.8 and pass environmental test conditions, including low-temperature endurance, high-temperature resistance, temperature cycling, thermal shock, humidity resistance, vibration, and shock stability (Chapter 6).

4.2.2   Facet passivation techniques

4.2.2.1   E2 process

This process was originally developed by IBM Laser Enterprise (now Avanex/Bookham = Oclaro), an independent business unit of IBM Research in Rüschlikon (Zurich), Switzerland, in the early 1990s, for high-power, highly reliable, single-mode 980 nm compressively-strained InGaAs/AlGaAs GRIN-SCH SQW ridge waveguide pump diode laser products for terrestrial and submerged optical communication systems applications. The E2 process is a follow-up development of an E1 process where "E" stands for enhanced (Gasser and Latta, 1992).

Essentially, the process consists of the following steps: a specially sized piece of material is cut out of a completely processed laser wafer, inserted and adjusted in a hinge-like mechanical xture, and then mounted in an ultrahigh-vacuum evaporation system. A scribe mark, which de nes the position of the cleave is adjusted to the rotation axis of the hinge. Scribing is usually performed outside the vacuum system with a diamond-tipped scriber. At a base pressure of $\sim 10^{-10}$ Torr cleaving is carried out by a 90° rotation of the hinge mechanism leaving the cleaved bar in one half of the hinge cleaver and the remaining material in the other half. Both freshly cleaved facet faces are immediately coated in situ with a thin amorphous Si (a-Si) lm by an electron-beam evaporator. In the original E2 process, the re ectivity coatings of the facets were performed outside the vacuum system in a separate deposition system. The coating for the front facet with a typical re ectivity of 10% is usually made of a single layer of Al_2O_3 whereas the back facet has a high re ectivity >90% realized by a stack of pairs of high and low refractive index layers a quarter wavelength thick, such as TiO_2 and SiO_2, respectively (cf. Section 1.3.8).

The thickness of the a-Si layer should be below 10 nm to avoid heating effects and the formation of a quantum state in the valence band of the Si well layer. The latter results from quantum mechanical calculations performed on a "particle in an asymmetric box" formed by the $GaAs/a-Si/Al_2O_3$ structure with bandgap energies of 1.42, 1.7, and 9.9 eV, respectively. Different band diagrams have been proposed for the $GaAs/a-Si/Al_2O_3$ system, including a type-II staggered band alignment for the GaAs/a-Si interface (Katnani and Margaritondo, 1983; Capasso and Margaritondo, 1987), and a type-I straddled band alignment for unstrained crystalline heterostructures (Tiwari and Frank, 1992). The existence of such quantum states in the a-Si layer would cause carriers to diffuse into the a-Si layer with the prospect of recombining nonradiatively (Tu et al., 1996).

Tu et al. (1996) applied a deviation to the original E2 process by executing the cleave while the samples were exposed to a ux of Si, thereby minimizing the time the facet was exposed to the residual gases in the vacuum chamber. The shorter exposure time led to higher threshold powers for COMD and relaxed requirements for the pressure in the vacuum system. Low surface recombination velocities $\sim 3 \times 10^4$ cm/s have been obtained for the a-Si/Al$_2$O$_3$ coating (Tu et al., 1996). The original E2 passivation principle has been dramatically improved over the years by IBM Corp. aimed at achieving higher throughput and lower cost. Fully automated systems with sophisticated in situ scribing and cleaving devices (Bauer et al., 1991) are now available, meeting industry standards of high throughput and low fabrication costs.

Diode lasers with E2 passivation technology have developed to state-of-the-art performance and reliability and offer pre-eminent laser products for most demanding high-power, single-mode, and ultrahigh-reliability applications. Kink-free powers >1.4 and 1.8 W rollover powers have been reported from narrow 980 nm InGaAs/AlGaAs lasers with ridge waveguides 4 μm wide and E2 passivated facets showing no COMD events and maximum power densities > 100 MW/cm^2 and ultralong lifetimes over 30 years at operating conditions above 1 W and 25 °C (Lichtenstein et al., 2004; Bookham, Inc., 2009).

4.2.2.2 Sulfide passivation

There are several studies on using sulfur treatment for facet passivation (Kamiyama et al., 1991; Lambert et al., 2006; Kawanishi et al., 1990). The treatment is usually performed chemically utilizing an ammonium sul de (NH$_4$)$_2$S solution in which the cleaved laser bars are submersed for a period of typically 5–10 min at 25 °C, followed by deionized water rinsing and blowing dry. These soak times give the maximum increase in COMD level.

Thus, the COMD power level of 110 mirrors of 780 nm AlGaAs high-power lasers has been doubled to 220 mW cw operation. Auger electron spectroscopy and x-ray photoelectron spectroscopy measurements showed that the sulfur treatment strongly reduces the surface oxides of Ga, Al, and As atoms and leaves a sul de layer of these constituent atoms. The improved COMD power level by the sul de passivation has been ascribed to a reduced surface recombination caused by the oxides. This nding is con rmed by measurements of the reverse leakage current, which has been found to be reduced for sul de passivated facets. A reduced reverse leakage current directly relates to the removal of conductive surface states and to the passivation of surface states that could trigger an oxidation process. Similar COMD power-level improvements have been obtained for AlGaInP visible diode lasers. In this case, transmission electron microscopy and energy dispersive x-ray measurements con rmed that most of the oxide at the mirror facets is replaced by sulfur after the treatment.

The quality of the sul de passivated facet tends to degrade quickly, requiring a post-deposited layer, though also stable operation of >2000 h has been achieved at higher temperature and power operation. SiN$_x$ encapsulation has proven to be an effective method for preserving the positive effects of the sul de passivation such as reduced surface recombination by inhibiting the oxidation process of the air-exposed semiconductor. SiN_x deposited by electron cyclotron resonance chemical vapor deposition has proven to be superior to that deposited by plasma-enhanced chemical vapor deposition (Hobson et al., 1995).

4.2.2.3   Reactive material process

A passivation layer comprising a reactive thin lm is deposited on the air-cleaved laser facet to getter oxygen, water, and other reactive contaminants. The lm consists of a highly reactive material such as the electropositive Al, which readily removes contaminants from the facet by reactive out-diffusion into this overlayer. Its layer thickness, however, must be controlled carefully to be suf ciently thick to react with an optimum amount of contaminants. On the other hand, it has to be suf ciently thin in order that the reactive material is substantially consumed in the oxidation or gettering process to render the thin layer, which is initially conductive, electrically nonconductive and therefore does not short-circuit the junction of the laser device (Tihanyi and Bauer, 1983).

Preferred Al layer thicknesses are in the range of 2.5 to 5.0 nm or 7.5 nm obtained from Auger spectroscopy depth-pro ling measurements on Al-coated air-cleaved GaAs/AlGaAs DH laser facets. The minimum thickness of ~2.5 nm is suf cient to getter any oxygen, oxide, or water present on the cleaved sample into the reactive Al lm in the form of electrically nonconducting Al_2O_3. The reduction in oxygen level at the laser facet is at least by a factor of ve. The empirically determined upper limit of ~5 nm is for devices with subsequent overcoatings used to modify the re ectivity. For devices with no re ectivity-modifying coatings, this thickness can be increased to ~7.5 nm by considering the surface oxidation of the passivation layer itself.

Due to the rapid and effective gettering actions of these layers, no shunting of the laser current through these coatings in the preferred thickness range has been observed. The passivation layer substantially improves the adhesion of any mirror overcoating for re ectivity modi cation. Other chemically reactive materials that can be used as thin passivation layers include Si, Ta, V, Fe, Mn, and Ti. The metal reactive materials may be deposited by sputtering using an Ar^+-ion beam in a vacuum chamber with an Ar partial pressure of ~6×10^{-5} Torr for ~10–20 s, whereas Si could be deposited either in a hydrogen environment or as a silane compound (Tihanyi and Bauer, 1983).

4.2.2.4 N^2IBE process

The native nitride ion beam epitaxy (N^2IBE; Nitrel® trademark of Comlase AB, Sweden) process belongs to the category of nitridization passivation technologies developed since the early 1980s and known for their excellent properties. These include the removal of surface states, the formation of a higher bandgap surface layer leading to a reduced optical absorption, the prevention of chemical contamination, and resistance to water, oxygen, and reoxidation.

The N^2IBE process is a facet passivation technology suitable for all material platforms covering the 400–1700 nm wavelength range and including materials such as InGaAs, AlGaAs, InAlGaAs, InGaAsP, and InGaAlN. It is a three-step process with the following consecutive steps (Silfvenius *et al.*, 2005):

- Cleaving of laser bars in ambient atmosphere.

- Nitrogen milling in a vacuum system to remove native oxides and impurities to smooth the facet surface.

- Chemical stabilization of and atomically sealing the exposed, pristine III–V surface by saturating dangling bonds and bonding of nitrogen ions with exposed group-III atoms to form near-surface nitride compounds, which have higher bandgaps than the crystal matrix and thus no discernible optical absorption for the laser radiation.

These nitride layers act as a diffusion barrier for the crystal matrix constituents as well as a protective layer against foreign contaminants such as air, moisture, and oxides. The passivation layer can be reinforced by growing an extraneous nitride layer on top of the native nitride layer. Careful management of the ion beam energy spectrum, its characteristics, and the exposure time allows processing to occur at energy levels suf cient to mill, clean, and react with the surface, while remaining below the energy levels that would otherwise damage the laser crystal. After completion of the nitride facet passivation, the AR and HR mirror coatings are deposited in the same vacuum system without breaking the vacuum. The nitride layers seal the surface from any further oxidation and hence enable the use of any coating material and deposition process that are most suitable from a performance and processing point of view. Figure 4.2 shows schematically the two-step passivation process in a vacuum.

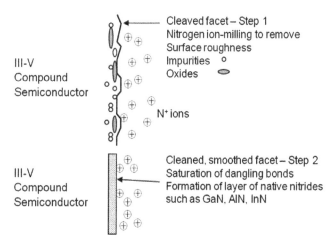

Figure 4.2   Schematic illustration of the two-step native nitride ion beam epitaxy passivation process comprising the rst step of nitrogen ion milling to remove surface roughness, impurities, and oxides from the air-cleaved facets, and the second step of saturating the dangling bonds and forming a layer of highly robust native nitrides on the facet surface.

The I^2IBE process is a cost-effective facet passivation technology with very high throughput of several hundred stacked bars passivated at the same time and delivers robust and reliable diode lasers exhibiting high COMD power levels, extremely low power degradation rates with no sudden failures in accelerated aging tests, and long lifetimes. Reliability tests on 980 nm InGaAs/AlGaAs QW lasers demonstrated COMD levels >25 MW/cm^2, typical accelerated power degradation rates of ~0.1% per 1000 h, and projected median lifetimes of 75 years at 25 °C (Silfvenius et al., 2004, 2003).

4.2.2.5 I-3 process

The I-3 process is a proprietary facet passivation procedure developed by Mitsubishi Chemical Corp. for the mass production of high-power, high-reliability, single-mode pump lasers at low cost and with high throughput. It consists of cleaving the facet in air, and thereby avoiding the complex, relatively slow, and cost-intensive facet cleaving process in high vacuum, followed by the novel three-step I-3 passivation process, which involves, in sequence (Horie et al., 1999):

1. Ion irradiation of the air-exposed facet.

2. Interlayer deposition.

3. Ion-assisted deposition of a dielectric coating layer.

The entire I-3 procedure is carried out in a vacuum chamber with the rst step consisting of low-energy (35 eV) Ar$^+$-ion irradiation for 90 s to sputter clean the air-cleaved facet without radiation damage (Ghandhi et al., 1982) and semiconductor composition change. In the second step, an a-Si layer 2 nm thick is deposited. The third step consists of Ar$^+$-ion assisted deposition of the AR and HR coatings, which use a single AlO$_x$ layer and AlO$_x$/a-Si multilayer stack, respectively. Figure 4.3 shows the side view of an I-3 passivated device.

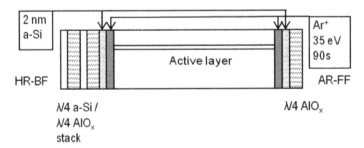

Figure 4.3 Schematic cross-section of an I-3 passivated diode laser including, rst, the Ar$^+$ ion milling cleaning process of the air-cleaved facet; second, the deposition of an amorphous Si interlayer 2 nm thick; and, third, the Ar$^+$ ion-assisted deposition of the AR/HR coating materials.

I-3 passivated 980 nm InGaAs/AlGaAs DQW lasers with buried-stripe lateral guides 2 µm wide and cavities 700 µm long (cf. Section 2.1.4.3) show excellent *P/I* characteristics with low threshold currents of ~20 mA, high slope ef ciencies of ~0.9 W/A, kink-free powers >300 mW, and thermal rollover >500 mW (Horie *et al.*, 1999). In addition, I-3 passivated devices are superior to non-I-3 passivated devices in accelerated aging life tests with no failures and stable operation at 250 mW cw and 50 °C over a test period of 3000 h. X-ray photoelectron spectroscopy (XPS) measurements con rm the signi cant improvement of aging characteristics compared to conventional coatings. I-3 treated facets are free of oxides whereas samples with conventional coatings show the presence of oxides. The damage-less removal of oxides and surface states reduces nonradiative recombination and surface current effects and therefore enhances the optical strength of the facet.

4.2.2.6   Pulsed UV laser-assisted techniques

In contrast to the preceding passivation technologies, N^2IBE and I-3, which ion mill the air-cleaved laser facets prior to the formation or deposition of a thin passivation layer, respectively, the technique to be described in this section employs a pulsed UV laser for cleaning the air-cleaved facets. After this cleaning of the facets, the concept consists of two steps (Kerboeuf *et al.*, 1999):

1. Oxide removal from the facets by pulsed laser-assisted etching (PLE).

2. Deposition of a thin passivating Si layer by pulsed laser-assisted deposition (PLD).

The essential parameters in both processes are the pulse energy density, pulse repetition rate, and number of pulses. Both processes use the same pulsed laser source (KrF excimer with wavelength 248 nm and pulse duration 25 ns). The beam is split into two optical paths, which are opened and closed by computer-controlled shutters. The same computer also controls the positioning of the laser bar for the PLE and PLD processes with normal incidence of the laser irradiation and normal exposure of the facet to the silicon plume, respectively. The time elapsing between completion of the PLE and beginning the PLD is a key parameter since gaseous contaminants due to the residual pressure in the vacuum chamber might condense on the freshly cleaned surfaces. Bear in mind that one atomic monolayer may be formed within a minute at a low pressure of only ~10^{-7} Torr, assuming a sticking coef cient equal to unity for the gaseous species. The elapsed time could be reduced to less than a second.

The oxide removal conditions were investigated on oxidized GaAs and AlGaAs test wafers, and fully PLE and PLD treated facets of 980 nm InGaAs/AlGaAs QW ridge lasers were aged in accelerated life tests to investigate the effect of this new passivation treatment. Various characterization techniques including scanning electron microscopy, atomic force microscopy, photoluminescence (PL; see Part III), and Raman scattering (see Part III) were employed to study potential modi cations of the surface morphology and sub-surface degradation induced by the PLE treatment (Kerboeuf *et al.*, 1999).

An upper limit of 180 mJ/cm^2 for the pulse energy density has been found for oxide removal on oxidized GaAs wafers. It should be recalled that 190 mJ/cm^2 is the value to reach the melting temperature at the GaAs surface (Vivet *et al.*, 1996). By proper adjustment of the laser irradiation conditions, surface oxide lms can be etched away without detectable degradation of the facet surface region. PL measurements showed that GaAs wafers could be fully deoxidized by applying 100 pulses, each of 150 mJ/cm^2 at a rate of 1 Hz.

Simulations on the thermal response of the exciting laser pulse showed that the return time to the initial surface temperature at around 25 °C is practically achieved after some milliseconds, which means that no heat accumulation occurs in the oxide removal processes.

Raman microprobing experiments revealed that the PLE process does not affect the lattice constant but induces compositional disorder, which leads to the broadening of Raman phonon modes, and to the activation of forbidden modes due to the relaxation of the momentum selection rules valid for the Raman scattering process (see Part III). The alloy disorder is sensitive to the speci c conditions of the PLE treatment. Thus, Raman spectroscopy can be considered a useful probing technique for monitoring the growth of compositional disorder at the laser facets during aging tests.

Two sets of laser devices were aged in air, at 50 °C, under automatic current control conditions of 350 mA and for a period of 10^3 h. The rst set had its facets PLE treated (80 mJ/cm^2, 10^4 pulses, 10 Hz) and PLD coated (5 nm Si). The second set had its facets only PLD passivated (5 nm Si) without PLE treatment. The lasers in both sets were conventionally coated with AR/HR lms. These rst tests unambiguously demonstrated the positive effect of the PLE treatment despite the fact that none of the steps involved in the passivation process had been fully optimized. Lasers without PLE suffered catastrophic degradation within 200 h, whereas PLE-treated lasers operated without any failure beyond 10^3 h (Kerboeuf *et al.*, 1999).

4.2.2.7 Hydrogenation and silicon hydride barrier layer process

Similar to the last passivation approaches, this technique cleaves the laser bars along natural (110) planes in an oxidizing air environment and subsequently places the material in a vacuum-controlled system for further processing without breaking the vacuum. However, the facet cleaning and deposition steps are different as given by:

- facet cleaning with either an ion-beam or a plasma of hydrogen, preferably with the addition of argon or xenon; and

- deposition of an amorphous hydrogenated silicon (a-Si:H) passivation layer preferably by sputtering a silicon target in the presence of hydrogen.

It is recommended that the cleaning and deposition steps are performed in the same chamber without moving the material and that the hydrogen plasma is maintained during these two processing steps. Preferably, the AR/HR coatings deposited over the passivation layer are also deposited in the same system without breaking the vacuum. There are actually two methods for cleaning the facets (Hu *et al.*, 2003).

The first method, which involves irradiating the facets with an ion-beam of hydrogen and either argon or xenon added in about equal proportions, is carried out at anode voltages of 30 V, which leads to ion energies in the range of 15 to 24 eV (50–80% of anode voltage). This gives an etching rate of $\sim$0.4 nm/min of the oxides and a low contamination of $\sim$0.2% of the facet due to sputtering of the stainless steel sample holder, which can be decreased further by coating the holder with a SiN layer. Applying a negative dc bias in the range -50 to -100 V to the sample during ion irradiation has the same effect as increased ion energy, but without the contamination problem by sputtering the sample holder. Once the oxides have been removed, the ion beam cleaning should stop to avoid degradation of the semiconductor surface due to the different sputtering rates for the individual components.

The alternative, second method, employs a hydrogen plasma produced by flowing between 2 and 60 sccm of H_2 (alternatively Ar/H_2 or Xe/H_2) through an electron cyclotron resonance (ECR) plasma source attached to the chamber held at a pressure of 10^{-4} Torr. The energy of the produced H^+-ions is in the range of 1 to some electronvolts with a maximum at 20 eV. To clean the facets takes between $\sim$10 and 60 min at an ECR power between 400 W and 1 kW, respectively. The sample is electrically grounded to minimize ion damage and the sample temperature is kept below 200 °C during ECR treatment. The effect of the hydrogen is to reduce the oxides formed by the semiconductor compound atoms and to fill the dangling bonds at the surface. Further positive effects are that hydrogen treatment releases surface strain and decreases interface charge density. The ECR procedure produces less damage, such as surface compositional contamination, structural change, doping, and lattice disorder (Hu *et al.*, 2003).

The deposition of the a-Si:H passivation layer is carried out immediately following the cleaning process by a sputtering method in the presence of a hydrogen plasma with the sample temperature kept below 200 °C and at a pressure in the range of 2×10^{-4} to 1×10^{-6} Torr. The method involves the ECR source and a silicon target in the same chamber. Si atoms are sputtered from the Si target by Ar^+-ions and directed onto the facet, while simultaneously the ECR-generated hydrogen plasma irradiates the facet surface. Both the silicon and hydrogen atoms will react and form a-Si:H.

The a-Si:H films are deposited to a thickness between 5 and 60 nm, have a bandgap energy of $\sim$1.6 eV, and therefore are transparent to laser emission wavelengths $\gtrsim$800 nm ($\lesssim$1.55 eV). Optical tests confirmed the transparency of the films, for example, at 980 nm. The electrical resistance is extremely high; its exact value could not be measured in a four-probe technique. The films are free of oxygen, have only H–Si–H bonding states, and have less stress than pure Si films. Further results from Raman spectroscopy measurements are that the films are amorphous and do not crystallize when irradiated with high power densities > 100 MW/cm^2 at 980 nm (see Chapter 8). Additional positive properties of the a-Si:H films due to the low-energy ions used in the deposition process include increased compactness, reduced pinhole density, and improved adhesion (Hu *et al.*, 2003).

Preliminary accelerated life tests, performed at 80 °C on 980 nm InGaAs/AlGaAs ridge lasers, operated at constant 300 mA drive current, demonstrated the relative optical robustness of the facets, passivated as described in this section. The optical power of the passivated lasers dropped gradually without sudden failure over ∼600 h, which is equivalent to 7000 h of room temperature operation. In contrast, lasers with a conventional pure Si barrier degrade rapidly and fail catastrophically at 220 h (Hu *et al.*, 2003).

4.2.3 Facet protection techniques

As discussed in Chapter 3, if the facet is not protected from contamination after cleaving in air, the laser will mainly degrade because of facet oxidation, which is enhanced by the emitted laser light and ambient moisture. However, in rare cases there may be also a positive effect in that the external differential quantum ef ciency increases due to a reduced mirror re ectivity caused by the growth of a uniform oxide layer. Nevertheless, facet oxidation and thus degradation can be easily suppressed by depositing dielectric protection coatings right after cleaving the laser chip. The most widely used materials for this purpose are SiO_2, Al_2O_3, SiN_x, and Ga_2O_3 having the required properties of a high laser-induced damage threshold, high chemical stability, good adhesion to semiconductor surfaces, and low optical absorption (see Section 1.3.8).

The effect of protective coatings on laser degradation modes such as COMD is different and depends on the material used. Thus, the COMD level of GaAs lasers with Al_2O_3 coatings is about twice that of lasers with SiO_2 coating lms (see Section 3.2.2.3). This can be mainly attributed to the higher thermal conductivity of Al_2O_3, which is about 20 times that of SiO_2 (Imai *et al.*, 1978). In addition, the thermal expansion coef cient of Al_2O_3 matches that of the GaAs crystal much better, and the interface between Al_2O_3 and the semiconductor is more stable under light irradiation (Fukuda and Wakita, 1980). In contrast, at the interface between SiO_2 and GaAs, oxidation of GaAs can take place and Ga atoms can out-diffuse into the SiO_2 under laser irradiation, which can generate vacancy point defects and hence nonradiative recombination leading to COMD. High-power 980 nm InGaAs/GaAs SCH-SQW ridge lasers 6 μm wide with lattice-matched InGaP cladding layers and low re ectivity (1%) front-facet Ga_2O_3 coatings have produced high peak cw power densities of 17 MW/cm^2, a value that is only limited by thermal rollover, but not by COMD (Passlack *et al.*, 1995). This is due to the very small absorption coef cient of ∼100 cm^{-1} of the Ga_2O_3 layer and a low front-facet interface recombination velocity of <10^5 cm/s, which assures a low local temperature rise at the front facet amounting to ∼9 K relative to the average bulk cavity temperature measured at 175 mW cw output power.

The protective function of the coating is usually combined with the function to modify the re ectivity of the end mirrors of the laser cavity. As discussed in Section 1.3.8, quarter-wavelength layers give minimal re ectivity, which approaches zero if the layer refractive index is equal to the square root of the semiconductor index assuming one for the index of air (see Equation 1.63). In contrast to this anti-re ection (AR) coating, half-wavelength thick layers are neutral to the re ectivity, keeping the re ectivity of the uncoated free-facet surface. AR coatings can increase the COMD power level. This effect, however, can be explained by considering the change in the ratio between the external and internal power of the laser beam (Ettenberg *et al.*, 1971). The power measured ex-facet is closer to the internal power (effective in COMD) in AR-coated lasers than in uncoated lasers, where the internal power is higher than the external power (see Equation 2.8). Therefore, the COMD power level of an AR-coated facet appears to be higher than that of an uncoated mirror, which is due more to optical effects than changes in interface properties. Neutral half-wavelength thick coatings can show improvements in COMD level by a factor of three and six for SiO_2 and Al_2O_3 coatings, respectively, which indicates real changes in the physical and chemical state of the coated facet surface, leading to a strong reduction of nonradiative recombination in the interface layer (Ladany *et al.*, 1977; Imai *et al.*, 1978; Hakki and Nash, 1974; Kappeler *et al.*, 1982).

Phase-shifting multilayer coatings controlling the position of nodes and anti-nodes of the standing wave inside the laser cavity relative to the mirror surface can have a positive effect on the COMD level (Eliseev, 1996; and references therein). A decrease in the eld strength at the facet can be obtained if the coating yields a negative amplitude re ectance shifting the anti-node position away from the facet surface by a quarter-wavelength, which is $\sim$60 nm for an 850 nm GaAs active medium. An effect on COMD can be expected if the degradation mechanism is taking place on the facet surface and is sensitively dependent on the distance along the cavity. A comparison between GaAs/AlGaAs DH 4 μm $\times$ 200 μm lasers with (i) uncoated mirrors, (ii) SiO_2 coatings (26.5% amplitude re ectivity), and (iii) multiple ZrO_2–SiO_2 layer pair coatings a quarter-wavelength thick (-28.3% amplitude re ectivity) led to average COMD power levels of (i) 85, (ii) 160, and (iii) 220 mW, respectively (Eliseev, 1996; and references therein). The higher value of the third group has been ascribed to the phase-shifting property of the multilayer coating stack.

4.3   Nonabsorbing mirror technologies

4.3.1   Concept

In general, there are two approaches for improving the COMD threshold, either to shift the laser emission to lower energies relative to the threshold absorption energy at the mirror facet or to increase the bandgap energy (absorption energy) of the facet material relative to the laser emission energy. In both approaches, the regions close to the facets become ideally transparent, that is, nonabsorbing to the laser light, and are called "windows" or "nonabsorbing mirrors" (NAMs). Window structures have a substantially reduced number of carrier pairs near the facet and the carrier recombination at the facet can be considered as negligible due to electric eld effects. Thinning the active layer can be employed to reduce the effective optical power density at the mirror facet. However, this approach yields only a limited COMD improvement but above all increases the threshold current density and deteriorates the temperature characteristics due to carrier over ow from the active layer into the cladding. These detrimental effects can be mitigated by thinning the active layer only near the facets and increased COMD levels can be obtained. We will discuss this approach in Section 4.4.3 below.

There are numerous methods to fabricate window layer type and nonabsorbing mirror structures. In the following, we describe some selected and proven techniques.

4.3.2 Window grown on facet

This approach involves the formation or regrowth of a thin wide-bandgap semiconductor material directly on the facet surface. It is rather simple and it allows selection of the proper window layer independent of the internal laser structure. To establish the highest integrity of the cleaved facet prior to the window layer, deposition is of critical signi cance.

4.3.2.1 ZnSe window layer

The window layer is a ZnSe layer 5 nm thick deposited on an essentially contamination-free facet surface. Preparing contamination-free facets can be achieved by cleaving in vacuum followed by in situ deposition or by air-cleaving followed by facet cleaning in vacuum with an appropriate method and in situ deposition of ZnSe. ZnSe has a bandgap energy of 2.75 eV corresponding to a 450 nm wavelength and can be grown on GaAs with low mechanical stress and a lattice-mismatch of only $\sim$0.27%. High crystal quality can be achieved at a low growth temperature of 300 °C. ZnSe can be formed on the mirror facets by any suitable deposition and growth technique. After completion of the ZnSe window layer formation, the facets are AR/HR coated.

ZnSe window lasers not only have a substantially improved resistance to COMD, but also exhibit a far lower gradual degradation rate of the laser output in accelerated aging tests. Thus, Al-free 980 nm InGaAs/InGaAsP/InGaP QW lasers passed without failure a series of stress tests including a "snap test" (500 mA dc for a short period, three times), a "purge test" (100 °C, 150 mA dc, 140 h), a post-purge "snap test" (same as rst test) and an aging test (85 °C, 300 mA dc, 10^3 h). In contrast, lasers with vacuum-cleaved facets but without ZnSe fail due to COMD during either the third or fourth test (Chand, 1997). The COMD power density level is 10 MW/cm^2, which is twice as high as that for facets without ZnSe windows. This gure is expected to be dramatically improved for depositing ZnSe windows on oxide-free facets (Syrbu et al., 1996).

4.3.2.2 AlGaInP window layer

Thin undoped $(Al_{0.7}Ga_{0.3})_{0.5}In_{0.5}P$ window layers were grown by low-pressure MOCVD on the cleaved facets of MBE-grown strained SQW AlGaInP ridge lasers 4 μm wide and 1200 μm long emitting at a wavelength of 680 nm. A kink-free maximum optical output power of $\sim$300 mW has been obtained, which is about twice as much as that of the conventional laser without window layer, where the power was limited by COMD. The threshold current of $\sim$100 mA and slope ef ciency of $\sim$0.75 W/A are the same as those of the conventional laser, which indicates that the window layer does not affect the laser properties. Slow-axis and fast-axis far- eld measurements up to 150 mW con rm the fundamental transverse mode operation. This technique applied to AlGaInP lasers appears to be promising, considering that these lasers are inherently very susceptible to facet degradation at high power (Watanabe *et al.*, 1994).

4.3.2.3   AlGaAs window layer

Maximum cw powers of 300 mW, limited only by thermal rollover, stable fundamental transverse mode operation up to 120 mW cw, and high reliability under 100 mW cw operation at 50 °C for >6000 h have been achieved for 830 nm $Al_{0.08}Ga_{0.92}As/Al_{0.31}Ga_{0.69}As$ DH lasers with active layers 60 nm thick, cavities 250 µm long, ridge waveguides, and $Al_{0.5}Ga_{0.5}As$ window layers 25 µm long at both facets. Lasers without a window structure have a COMD limited output power of $\sim$160 mW, which is about half that of window lasers. Threshold current and external quantum ef ciency are the same for both laser types, which is due to the fact that in the window region there is no generation of carriers by injection or light absorption, which could lead to free-carrier losses (Naito *et al.*, 1989). Extensive life tests showed that the degradation rate is not dependent on output power, but only on current and temperature. This is attributed to the strong suppression of mirror degradation. The thermal activation energy of the degradation rate is high at $\sim$0.85 eV, and lifetimes $>10^5$ h have been estimated for 100 mW cw operation at 25 °C (Naito *et al.*, 1991).

The window structure was prepared by etching down to the lower $Al_{0.31}Ga_{0.69}As$ guide layer and regrowing by MOCVD the high AlAs content window layer, which also serves as the upper cladding in the center part of the laser. Prior to the regrowth, the etched wafer is cleaned at 750 °C under AsH_3 ow for several minutes. A small V/III ratio of $\sim$10 in the regrowth was chosen for suppressing defect formation. Laser chips are formed by cleaving and are AR/HR coated and mounted junction-side down on Si heat sinks In solder bonded onto copper blocks (Naito *et al.*, 1989).

4.3.2.4   EMOF process

EMOF, which stands for epitaxial mirror on facet, is a proprietary window layer process originally developed by the startup Spectracom, acquired by ADC Telecommunications, Inc. in 1999. Facets are cleaved along (110) planes under vacuum and a nonabsorbing epitaxial mirror layer is deposited simultaneously in a batch process on each cleaved bar placed in a cassette. The nature of the material has not been disclosed, but it is a wide-bandgap material grown by MBE at low temperature to a thickness that achieves 9% re ectivity for the front-facet mirror. In reliability tests of 980 nm InGaAs GRIN-SCH SQW ridge lasers 4 µm wide, no power dependence of the degradation rate could be detected. Extremely low degradation rates of 4×10^{-7} per hour after 5000 h for 350 mW cw ex-facet have been achieved (Whitaker, 2000).

4.3.2.5   Disordering ordered InGaP

Window-type regions in visible InGaAlP diode lasers can be produced by employing a property of the GaInP active material to change its bandgap as a function of the degree of ordering of the alloy crystallography. $Ga_xIn_{1-x}P$ exists in two phases, ordered and disordered, which differ in the atomic arrangement of group-III atoms. In the ordered phase, a monolayer superlattice consisting of a periodic arrangement of Ga-rich planes and In-rich planes is naturally formed under special epitaxial conditions. The bandgap energy increases by typically 50 meV when changing from the ordered phase to the disordered phase (Gomyo *et al.*, 1987). The ordered phase grows on a standard (100) substrate whereas the disordered phase grows on a substrate misoriented relative to the (100) plane. Thus, a window laser can be formed by growing GaInP on a structured substrate with the center section on a (100) plane forming the cavity and the inclined end sections misoriented relative to (100) forming the window layers with higher bandgap energy (Minagawa and Kondow, 1989).

This feature of bandgap engineering can also be used for designing index-guided transverse lateral waveguides, a technique we want to mention here without further discussion, only to complete the list of relevant approaches discussed in Section 2.1.4.

The phase change from the ordered to random disordered alloy can also be achieved by impurity-induced disordering by diffusing Zn into the ordered GaInP regions. Various kinds of Zn diffusion techniques have been realized including solid phase diffusion using ZnO lm (Arimoto *et al.*, 1993), vapor phase diffusion of Zn in a closed tube (Ueno *et al.*, 1990), and selectively enhanced Zn diffusion into the active layer in the mirror region by an n-GaAs capping layer (Itaya *et al.*, 1991). The latter approach yielded a COMD-free maximum power >80 mW cw and 400 mW for pulsed operation of gain-guided lasers with active layers 60 nm thick, stripes 7 μm wide, window regions 20 μm long, p-side down mounted con guration, and without facet coatings. The power of 80 mW is limited by thermal saturation and is ve times as high as that of nonwindow structure lasers. Fundamental transverse mode high-power cw operation of >150 mW has been realized in compressively-strained 670 nm AlGaInP/GaInP DQW window structure lasers without degradation. Stable operation of >1500 h has been observed for 50 mW operation at 50 °C (Arimoto *et al.*, 1993).

4.3.3   Quantum well intermixing processes

4.3.3.1   Concept

Quantum well intermixing (QWI), also called QW disordering, involves the interdif-fusion of constituent atoms across the interface of a well–barrier structure resulting in a controlled modi cation of the material composition (see Sections 2.1.4.2 and 3.1.1.3). This causes a change in the con nement pro le and subband edge struc-ture in the QW leading to a blueshift of the effective bandgap energy and thus a modi cation of the refractive index.

The degree of intermixing depends on the technique applied, its parameters such as the type and concentration of impurities to enhance the interdiffusion, and the process temperature and time. The presence of point defects like vacancies or interstitials in the structure is mandatory for the QW disordering process to occur. These point defects diffuse through the structure and thereby allow the atoms, for example, group-III elements Al and Ga in an AlGaAs/GaAs QW structure, to hop from one lattice site to another across the interface, resulting in the intermixing of the well material with the barrier material.

Thermal annealing alone can lead to interdiffusion of constituent atoms for sufficiently high temperatures and long times. Typical temperatures are >800 °C and >600 °C for InGaAs/GaAs and InGaAs/InP QW structures, respectively. QWI, however, can take place at lower temperatures in the presence of point defects. Key technologies for introducing point defects include impurity-induced disordering, impurity-free vacancy diffusion, and to a lesser extent laser-induced annealing. QWI can be localized to selected regions so that bandgap engineering is only effective in these selected areas of the QW structure.

In Section 2.1.4.2, we described how this effect could be used to produce low-loss transverse lateral waveguide structures. Furthermore, QWI can also be used in realizing NAMs by producing the compositional disordering of QW or DH laser structures at the ends of the laser cavity, where they act as a window for laser emission and a barrier for electron–hole pairs. Figure 4.4 illustrates the principle of a QWI-NAM. As described previously, there are many different types of intermixing techniques, which are discussed in a series of review articles (e.g. Li, 2000, 1998; Yu and Li, 1998). In the following, we discuss some relevant methods along with their key properties and results achieved in NAM applications.

4.3.3.2   Impurity-induced disordering

Ion implantation and annealing

Ion implantation has been employed to enhance the disordering of QW structures by using implants such as Si and subsequent high-temperature annealing. High-temperature annealing is applied to anneal the implantation damage and accelerate the layer disordering process. The ion implantation process can be made effectively impurity-free if a matrix element is used as the implanted species, thus avoiding a potential degradation of the laser performance. Examples are P^+ implants in InGaAsP/InP lasers (Noël et al., 1996) and As^{4+} ions implanted in InGaAs/GaAs QW lasers (Charbonneau et al., 1995). In addition, nonelectrically active impurity implantation, such as B, F (O'Neill et al., 1989), Ar (Myers et al., 1990), and N (Hashimoto et al., 2000) implants used for the impurity-induced interdiffusion process, avoid the problems linked to free-carrier absorption and unintentional doping of the window layer. The latter would lead to conductivity and hence leakage current and increased nonradiative recombination, which would diminish the COMD robustness of the facet.

Crucial parameters are implant species, ion energies, doses and uxes, post-implantation annealing temperature, time, and capping layer material used in the annealing process. Implantation of low-energy ions along particular crystallographic

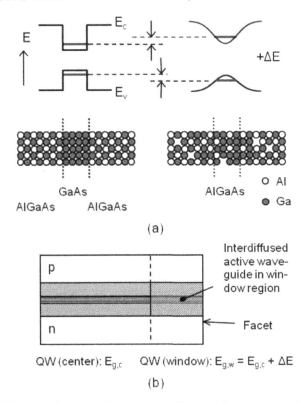

Figure 4.4 Schematic diagrams of the quantum well intermixing process and its application in nonabsorbing mirror design. (a) Simpli ed energy band structure of a GaAs/AlGaAs quantum well as an illustrating example, before (left) and after (right) intermixing. The nal increase of the quantum well bandgap energy after intermixing and only group-III atoms participating in the process are shown. (b) Simpli ed illustration of a quantum well intermixed nonabsorbing mirror structure.

directions, for example, perpendicular to the (100) plane of the laser wafer, can lead to strong ion channeling effects. Once an ion is in a channel, small-angle scattering events from atoms that line the walls of the channel enable the ion to steer quite a long distance along the channel before coming to rest due to electronic drag forces or a sharp collision causing the ion to exit the channel. However, if this depth exceeds the projected ion range then the channeling effect has to be minimized or eliminated. Usually this can be achieved by a small tilt and rotation of the wafer relative to the ion beam direction leading to an off-axis implantation direction of a few degrees. Ions have to be implanted in the vicinity of the QW active layer in order for the intermixing to occur due to the small diffusivity of vacancies and interstitials in material systems such as InGaAs/GaAs and GaAs/AlGaAs at temperatures <850 °C.

NAM structures in 980 nm InGaAs/InGaAsP/InGaP QW lasers were produced by implanting Si ions at energies in the range of 100 to 350 keV and doses of

10^{12} to 10^{13} ions/cm^2 with a post-implantation annealing under H$_2$ atmosphere (Hiramoto *et al.*, 1999). The blueshift of the bandgap energy increases with increasing implantation energy and amounts to $\sim$33 meV and 73 meV for 100 keV and 350 keV, respectively, obtained for a dose of 10^{13} ions/cm^2. The sensitivity of the bandgap energy blueshift on dose is $\sim$1.5 meV per 10^{12} ions/cm^2. The length of the window region forming the NAM is 30 μm. The threshold current of $\sim$20 mA and kink-free power of >300 mW cw are the same for lasers with and without a NAM structure, indicating the absence of scattering and absorption losses in the window structure.

Reliability tests performed at a constant power of 200 mW at 50 °C showed an increase in drive current of $\sim$6% over 1000 h for lasers without NAM, in contrast to lasers with NAM showing practically no change in drive current under the same conditions. In addition, from life tests performed on lasers with NAMs at 240 mW cw and 50 °C over 1000 h, and by assuming a thermal activation energy of 0.4 eV, a lifetime of 2×10^5 h at 25 °C and 20% current increase was estimated (Hiramoto *et al.*, 1999).

Superior performance and reliability results have been achieved for 980 nm InGaAs/AlGaAs QW ridge lasers with Si ion-implanted and annealed NAM structures (Yamamura *et al.*, 2000). These lasers have on top of the window region an H$^+$-implanted high-resistive region for avoiding current injection into the NAM region (see Section 4.4.1 below). NAM devices with cavities 900 μm long and 4%–AR/90%–HR coatings have >300 mW kink-free powers, maximum powers limited only by a thermal rollover of >600 mW, and are COMD-free at 200 mW automatic power control (APC) and 50 °C over the aging test time of 10^4 h. In contrast, the COMD level of lasers without NAM decreases proportional to the logarithm of aging time at 50 °C–100 mW–APC. For NAM lasers very low degradation rates of 10^{-6} per hour have been obtained at 50 °C–150 mW–APC. The wear-out mean time to failure (MTTF, see next Chapter 5 for details) could be modeled from the accelerated life tests as

$$\text{MTTF} \propto \exp\{E_a / k_B T\} \times I^{-n} \tag{4.2}$$

where $E_a = 0.70$ eV and $n = 6.9$ are   tting parameters to the experimental MTTF data obtained from the life tests carried out on groups of samples at different effective junction temperatures T and drive currents I. For 25 °C–250 mW–APC very long MTTFs of wear-out of $\sim$7.5 × 10^6 h, very low cumulative failure rates for wear-out, and sudden failures of 0.4% and 5.6%, respectively, have been estimated from the long-term reliability tests (Yamamura *et al.*, 2000). The various reliability terms used above will be discussed in Chapter 5.

Selective diffusion techniques

Diffusion techniques have been widely used in both lattice-matched QW structures grown on GaAs and InP substrates and strained QW structures by selectively diffusing a certain impurity into the QW active region near the facet to enhance the interdiffusion of group-III elements across the QW interface, thereby disordering the QW structure required for the formation of the NAM window. Mainly Zn (p-type ion) has been used in the diffusion process, but also to a lesser extent Si, Sn, and S (n-type ions). The window fabrication methods based on diffusion are in principle easy to implement and are sensitively dependent on annealing temperature and annealing time (Itaya *et al.*, 1996).

Watanabe *et al.* (1999) have developed a well-controlled Zn diffusion process by depositing a highly Zn-doped GaAs layer on top of the InGaP/InGaAlP MQW-SCH laser structure emitting in the 660 nm wavelength band. Annealing the laser wafer in a hydrogen ambience, typically at 620 °C for 40 min, enhances the localized diffusion of Zn atoms through a SiO_2 aperture down into the n-cladding layer through the MQW-SCH, resulting in the QW disordering. This technique avoids uncontrolled parasitic Zn diffusion effects, which could possibly impact laser operational characteristics and reliability.

A large blueshift of the bandgap energy of 156 meV can be achieved, which is mainly due to disorder by the Zn diffusion and not due to disorder of the natural ordered InGaP superlattice (see Section 4.3.2.5 above). The lasers have cavity lengths of 800 μm, which includes 2×20 μm for window regions at both ends, ridge widths of 5 μm, and mirror re ectivities of 10%–AR and 93%–HR. NAM lasers fabricated by this method operate COMD-free up to 180 mW cw at 25 °C, limited only by thermal rollover. This is an improvement by a factor of ∼3 compared to lasers without NAM. COMD power levels of 50 and 30 mW cw have been achieved at high operating temperatures of 85 and 95 °C, respectively. NAM devices have been operating without COMD for >1200 h at 50 mW cw and 70 °C. A MTTF (time when 63.2% of population failed, see Chapter 5) of $>3.7 \times 10^4$ h could be predicted for a failure criterion of 20% increase of drive current relative to its initial value in the APC tests (Watanabe *et al.*, 1999).

Ion beam intermixing

Ion beam intermixing is an alternative process for QW disordering where implantation of a low-mass ion such as Si^+, Be^+, B^+, Bi^+, or O^+ is carried out at an elevated temperature. This elevated temperature, however, is lower than the annealing temperature used in impurity-induced disordering processes and requires no high-temperature post-implantation annealing to enhance the compositional disordering. It has been found that implantation at elevated temperatures (referred to the ion beam mixing process) usually leads to much more ef cient disordering compared to implantation at room temperature followed by annealing at the same elevated temperature. A typical example is an ion beam mixing process using oxygen ions with a dose of 1×10^{13} ions/cm^2 at 550 °C (Pappert *et al.*, 1992, 1994). Ion beam intermixing is a powerful technology for fabricating opto- and microelectronic devices. In particular, there are numerous reports on engineering the bandgap energies of QW structures by ion beam intermixing to produce NAMs and transverse lateral waveguiding in diode lasers (Konig *et al.*, 1998; Goldberg *et al.*, 1996; Elenkrig *et al.*, 1995).

4.3.3.3   Impurity-free vacancy disordering

Impurity-free vacancy disordering (IFVD) also results in an enhanced QW inter-diffusion process (Li, 2000, 1998; Yu and Li, 1998; and references therein). The mechanism of IFVD requires the deposition of a dielectric cap layer on the surface of the QW sample and then annealing, usually a rapid thermal annealing (RTA) process at high temperature around 850–900 °C for $\sim$30–200 s. The challenge is to develop caps that control the amount of intermixing. SiO_2 caps are employed for both AlGaAs/GaAs and InGaAs/GaAs QW disordering, whereas Si_3N_4 caps are used to promote IFVD of InGaAs/InP QW structures. The latter produces good discrimination between disordered and as-grown regions. However, these layers are usually highly strained, which adversely affects the disordering process beneath the cap. SrF_2 caps deliver an enhanced lateral discrimination of the intermixing process. The blueshift of the energy gap increases with increasing annealing temperature and annealing time. For SiO_2 caps 200 nm thick e-beam evaporated on GaAs/AlGaAs QW structures, the blueshift for an RTA time of 90 s amounts to 37 and 132 meV obtained at RTA temperatures of 910 and 960 °C, respectively. SrF_2 caps 200 nm thick thermally evaporated deliver much lower blueshifts by a factor of $\sim$5 under the same conditions (Hofstetter *et al.*, 1998).

The microscopic origin of the intermixing process involves the out-diffusion of Ga atoms into the cap layer and the generation of vacancies on the group-III sublattice at the surface. The great affinity of the cap material for Ga is due to both the high diffusion coefficient of Ga and high solubility of Ga in the dielectric layer. These vacancies then diffuse through the structure during the RTA process, and their presence at the well–barrier interfaces enhances the anneal-driven interdiffusion of the group-III atoms. The degree of intermixing depends on the encapsulating dielectric material, its thickness and porosity, deposition and thermal treatment conditions, and stress level at the cap layer/semiconductor interface due to the large difference of thermal expansion coefficients.

The strength of IFVD is that it is free from the influence of additional carrier-related impurities, and avoids implantation-induced device damage. In addition, a laterally selective interdiffusion process, which locally modifies thickness, composition, or porosity of the dielectric cap layer, can achieve different on-chip wavelength shifts, which are required, for example, for the fabrication of photonic integrated circuits. Moreover, the diffusion of Ga vacancies depends highly on the stress field distribution in the sample, which is demonstrated in the following example. With the deposition of Ga_xO_y on top of the dielectric cap a significant suppression of interdiffusion during annealing can be achieved, which can be ascribed to the modification of stress distribution by the large expansion coefficient of Ga_xO_y. This effect can be employed to control and selectively achieve bandgap engineering through IFVD (Fu *et al.*, 2002).

COMD power threshold levels, improved by a factor of 2.6 for 860 nm GaAs/AlGaAs DQW-SCH ridge lasers 3 μm wide and 1000 μm long with NAM regions 60 μm long based on the IFVD technology, have been reported by Walker *et al.* (2002). Sputtered SiO_2 caps 50 nm thick, annealed at 875 °C for 60 s, produced a blueshift of 70 meV. A dielectric current injection window in the top p-contact ensures that current injection is inhibited within the NAM region. The lasers were not AR/HR coated and were not bonded/heat sunk to avoid possible facet contamination issues. Hence, to avoid thermal artifacts under cw test conditions, the lasers were pulse tested (400 ns, 1 kHz), which allowed a comparison of the relative facet performance. Thus, 220 mW and 580 mW have been achieved for COMD levels of lasers without NAM and with NAM, respectively. Both laser types have nearly the same threshold currents and differential quantum ef ciencies, indicating that no detrimental side effects are caused by the NAM region. It is expected that these COMD levels increase by using high-quality AR/HR coatings (Walker *et al.*, 2002).

4.3.3.4 Laser-induced disordering

The functioning of laser-induced QWI results from the heating effect of a strong laser irradiation leading to an increase in the concentration and diffusion coef cient of vacancies and as a consequence to the disordering of the QW structure. This process requires high power densities to melt the material, but the quality of the recrystallized material might be poor. Moreover, the spatial resolution for a cw irradiation is poor due to lateral heat conduction.

This issue can be eliminated by using the laser in high-energy pulsed mode. Thus, the disruption of the lattice due to rapid thermal expansion results directly in a localized increase in the required point defects. The high-energy pulse could introduce thermal shock damage. However, intermixing of GaAs/AlGaAs MQW structures with a KrF excimer laser operated at a wavelength of 248 nm with pulses of 20 ns duration and 220 mJ/cm^2 energy density showed a complete intermixed alloy with no detected residual damage, as found by Raman spectroscopy and sputter Auger pro ling. For energy densities exceeding $\sim$400 mJ/cm^2 visible damage could be observed (Ralston *et al.*, 1987). NAMs have been produced by irradiating the cleaved facets of AlGaAs/GaAs lasers with a pulsed excimer laser. The wavelength and pulse width were 306 nm and 17 ns, respectively, and the annealing powers varied in the range of 60 to 200 mJ. The COMD level reaches a maximum at $\sim$60 mJ and is three times higher than that of the untreated facets. Beyond 60 mJ it drops rapidly to the value of the untreated facet. The threshold current density is not changed relative to that of the as-grown laser in the investigated power range up to 200 mJ, indicating that the facet is not damaged by the excimer laser treatment (Lim *et al.*, 1994).

An alternative laser disordering technique is photon absorption-induced disordering (PAID) (McKee *et al.*, 1994). The process makes use of the poor thermal stability of the InGaAsP material system and relies on the generation of a high photo-created carrier density through bandgap-dependent absorption of the incident cw laser radiation within the active layer region of a multilayer structure. Subsequently, heat is generated by carrier cooling and nonradiative recombination causing the QWI process to occur. The PAID process offers some advantages including high blueshifts >100 meV in standard MQW structures, impurity-free processing, much lower power densities than other cw techniques, absence of a melt-phase, and layer composition selectivity by tuning the laser energy to the bandgap energy of the selected layer. As mentioned above, the poor spatial resolution due to lateral heat ow at cw laser irradiation can be improved to below 20 μm by using a pulsed laser. The absorption of pulsed high- uence photons in a QW structure locally increases the density of point defects, which enhance the QWI process in a subsequent high-temperature annealing treatment (Shin *et al.*, 1998).

4.3.4 Bent waveguide

Nonabsorbing etched-mirror structures based on a bent-waveguide concept, in which the guided laser beam exits from the active waveguide layers into the nonabsorbing cladding layer with higher bandgap energy, have been demonstrated for AlGaAs/GaAs SQW GRIN-SCH ridge lasers 3.5 μm wide (Gfeller *et al.*, 1992a, 1992b). The bent-waveguide structure in the mirror regions is formed by overgrowing a wet-etched channel 1 μm deep with 22° sloped walls in the n^+-GaAs (001) substrate by MBE. Exact wafer orientation is necessary to obtain symmetrical growth conditions on both channel slopes. The resulting kink in the active waveguide then allows decoupling of the beam into the cladding. Simulations showed that the beam is almost completely decoupled without distortion provided the slope angle of the bent waveguide is >20° and the NAM section length is <15 μm for a channel depth of 1 μm, otherwise the beam recouples partially into the active waveguide.

Pulsed COMD power levels are four times higher for NAM lasers with ∼300 mW compared to conventional etched-mirror lasers with ∼70 mW. In general, cw-operated etched-mirror lasers with NAM sections are free of COMD. A maximum, thermal rollover-limited COMD power level of 165 mW cw has been achieved with fundamental transverse lateral mode operation up to 80 mW. Due to beam divergence in the short NAM section a minor mirror coupling loss occurs, which causes an acceptable increase in the threshold current of, for example, ∼12% for a NAM section 5 μm long relative to a conventional laser. The far- eld patterns are not disturbed by the presence of the NAM section. Typical slow-axis and fast-axis beam divergence angles are 11° and 23°, respectively. Figure 4.5 shows schematically the NAM section and near- eld and far- eld patterns taken at two positions, one at a conventional facet outside the NAM section and the other where the beam partially recoupled into the active waveguide.

Mirror temperatures measured for different NAM section lengths re ect the overlap of the optical eld intensity with the absorbing waveguide pro le and are anticorrelated with the COMD levels. Typical temperature rises are 50 K at 30 mW cw for conventional etched mirrors compared to only 20 K in the NAM section. The origin of the remaining temperature rise of 20 K may be due to nonradiative recombination of diffused carriers from the pumped laser section or to optical power absorption at surface defects of the etched mirrors. Details of these temperature and COMD measurements including the measurement technique (Epperlein, 1993a, 1997) will be discussed in Part III.

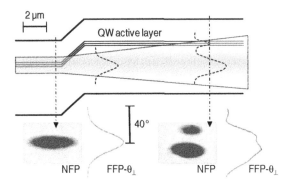

Figure 4.5 Scheme of a bent-waveguide nonabsorbing mirror (NAM) structure, illustrating the ared propagation of the beam (pro le shown as light grey) in the unguided NAM region and the anticipated mode pro les at locations without and with partial recoupling of the beam into the displaced active waveguide. Experimental veri cation of the beam recoupling effect by near- eld patterns (black) and transverse vertical far- eld patterns at two distinct locations, one far (>10 μm) in the NAM section with recoupling and the other at a conventional facet. Approximate scale along cavity axis only.

4.4 Further optical strength enhancement approaches

4.4.1 Current blocking mirrors and material optimization

4.4.1.1 Current blocking mirrors

The function of current blocking mirrors is to avoid the injection of electron–hole pairs near the facet surface. In this way, facet heating caused by nonradiative surface recombination and surface currents can be minimized and hence the positive feedback cycle to COMD interrupted. The use of a high-resistivity material for realizing the noninjection region is an effective approach, though also other techniques have been developed (Figure 4.6), which can be applied in combination with one of the NAM structures discussed above.

Rinner *et al.* (2003) reported a facet temperature reduction by a factor of 3–4 and increased COMD levels of InGaAs/AlGaAs SQW lasers junction In-soldered down on copper heat sinks with SiN current blocking layers 30 μm long at the front facet (Figure 4.6a). The "aspect ratio" of 15 between the blocking layer length of 30 μm and the total thickness 2 μm of all p-layers above the active layer is suf cient to prevent current spreading toward the facet. In addition, the highly doped p-cap layer is removed in the SiN current blocking region.

Herrmann *et al.* (1991) developed a segmented contacts approach. Here the top electrode was segmented in three parts consisting of a gain section 340 μm long and two mirror contact sections each 20 μm long of ridge (Al)GaAs QW lasers 4 μm wide. The in uence of a separately controllable potential in the mirror region on the temperature was studied. There is an optimum potential of 0 V, easily obtainable by grounding the mirror contacts, leading to a substantial temperature reduction of the

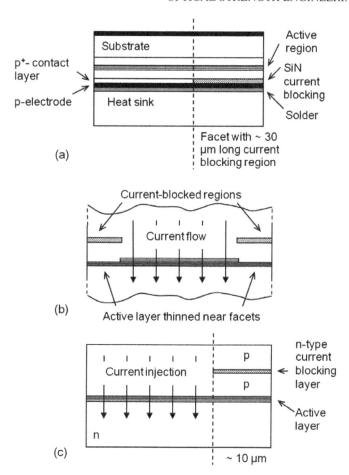

Figure 4.6   Current blocking mirror schemes. (a) SiN current blocking layer on top of the epitaxial laser structure lined up with the mirror edge and extending 30 μm into the laser. (b) Current-blocked regions and in addition thinned active layer sections near cavity facets with the latter for enlarging the mode spot size and bandgap energy to increase the COMD power level. (c) Current blocking facet comprising a p–n–p structure formed above the active layer near the facet.

mirrors by a factor of ∼3 for 10 mW output power and in relation to a mirror contact potential of ±0.4 V. The mirror temperatures were measured by Raman scattering (see Chapter 8).

Another approach deals with current-blocking regions near both facets in combination with a thinning of the active layer in the vicinity of the facets (Figure 4.6b) to enlarge the near- eld spot size without signi cant increase of the operating current (Shibutani *et al.*, 1987). The combined effect of both methods led to higher COMD levels by a factor of 1.4 and 50% lower degradation rates compared to the values of conventional lasers.

A different approach was realized by introducing a novel p–n–p current-blocking facet structure above the active layer (Figure 4.6c) in 980 nm InGaAs/GaAs DQW buried-ridge lasers with ridge mesa widths $\lesssim 4$ μm and cavity lengths of 1200 μm (Igarashi *et al.*, 2000). COMD power levels of uncoated lasers increased from 230 mW for as-cleaved conventional facets to 360 mW for the novel current blocking facets. The maximum output power is 760 mW cw for lasers with current-blocking and AR/HR-coated facets. Accelerated life tests at 250 mW APC and 50 °C yielded a degradation rate of 6.5×10^{-7} per hour and an estimated median lifetime of $>3 \times 10^{5}$ h at a 20% drive current increase. There is no difference between the *P/I* characteristics of lasers before and after the aging tests indicating the robustness of the p–n–p current blocking facet structure.

4.4.1.2 Material optimization

An optimized material is a crucial parameter in developing and fabricating diode lasers with high optical strengths in the laser cavity and at the laser mirrors. Contributing factors include the following:

- Laser structure and type, device geometry, and layout.
- Growth procedures, processing details, material properties.
- Mirror passivations, coatings, and re ectivities.
- Heat dissipation, device mounting, heat-sinking techniques.
- Measurement techniques, conditions, and accuracy.

This variety of factors actually makes impossible a valid comparison of performance and reliability data including maximum output powers, COMD power thresholds, and laser lifetimes of the various material systems.

In Section 1.1.5, we discussed the properties of some key compound material systems used for semiconductor lasers in the lasing wavelength range roughly between 0.4 and 1.6 μm. In spite of the aforementioned uncertainties and ambiguities when comparing device functionality data in the different wavelength regimes, we have extensively researched and discussed laser performance in this chapter, with a focus on maximum optical output powers and whenever possible COMD power densities, as well as life-test data of edge-emitting lasers in the wavelength range of ∼0.4 to 1.6 μm.

Section 2.1.3.3 in particular, but also Sections 2.1.3.4 and 2.1.3.5, discuss the most representative data of these properties in each material system and wavelength category. The elaborate discussions in these sections should be read in conjunction with the maximum power data listed in Tables 1.2–1.5. We want to refer to these sections regarding aspects of material and device optimization in the various wavelength bands.

4.4.2   Heat spreader layer; device mounting; and number of quantum wells

4.4.2.1   Heat spreader and device mounting

By depositing a thick Au layer onto the top electrode layer of a ridge laser, a signi - cant effect on the effective operating facet temperature and hence facet degradation performance can be achieved (Epperlein, 1993a; Epperlein and Bona, 1993a). Au is usually deposited by an electroplating technique to a thickness of ~ 5 μm. The cooling ef ciency depends sensitively on the recess distance of the Au layer from the mirror edge (see Figure 9.5b) and the mounting con guration of the laser chip on a heat sink (junction-side up, JSU; junction-side down, JSD). In all investigated cases the top electrode and p-contact layers are not recessed from the facet mirror edge.

Facet heating decreases in the sequence (i) 10 μm recess, JSU, (ii) 10 μm recess, JSD, and (iii) <1 μm recess, JSU.

In particular, facet heating is lowest for lasers with an aligned (<1 μm recess) heat spreader and JSU mounting and that is by a factor of ~ 1.5 relative to lasers with a 10 μm recessed heat spreader and JSD mounting, and by a factor of ~ 2.5 relative to lasers with a 10 μm recessed heat spreader but with JSU mounting. The ef cient cooling with an aligned heat spreader relative to a recessed heat spreader can be explained by taking into account the fact that the characteristic decay length of the facet mirror temperature into the cavity is ~ 6 μm (at 1/e point) (Epperlein, 1997). This means that the cooling effect of the heat spreader decreases with increasing recess distance from the mirror edge and is practically irrelevant for distances $\gtrsim 10$ μm. In Part III, we will discuss the temperature measurement technique along with further experimental data.

4.4.2.2   Number of quantum wells

The effects of number of active QWs on laser performance and reliability are manifold and have been discussed thoroughly in Section 2.1.3.4. We refer to this section and Section 7.4.6.3 for details. In general, lasers with MQWs in the active region have higher differential gain and transverse vertical optical con nement factor, which leads to lower threshold current density, weaker temperature dependence of laser characteristics, and higher characteristic temperature coef cient T_0 (see Section 1.3.7.3). If the losses are low, then a SQW is suf cient to supply the required gain for threshold; for higher losses, however, more active wells will be required (see Section 1.3.5.1).

More QWs in the active region, however, may also lead to the generation of mis t dislocations due to the increased total active layer thickness (Namegaya *et al.*, 1994) and stronger facet heating with the consequence of lowering COMD levels (Epperlein, 1997; Epperlein and Bona, 1993a, 1993b).

Figure 4.7 shows the facet temperature rise ΔT and COMD power level measured on red-emitting compressively-strained GaInP/AlGaInP QW GRIN-SCH lasers with cleaved, uncoated facets and ridge waveguides 4 μm wide and 500 μm long as a

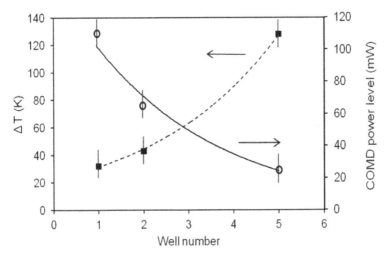

Figure 4.7 Experimental facet temperature rises ΔT at 15 mW cw and catastrophic optical mirror damage (COMD) power levels of compressively strained GaInP/AlGaInP GRIN-SCH QW lasers with cleaved, uncoated facets, and ridge waveguides 4 μm wide and 500 μm long as a function of the number of active quantum wells (QWs). AlGaInP/AlGaAs claddings for devices with one and two active QWs; AlGaInP claddings for devices with ve QWs. (Selected data adapted in modi ed form from Epperlein and Bona, 1993a; Epperlein, 1997.)

function of the number of active QWs. The temperature rise, increasing with the number of wells, is due to the cumulative effect of heat generation at the facet surface caused by nonradiative recombination processes in the individual QWs. The temperatures are directly proportional to the number of active QWs of up to two for lasers with mixed AlGaInP/AlGaAs claddings, whereas lasers with ve QWs and AlGaInP claddings show a stronger dependence due to the higher thermal and electrical resistivity of AlGaInP leading to higher temperatures (see Chapters 7 and 9 for details). The higher temperature then leads to a lower optical strength of the mirror facet with a reduced output power level at COMD.

Such an anti-correlation between ΔT and COMD has also been found independently on 830 nm AlGaAs QW GRIN-SCH lasers as a function of the degree of lattice disorder (defect level) in the laser facet. These lasers had dry-etched facets with passivation coatings half-wavelength thick and ridge waveguides 5 μm wide and 720 μm long (Epperlein *et al.*, 1993; Epperlein, 1993b, 1997). We will discuss all these signi cant world- rst diagnostic ndings including the measurement approach and technique in Part III.

4.4.3   Mode spot widening techniques

In this last section of the chapter, we discuss methods to expand the mode spot size at the facet aimed at reducing the local output power density and hence increase the optical strength of the facet mirror dramatically by enhancing the power level at which COMD occurs. This aspect of *facet load-off* includes:

- low optical con nement structures in the transverse vertical direction;
- thin active layer and tapered con gurations; and
- ared waveguide structures in the transverse lateral direction.

A great variety of techniques for increasing the mode spot in the diode laser near- eld have been discussed in detail in Section 2.1.3.5. These techniques for expanding the mode size in the transverse vertical direction include:

- thin waveguides;
- broad waveguides;
- low refractive index mode puller layers;
- optical traps;
- spread index or passive waveguides;
- leaky waveguides;
- spot-size converters; and
- photonic bandgap crystals.

We refer to Section 2.1.3.5 for further detailed information regarding technological aspects and device properties.

Another technique is based on the fact that DH lasers with thinner active layers are more resistant to COMD failure and QW lasers are more robust to COMD than DH lasers (see Section 3.2.2.3). This can be understood primarily in terms of the smaller optical con nement factor (larger mode) in thin active layers and hence smaller heat-generating area at the facet. Based on this effect, a thin-tapered-thickness (T^3) active layer con guration of a DH laser was developed with thinned active layers at a length of 100 µm at both ends of the 350 µm long cavity of 780 nm AlGaAs diode lasers with otherwise ordinary and optimized active layer thickness in the bulk of the cavity (Shima *et al.*, 1990). Thus, the mode spot at the facet is expanded relative to that inside the bulk of the cavity, which reduces the local power density at the same overall output power. The expansion of the mode leads also to narrower far
 elds with beam divergence angles of 11 and 15° in the transverse lateral and vertical directions, respectively. The thinned tapered active layer regions have been realized by a selective area growth method (see also Section 2.1.3.5, "Spot size converters"). AR/HR-coated devices yielded COMD-free operation up to 160 mW cw and failure-free life tests over 10^4 h at 70 °C, 30 mW APC, and over 2000 h at 50 °C, 50 mW APC aging conditions.

Finally, another way to increase the output aperture size at the facet is to use a
 ared lateral waveguide con guration. In this case, the lateral waveguide is tapered along the cavity axis and the mode spot size grows steadily up to the cavity end

and lls the increasing cross-section of the waveguide due to light diffraction. These lasers, known as tapered lasers and unstable resonator lasers, have been discussed in depth in Sections 2.4.4 and 2.4.3. For detailed information regarding technological details, laser device properties, and laser performances, the reader should refer to these sections.

References

Arimoto, S., Yasuda, M., Shima, A., Kadoiwa, K., Kamizato, T., Watanabe, H., Omura, E., Aiga, M., Ikeda, K., and Mitsui, S. (1993). *IEEE J. Quantum Electron.*, **QE-29**, 1874.

Bauer, T. L., Cavaliere, W. A., Linnell, D. C., and Ruckel, R. R. (1991). *US Patent* No. 5154333.

Bookham, Inc. (now part of Oclaro, Inc.) (2009). *High power laser diodes*. www.oclaro.com.

Boroditsky, M., Gontijo, I., Jackson, M., Vrijen, R., Yablonovitch, E., Krauss, T., Cheng, C., Scherer, A., Bhat, R., and Krames, M. (2000). *J. Appl. Phys.*, **87**, 3497.

Capasso, F. and Margaritondo, G. (eds.) (1987). *Heterojunction Band Discontinuities*, North-Holland, Amsterdam.

Chand, N. (1997). *US Patent* No. 5665637.

Charbonneau, S., Poole, P., Piva, P., Aers, G., Koteles, E., Fallahi, M., He, J., McCaffrey, J., Buchanan, M., Dion, M., Goldberg, R., and Mitchell, I. (1995). *J. Appl. Phys.*, **78**, 3697.

Elenkrig, B., Yang, J., Cassidy, D., Bruce, D., Lakshmi, B., and Champion, G. (1995). *IEEE Proceedings of the 7th Conference on Indium Phosphide and Related Materials*, 612.

Eliseev, P. G. (1996). *Prog. Quantum Electron.*, **20**, 1.

Epperlein, P. W. (1993a). *Jpn. J. Appl. Phys.*, **32**, 5514.

Epperlein, P. W. (1993b). *Conference on Lasers and Electro-Optics, CLEO/QELS 1993, OSA Tech. Dig. Ser.*, **11**, paper CTuN13, 158.

Epperlein, P. W. (1997). *Proc. SPIE*, **3001**, 13.

Epperlein, P. W. and Bona, G. L. (1993a). *Appl. Phys. Lett.*, **62**, 3074.

Epperlein, P.W. and Bona, G. L. (1993b). *Conference on Lasers and Electro-Optics, CLEO/QELS 1993, OSA Tech. Dig. Ser.*, **11**, paper CThQ4, 478.

Epperlein, P. W., Buchmann, P., and Jakubowicz, A. (1993). *Appl. Phys. Lett.*, **62**, 455.

Ettenberg, M., Sommers, Jr., H., Kressel, H., and Lockwood, H. (1971). *Appl. Phys. Lett.*, **18**, 571.

Fu, L., Wong-Leung, J., Deenapanray, P., Tan, H., Jagadish, C., Gong, B., Lamb, R., Cohen, R., Reichert, W., Dao, L., and Gal, M. (2002). *J. Appl. Phys.*, **92**, 3579.

Fukuda, M. and Wakita, K. (1980). *Jpn. J. Appl. Phys.*, **19**, 1969.

Gasser, M. and Latta, E. E. (1992). *US Patent* No. 5144634.

Gfeller, F. R., Buchmann, P., Epperlein, P. W., Meier, H. P., and Reithmaier, J. P. (1992a). *J. Appl. Phys.*, **72**, 2131.

Gfeller, F. R., Buchmann, P., Epperlein, P. W., Meier, H. P., and Reithmaier, J. P. (1992b). *Proceedings of the IEEE 13th International Semiconductor Laser Conference, Conf. Dig.*, 222.

Ghandhi, S., Kwan, P., Bhat, K., and Borrego, J. (1982). *IEEE Electron Device Lett.*, **3**, 48.

Goldberg, R., Piva, P., Mitchell, I., Poole, P., Fafard, S., Dion, M., Buchanan, M., Feng, Y., and Charbonneau, S. (1996). *IEEE Proceedings of the Conference on Optoelectronic and Microelectronic Materials and Devices*, 118.

Gomyo, A., Suzuki, T., Kobayashi, K., Kawata, S., Hino, I., and Yuasa, T. (1987). *Appl. Phys. Lett.*, **50**, 673.

Hakki, B. and Nash, F. (1974). *J. Appl. Phys.*, **45**, 3907.

Hashimoto, J., Ikoma, N., Murata, M., and Katsuyama, T. (2000). *IEEE J. Quantum Electron.*, **QE-36**, 971.

Herrmann, F. U., Beeck, S., Abstreiter, G., Hanke, C., Hoyler, C., and Korte, L. (1991). *Appl. Phys. Lett.*, **58**, 1007.

Hiramoto, K., Sagawa, M., Kikawa, T., and Tsuji, S. (1999). *IEEE J. Sel. Top. Quantum Electron.*, **5**, 817.

Hobson, W., Ren, F., Mohideen, U., Slusher, R., Schnoes, M., and Pearton, S. (1995). *J. Vac. Sci. Technol. A*, **13**, 642.

Hoffman, C. A., Gerritzen, H. J., and Nurmikko, A. W. (1980). *J. Appl. Phys.*, **51**, 1603.

Hofstetter, D., Maisenhölder, B., and Zappe, H. P. (1998). *IEEE J. Sel. Top. Quantum Electron.*, **4**, 794.

Horie, H., Ohta, H., and Fujimori, T. (1999). *IEEE J. Sel. Top. Quantum Electron.*, **5**, 832.

Hu, M., Kinney, L., Onyiriuka, E., Ouyang, M., and Zah, C. (2003). *US Patent* No. 6618409.

Hu, S. Y., Corzine, S. W., Law, K. K., Young, D. B., Gossard, A. C., Coldren, L. A., and Merz, J. L. (1994). *J. Appl. Phys.*, **76**, 4479.

Igarashi, T., Fukagai, K., Chida, H., Miyazaki, T., Horie, M., Ishikawa, S., and Torikai, T. (2000). *Proceedings of the IEEE 17th International Semiconductor Laser Conference, Conf. Dig.*, 25.

Imai, H., Morimoto, M., Sudo, H., Fujiwara, T., and Takusagawa, M. (1978). *Appl. Phys. Lett.*, **33**, 1011.

Itaya, K., Ishikawa, M., Hatakoshi, G., and Uematsu, Y. (1991). *IEEE J. Quantum Electron.*, **QE-27**, 1496.

Itaya, K., Mondry, M., Floyd, P., Coldren, L., and Merz, J. (1996). *J. Electron. Mater.*, **25**, 565.

Jastrzebski, L., Lagowski, J., and Gatos, H. C. (1975). *Appl. Phys. Lett.*, **27**, 537.

Kamiyama, S., Mori, Y., Takahashi, Y., and Ohnaka, K. (1991). *Appl. Phys. Lett.*, **58**, 2595.

Kappeler, F., Mettler, K., and Zschauer, K. H. (1982). *IEE Proc.*, **129**, Pt. 1, 256.

Katnani, F. and Margaritondo, G. (1983). *Phys. Rev.*, **B28**, 1944.

Kawanishi, H., Ohno, H., Morimoto, T., Kaneiwa, S., Miyauchi, N., Hayashi, H., Akagi, Y., Hijikata, T., and Nakajima, Y. (1990). *Proc. SPIE*, **1219**, 309.

Kerboeuf, S., Bettiati, M., Gentner, J., Belouet, C., Perriere, J., Jimenez, J., and Martin, E. (1999). *J. Electron. Mater.*, **28**, 83.

Konig, H., Mais, N., Ho ing, E., Reithmaier, J., Forchel, A., Mussig, H., and Brugger, H. (1998). *J. Vac. Sci. Technol. B*, **16**, 2562.

Ladany, I., Ettenberg, M., Lockwood, H., and Kressel, H. (1977). *Appl. Phys. Lett.*, **30**, 87.

Lambert, R., Ayling, T., Hendry, A., Carson, J., Barrow, D., McHendry, S., Scott, C., McKee, A., and Meredith, W. (2006). *IEEE J. Lightwave Technol.*, **24**, 956.

Li, E. H. (1998). *Proc. SPIE*, **3491**, 432.

Li, E. H. (ed.) (2000). *Semiconductor Quantum Wells Intermixing*, Gordon and Breach, Singapore.

Lichtenstein, N., Manz, Y., Mauron, P., Fily, A., Arlt, S., Thies, A., Schmidt, B., Muller, J., Pawlik, S., Sverdlov, B., and Harder, C. (2004). *Proceedings of the IEEE 19th International Semiconductor Laser Conference, Conf. Dig.*, 45.

Lim, G., Lee, J., Park, G., and Kim, T. (1994). *Proceedings of the IEEE 14th International Semiconductor Laser Conference, Conf. Dig.*, 155.

McKee, A., McLean, C., Bryce, A., De La Rue, R., Marsh, J., and Button, C. (1994). *Appl. Phys. Lett.*, **65**, 2263.

Minagawa, S. and Kondow, M. (1989). *Electron. Lett.*, **25**, 758.

Myers, D. R., Lee, K., Hausken, T., Simes, R. J., Ribot, H., Laruelle, F., and Coldren, L. A. (1990). *Appl. Phys. Lett.*, **57**, 2051.

Naito, H., Kume, M., Hamada, K., Shimizu, H., and Kano, G. (1989). *IEEE J. Quantum Electron.*, **QE-25**, 1495.

Naito, H., Kume, M., Hamada, K., Shimizu, H., Kazumura, M., Kano, G., and Teramoto, I. (1991). *IEEE J. Quantum Electron.*, **QE-27**, 1550.

Namegaya, T., Matsumoto, N., Yamanaka, N., Iwai, N., Nakayama, H., and Kasukawa, A. (1994). *IEEE J. Quantum Electron.*, **QE-30**, 578.

Noël, J. P., Melville, D., Jones, T., Shepherd, F. R., Miner, C. J., Puetz, N., Fox, K., Poole, P. J., Feng, Y., Koteles, E. S., Charbonneau, S., Goldberg, R. D., and Mitchell, I. V. (1996). *Appl. Phys. Lett.*, **69**, 3516.

O'Neill, M., Bryce, A., Marsh, J., De La Rue, R., Roberts, J., and Jeynes, C. (1989). *Appl. Phys. Lett.*, **55**, 1373.

Pappert, S. A., Xia, W., Zhu, B., Clawson, A. R., Guan, Z. F., Yu, P. K., and Lau, S. S. (1992). *J. Appl. Phys.*, **72**, 1306.

Pappert, S. A., Xia, W., Jlang, X. S., Guan, Z. F., Zhu, B., Liu, Q. Z., Yu, L. S., Clawson, A. R., Yu, P. K., and Lau, S. S. (1994). *J. Appl. Phys.*, **75**, 4352.

Passlack, M., Bethea, C., Hobson, W., Lopota, J., Schubert, E., Zydzik, G., Nichols, D., Jong, J., Chakrabarti, U., and Dutta, N. (1995). *IEEE J. Sel. Top. Quantum Electron.*, **1**, 110.

Ralston, J., Moretti, A., Jain, R., and Chambers, F. (1987). *Appl. Phys. Lett.*, **50**, 1817.

Rinner, F., Rogg, J., Keleman, M., Mikulla, M., Weimann, G., Tomm, J., Thamm, E., and Poprawe, R. (2003). *J. Appl. Phys.*, **93**, 1848.

Shibutani, T., Kume, M., Hamada, K., Shimizu, H., Itoh, K., Kano, G., and Teramoto, I. (1987). *IEEE J. Quantum Electron.*, **QE-23**, 760.

Shima, A., Matsubara, H., and Susaki, W. (1990). *IEEE J. Quantum Electron.*, **QE-26**, 1864.

Shin, J., Gurtler, S., Chang, Y., and Yang, C. (1998). *Appl. Phys. Lett*, **72**, 2808.

Silfvenius, C., Blixt, P., Lindstrom, C., and Feitisch, A. (2003). *Laser Focus World*, **39** (11), 69.

Silfvenius, C., Blixt, P., Lindstrom, C., and Feitisch, A. (2004). *Proc. SPIE*, **5336**, 132.

Silfvenius, C., Sun, Y., Blixt, P., Lindstrom, C., and Feitisch, A. (2005). *Proc. SPIE*, **5711**, 189.

Swaminathan, V., Freund, J. M., Chirovsky, L. M., Harris, T. D., Kuebler, N. A., and D'Asaro, L. A. (1990). *J. Appl. Phys.*, **68**, 4116.

Syrbu, A., Yakovlev, V., Suruceanu, G., Mereutza, A., Mawst, L., Bhattacharya, A., Nesnidal, M., Lopez, J., and Botez, D. (1996). *Conference on Lasers and Electro-Optics, CLEO/QELS 1996, OSA Tech. Dig. Ser.*, paper CTUC4, 78.

Sze, S. M. (1981). *Physics of Semiconductor Devices*, 2nd edn, John Wiley & Sons, Inc., New York.

Tihanyi, P. and Bauer, R. (1983). *US Patent* No. 4656638.

Tiwari, S. and Frank, D. J. (1992). *Appl. Phys. Lett.*, **60**, 630.

Tu, L. W., Schubert, E. F., Hong, M., and Zydzik, G. J. (1996). *J. Appl. Phys.*, **80**, 6448.

Ueno, Y., Fujii, H., Kobayashi, K., Endo, K., Gomyo, A., Hara, K., Kawata, S., Yuasa, T., and Suzuki, T. (1990). *Jpn. J. Appl. Phys.*, **29**, L1666.

Vivet, L., Dubreuil, B., Gibert-Legrand, T., and Barthe, M. (1996). *J. Appl. Phys.*, **79**, 1099.

Walker, C., Bryce, A., and Marsh, J. (2002). *IEEE Photonics Technol. Lett.*, **14**, 1394.

Wang, C., Shiau, D., Donetsky, D., Anikeev, S., Belenky, G., and Luryi, S. (2005). *Appl. Phys. Lett.*, **86**, 101910.

Watanabe, M., Tani, K., Sasaki, K., Nakatsu, H., Hosada, M., Matsui, S., Yamamoto, O., and Yamamoto, S. (1994). *Proceedings of the IEEE 14th International Semiconductor Laser Conference, Conf. Dig.*, 251.

Watanabe, M., Shiozawa, H., Horiuchi, O., Itoh, Y., Okada, M., Tanaka, A., Gen-ei, K., Shimada, N., Okude, H., and Fukuoka, K. (1999). *IEEE J. Sel. Top. Quantum Electron.*, **5**, 750.

Whitaker, T. (2000). *Compound Semicond.*, **6**, 52.

Wittry, D. B. and Kyser, D. F. (1966). *J. Phys. Soc. Jpn, Suppl.*, **21**, 312.

Wolfe, C. M., Holonyak, Jr., N., and Stillman, G. E. (1989). *Physical Properties of Semiconductors*, Prentice-Hall, Englewood Cliffs, NJ.

Yamamura, S., Hanamaki, Y., Kawasaki, K., Shigihara, K., Nagai, Y., Nishinura, T., and Omura, E. (2000). *Proceedings of the Optical Fiber Communication Conference, OFC 2000, OSA Tech. Dig.*, 394, paper ThK3-1.

Yu, P. Y. and Cardona, M. (2001). *Fundamentals of Semiconductors*, 3rd edn, Springer-Verlag, Berlin.

Yu, S. F. and Li, E. H. (1998). *IEEE J. Sel. Top. Quantum Electron.*, **4**, 723.

Chapter 5

Basic reliability engineering concepts

Semiconductor Laser Engineering, Reliability and Diagnostics: A Practical Approach to High Power and Single Mode Devices, First Edition. Peter W. Epperlein.
© 2013 John Wiley & Sons, Ltd. Published 2013 by John Wiley & Sons, Ltd.

Introduction

In the previous chapters, we described aspects of design and fabrication technology aimed at maximizing the performance and reliability of diode lasers. Reliability-related aspects included optical strength of laser materials, optimum design of transverse vertical and lateral waveguide structures, and optically robust mirror technologies. Reliability has to be an integral part of all phases of the laser product cycle, from product proposal, through design and fabrication, to use by the customer. Only then can a high-quality laser product be assured of meeting the speci cations in a speci ed environment for the required life. Hence, laser reliability can be de ned as the probability that a laser device will perform within speci ed performance limits for a speci ed length of time under the stated conditions without failure. On the other hand, failure can be de ned as the point when the laser does not comply with any of these requirements.

This chapter focuses on the mathematical methods employed in analyzing laser product reliability and failure data. Usually these data are available through various specially designed life tests carried out in the laboratory, but also through in- eld use with the objective of obtaining the most realistic values of parameters that characterize reliability, failure rates, and projected lifetimes. These topics will be discussed in the context of Chapter 6 next dealing with a diode laser reliability assessment program in detail.

Reliability mathematics and statistical methods and analysis play a central part in evaluating reliability test data in order to predict the reliability performance of a laser product during service. Predicting with some degree of con dence is very dependent on correctly de ning a number of parameters. For example, choosing the correct failure distribution, also called the life distribution, that matches the test data is of primary signi cance, otherwise the conclusions drawn from the evaluations will not be reliable. The con dence that depends on the sample size must be adequate to make correct decisions; or failure rate evaluations must be based on large enough and relevant populations to truly re ect practical normal usage in the eld. The discussion in the next chapter will also cover these aspects.

This chapter describes a minimum set of basic reliability terms and concepts needed to set up a meaningful reliability test program (Chapter 6) aimed at delivering correct product reliability results re ecting the robustness of a design and the quality of its realization. It consists essentially of ve sections related to single-device terms, and a nal section dealing brie y with system-related reliability aspects.

The rst section treats fundamental topics such as probability density function, cumulative distribution function, reliability function, hazard function or failure rate, and the bathtub failure rate curve. The next section is on failure distributions including the lognormal, Weibull, and exponential, and the application of the rst four functions in the rst section to these life distributions. The third and fourth sections describe reliability data interpretation terms, which include life-test data plotting, censoring data, mean time to failure, con dence limits, and reliability estimations. In the fth section, terms, aspects, and models of accelerated reliability testing will be discussed.

This chapter provides only a basic understanding of reliability engineering concepts, and the selection of material was dictated by what was needed in the context of this book. More extensive and detailed treatments of the subject can be found in the texts by, for example, Elsayed (1996), Tobias and Trindade (1986), Condra (1993), Nash (1993), Kececioglu (1991), and O'Connor (2002).

5.1   Descriptive reliability statistics

5.1.1   Probability density function

The failure probability density function (PDF), designated as $f(t)$, describes the probability that a device will fail at a given time t. Thus, $f(t)dt$ is the fraction of failures of a population occurring in an in nitely small interval dt. The general mathematical expression of the PDF is

$$\int_{0}^{\infty} f(t)\,dt = 1; \quad 0 \le f(t) \le 1 \quad \text{for } 0 \le t \le \infty. \tag{5.1}$$

A qualitative illustration of the PDF is shown in Figure 5.1a for an arbitrary distribution. The PDFs for the lognormal, Weibull, and exponential failure distributions have their own speci c forms that will be given in Section 5.2 below.

5.1.2   Cumulative distribution function

The cumulative distribution function (CDF), designated $F(t)$, describes that a random device drawn from the population will have failed by a given time t. The CDF for a population is called a life distribution. It is illustrated qualitatively in Figure 5.1b for an arbitrary failure distribution. The CDF is related to the PDF via the following relationship:

$$F(t) = \int_{0}^{t} f(t)\,dt; \quad 0 \le t < \infty \tag{5.2}$$

where $F(t)$ is the probability that a failure will occur before time t and the PDF indicates the velocity of the increase in $F(t)$ with time and is given by

$$f(t) = dF(t)/dt. \tag{5.3}$$

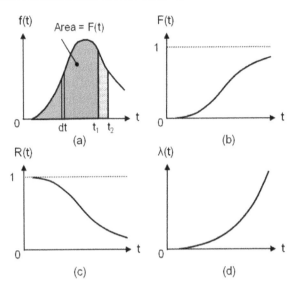

Figure 5.1   Schematic representation of the (a) probability density function $f(t)$, (b) cumulative distribution function $F(t)$, (c) reliability function $R(t)$, and (d) failure rate $\lambda(t)$ or hazard rate $h(t)$ as a function of time t for an unspeci ed life distribution. The area in (a) under the $f(t)$ curve shaded up to a time t corresponds to the cumulative failure distribution $F(t)$ at that time.

The CDF is also called the failure distribution function or unreliability function, which is linked to the reliability function $R(t)$ as discussed in the next section. The CDF begins at zero and starts to increase at the same point in time as the PDF, and since it is cumulative, it is always increasing by approaching a maximum of 1 or 100%.

In contrast to the rst de nition given at the beginning of this section, the CDF can also be interpreted as the percentage of all devices in the population, which failed by time t. $F(t)$ is the shaded area under the PDF in Figure 5.1a and is the probability of a device failing by time t. The area under $f(t)$ between two vertical lines drawn at time t_1 and time $t_2 > t_1$ in Figure 5.1a corresponds to the probability $F(t_2) - F(t_1)$ of a device surviving to time t_1, but then failing in the interval between t_1 an t_2. $F(t_2) - F(t_1)$ is also the percentage of the entire population, which fails in the interval $t_2 - t_1$.

5.1.3   Reliability function

The reliability function $R(t)$ represents the probability that a device will not have failed by a given time t or, in other words, $R(t)$ gives the percentage of devices operating until a time t without failure. In its general form it is therefore de ned by

$$R(t) = 1 - F(t) = \int_t^\infty f(t)\,dt. \tag{5.4}$$

The value of $R(t)$ is one at time zero and zero at infinite time. If there are n identical devices operating in a population, then $nF(t)$ is the expected number of failures by the time t and $nR(t)$ is the number expected of not having failed.

Example for illustration: Suppose $F(t) = 1 - (1 + 0.01t)^{-2}$. Then the probability of failing, for example, between 100 and 400 h, is $F(400) - F(100) = 0.96 - 0.75 = 0.21$. Or, the fraction of a population surviving past 1000 h is $R(1000) = 1 - 1 + (1 + 10)^{-2} \cong 0.008$. Expected failures out of a population of, let us say, 100 in the first 100 h, are $100 \times 0.75 = 75$, and in the next 300 h a further $100 \times 0.21 = 21$ devices are expected to fail.

There are two useful rules (Tobias and Trindade, 1986):

1. The probability that N independent and identical devices each with a reliability $R(t)$ will survive after time t is $[R(t)]^N$.

2. The probability that at least one of such devices will fail by time t is $1 - [R(t)]^N = 1 - [1 - F(t)]^N$ (see also Section 5.6.1).

5.1.4 Instantaneous failure rate or hazard rate

The hazard rate or failure rate is the probability that a device will fail during the next time interval Δt provided that it is functioning at the beginning of Δt and is given by

$$\overline{\lambda(t)} = \frac{N_f}{N_0 \Delta t} \tag{5.5}$$

where $\overline{\lambda(t)}$ is an estimate of the average failure rate during Δt, N_f is the number of failures during Δt, and N_0 is the number of good devices entering Δt. We can also get an expression for the hazard rate by calculating the probability of failing after surviving up to time t in an interval Δt, and divide by Δt to convert this probability into a rate:

$$\lambda(t) \text{ or } h(t) = \frac{F(t + \Delta t) - F(t)}{R(t)} \frac{1}{\Delta t} \rightarrow \frac{f(t)}{R(t)} \quad \text{for } \Delta t \rightarrow 0 \tag{5.6}$$

where we used Equation (5.3) for the derivative of $F(t)$ for letting Δt approach zero and Equation (5.6) then gives the instantaneous failure rate, also called the failure rate or hazard rate. It has units of failures per unit time, is not a probability, can have values > 1, and in general is a function of time. The most common units will be discussed below.

5.1.5 Cumulative hazard function

Analogously to Equation (5.2), the hazard function $h(t)$ can be integrated to yield the cumulative hazard function $H(t)$ (Tobias and Trindade, 1986)

$$H(t) = \int_0^t h(t)\, dt \tag{5.7}$$

where the integral can be expressed in closed form as

$$H(t) = -\ln R(t) \tag{5.8}$$

which can be verified by taking derivatives of both sides to obtain $h(t) = f(t)/R(t)$. By taking the natural anti-logarithms of both sides in Equation (5.8), we obtain a very useful expression linking CDF with failure rate:

$$F(t) = 1 - \exp\{-H(t)\} = 1 - \exp\left\{-\int_0^t h(t)\,dt\right\}. \tag{5.9}$$

By using Equation (5.4), the reliability function $R(t)$ can be linked to the hazard function

$$R(t) = \exp\{-H(t)\} = \exp\left\{-\int_0^t h(t)\,dt\right\}. \tag{5.10}$$

These equations show that, by knowing any quantity from $F(t), f(t), h(t),$ or $H(t)$, the others can be calculated. $H(t)$ can play an important role in graphical plotting methods for estimating life distribution parameters from failure data (see Sections 5.3.1.2 and 5.3.1.3).

5.1.6   Average failure rate

Since the hazard rate $h(t)$ (or $\lambda(t)$) changes over time, it may be useful to define a single average failure rate (AFR) characterizing the failure rate over a certain interval between times t_1 and t_2 by using (Tobias and Trindade, 1986)

$$\text{AFR}(t_1, t_2) = \frac{\int_{t_1}^{t_2} h(t)\,dt}{t_2 - t_1} = \frac{H(t_2) - H(t_1)}{t_2 - t_1} = \frac{\ln R(t_1) - \ln R(t_2)}{t_2 - t_1} \tag{5.11}$$

which simplifies to

$$\text{AFR}(T) = \frac{H(T)}{T} = \frac{-\ln R(T)}{T} \cong \frac{F(T)}{T} \text{ for small } F(T) \tag{5.12}$$

if the time interval is from 0 to T.

5.1.7   Failure rate units

Failure rates are usually very small numbers, if expressed in units of failures per hour, which may be inconvenient, as an *example may illustrate*: if $N_f/N_0 = 2 \times 10^{-1}$ devices fail in $\Delta t = 10^4$ h, then the failure rate would be $\lambda = 2 \times 10^{-5}$ h^{-1}.

A more easily usable scale would be percent per thousand hours (%/k), which means 1 failure for each 100 units operating for 1000 h. The factor would then be 10^5

to convert h^{-1} units to %/k units and the above example would translate to a failure rate of 2%/k.

However, a now well-established and widely used scale for very small failure rates is parts per million per thousand hours (ppm/k), which means 1 failure is expected out of 1 million devices operating for 1000 h. A more popular name for ppm/k is FIT, which stands for "failure in time." In simple terms, it equals the number of failures in 10^9 device hours. The factor is 10^9 to convert h^{-1} units to FIT units. For example, if 0.1% of a population fails in 10^4 h, then the failure rate is 100 FITs. We will use this unit in all forthcoming failure rate data.

In summary, the conversion of equivalent failure rates in different units is as follows:

- Failures per hour $\times 10^5 = $ %/k.

- Failures per hour $\times 10^9 = $ FIT.

- %/k $\times 10^4 = $ FIT.

5.1.8   Bathtub failure rate curve

Diode lasers are generally nonrepairable devices and cannot be reused after failure. Figure 5.2 shows schematically the failure rate of nonrepairable devices as a function of operating time. The typical shape of the curve, known as the bathtub curve, consists essentially of three regions.

The initial region has a decreasing failure rate and is known as the early failure or infant mortality period. During this period, the weak parts that were marginally functional are weeded out. The failures in this region are usually due to serious defects introduced during the manufacturing process. They are often called intrinsic failures, because they result from causes within the device. If defects cannot be eliminated at source, then eliminating this region is accomplished by screening the devices in so-called burn-in stress tests, such as high-temperature, electrical overstress, or temperature cycling tests (see Chapter 6). Devices passing these screening tests are

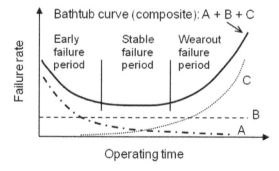

Figure 5.2   Simpli ed illustration of the bathtub curve for failure rates composed schematically of the early failure period (A), stable failure period (B), and wearout failure period (C).

usually ready for shipment. These devices are ultimately subject to the next two degradation phases of random and wear-out failures. As we will see further down, the initial early failure rate region can be modeled by a Weibull hazard function.

The long, fairly at region is called the stable failure period because here failures seem to occur in a random fashion at a relatively constant rate enabling a mean time between failures (MTBF) to be expressed, as we will discuss further below. Most of the useful life of a device should take place in this region of the curve and reliability testing is conducted here to determine values for the failure rate. Defects not as severe as those in the infant mortality region cause failures here, but predominantly random environmental or operating events, which can overstress a device. Failures here are called extrinsic, because they result from events external to the device. Mathematically, the constant failure rate of the exponential life distribution is perfectly suited to model such random failures (see Section 5.2).

The nal part of the curve is characterized by an increasing failure rate and is called the wear-out failure regime. This is the region, in which a major failure mechanism progresses to the point, where it can cause failure of all the surviving devices. Here microscopic defects grow over time, shortening the useful device life dramatically. It is possible, but unlikely, that more than one failure distribution will be involved in the wear-out region. The failures are usually considered to be intrinsic. A major objective of manufacturing high-reliability diode lasers and supportive reliability tests is to shift the onset of the wear-out region far enough in time so that it does not impact the speci ed life of the laser product. Wear-out failure can be combated by developing robust laser designs, damage-resistant materials, and defect-free laser wafer growth and processing. The wear-out regime can be best tted by the statistics of the lognormal and Weibull life distributions (see next section).

5.2 Failure distribution functions – statistical models for nonrepairable populations

5.2.1 Introduction

There are many statistical probability distribution functions, including the normal, lognormal, Weibull, and exponential functions. The normal distribution is not as common in reliability work as the others. The two-parameter continuous normal, or Gaussian, distribution is ideally suited for dealing with the statistics of random events, numbers, and phenomena. It is also frequently used to model measurement errors of almost any kind. Many populations encountered in industrial applications have bell-shaped symmetrical distributions that can be nicely modeled by the normal distribution. However, it rarely ts the failure distributions empirically determined in reliability work. The vast majority of reliability results have been modeled by the lognormal, Weibull, or exponential failure distribution. Therefore, we will focus on these functions and will not include the normal distribution in the discussions below.

There are some signi cant parameters in reliability terminology, including t_{50}, t_{16}, t_{75}, which designate the times by which 50%, 16%, 75% of the devices in a population have failed. The location parameter, also called measures of central values or measures of location of a distribution, locates it in time. The shape parameter, also called measures of variation, gives a quantitative measure of the shape or spread of a distribution. The relevant parameters for each distribution will be given below.

5.2.2 Lognormal distribution

5.2.2.1 Introduction

The lognormal is not a separate distribution function; it can simply be derived by taking the natural logarithms of all data points and analyzing the transformed data as a normal distribution. If a normal distribution of the random variable x has the mean μ and standard deviation σ then $t_f = e^x$ is the random failure time variable of a lognormal distribution with the quantities $t_{50} = e^\mu$ and σ.

Alternatively, starting with a lognormal distribution population of variables t_f and with a median t_{50} and shape parameter σ, then the population of logarithmic failure times $x = \ln t_f$ is normal with a mean value of $\mu = \ln t_{50}$ and standard deviation σ. The logarithm of a lognormal distribution is a normal distribution. The parameter σ can be considered as a shape parameter for the lognormal distribution. It is not the standard deviation of the population failure times, but is, in units of logarithmic time, the standard deviation of the population of logarithmic failure times   tted by a normal distribution.

5.2.2.2 Properties

The PDF $f(t)$ for the lognormal distribution is given by

$$f(t) = \frac{1}{\sigma t \sqrt{2\pi}} \exp\left\{ -\frac{1}{2} \left(\frac{\ln t - \mu}{\sigma} \right)^2 \right\} \tag{5.13}$$

while the CDF $F(t)$ has the form

$$F(t) = \frac{1}{\sigma \sqrt{2\pi}} \int_0^t \frac{1}{x} \exp\left\{ -\frac{1}{2} \left(\frac{\ln x - \mu}{\sigma} \right)^2 \right\} dx. \tag{5.14}$$

The reliability function is

$$R(t) = 1 - F(t) \tag{5.15}$$

and the failure rate is

$$\lambda(t) = \frac{f(t)}{1 - F(t)}. \tag{5.16}$$

The lognormal is, like the normal distribution, a two-parameter function with the important quantities (Condra, 1993) listed in Table 5.1.

The shape parameter σ strongly in uences the shape of the PDF $f(t)$ and CDF $F(t)$. Figure 5.3 shows the curves calculated for a series of σ values. As we will see

Table 5.1 De nitions and formulae of important lognormal distribution parameters.

Parameter	Definition
Mean (arithmetic average)	$t = \exp\{\mu + (\sigma^2/2\}$
Median (t_{50})	$t = e^{\mu}$
Mode (peak location of $f(t)$)	$t = \exp\{\mu - \sigma^2\}$
Location parameter	e^{μ}
Shape parameter	σ
s (estimate of σ)	$\ln(t_{50}/t_{16})$

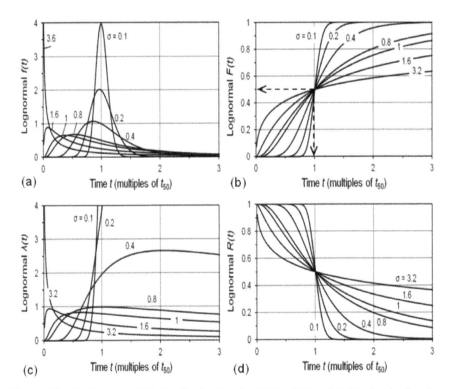

Figure 5.3   The lognormal life distribution function. Plots of (a) probability density function $f(t)$, (b) cumulative distribution function $F(t)$, (c) failure rate $\lambda(t)$, and (d) reliability function $R(t)$ calculated as a function of time t (in units of median time t_{50}) and at different values of the shape parameter σ. The dashed lines in (b) indicate the 50% failure point corresponding to the median time t_{50}.

in the next section, their shapes resemble those calculated for a Weibull distribution. This similarity indicates that both functions can t the same set of experimental data. Whether a lognormal or a Weibull will work better for a set of data can be decided from the shape of the histogram of the logarithm of data. If the shape is symmetrical and bell-like, the lognormal will match the data, whereas if the histogram is left skewed, the t of the data is better done with a Weibull (Tobias and Trindade, 1986).

Also shown in Figure 5.3 are the reliability function $R(t)$ derived from $R(t) = 1 - F(t)$ and the failure rate function $\lambda(t)$ calculated by using $\lambda(t) = f(t)/(1 - F(t))$ for the same set of σ values. In Section 5.3 it is shown, how to extract values for the shape parameter σ and the median time t_{50}. The shapes of these curves are also very similar to the shapes taken on by the Weibull failure rate (see next section). Failure rate curves with large values of σ are similar in shape to curves with small β values of the Weibull distribution. For large $\sigma > 2$ values the failure rate practically decreases throughout life with a strong initial rate decreasing with time. Typical high values of $\sigma \sim 4$ indicate early manufacturing problems leading to failures not under control. With increasing maturity of manufacturing, the σ value approaches one, which re ects roughly a constant failure rate. However, small σ values $\lesssim 0.5$ describe a failure rate increasing with time, which represents the wear-out failure regime. Thus, the lognormal distribution can describe all three failure rate regions: namely, that of early failures, stable failures, and wear-out failures.

5.2.2.3 Areas of application

The model underlying the lognormal distribution is called a *multiplicative growth model*. In essence, this model says that at any instant of time, the process undergoes a random increase of degradation that is proportional to its present state. The multiplicative effect of all these random growth processes ultimately builds up to failure (Tobias and Trindade, 1986). Whenever a multiplicative degradation process is occurring, in which a defect gets progressively worse due to accumulated damage, the lognormal model tends to apply. The main requirement is that the change in the degradation process at any time is a small random proportion of the accumulated degradation up to that time. Examples of such gradual degradation processes involve many failure mechanisms, including diffusion effects, corrosion processes, chemical reactions, oxide growth, and crack growth propagation in electrical and optical semiconductor devices.

5.2.3 Weibull distribution

5.2.3.1 Introduction

The Weibull is a three-parameter distribution function. By adjusting the shape parameter β and the timescale parameter η, which is also called the characteristic life parameter α, a large variety of functional behavior can be created to t a wide range of experimental life-test data. Both β and η must be >0 and the distribution is a life distribution de ned only for positive times t. The location or time delay parameter γ is only used to move the distribution forward or backward in time. If the distribution begins at time $t = 0$, then $\gamma = 0$, and the distribution becomes the two-parameter Weibull, which is very common in reliability work.

The Weibull distribution developed by Weibull (1951) has proved to be a exible and successful model for many product failure mechanisms. It applies to both decreasing failure rates, typical of early failures, and increasing failure rates describing the long-term wear-out regime. In addition, it can handle the stable failure period regime for $\beta = 1$, when the Weibull reduces to a standard exponential distribution with constant failure rate $\lambda = 1/\eta$.

5.2.3.2   Properties

The Weibull PDF $f(t)$ is

$$f(t) = \frac{\beta}{\eta} \left(\frac{t - \gamma}{\eta} \right)^{\beta - 1} \exp \left\{ - \left(\frac{t - \gamma}{\eta} \right)^{\beta} \right\} \tag{5.17}$$

and the CDF $F(t)$ is given by

$$F(t) = 1 - \exp \left\{ - \left(\frac{t - \gamma}{\eta} \right)^{\beta} \right\} . \tag{5.18}$$

The Weibull reliability function can be written as

$$R(t) = \exp \left\{ - \left(\frac{t - \gamma}{\eta} \right)^{\beta} \right\} \tag{5.19}$$

and the failure or hazard rate function is

$$\lambda(t) = \frac{\beta}{\eta} \left(\frac{t - \gamma}{\eta} \right)^{\beta - 1} = \frac{\beta}{(t - \gamma)} \left(\frac{t - \gamma}{\eta} \right)^{\beta} . \tag{5.20}$$

In addition to the above major Weibull formulas, some derived expressions for useful parameters (Tobias and Trindade, 1986) are listed in Table 5.2 for $\gamma = 0$ for simpli cation.

The shape parameter β plays a dominating role in determining the shape of the $f(t)$ and $\lambda(t)$ functions and therefore in tting a broad range of reliability data. In Figure 5.4, we calculated a set of curves of all four functions $f(t)$, $F(t)$, $R(t)$, and $\lambda(t)$ for a series of β values. For $0 < \beta < 1$, $f(t)$ and $\lambda(t)$ are exponentially decreasing from in nity toward zero making this type of Weibull an ideal model for early failure mechanisms typical of the initial part of the bathtub curve. For $\beta = 1$, $f(t)$ is

Table 5.2 Definitions and formulae of important Weibull distribution parameters.

Parameter	Definition
Scale parameter η (= characteristic life α) at:	$F(\eta) = 0.632$ (63.2% of population fails by η point independent of value of β)
AFR(t_1, t_2)	$[(t_2/\eta)^\beta - (t_1/\eta)^\beta]/(t_2 - t_1)$
Median (t_{50})	$\eta(\ln 2)^{1/\beta}$
Mode (peak location of $f(t)$ for $\beta > 1$ at:)	$\eta(1 - 1/\beta)^{1/\beta}$
System characteristic life α_s when n components have independent Weibull distributions with same β each but different α	$\alpha_s = \left(\sum_{i=1}^{n} \dfrac{1}{\alpha_i^\beta}\right)^{-1/\beta}$
Shape parameter β	$\beta = \ln[-\ln(1 - F(t))]/[\ln(t/\eta)]$
Time t	$t = \eta[-\ln(1- F(t))]^{1/\beta}$
Parameter η (= α)	$\eta = t/[-\ln(1 - F(t))]^{1/\beta}$

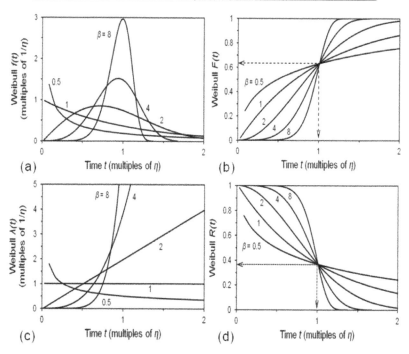

Figure 5.4 The Weibull life distribution function. Plots of (a) probability density function $f(t)$, (b) cumulative distribution function $F(t)$, (c) failure rate $\lambda(t)$, and (d) reliability function $R(t)$ calculated as a function of time t (in units of timescale parameter η) and at different values of the shape parameter β. The dashed lines in (b) indicate the 63.2% failure point corresponding to the timescale parameter μ.

exponentially decreasing from $1/\eta$ and the Weibull reduces to a standard exponential distribution with constant failure rate $\lambda(t) = 1/\eta$. For $\beta > 1$, $f(t)$ starts at zero (for $\gamma = 0$) and rises to a peak at $\eta(1 - 1/\beta)^{1/\beta}$ and then decreases toward zero with increasing time. $\lambda(t)$ starts also at zero and increases with time at a rate dependent on the value of β. For $\beta = 2$, $\lambda(t)$ is linearly increasing and the $f(t)$ distribution is also known as the Rayleigh distribution. For $\beta \geq 3$, the $f(t)$ function has a bell-shaped appearance narrowing in width with β, and the failure rate increases rapidly with time, which represents a useful model for wear-out failure mechanisms at the long-term end of the bathtub curve.

5.2.3.3 Areas of application

As described above, the Weibull distribution is highly exible and can be used to describe all three regions of the bathtub curve. It applies to a wide range of failure phenomena. In particular, the Weibull model appears to be applicable in cases where the weak link or the rst of many aws propagates to failure. Examples are dielectric breakdown, capacitor failures, or fracture in ceramics. In general, fatigue tests at constant alternating stress, early failures in environmental stress screening, and any set of items with defects with a range of severity are typically described by Weibull distributions.

However, as outlined in Section 5.2.2.3, degradation processes linked to chemical reaction, corrosion, migration, or diffusion, typical of many semiconductor failure mechanisms, are better matched by the lognormal than the Weibull distribution on the basis of the multiplicative random failure growth model. The Weibull distribution tends to provide a better t to short time failures in the infant mortality regime, whereas the lognormal model is better in predicting longer lifetimes in the wear-out failure regime.

5.2.4 Exponential distribution

5.2.4.1 Introduction

The exponential distribution is a special case of the Weibull distribution occurring when the Weibull shape parameter $\beta = 1$, which describes a constant failure rate life distribution. From a reliability point of view, the exponential distribution usually does not represent a single mechanism. Instead, it is the sum of several other distributions. Such a *mixed-mode failure mechanism* may govern most products and therefore the exponential distribution has been the preferred basis for describing, modeling, and predicting product reliability.

Equipment repeatedly repaired or device populations with replacements of failed devices may contain a variety of components in a variable state of wear. This may lead to an admixture of varying projected lifetimes and thus conspire to yield a roughly time-independent failure rate. In simple terms, the addition of a decreasing infant mortality rate and an increasing wear-out failure rate leads to an approximately constant failure rate over a certain period of time.

5.2.4.2   Properties

The PDF $f(t)$, CDF $F(t)$, reliability function $R(t)$, and failure rate $\lambda(t)$ or hazard rate $h(t)$ function for the exponential distribution are given by

$$f(t) = \lambda \exp\{-\lambda t\} \tag{5.21}$$

$$F(t) = 1 - \exp\{-\lambda t\} \tag{5.22}$$

$$R(t) = 1 - F(t) = \exp\{-\lambda t\} \tag{5.23}$$

$$\lambda(t) = h(t) = \frac{f(t)}{R(t)} = \frac{\lambda \exp\{-\lambda t\}}{\exp\{-\lambda t\}} = \lambda \tag{5.24}$$

where λ is the single unknown parameter that de nes the exponential distribution. The units for λ are failures per unit time. This means that, when failure rates are expressed, for example, in percent per thousand hour units or in FITs, these units have to be converted rst to failures per unit time before making any calculations using the exponential formula. Plots of the above four functions are shown in Figure 5.5 as a function of time.

By using Equations (5.11) and (5.23) the average failure rate between time t_1 and time t_2 can be calculated as

$$\text{AFR}(t_1, t_2) = \lambda. \tag{5.25}$$

The time t corresponding to any percentile of failure can be obtained by inverting the formula for $F(t)$. The inversion of Equation (5.22) gives

$$t = -\ln(1 - F(t))/\lambda. \tag{5.26}$$

Example for illustration: A certain type of diode laser has a constant failure rate of 1000 FITs $= 1 \times 10^{-6}$ h^{-1}. Then the probability that one of these lasers fails before 16 kh of use is calculated as: $F(16\,\text{kh}) = 1 - \exp\{-0.000\,001 \times 16\,000\} = 0.016 = 1.6\%$. Moreover, the waiting time for 1% failures is calculated as $t = -\ln(1 - 0.01)/1 \times 10^{-6}$ h$^{-1} = 10\,050$ h according to Equation (5.26).

The mean time to failure (MTTF) for the exponential distribution can be de ned as

$$\text{MTTF} = \int_0^\infty t f(t)\, dt = \int_0^\infty t \lambda \exp\{-\lambda t\} dt = 1/\lambda. \tag{5.27}$$

Equation (5.27) is the average time to failure, but it is not the time when half of the device population will have failed. This time is the median time to failure t_{50} de ned to be the time when the $F(t)$ function rst reaches the value 0.5 and is derived for the exponential distribution from

$$F(t_{50}) = 0.5 = 1 - \exp\{-\lambda t_{50}\}$$

as

$$t_{50} = \ln 2/\lambda \cong 0.693/\lambda. \tag{5.28}$$

The time t_{50} is less than the MTTF and

$$F(\text{MTTF}) = 1 - \exp\{-\lambda/\lambda\} = 1 - \exp\{-1\} \cong 0.632 \tag{5.29}$$

shows that $\cong 63.2\%$ of a population will have failed by MTTF $= 1/\lambda$ (see Figure 5.5b).

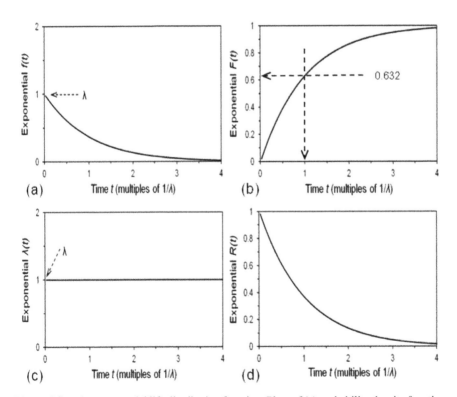

Figure 5.5 The exponential life distribution function. Plots of (a) probability density function $f(t)$, (b) cumulative distribution function $F(t)$, (c) failure rate $\lambda(t)$, and (d) reliability function $R(t)$ calculated as a function of time t (in units of mean time to failure $1/\lambda$). The dashed lines in (b) indicate the 63.2% failure point corresponding to the mean time to failure.

Besides the constant failure rate, the exponential distribution has another key feature, that of *lack of memory*, which means that there is no process of aging, wearing out, or degrading involved with time, but failure is happening by chance at the same constant rate without any link to the history of the item. This feature says that the conditional probability of failure in some time interval Δt given survival up to the start of that interval is the same as the probability of a new unit failing in its rst Δt hours (Tobias and Trindade, 1986).

This concept implies that the information gained from testing x units for y hours is equivalent to the information gained from testing y units for x hours or any other partition as long as the accumulated power-on hours product $x \times y$ is the same. In addition, failed units can be replaced and the test continued without affecting the test result in spite of the fact that these units have a different degree of aging than the nonreplaced ones. Further consequences of the *lack of memory* property are that the failure rate and average failure rate are the same (see above), and the mean time between failures (MTBF) for a repairable population of exponential distribution is also $1/\lambda$, like that for the MTTF referring to a system of nonrepairable devices.

The best estimate for the constant failure rate parameter λ of a complete set of samples of an exponential distribution can be expressed as

$$\bar{\lambda} = \frac{\text{number of failures}}{\text{total unit test hours}} = \frac{1}{\text{sample mean time to failure}} \tag{5.30}$$

where the denominator is obtained by adding up the test hours of all units, failed and completed ones without failing, and the expression reduces to the inverse of the sample mean, similar to $\lambda = 1/\text{MTTF}$. We will present below the de nition of censoring test data and how censoring affects Equation (5.30).

5.2.4.3   Areas of application

From a statistical point of view, the exponential model is only applicable to failures that are totally random or at least random in the intended application life with no signi cant wear-out mechanisms and low expectation of early defect failures. Exponential methods can be applied to the constant failure rate section but also piecewise to other constant exponential portions of the failure rate curve by approximating it piecewise by straight lines and using the average failure rate concept within these quasi-linear sections. Testing times then need to be con ned to these regions with nearly constant failure rate.

The exponential model applies to all processes where the probability of failure is independent of past history, or an item is as good the moment before failing as it was at the time it was put into use. Examples are blowing an electrical fuse, sudden breakdown due to mechanical overload, or electrostatic discharge damage of an electronic item; all of them have an external cause of failure.

In Chapter 6, we will discuss the process of proper experimental planning of reliability tests by using the exponential distribution, where the planning dif culty is not present and the planning process is more straightforward compared to the other life distributions (Tobias and Trindade, 1986). Thus, the exponential model is a useful trial model in the experimental planning stage by giving early consideration to sample sizes and durations of the reliability tests.

5.3 Reliability data plotting

5.3.1 Life-test data plotting

In the previous section we discussed analytical techniques for the lognormal, Weibull, and exponential life distributions. As demonstrated in the various gures above, all the major forms of reliability distributions including $f(t)$, $F(t)$, $R(t)$, and $\lambda(t)$ can be plotted, where, however, the cumulative distribution function $F(t)$ is the most practical one. Required is to have simple graphical procedures that allow checking of the applicability of a distribution model to the experimental reliability data. This implies having a simple linear plot of the number of failures versus time, from which relevant reliability parameters can be quickly and easily derived. The task is to linearize the various nonlinear reliability functions. In the following sections, we will discuss relevant practical approaches for plotting reliability data.

5.3.1.1 Lognormal distribution

A lognormal probability plot is a plot of Equation (5.14). There are different options for generating a probability plot. However, we will discuss here only the standard approach by plotting the cumulative percent failures versus the failure times on (electronic) lognormal probability paper. To illustrate the concept, we consider the failures of a collection of InGaAs/AlGaAs QW diode lasers age accelerated at high temperature and high constant current operation over a long period of time. There are four steps involved in generating the probability plotting:

1. Rank order the failure times from the smallest to the largest value. For example, if the rank order of a failure is i and the total number of failures is j, then i ranges from 1 to j. In the example of Table 5.3, $j = 15$.

2. Estimate the cumulative percent for each failure data point. There are several equations commonly used for this. However, statistically the most accurate concept for estimating the population CDF or $F(t_i)$, called median ranks, is de ned by

$$F(t_i) = \frac{i - 0.3}{N_0 + 0.4} \times 100 \, (\text{cumulative}\%) \, ; i = 1, 2, 3, \ldots, j \leq N_0 \qquad (5.31)$$

where N_0 is the sample size at the beginning of the test, which in our example in Table 5.3 is equal to the number of total failures. The other methods yield slightly different $F(t_i)$ values by a couple of percent at the start and end

Table 5.3 Estimation of cumulative distribution function $F(t)$ by using median ranks (col. (c)) of failures in highly accelerated aging tests of InGaAs/AlGaAs diode lasers. Transformation of $F(t)$ data used for Weibull probability plotting (col. (d)) and exponential probability plotting (col. (e)) as a function of failure times t_i (col. (a)).

(a)	(b)	(c)	(d)	(e)
Failure time t_i [h]	Failure count i	Median rank $F(t_i) = \dfrac{i-0.3}{N_0+0.4}$ $N_0 = 15$	$-\ln[1-F(t_i)]$	$1/[1-F(t_i)]$
102	1	0.045 (4.5%)	0.046	1.048
250	2	0.110 (11.0%)	0.117	1.124
400	3	0.175 (17.5%)	0.192	1.212
660	4	0.240 (24.0%)	0.274	1.316
800	5	0.305 (30.5%)	0.364	1.439
970	6	0.370 (37.0%)	0.462	1.587
1300	7	0.435 (43.5%)	0.571	1.770
1350	8	0.500 (50.0%)	0.693	2.000
1600	9	0.565 (56.5%)	0.832	2.299
2000	10	0.630 (63.0%)	0.994	2.703
2200	11	0.695 (69.5%)	1.187	3.279
2600	12	0.760 (76.0%)	1.427	4.167
3000	13	0.825 (82.5%)	1.743	5.714
3800	14	0.890 (89.0%)	2.207	9.091
4500	15	0.955 (95.5%)	3.101	22.222

of the rank order range. Given the inaccuracies generally experienced in data collection and evaluation, these differences hardly have any practical in uence on the nal test results and conclusions.

3. Plot each cumulative failure percent (column (c) in Table 5.3) versus its time to failure (column (a) in Table 5.3). Figure 5.6 shows the plot for our example.

4. Draw a best- t straight line through the plotted data points. In the gure this was achieved by an electronic least squares t.

The gure shows quite some scatter of the data around the least squares t straight line with major deviations toward median to longer times. This indicates that the lognormal model does not t the data well. Nevertheless, by using the de nitions in Section 5.2.2 and Table 5.1 for the median time t_{50} and the shape parameter σ the plot yielded 1153 h and 1.166 for these parameters, respectively. The relatively short median time is a consequence of the highly accelerated life tests. We will discuss this

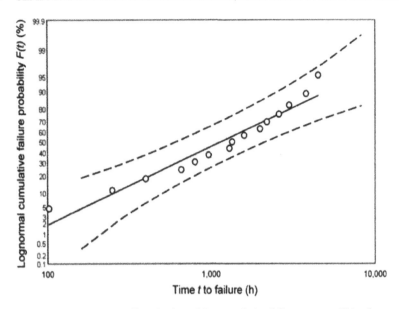

Figure 5.6   Lognormal probability plotting of the cumulative failure percent $F(t)$ values versus failure times calculated in Table 5.3 for the failure events in highly accelerated aging tests on InGaAs/AlGaAs diode lasers. The solid line in the lognormal probability graph represents the least-squares t and the dashed lines the calculated 90% two-sided con dence limits.

topic in Section 5.5. According to the discussion in Section 5.2.2.2 and Figure 5.3, such low values of σ of around one re ect roughly a constant failure rate behavior and it may well be that the data are better matched by a Weibull or exponential failure distribution model (see the following sections).

Finally, the dashed lines in the gure are the 90% two-sided con dence bounds with 95% one-sided lower con dence limit and 95% one-sided upper con dence limit of the sample data for the selected distribution model. There are standard mathematical procedures for evaluating con dence limits, but a discussion of them would be beyond the scope of this text. For this information the reader is referred to relevant references including Epstein and Sobel (1953), Jordan (1984), Sudo and Nakano (1985), Nash (1993), O'Connor (2002), Tobias and Trindade (1986), and Dummer and Grif n (1966). We will, however, return to the topic brie y in Section 5.4 below, and discuss the con dence limits on MTTF estimates by giving the relevant mathematical expressions.

5.3.1.2   Weibull distribution

A Weibull probability plot is a plot of Equation (5.18). To realize this, we want to discuss two methods. The rst comprises the standard procedure by plotting the $F(t_i)$ data versus failure times t_i on Weibull probability graph paper. The second method comprises a linearization of Equation (5.18) with subsequent plotting of the

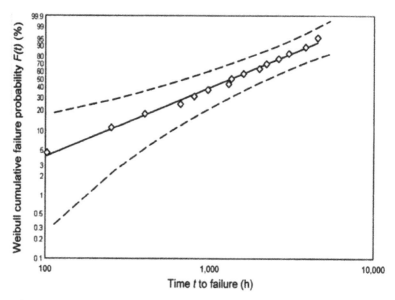

Figure 5.7 Weibull probability plotting of the cumulative failure percent $F(t)$ values versus failure times calculated in Table 5.3 for the failure events in highly accelerated aging tests on InGaAs/AlGaAs diode lasers. The solid line in the Weibull probability graph represents the least-squares t and the dashed lines the calculated 90% two-sided con dence limits.

transformed data. We use the same example and highly accelerated failure data as in the previous section (see Table 5.3) to illustrate both methods.

Figure 5.7 shows a plot of the cumulative percent failures $F(t_i)$ versus failure times t_i on Weibull graph paper. The data are matched by the Weibull model much better than by the lognormal distribution shown in Figure 5.6. Using the de nitions in Table 5.2 for relevant Weibull parameters and the plot in Figure 5.7, the timescale parameter η, also called the characteristic life parameter α (time at $F = 63.2\%$), has been derived to 1890 h (highly accelerated) and the important shape parameter β to 1.085. The value of the β parameter close to one indicates a nearly constant failure rate, which can actually be ascribed to the exponential distribution. However, as the β value is slightly above one, a weak wear-out component with a failure rate increasing in time might be involved in the aging process.

The second method is based on rewriting Equation (5.18) in the form ($\gamma = 0$ for simplicity) (Tobias and Trindade, 1986)

$$1 - F(t) = \exp\left\{-(t/\eta)^\beta\right\} \qquad (5.32)$$

and taking natural logarithms of both sides twice to get

$$\ln\left\{-\ln[1 - F(t)]\right\} = \beta \ln t - \beta \ln \eta. \qquad (5.33)$$

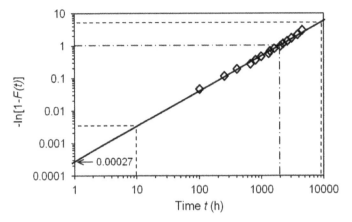

Figure 5.8 Weibull probability plotting of the transformed cumulative failure percentage variable $-\ln[1 - F(t)]$ versus failure times calculated in Table 5.3 for the failure events in highly accelerated aging tests on InGaAs/AlGaAs diode lasers. Dashed lines represent x, y coordinates of points on the anticipated straight-line least-squares t used to calculate the shape parameter β from the slope of the straight line. The intercept is used to calculate a value for the timescale parameter η. The dotted–dashed lines give the coordinates for deriving a value for η from the condition of the transformed variable $-\ln[1 - F(\eta)] = 1$ for $t = \eta$.

Obviously, in a $\ln\{-\ln[1 - F(t)]\}$ versus $\ln t$ plot, Equation (5.33) represents a straight line of slope β and intercept $-\beta \ln \eta$.

Figure 5.8 shows the experimental dependence of $-\ln[1 - F(t_i)]$ versus t_i in a log–log plot for our example by using the data listed in Table 5.3 and the median ranks to estimate $F(t_i)$. The data t reasonably well on a straight line, con rming also in this representation that the Weibull distribution can describe the data. To estimate the slope we pick, say, $t_1 = 10$ h and $t_2 = 9000$ h, which correspond to $y_1 = 0.0031$ and $y_2 = 5$, respectively. The intercept is found at $t = 1$ (where $\ln 1 = 0$) and amounts to 0.000 27. From these data average values for the shape parameter β and timescale parameter η (characteristic life α) have been calculated as

$$\bar{\beta} = \frac{\ln(5/0.0031)}{\ln(9000/10)} \cong 1.086 \pm 0.024 \tag{5.34}$$

$$\bar{\eta} = 0.00027^{-1/1.086} \cong (1932 \pm 130)\,\text{h}. \tag{5.35}$$

Within the measurement errors, these results are in excellent agreement with the results derived from the probability paper plot in Figure 5.7.

Another procedure can be developed by replacing the left-hand side of Equation (5.33) by $\ln H(t)$, where $H(t)$ is the cumulative hazard expressed in Equation (5.8) and which can be linked to $F(t)$ by using Equation (5.4). A plot of $H(t)$ versus t on log–log paper would then yield again the shape parameter β from the slope of the straight line and the timescale parameter η from the intercept if the assumed Weibull model applies. The task is, however, to de ne an appropriate method for estimating $H(t)$, which will be the reverse rank procedure and not the median ranks used so far. We will not describe the procedure here, but will give a detailed account of the $H(t)$ versus t plotting in the next section dealing with exponential probability plotting.

Finally, we should mention a quick method to estimate a value for η. From Equation (5.32) we obtain $F(t = \eta) = 1 - e^{-1}$ or $-\ln[1 - F(\eta)] = 1$ for $t = \eta$. From Figure 5.8 we then get $\eta \cong 2000$ h, which is within reading accuracy in good agreement with the values determined by the other methods described above.

5.3.1.3 Exponential distribution

As discussed above, the exponential distribution is a special case of the Weibull distribution that occurs when the Weibull shape parameter $\beta = 1$. A probability plot of the exponential distribution is a plot of $F(t)$ expressed as in Equation (5.22). Again, the graphical analysis is based on the concept of transforming the data in such a way that straight lines can be expected when the data are plotted (Tobias and Trindade, 1986). Two procedures will be discussed.

First, rewriting Equation (5.22) and taking natural logarithms of both sides, we get nally

$$- \ln[1 - F(t)] = \ln \frac{1}{1 - F(t)} = \lambda t. \tag{5.36}$$

By plotting $\ln\{1/[1 - F(t)]\}$ versus time t on linear–linear paper or equivalently $1/[1 - F(t)]$ versus time on log–linear paper, the failure data should fall on a straight line if the exponential distribution applies. The linear t has a slope λ and an intercept at zero ($\ln 1 = 0$) for $t = 0$.

Figure 5.9 shows $1/[1 - F(t)]$ versus time t for the data of our example (Table 5.3) by again using median ranks for calculating the $F(t_i)$ values. The data up to $\sim$2500 h t a straight line with the origin at zero indicating an exponential failure distribution with constant failure rate. However, at longer failure times the data deviate from the initial straight line due to higher $F(t)$ values, leading to a failure rate increasing with time. This indicates that the aging process (strongly accelerated) at the longer times may have already entered the wear-out regime, which is better described by the Weibull model con rmed by $\beta \cong 1.085$, which is slightly higher than $\beta = 1$ required for a pure exponential distribution. The better match of the data by a Weibull model is also illustrated in Figures 5.7 and 5.8. From the slope of the least-squares t line in Figure 5.9 an average value of the failure rate has been calculated as

$$\bar{\lambda} = \text{slope} = \frac{\ln 7 - \ln 1}{3500 - 0} \cong 5.6 \times 10^{-4} \text{h}^{-1} \tag{5.37}$$

and the MTTF is given by

$$\text{MTTF} = 1/\bar{\lambda} \cong 1785 \text{ h.} \tag{5.38}$$

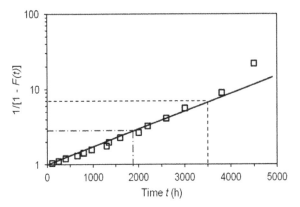

Figure 5.9 Exponential probability plotting of the transformed cumulative failure percentage $1/[1 - F(t)]$ values versus failure times calculated in Table 5.3 for the failure events in highly accelerated aging tests on InGaAs/AlGaAs diode lasers. Dashed lines represent x, y coordinates of the point on the anticipated straight-line least-squares t used to calculate the failure rate constant λ from the slope of the straight line. The dotted–dashed lines give the coordinates for deriving a value for the mean time to failure from the condition of the transformed variable $1/[1 - F(\text{MTTF} = 1/\lambda)] = e = 2.7183$.

A quick estimation of the failure rate and mean life MTTF can be found by considering that $F(t = 1/\lambda) = 1 - e^{-1} = 0.632\ 12$ for $t = 1/\lambda$. Thus, a value for the MTTF can be obtained by determining the time corresponding to $F(1/\lambda) = 0.632\ 12$, or, equivalently, where the transformed variable $1/[1 - F(\text{MTTF} = 1/\lambda)] = e = 2.7183$. From Figure 5.9 a value of MTTF $\cong$ 1850 h can be read off. Given the limited reading accuracy, this quickly estimated value has to be considered as in excellent agreement with the value above. It also agrees with the characteristic life value determined in the Weibull plots above. In both independent distribution models, the time was determined at the cumulative distribution density $F(t) = 63.2\%$ in the Weibull and exponential probability plots.

A second alternative method, which is different to all the procedures discussed so far, employs the cumulative hazard function $H(t)$. According to Equation (5.8), $H(t)$ of any distribution is related to $F(t)$ by

$$H(t) = -\ln R(t) = -\ln[1 - F(t)] \tag{5.39}$$

which, by using Equation (5.22) for $F(t)$ of the exponential distribution, leads to

$$H(t) = -\ln \exp\{-\lambda t\} = \lambda t. \tag{5.40}$$

Equation (5.40) varies linearly with time, and data plotted as $H(t_i)$ versus failure time t_i should fall on a straight line with slope λ and intercept zero if the exponential model ts the data. However, the task is to nd the appropriate method for estimating $H(t)$, which is different to the median ranks procedure described in Section 5.3.1.1 for calculating $F(t)$ according to Equation (5.31).

We adopt a method described by Tobias and Trindade (1986): at each time t, $H(t)$ is the sum of the individual hazard terms found by dividing the number of failures by the number of units surviving previously before that time t. For exact, unique times to failure, the number of failures at each failure time is always one. Therefore, a simple process is to order the failure times from lowest to highest, and then associate with each failure time the reverse rank starting with the initial sample (failed devices) size. For each time, the $H(t)$ estimate is just the sum of the individual hazard values (reciprocals of the reverse ranks), since there is one failure for each of the reverse ranks. Table 5.4 gives the various steps involved in the procedure by using the failures occurring in our example used throughout Section 5.3.

Table 5.4 Estimation of cumulative hazard function $H(t)$ (col. (e)) by using individual hazard values (col. (d)) calculated by applying reverse ranks (col. (c)) of failures in highly accelerated aging tests of InGaAs/AlGaAs diode lasers; 15 failures in an initial sample size of 15. Cumulative hazard values (col. (e)) used in an alternative method for exponential probability plotting versus failure times t_i (col. (a)).

(a)	(b)	(c)	(d)	(e)
Failure time t_i [h]	Failure count i	Reverse rank k	Hazard value $100 \times (1/k)$ [%]	Cumulative hazard [%]
102	1	15	6.66	6.7
250	2	14	7.14	13.8
400	3	13	7.69	21.5
660	4	12	8.33	29.8
800	5	11	9.09	38.9
970	6	10	10.00	48.9
1300	7	9	11.11	60.0
1350	8	8	12.50	72.5
1600	9	7	14.28	86.8
2000	10	6	16.66	103.5
2200	11	5	20.00	123.5
2600	12	4	25.00	148.5
3000	13	3	33.33	181.8
3800	14	2	50.00	231.8
4500	15	1	100.00	331.8

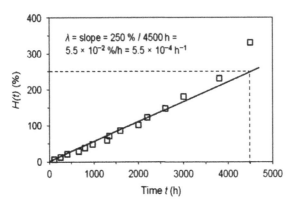

Figure 5.10 Exponential probability plotting of the cumulative hazard function $H(t)$ versus failure times calculated in Table 5.4 for the failure events in highly accelerated aging tests on InGaAs/AlGaAs diode lasers. Dashed lines represent x, y coordinates of the point on the anticipated straight-line least-squares t used to calculate the failure rate constant λ from the slope of the straight line.

The $H(t)$ versus t plot, shown in Figure 5.10, appears to t reasonably well a straight line in a linear–linear representation with intercept zero. From its slope the average failure rate has been derived as 5.5×10^{-4} h^{-1} and thus the mean time to failure MTTF $= 1/\lambda$ amounts to 1818 h. Both values are in excellent agreement with the values obtained from the different procedures described above and in Equation (5.36) and Figure 5.9.

5.4 Further reliability concepts

5.4.1 Data types

In general, reliability data fall into categories including exact failure times with censoring and readout time data. The de nitions are brie y as follows.

5.4.1.1 Time-censored or time-terminated tests

These are terminated after a certain time, regardless of how many of the units under test have failed. For example, if N_0 units are tested for a xed planned duration of T, and r of them failed at exact times $t_1 \leq t_2 \leq t_3 \leq \cdots \leq t_r \leq T$, then at the end of the test, there are $N_0 - r$ survivors. All that is known about these survivors is that they will fail at times beyond T. The number of failures is unknown with this kind of testing. This may cause a problem in the precision of failure rate estimates, which depend on the number of failures. An unsuitable choice of sample size and/or test conditions may result in an insuf cient number of failures and hence information from the tests.

5.4.1.2   Failure-censored or failure-terminated tests

These are terminated after a certain number of failures r have occurred. Since r is specified in advance, one knows how much data the test will yield, which is an advantage. However, the drawback with this kind of testing is that the length of test time is open ended. For practical reasons, failure-determined testing is less common and usually ruled out in favor of time-determined testing.

5.4.1.3   Readout time data tests

These data are collected from tests, in which all units are tested at specified intervals. For example, at time T_1 the readout finds r_1 failures, then $N_0 - r_1$ units go back on test, at time T_2, after $T_2 - T_1$ time of test, the readout may find r_2 failures, then $N_0 - r_1 - r_2$ units go back on test, and so forth until the last readout at the end of the test at time $T_k = T$. The readout times and the end of test time are predetermined. The number of failures is only known when the readout has been taken and the exact times of failures are never known. The same problem as with time-censored tests is that the test may end before a sufficient number of failures are available. Moreover, not knowing the exact failure times and a possible spreading of the failures over only a few intervals may strongly affect the precision of the test results. Nevertheless, this kind of testing is popular, because it does not require sophisticated and expensive equipment to perform in situ monitoring for detecting the exact failure times.

5.4.2   Confidence limits

The exact failure rate or failure time of a product such as a diode laser predicted from reliability tests can never be known. However, there are mathematical models that describe the relationship between the observed and actual values. The most frequently used and most accurate model for calculating failure times and failure rates from a relatively small number of failures is the one that uses a χ^2 (chi-squared) distribution with upper (u) and lower (l) limits.

We will discuss the confidence limits on MTTF estimates for two cases: (i) for two-sided estimations in a fixed-time test plan, that is, for time-terminated tests with r failures; and (ii) for tests with no failures observed. The corresponding expressions for the failure rate can be easily derived from $\lambda = 1/\text{MTTF}$. Note that $\lambda(u) = 1/\text{MTTF}(l)$ and $\lambda(l) = 1/\text{MTTF}(u)$.

The general expression for the upper and lower confidence limits of an estimate of MTTF is given by (Condra, 1993)

$$\text{MTTF}(u, l) = \frac{2t_a}{\chi^2(\gamma; dF)} \tag{5.41}$$

where t_a is the total number of device hours. It is simply the sum of the number of hours each unit operates up to its point of removal for any cause. Cause of removal can be any failure, removal for nontest reasons, or termination of the test. The confidence level (CL) is given by CL $= 100 \times (1 - \alpha)\%$; for example, for CL $= 95\%$, $\alpha = 0.05$ or for CL $= 80\%$, $\alpha = 0.2$.

A value for the parameter γ is a function of α, and the parameter dF gives the number of degrees of freedom. Both parameters depend on the upper and lower CL limits and type of test, that is, two-sided or one-sided failure-terminated or time-terminated tests with r failures or a test without failures observed. γ and dF are for the lower and upper limits of MTTF for the two-sided, time-terminated tests as follows (Condra, 1993; NIST/SEMATECH, 2010):

$$\gamma = \frac{\alpha}{2} \quad \text{and} \quad dF = 2(r+1) \quad \text{for MTTF (l)} \tag{5.42a}$$

$$\gamma = 1 - \frac{\alpha}{2} \quad \text{and} \quad dF = 2r \quad \text{for MTTF (u)} \tag{5.42b}$$

which results in

$$\text{MTTF (l)} = \frac{2t_a}{\chi^2\left(\dfrac{\alpha}{2}; 2r+2\right)} \tag{5.42c}$$

$$\text{MTTF (u)} = \frac{2t_a}{\chi^2\left(1 - \dfrac{\alpha}{2}; 2r\right)}. \tag{5.42d}$$

For tests with no failures observed on N_0 units tested for T hours, the MTTF is given by a one-sided estimate, which can be generalized as follows to derive values for any CL percentile (Tobias and Trindade, 1986):

$$\text{MTTF (l)} = \frac{2N_0T}{\chi^2(\alpha; 2)} = \frac{N_0T}{-\ln \alpha}. \tag{5.43}$$

It is tempting to argue that if no failures occur in a test, the MTTF is in nite. However, it has to be considered that a failure might occur just after termination of the test. An estimate of the upper limit is obviously meaningless.

Therefore, a realistic estimate can be obtained by the lower one-sided CL for a failure-terminated test, in which one failure takes place. For this case, $dF = 2r$ or 2 for $r = 1$, and $\gamma = \alpha$, leading to an expression for chi-squared of $\chi^2(\alpha; 2)$ (Condra, 1993). Equation (5.43) yields, for example, MTTF(l) $= (N_0T)/2.3026$ for CL $= 90\%$ by using the value for χ^2 from tables (NIST/SEMATECH, 2010) divided by two in the rst term or using $\alpha = 0.1$ in the second term.

Example for illustration: An example may illustrate the case expressed in Equations (5.42c) and (5.42d). We take $r = 15$ failed diode lasers used throughout Section 5.3 above. The accumulated device hours are $t_a = 25\,532$ h. To know the lower and upper limits on the MTTF with 90% con dence, the following values apply: $t_a = 25\,532$ h, CL $= 1 - \alpha = 0.9$, $\alpha = 0.1$, $\alpha/2 = 0.05$, $1 - (\alpha/2) = 0.95$, and $r = 15$. Using Equations (5.42c) and (5.42d) and chi-squared values from relevant tables (e.g. NIST/SEMATECH, 2010), the lower and upper 90% con dence limits on the MTTF are calculated as

$$\text{MTTF (l)} = \frac{2 \times 25532 \text{ h}}{\chi^2 (0.05; 32)} = \frac{51064 \text{ h}}{46.194} = 1105 \text{ h} \tag{5.44a}$$

$$\text{MTTF (u)} = \frac{2 \times 25532 \text{ h}}{\chi^2 (0.95; 30)} = \frac{51064 \text{ h}}{18.493} = 2761 \text{ h.} \tag{5.44b}$$

The two values in Equations (5.44a) and (5.44b) give the 90% con dence interval that contains the true value with 90% con dence after the same tests were repeated many times and the same method was used over and over again to construct an interval for the MTTF. The value of $\sim$1850 h averaged over the values determined by the various methods in Section 5.3 ts nicely in the calculated 90% con dence interval.

5.4.3   Mean time to failure calculations

As described in Section 5.2.4, when data come from an exponential distribution there is only one parameter required to estimate the failure rate function, which is then a constant with time. In addition, equivalently for this case, only the failure rate is constant and the failures are caused by random events; the MTTF is the reciprocal of the constant failure rate parameter λ quanti ed in Equation (5.30). This facilitates easy MTTF calculations. Equation (5.27) de nes the MTTF for the exponential distribution. It is simply the sum of all failure times divided by the number of failures and applies for nonrepairable items. In contrast, the MTBF is de ned as the sum of time intervals of consecutive failure times divided by the number of failures.

Equation (5.30) can be written as

$$\text{MTTF} = t_a/r \tag{5.45}$$

for an estimate of the MTTF, where t_a is the total number of device hours of all devices removed by failure, nontest reasons, or test termination and r is the number of relevant failures. For time-terminated tests, in which failed items are not replaced, as is generally the case for diode laser testing, Equation (5.45) yields

$$\text{MTTF} = \frac{1}{r} \sum_{i=1}^{r} t_i + (N_0 - r) T \tag{5.46}$$

where T is the pre- xed end of test time, and t_i are the exact failure times of the r units that fail before termination of the test. Similarly, for failure-terminated tests without replacement we obtain

$$\text{MTTF} = \frac{1}{r} \sum_{i=1}^{r} t_i + (N_0 - r) t_r \tag{5.47}$$

where the test ends at the rth failure time. See Section 5.4.1 regarding time- and failure-terminated tests.

5.4.4 Reliability estimations

The con dence limits on reliability estimates for exponentially distributed failures can be obtained by using the method in Section 5.4.2 above and the expression

$$R = \exp\{-\lambda t\} = \exp\{-t/\text{MTTF}\}. \tag{5.48}$$

If we take the lower and upper estimates for the MTTF of 1105 h and 2761 h, respectively, calculated at the 90% con dence level for our diode laser test example in Section 5.4.2, then the lower and upper reliability limits can be estimated as

$$R(l) = \exp\{-t/1105\} \quad \text{and} \quad R(u) = \exp\{-t/2761\}. \tag{5.49}$$

Thus, we can estimate with 90% con dence that the reliability of these highly stressed (temperature and current) lasers will be between 0.913 and 0.964 over the rst 100 h or between only 0.405 and 0.696 over 1000 h.

Another example is the reliability calculation of a system with N components, such as lasers in an optical ber communication system, each of which has a constant failure rate λ_i for $i = 1, 2, \ldots, N$ within an exponential failure distribution. Considering in addition that each component must operate for the system to work, we can calculate the system reliability R_s by using

$$R_s = \exp\left\{-\sum_{i=1}^{N} \lambda_i t\right\} \tag{5.50}$$

where the failure rate of the system is the sum of the failure rates of its components. Let us consider a system with MTTF $= 10$ years $= 8.76 \times 10^4$ h with 1000 lasers all with the same failure rate λ, which can then be calculated according to

$$R_s = \exp\{-\lambda_s t\} = \exp\{-t/\text{MTTF}\} = \exp\{-t/8.76 \times 10^4\}$$
$$= \exp\left\{-\sum_{i=1}^{1000} \lambda_i t\right\} = \exp\{-1000\lambda t\}. \tag{5.51}$$

From Equation (5.51) a very low failure rate $\lambda = 1/(8.76 \times 10^4 \times 10^3)\,\text{h}^{-1} \cong 11$ FITs has been obtained for the individual components demonstrating the extremely high requirements demanded from the reliability of the individual components in a system.

5.5 Accelerated reliability testing – physics–statistics models

5.5.1 Acceleration relationships

Under normal use conditions, it is not possible to obtain a reasonable amount of test data within a reasonable test time if the components have high reliability and long lifetimes. Therefore, tests are performed at much higher stress conditions to get failure data at relatively small test sample sizes and practical test times and that can then be tted to life distribution models. Typical stresses include temperature, current, optical power, voltage, and humidity taken singly or in combination (see below). For example, raising the temperature of the item under test can be understood in terms of accelerating the aging mechanisms that are temperature dependent following the Maxwell–Boltzmann law.

It is crucial that the right levels of increased stress are chosen so that the failure mechanisms are not changed at higher stresses, but only the times to failure are shortened. Such a process represents true acceleration. The acceleration factor (AF) is de ned as the ratio of the failure time at the use test condition (u) to that at an elevated stress condition (s). Under a linear acceleration assumption, every test failure time and distribution function is multiplied by the same AF constant value to obtain projected results during use. The key equations relating the random time to failure, PDF $f(t)$, CDF $F(t)$, and failure rate function at use condition to a stress condition can be written (Tobias and Trindade, 1986) in their most general form for true and linear acceleration conditions as follows:

$$t_u = \text{AF} \times t_s \tag{5.52}$$

$$f_u(t) = \left(\frac{1}{\text{AF}}\right) f_s\left(\frac{t}{\text{AF}}\right) \tag{5.53}$$

$$F_u(t) = F_s\left(\frac{t}{\text{AF}}\right) \tag{5.54}$$

$$\lambda_u(t) = \left(\frac{1}{\text{AF}}\right) \lambda_s\left(\frac{t}{\text{AF}}\right). \tag{5.55}$$

5.5.1.1 Exponential; Weibull; and lognormal distribution acceleration

In the following, we will see how these four equations are transformed when applied to the three life distributions.

Assuming Equation (5.22) for the CDF of the exponential distribution is also valid at high stresses, leading to failures occurring with a constant failure rate λ_s, we get

$$F_u(t) = F_s\left(\frac{t}{\text{AF}}\right) = 1 - \exp\left\{-\lambda_s \frac{t}{\text{AF}}\right\} = 1 - \exp\left\{-\frac{\lambda_s}{\text{AF}} t\right\} = 1 - \exp\{-\lambda_u t\}. \tag{5.56}$$

This means that the F function at use conditions remains exponential with the new failure rate parameter $\lambda_u = \lambda_s/\text{AF}$. Equation (5.56) implies that if an exponential model ts the data at a certain stress condition, then it also matches at any other stress condition as long as true and linear acceleration conditions are met. The simple inverse relationship of the failure rate with the acceleration factor applies only for the exponential life distribution.

If the data at stress condition can be matched by $F_s(t)$ of a Weibull distribution with shape parameter β_s and timescale parameter η_s, Equation (5.18) can be written ($\gamma = 0$ for simplicity) as

$$F_s(t) = 1 - \exp\left\{-(t/\eta_s)^{\beta_s}\right\}$$ (5.57)

and transforms to use stress as

$$F_u(t) = F_s\left(\frac{t}{AF}\right) = 1 - \exp\left\{-\left(\frac{t}{AF} \times \frac{1}{\eta_s}\right)^{\beta_s}\right\} = 1 - \exp\left\{-\left(\frac{t}{\eta_u}\right)^{\beta_u}\right\}$$ (5.58)

where $\eta_u = \eta_s \times AF$ and $\beta_u = \beta_s = \beta$.

This shows that if the Weibull model is valid at one stress level, then it is also valid at another stress condition provided there is true and linear acceleration. It is crucial that the shape parameter β remains the same. This means that, if there is Weibull acceleration in a test, then it is expected that the shape parameter will be the same for all the cells with different stress levels (corresponding to the slope of the different parallel straight lines of the Weibull data plots, see Figures 5.7 and 5.8). If this is not the case, then either the Weibull distribution is the wrong model to t the data or the true and linear acceleration condition is not met.

By using Equations (5.20) and (5.55) the relationship between the Weibull failure rate at stress and use conditions can be obtained as

$$\lambda_u(t) = \frac{1}{AF}\frac{\beta}{\eta_s}\left(\frac{t}{AF \times \eta_s}\right)^{\beta-1} = \frac{1}{AF^\beta}\frac{\beta}{\eta_s}\left(\frac{t}{\eta_s}\right)^{\beta-1} = \frac{1}{AF^\beta}\lambda_s(t).$$ (5.59)

Equation (5.59) shows that the failure rate is multiplied by $1/AF$ only when $\beta = 1$ and the data t an exponential distribution.

Finally, it can be shown that by applying Equation (5.54) to the lognormal CDF F_s at stress conditions with the characteristic parameters $t_{50,s}$ and σ_s (see Section 5.2.2.2), again a lognormal distribution can be found at use conditions with $\sigma_u = \sigma_s = \sigma$ and $t_{50,u} = AF \times t_{50,s}$, where σ_u and $t_{50,u}$ are the use parameters for the lognormal shape parameter and median time, respectively. The reader is referred to Tobias and Trindade (1986) for details on making the acceleration transformation of the timescale given by $F_u(t) = F_s(t/AF)$. As with the Weibull, true linear acceleration does not change the type of distribution. It is also expected that different stress cells of data will yield the same shape parameter.

5.5.2   Remarks on acceleration models

In the previous section, we saw that the acceleration factor between stresses can be calculated by knowing the failure rate constants, scale parameters, or the median times at two stress levels. If the acceleration factor is already known, then the results from the stress tests can be converted to use condition reliability projections. However, if the acceleration factor is not known, then the stress data have to be used to t an appropriate acceleration model, which can be used to extrapolate to use conditions. Different failure mechanisms may be involved in the accelerated aging process, which follow different life distributions and may also have different acceleration models. Therefore, accelerated aging tests have to be designed carefully to produce unique data from only one failure mode, which may be accelerated by more than one stress factor.

In the following we discuss three general well-established model forms including (i) the Arrhenius model describing thermally activated mechanisms, (ii) the inverse power law describing the life of an item inversely proportional to an applied stress, and (iii) the Eyring model considering a temperature term and additional nonthermal stress terms.

5.5.2.1   Arrhenius model

The Arrhenius model has been used with great success to describe thermally activated mechanisms regardless of the underlying life distributions – lognormal, Weibull, or exponential. It describes the temperature dependence of features such as median time t_{50}, timescale parameter η (characteristic life parameter α), or MTTF $= 1/\lambda$ or any other percentile of a life distribution as a rate equation

$$R = R_0 \exp\{-E_a/k_B T\} \tag{5.60}$$

where R is the reaction rate, R_0 is a constant, E_a is the activation energy in units of eV, $k_B = 8.617 \times 10^{-5}$ eV/K is the Boltzmann constant, and T is the reaction temperature in units of K. If this relationship holds, then the time to failure is inversely proportional to the reaction rate. For convenience we take $t_{50,f}$ which can be written as

$$t_{50,f} = A \exp\{E_a/k_B T\} \tag{5.61}$$

where A is a constant and $t_{50,f}$ is the time to reach 50% failures. The acceleration factor (AF) between temperature T_1 and temperature $T_2 > T_1$ is then given by

$$AF = \frac{R_2(T_2)}{R_1(T_1)} = \frac{t_{50,1}(T_1)}{t_{50,2}(T_2)} = \exp\left\{\frac{E_a}{k_B}\left(\frac{1}{T_1} - \frac{1}{T_2}\right)\right\} \tag{5.62}$$

and is independent of the chosen cumulative percentage of failure. Equation (5.62) shows that, knowing E_a, the acceleration factor between any two temperatures can then be calculated. Conversely, knowing the acceleration factor, the activation energy E_a can be calculated from

$$E_a = k_B(\ln AF)\frac{1}{\left(\dfrac{1}{T_1} - \dfrac{1}{T_2}\right)}. \tag{5.63}$$

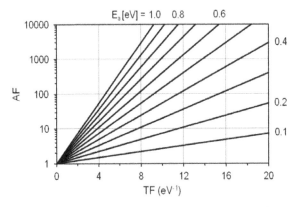

Figure 5.11 Acceleration factor AF versus parameter TF calculated for a range of thermal activation energies E_a. TF is de ned as in Equation (5.64) and measured in units of $(eV)^{-1}$, and the temperatures $T_2 > T_1$ are in Kelvin (K). AF can be read from the graph for a given E_a and pair of temperatures yielding TF.

Equation (5.63) is valid for any life distribution. A useful plot of AF versus TF for a range of E_a values (Figure 5.11) can be generated from Equation (5.62) where

$$\text{TF} = k_B^{-1}\left(T_1^{-1} - T_2^{-1}\right) = 11605 \times \left(T_1^{-1} - T_2^{-1}\right)\left[eV^{-1}\right]. \qquad (5.64)$$

TF is measured in units of $(eV)^{-1}$ and the temperatures T are in units of Kelvin (K), which are obtained by adding 273.16 to the temperature in degrees Celsius (°C).

Example for illustration: We apply the Arrhenius formalism to our example used throughout Section 5.3. Diode lasers were highly stressed at high temperatures of 120 °C $\triangleq$ 393 K heat sink temperature resulting in TF = 9.4 $(eV)^{-1}$ for a use temperature of 25 °C $\triangleq$ 298 K, which leads to AF = 410 read off Figure 5.11 or calculated from Equation (5.62) for $E_a = 0.64$ eV. This activation energy has been determined separately and is a typical value for the stressed InGaAs/AlGaAs QW lasers in the wear-out regime. The timescale parameter (at the failure percentile $F = 63.2\%$) at use conditions η_u can then be calculated using the Weibull model, which ts the failures well (see Section 5.3.1.2). It amounts to $\eta_u = \eta_s \times \text{AF} = 7.7 \times 10^5$ h $\cong 88$ years using $\eta_s = 1890$ h and AF = 410.

Using the available data, we can also calculate the probability that the lasers survive, say, 10 years. The failure probability is given by Equation (5.58) and is calculated as $F_u(t = 10 \text{ years} = 8.76 \times 10^4 \text{ h}) \cong 9 \times 10^{-2}$ by using AF = 410, $\beta_s = 1.085$, and $\eta_s = 1890$ h. Thus, the survival probability of the lasers after 10 years is $1 - F(10 \text{ years}) \cong 0.91$ (~91%).

Alternatively to Equation (5.63), a procedure based on more than two temperature cells can be derived from the key Arrhenius formula in Equation (5.61) and can yield, in principle, a higher accuracy in estimating E_a. Equation (5.61) is linearized by writing it in logarithmic form

$$\ln t_{50,f} = \ln A + E_a \times \left(\frac{1}{k_B T} \right). \tag{5.65}$$

A plot of the data points of all the temperature cells in the form ($\ln t_{50,f}$) versus ($1/k_B T$) should show up as a straight line. The activation energy E_a is determined from the slope of the line tted using least-squares. This procedure holds for any other percentile of life distribution and affects only the constant A but leaves E_a unchanged.

5.5.2.2 Inverse power law

The inverse power rule model is used when the time to failure is inversely proportional to an applied stress S. In the next section, we will see that it is a very simpli ed form of the more general Eyring model. As with the Arrhenius model, the underlying failure rate distribution can be lognormal, Weibull, or exponential. The general form of the inverse power law is

$$t_f = \frac{A}{S^n} \tag{5.66}$$

where the time to failure t_f can be, as with the Arrhenius model above, the median time t_{50}, timescale parameter η (characteristic life parameter α), or MTTF $= 1/\lambda$ or any other percentile of a life distribution. A is a constant typical for the laser type, S is the applied stress, and n is an exponent characteristic of the laser device. The types of stress S can be voltage, current, optical power, temperature cycling, or mechanical vibration, where the latter two can cause fatigue due to alternating stress. In the next section, we give values for n derived from typical combined temperature/current stress tests on diode lasers.

Many different forms of the inverse power law have been developed for various applications including models for mechanical failure due to cracks and material fatigue or deformation. These models often have terms relating to cycles of stress or changes in temperature or frequency of use. We will discuss such product tests carried out under environmental stress conditions in detail in Chapter 6. One of the most common is the Cof n–Manson Law (Cof n, 1954; Manson, 1965) for fatigue testing de ned as

$$N_f = A \left(\frac{1}{\Delta \epsilon_p} \right)^B \tag{5.67}$$

where N_f is the number of cycles to failure, A and B are material constants, and $\Delta \epsilon_p$ is the plastic strain range. Equation (5.67) has been applied to many different situations. It applies also to both isothermal mechanical fatigue cycling and to fatigue due to mechanical stresses resulting from thermal cycling. A simpli ed acceleration factor can be de ned as

$$AF = \frac{N_{f,u}}{N_{f,s}} = \left(\frac{\Delta\epsilon_s}{\Delta\epsilon_u}\right)^B \tag{5.68}$$

if the total applied stress is much higher than the elastic strain range for a fatigue test and where subscripts u and s again denote use and stress conditions, respectively. The strain ranges $\Delta\epsilon$ can be due to, for example, displacement in bending or elongation in tension. A similar expression can be written for fatigue testing in temperature cycling

$$AF = \frac{N_{f,u}}{N_{f,s}} = \left(\frac{\Delta T_s}{\Delta T_u}\right)^B \tag{5.69}$$

where ΔT_s and ΔT_u are the applied temperature cycling ranges under stress and use conditions, respectively. From the slope of log–log plots, a value for the constant B can be obtained, which is in the range of $\sim$2 to 12 with typical numbers of 2–3 for solder joints and 12 for passivation layers. The inverse power law applies also to the relative humidity (RH) as a stress factor and, similar to Equation (5.67), the laser lifetime, such as t_{50}, is proportional to $(RH)^{-B}$.

5.5.2.3 Eyring model

In general, the Eyring model contains at least two stress terms, one of which is the temperature stress. For example, for a second stress it takes the form

$$t_f = AT^\alpha \exp\left\{\frac{E_a}{k_B T}\right\} \exp\left\{\left(B + \frac{C}{T}\right) S_1\right\} \tag{5.70}$$

where A, α, B, and C are constants and S_1 is a nonthermal stress, which can be of almost any type, and which exists in combination with temperature (Tobias and Trindade, 1986). If a second nonthermal stress S_2 is needed, then a third exponential term similar to the second one but with the constants D and E has to be added to Equation (5.70). The rst term models the effect of temperature and except for the T^α factor is the same as the Arrhenius model. This factor can be included in the A constant without changing the practical value of Equation (5.70) if $\alpha \sim 0$ or the temperature application range of the model is small. This means that the successful Arrhenius model is a useful simpli cation of the Eyring model, which was theoretically derived on the basis of chemical reaction rate theory and quantum mechanics where E_a represents the energy required to lift an electron to the state where the processes of chemical reaction, diffusion, or migration can occur (Eyring et al., 1941).

The Eyring model is dif cult to work with because, due to its complexity, it requires at least as many separate experimental stress cells as there are unknown constants. As an example, a two-stress model requires a minimum of ve cells where 1–2 cells have to be added to examine the adequacy of the model t. In addition, the design and execution of such a set of experiments are crucial. We will describe an example below. The two-stress Eyring model can be reduced to a simple acceleration equation with three unknown parameters by making the substitutions $\alpha = 0$, $C = 0$, and $S_1 = -\ln S$ where the stress S can be realistically voltage V, current I, optical power P, or relative humidity RH

$$t_f = A \frac{1}{S^B} \exp\{E_a/k_B T\}. \tag{5.71}$$

The exponent B has the same meaning as that of n in Equation (5.66) and Equation (3.5).

An example may illustrate the determination of the parameters of a two-stress Eyring model based on Equation (5.71) with the stress parameters of laser operating temperature and drive current. The matrix of the multi-cell life tests consists of three cells with different temperatures but equal current levels and at least two cells with the same temperatures but different current levels. The rst set of cells can be used to determine E_a according to the procedure described in Section 5.5.2.1 and Equation (5.65), since the current is constant and the temperature portion of the model is Arrhenius. Alternatively, taking the ratio of the t_f for any pair of this set cancels out the inverse current portion and constant A and by taking the logarithms of both sides yields a simple equation for E_a. This latter approach can also be applied to any pair of cells of the second set where the Arrhenius portion cancels out and the constant B can then be calculated. If the failures are not evenly distributed across the cells, but concentrated on a few cells, the estimates have to be weighted by the number of failures. A more effective way to t the acceleration model (Equation 5.71) would be to use a multiple regression program on the t_f of all the cells simultaneously by inputting the data in a linearized form of Equation (5.71) generated by taking the logarithms of both sides. A weighted regression evaluation can provide a signi cant improvement in achieving estimates with less deviation from the true values.

We applied this estimation procedure to the multi-cell life tests of a matrix of accelerated tests covering a range of drive currents and temperatures applied to the InGaAs/AlGaAs lasers used throughout Section 5.3 above. In the sudden failure rate regime matched by the Weibull model, a thermal activation energy $E_a = 0.65$ eV and current exponent B (equivalent to n) $= 0.76$ was obtained, whereas in the wear-out regime matched by a lognormal function, $E_a = 0.64$ eV and $B = 1.17$ (Epperlein et al., 2001, unpublished). Accelerated aging tests on InGaAs/AlGaAs lasers with a modi ed structure have yielded a higher current acceleration with exponent B up to three. We conclude that, in general, current exponents B are in the range of ~ 0.7 to 3 for current acceleration aging tests on diode lasers.

The total acceleration factor for Equation (5.71) with temperature T and current I as stress factors is

$$\text{AF} = \frac{t_{f,u}}{t_{f,s}} = \left(\frac{I_s}{I_u}\right)^B \exp\left\{\frac{E_a}{k_B}\left(\frac{1}{T_u} - \frac{1}{T_s}\right)\right\} \tag{5.72}$$

where, in this case, the acceleration factor of the Eyring model is just the product of the acceleration factors of the inverse power law for current and the Arrhenius model for temperature. Using $T_u = 298$ K, $T_s = 393$ K, $E_a = 0.64$ eV, $I_s = 500$ mA, and

$I_u = 250$ mA (for 200 mW cw output power) for our laser example, we obtained $AF = 1.6 \times 410 = 656$ for $B = 0.7$ and $AF = 8 \times 410 = 3280$ for $B = 3$. This shows that the temperature dominates the current in the acceleration process for the entire, practically relevant current exponent B.

5.5.2.4 Other acceleration models

The three general acceleration models discussed above are certainly the most widely used models in accelerated testing. However, there may be situations where other forms of the applied stress S have to be used to match the actual failure mechanism accurately. Typical variants include

$$A + B \ln S \quad \text{or} \quad A + BS \quad \text{or} \quad 1/(A + BS). \tag{5.73}$$

The respective form depends on the laser type, its speci cations, and operating conditions and usually can be determined in an iterative process by tting the model to the experimental test data.

Moreover, in the preceding discussions on the common acceleration models, we assumed that the applied stresses are well de ned and constant in time. However, in reality, this is usually not the case, and the operating environment may consist of many different stresses varying in relative importance, range, and relative importance during use. Such a situation can be handled by testing sequentially under the various conditions or by testing under a set of conditions applied simultaneously, though this complicates the test data analysis due to the complexity of the test conditions. There is, however, a most widely used cumulative damage model for fatigue failures caused by combined stresses.

This is the Linear Damage Rule or Miner's Rule (Miner, 1945; ReliaSoft Corporation, 2010), which is probably the simplest cumulative damage model. It de nes a damage fraction, D, as the fraction of life used up by a series of events. Failure is then predicted to occur when

$$\sum_{i=1}^{j} D_i = X \tag{5.74}$$

where i is the index of each set of applied load cycles at constant stress level S_i, D_i is the damage fraction accumulated during the load cycles of interval i at S_i, and X is the damage criterion which is a constant. The damage fraction D_i at the ith stress level S_i is equal to the number of applied cycles n_i at a speci c applied stress level S_i divided by the number N_i of cycles to failure at that level. That is, n_i/N_i is the fraction of an item's useful life, which is used up at each level of applied stress. Equation (5.74) can then be rewritten as

$$\sum_{i=1}^{j} \frac{n_i}{N_i} = X \tag{5.75}$$

where X is the fraction of life consumed by exposure to the cycles at all the different stress levels S_i with $i = 1, 2, \ldots, j$. In general, failure is predicted to occur when $X \geq 1$. Instead of cycles we can also use lifetimes, and Equation (5.75) is then

$$\sum_{i=1}^{j} \frac{t_i}{T_i} = X \qquad (5.76)$$

where t_i is the time at the ith stress level, T_i is the permissible lifetime consumed at the ith stress level, and t_i/T_i is again the fraction of useful life used up at the ith stress level. If we assume the fraction of time α_i at each stress level S_i is known, then $t_i = \alpha_i \times T$, where T is the total lifetime resulting from all cycles applied. With $X = 1$ we obtain from Equation (5.76) for the resultant life

$$T = 1 / \sum_{i=1}^{j} \alpha_i / T_i. \qquad (5.77)$$

Miner's Rule is a useful approximation in many applications. However, it has several major shortcomings, which include: (i) the assumption of a simple linear life–stress relationship, which may not be true in some real-world applications; (ii) ignorance of possible effects of the application sequence of the individual stress cycles; and (iii) the assumption that the rate of damage accumulation is independent of the stress level. Despite these limitations, Miner's Rule is very popular in fatigue testing, since it is both simple and obviously more sophisticated methods do not always yield more accurate predictions. It can also be applied to other failure mechanisms, if properly de ned. A more detailed discussion of Miner's Rule is beyond the scope of this text.

5.5.2.5 Selection of accelerated test conditions

Expected failure mechanisms usually determine the selection of the conditions for accelerated testing. This means considering the various ambient and operating conditions for the product, and probing which mechanisms emerge as dominant during testing at accelerated levels of these conditions.

It may well be, for example, that a product exposed to elevated temperatures might be affected by several thermally activated failure mechanisms, which might have different activation energies E_a, but which one of these will be effective might be unknown. Hidden failure mechanisms are a risk of predicting failures with the Arrhenius equation. For example, a stress test at high temperature may reveal a failure mechanism with high activation energy, but cannot detect a failure mechanism with low activation energy. In this case, a product will fail at low use temperature in a much shorter time due to the mechanism with low activation energy. Unanticipated thermally activated failures might also contribute to inconclusive test results leading to unrealistic predictions and ambiguous further evaluations. Moreover, because E_a is in the exponent in the Arrhenius equation, it has to be known precisely, otherwise enormous uncertainties in the calculated acceleration factor and extrapolated lifetime values will result. In addition, it is crucial to stay within the temperature range of applicability of the Arrhenius equation. If the stress temperature is so high that the properties of the sample are different from those at use temperature, then the test results cannot be extrapolated to use temperature.

The proper selection of time and temperature as test conditions will help to detect every mechanism effectively present in the tests. Without further elaboration, it should be mentioned that a nite element analysis could make it possible to nd the locations of maximum applied stress and to determine the level of stress at these locations.

A well-established way out of this dilemma is to expose the samples to well-de ned and controlled accelerated lifetime tests and environmental stress tests, which may cause failures during the use time in the eld. Typical stress factors can be categorized as follows:

- Accelerated lifetime tests – including temperature, current, and optical power as stress factors.

- Endurance tests – including slow temperature cycling, fast temperature cycling, low-temperature storage, high-temperature storage.

- Mechanical integrity tests – including wire bond pull, ber pull, die shear, vibration, mechanical shock, thermal shock, solderability, package hermeticity.

- Other tests – including humidity stress, electrostatic discharge (ESD).

In Chapter 6, we will elaborate on some of these stress factors and their typical test conditions for diode laser devices. When these tests are carried out properly and completely, then actually any failures that might occur in use should be detected in these tests, even if the underlying failure mechanisms are not taken into account in the design phase of the tests.

5.6 System reliability calculations

5.6.1 Introduction

So far, we have discussed in this chapter the reliability of an individual device. In practice, however, we have to deal also with multi-device structures that operate together in a system. Relevant questions in this respect include:

- How can the total system performance be predicted?

- How does the design of the system affect reliability?

- How can the design be made so that the system operates reliably even if some of the components fail?

- How can the system reliability by improved by redundancy?

To calculate in the following the reliability of the simplest types of systems, we need two rules (Tobias and Trindade, 1986) (see also Section 5.1.3):

- The *Multiplication Rule*: The probability that several independent events will all occur is the product of the individual event probabilities. Accordingly, the probability that N independent identical units, each with reliability $R(t)$, will survive after time t is $[R(t)]^N$.

- The *Complement Rule:* The probability that an event will not occur is one minus the probability of the event occurring. Accordingly, the probability that at least one of N independent identical units fail by time t is given by $1 - [R(t)]^N = 1 - [1 - F(t)]^N$.

Both these expressions show how, in this simple case, system reliability is built up by the reliability of individual units in a bottom-up approach. Within the scope of this text, we can only discuss two simple system reliability methods, so we will show that the lower and upper bounds for system reliability are given by the series and parallel models, respectively.

5.6.2   Independent elements connected in series

A system composed of N independent elements connected in series fails when any one of the elements fails. If $R_i(t)$ is the reliability or probability of survival of a given element, then the system reliability function $R_s(t)$ is given according to the Multiplication Rule as

$$R_s(t) = \prod_{i=1}^{N} R_i(t) \qquad (5.78)$$

and the CDF (see Equation 5.4) or unreliability function $F_s(t)$ of the system is given by the Complement Rule as

$$F_s(t) = 1 - \prod_{i=1}^{N} [1 - F_i(t)]. \qquad (5.79)$$

Example for illustration: An example may illustrate the effect of Equation (5.79). Let us take a 72-pin memory module where the module is the system and the pins are the elements. We want to calculate the probability of failure of the module by assuming that the probability of failure is 1×10^{-5} per pin when inserting the module into a mating connector. Equation (5.79) applies, when we further assume that a single pin failure leads to failure of the module. The equation can be written as $F_s = 1 - (1 - F_i)^N$ where $N = 72$ is the number of pins and $F_i = 1 \times 10^{-5}$ is the unreliability of a single pin. The result is $F_s = 7.2 \times 10^{-4}$, which means that the module is predicted to fail with a probability of about 0.07%, or in 10^4 plug-in events there will be about 7 failures.

The system failure rate λ_s is the sum of the individual failure rates λ_i. To show this, we use the fact that the failure rate function is de ned as the negative derivative of the natural logarithm of the reliability function ($dH(t)/dt = h(t) = \lambda(t) = -d[\ln R(t)]/dt$, which can be obtained by taking the derivatives of both sides of Equation 5.8). From Equation (5.78) we obtain

$$- \ln R_s (t) = - \ln \prod_{i=1}^{N} R_i (t) = \sum_{i=1}^{N} - \ln R_i(t) \qquad (5.80)$$

and the system failure rate is then

$$\lambda_s (t) = \frac{d[- \ln R_s(t)]}{dt} = \sum_{i=1}^{N} \frac{d[- \ln R_i(t)]}{dt} = \lambda_1 (t) + \lambda_2 (t) + \cdots + \lambda_N (t). \qquad (5.81)$$

From Equation (5.81) it follows that the reliability of a series system is always worse than that of the weakest system element.

5.6.3   Parallel system of independent components

The active redundant system consisting of N independent elements connected in parallel continues to operate until the last element fails. This model is the other extreme of the series model where all elements must operate. Parallel systems offer big advantages in reliability, in particular in early life when the failure function F is small, as we will see below. To design-in redundancy is crucial in applications where excellent reliability and low front-end failure rates have paramount priority. Such applications include computer systems in aerospace/space transportation systems and laser systems deployed in transoceanic cables for optical communication.

The probability that the system fails by time t is the probability that all N independent components have failed by time t. The corresponding system unreliability $F_s(t)$ and reliability $R_s(t)$ functions are then given by

$$F_s (t) = \prod_{i=1}^{N} F_i (t) \qquad (5.82)$$

$$R_s (t) = 1 - \prod_{i=1}^{N} [1 - R_i (t)] = 1 - \prod_{i=1}^{N} F_i (t). \qquad (5.83)$$

Without showing the actual calculation here, it can be stated that the failure rates are no longer additive. In fact, the system failure rate in a parallel (redundant) con guration is smaller than the smallest component failure rate.

It can also be shown (Tobias and Trindade, 1986) that if a single component with an unreliability F and failure rate λ is replaced by N components in parallel, then the system failure rate is reduced or the ratio of the system to component failure rates is lowered by the factor

$$k = \frac{\lambda}{\lambda_s} = \frac{1 + F + F^2 + \cdots + F^{N-1}}{NF^{N-1}}. \tag{5.84}$$

For small F values in the product's early-life regime the improvement is about a factor of $(1/NF^{N-1})$.

Example for illustration: An example may illustrate this effect. For $F = 0.01$ and $N = 2, 4$, and 6, we obtain $k \cong 50, 2.5 \times 10^5$, and 1.7×10^9, respectively, and for $F = 0.1$ and $N = 2, 4$, and 6, $k \cong 5, 2.5 \times 10^2$, and 1.7×10^4, respectively. These results clearly demonstrate that failure rate reduction and redundancy is greatest early in life when F is small. It also shows that the improvement factor can already be very high for small N and increases much faster with N when the F values are small. In addition, it drops to much lower values for larger F values and increases for example by a factor of 10, where its increase rate with N is also slowed down by at least two orders of magnitude. In summary, redundancy makes a large difference early in life when F is small, and much less of a difference later on.

There are many other redundancy models different to the parallel model, which assumes the redundant components are operating all the time. Such systems would have, for example, a certain portion required to be operable out of all N components or have the redundant components in a standby mode, only being activated when switched on by a failure in the primary component. However, their discussion is beyond the scope of this text and the reader is referred to relevant texts including Tobias and Trindade (1986), Condra (1993), and Joyce and Anthony (1988).

References

Cof n, Jr., L. F. (1954). *Transactions of the ASME*, **76**, 931.

Condra, L. W. (1993). *Reliability Improvement with Design of Experiments*, Marcel Dekker, New York.

Dummer, G. W. and Grif n, N. B. (1966). *Electronics Reliability – Calculation and Design*, Pergamon Press, Oxford.

Elsayed, A. E. (1996). *Reliability Engineering*, Addison Wesley Longman, Reading, MA.

Epperlein, P. W., Hawkridge, A. R., and Skeats, A. (2001). For further information see www.pwe-photonicselectronics-issueresolution.com.

Epstein, B. and Sobel, M. (1953). *J. Am. Stat. Assoc.*, **48**, 486.

Eyring, H., Glasstones, S., and Laidler, K. J. (1941). *The Theory of Rate Processes*, McGraw-Hill, New York.

Jordan, A. S. (1984). *Microelectron. Reliab.*, **24**, 101.

Joyce, W. B. and Anthony, P. J. (1988). *IEEE Trans. Reliab.*, **37**, 299.

Kececioglu, D. (1991). *Reliability Engineering Handbook*, Vols. 1 and 2, Prentice-Hall, Englewood Cliffs, NJ.

Manson, S. S. (1965). *Exp. Mech.*, **5**, 193.

Miner, M. A. (1945). *J. Appl. Mech.: Trans. ASME*, **67**, A159.

Nash, F. R. (1993). *Estimating Device Reliability: Assessment of Credibility*, Kluwer Academic, Boston, MA.

NIST/SEMATECH (2010). *e-Handbook of Statistical Methods, Reliability*. www.itl.nist.gov/div898/handbook/.

O'Connor, P. D. T. (2002). *Practical Reliability Engineering*, 4th edn, John Wiley & Sons, Ltd, Chichester.

ReliaSoft Corporation (2010). *Reliability HotWire*, Issue 116, October. www.weibull.com/hotwire/issue116/hottopics116.htm.

Sudo, H. and Nakano, Y. (1985). *Microelectron. Reliab.*, **25**, 525.

Tobias, P. A. and Trindade, D. C. (1986). *Applied Reliability*, Van Nostrand Reinhold, New York.

Weibull, W. (1951). *J. Appl. Mech.*, **18**, 293.

Chapter 6

Diode laser reliability engineering program

Introduction

The main part of this chapter is dedicated to the detailed description of a typical laser reliability test program required for achieving qualification of a diode laser product. In the first part, some up-front activities are addressed, which are required to implement an efficient reliability engineering program. These include fostering a culture of reliability in the organization, the definition of reliability specifications, and product failures. A crucial necessity is also to conduct a risk assessment by implementing a failure modes and effect analysis program at the beginning of the product design.

Semiconductor Laser Engineering, Reliability and Diagnostics: A Practical Approach to High Power and Single Mode Devices, First Edition. Peter W. Epperlein.
© 2013 John Wiley & Sons, Ltd. Published 2013 by John Wiley & Sons, Ltd.

Critical laser parameters for long-term stability and reliability will be determined. Essential for successful reliability investigations are the test preparations, which include procuring effective test equipment and defining sample sizes, and test durations of the experiments. The central portion of this chapter deals extensively with the technical aspects of the individual building blocks of the reliability engineering program including specifications, conditions, and results of the various test procedures for the laser chip, subcomponents, and module. We will also describe how the collection, analysis, and reporting of reliability data play an integral part in the program. In the final sections, the advantages and benefits of a reliability growth program and a reliability engineering program will be described, including a reliability cost model for determining an optimum reliability point by balancing the initial and post-production costs.

6.1 Reliability test plan

6.1.1 Main purpose; motivation; and goals

The implementation of a reliability engineering program (REP) adds great value to an organization's assets and success. To succeed in today's highly competitive and technologically complex industrial laser markets with increasing performance and reliability demands, knowledge of product reliability is mandatory, as is the ability to control it in order to produce products at an optimum level of reliability. In real terms, this means yielding the minimum life-cycle costs for the user and minimizing the manufacturer's life-cycle costs of such a product without compromising the product's reliability and quality.

This complex goal can be achieved by the implementation of an effective and efficient REP. However, a condition for success is to have a culture of reliability in the organization that requires everybody involved in the planning, concept and design, production and delivery of a product to understand the need for a healthy REP for the organization's success. An essential step in this direction is to have the support of the organization's senior management, which can usually be obtained by emphasizing the financial benefits resulting from a successful, living REP in the form of increased customer goodwill and satisfaction, positive image and favorable reputation, increased sales revenue and positive impact on future business, and enhanced competitive advantage. Equally important are the support and understanding of the technical personnel responsible for the implementation and operation of the REP. Only then can the REP contribute to the success of the organization's high-quality and high-reliability products.

Reliability engineering provides the theoretical and practical tools with which the capability of products can be specified, designed-in, predicted, tested, and demonstrated with regard to the performance of their desired functions for the required periods of time without failure, in specified environments, and with the desired confidence. It covers all aspects of a product's life from its conception, subsequent design, and production through to the end of its practical use life. A highly reliable product is as good as the inherent reliability of the product and the quality of the manufacturing processes. Reliability is the most important quality factor of a product or, in other terms, quality is based on reliability.

The REP is an essential component of a good product life-cycle management program comprising the following goals:

- To evaluate the inherent reliability of a product or manufacturing process.

- To pinpoint potential areas for reliability improvement.

- To identify the most likely failures as early as possible in the product life cycle.

- To identify appropriate actions to eliminate failures or mitigate the effects of these failures.

The overriding goal is to use the reliability information at the highest level to assess the financial impact of the reliability of the products and to improve the overall product reliability, leading to decreased overall lifetime costs and consequently to an improvement in the financial strength of the organization. However, a proper balance must be struck between reliability and other business aspects such as time to market, manufacturing costs, sales, product features, and customer satisfaction.

6.1.2   Up-front requirements and activities

Important foundations for an REP are the clear definitions of reliability specifications for a product and what constitutes a failure. It is also critically important to provide a methodical way of examining the proposed design of a product for possible ways in which failure can occur. We will discuss the basics of the failure modes and effect analysis approach revealing the criticality of failure modes.

6.1.2.1   Functional and reliability specifications

The definition of reliability is the ability of a product to perform its intended mission under given conditions for a specified time without failing. This means that clear, unambiguous, and detailed reliability specifications must be agreed in line with the conditions in the reliability definition addressing failure rate, effective use time, usage limitations, and operating environment. An example from the highly demanding applications of pump lasers in submerged optical communication systems may illustrate the situation:

- Wear-out failure rate at 200 mW single-mode power in fiber: $<0.2\%$ at 25 °C, with 95% upper confidence limit over 27 years.

- Sudden failure rate at 200 mW single-mode power in fiber: $<2\%$ at 25 °C, with 95% upper confidence limit over 27 years (≈ 80 FITs).

Setting clearly specified reliability goals requires the involvement of the customer and design, manufacturing, and quality engineers from the manufacturer's organization. Of course, the ideal situation is when the organization has solid reliability expertise and the background required to define realistic goals and specifications to be met by the product under use operation. In the formulation process of the reliability specifications, it is also essential to consider financial aspects such as planning for warranty and production part costs, so that in the end a proper balance of realistic reliability performance expectations and financial issues can be achieved.

6.1.2.2 Definition of product failures

Another critical issue is the definition of what constitutes a failure, which can be decisive as to whether a product meets the reliability tests. It is imperative to agree commonly accepted definitions of product failures across all parties involved in the product development processes and for various different reasons. These groups may have different definitions of product failure, which may yield radically different results from a test. Universally agreed failure definitions may have the benefit of minimizing the tendency to rationalize away certain failures, particularly in the early stages of the product development. This may result in a lower risk of releasing products into the field with poorly defined but very real failure modes. Communications and the management of the REP may also become more effective and error-free. In the case of a complex product with a number of distinct failure modes, the implementation of a multi-tiered failure definition structure may ease the assessment and analysis of the different failures by logging them under individual codes into a database.

Typical failure criteria for pump diode lasers are, for example, a 5% reduction in optical power ex-facet during accelerated life tests at high temperature and power or 50% increase in threshold current at the rated power.

6.1.2.3 Failure modes, effects, and criticality analysis

The most commonly used form of risk analysis is the failure modes, effects, and criticality analysis (FMECA). It is formalized in the standard MIL-STD-1629 (e.g., ReliaSoft Corporation, 2011a) as a structured analysis of potential failure modes and their effects on the product. The aim is to reduce or eliminate failures in products or processes by identifying and applying appropriate corrective actions. In this way, failures are prevented from reaching the customer and hence the highest yield, quality, and reliability can be assured and the cost of quality can be reduced both in the organization and at the customer.

There is a design FMECA, used by product design engineers, which addresses potential product failures, and a process FMECA, used by process design engineers, which addresses potential manufacturing process failures, which ultimately could lead to product failures. Both types are independent of each other and can be produced separately.

Conducting an FMECA is a multifunctional team effort requiring the input of many disciplines including design, manufacturing, testing, quality, reliability, failure analysis, packaging, and marketing. One of the most important factors for the successful implementation of an FMECA program is timing. It is meant to be a *before-the-event* action and not an *after-the-fact* exercise. To achieve its greatest value, it must be done *before* a design or process failure mode is unknowingly designed into the product. It would be best placed as an integral part of the product/process development in the detailed design development phase *before* the pre-production phase and design approval. An FMECA offers various benefits including:

- It ensures that all possible failure modes and their effects on the fitness-for-use of the product have been taken into account.

- It identifies potential failure mechanisms and the relative severity of their effects.

- It records any improvement due to any necessary corrective action taken.

- It identifies design alternatives with high functionality potential during the early phase of product development.

- It assures the customer about a product with a minimized failure potential and thus increasing yield, quality, and reliability, thereby reducing the cost of the product.

All aspects of an FMECA are recorded on a form (Table 6.1). Note that the failure mode is the symptom of the failure as distinct from the cause of failure.

Various approaches are used to show the criticality of the failure modes. For each identified potential failure mode, an estimate is made of its cause and the likely effect on the fitness-for-use of the product. A risk priority number (RPN) is calculated as the product of ratings on frequency of occurrence (O), severity (S), and likelihood of detection (D) for each identified failure mode. A value of 1 (low risk) to 10 (high risk) is assigned to O, S, and D, so the RPN can be between 1 (low risk) and 1000 (high

Table 6.1 Typical failure modes, effects, and criticality analysis (FMECA) form for risk analysis.

Failure Modes, Effects and Criticality Analysis (FMECA) Worksheet															
Type: Device:			Team: Leader:				Report no.: Sheet no.: Date:								
Design FMECA									Action results						
Running No.	Item / Function	Potential failure mode	Potential effect of failure	Potential cause of failure	S (1 – 10)	O (1 – 10)	D (1 – 10)	RPN (1 – 1000)	Recommended corrective action	Responsible person	Completion date	S (1 – 10)	O (1 – 10)	D (1 – 10)	RPN (1 – 1000)

risk). MIL-STD-1629 gives a description of criteria allocated to each risk number. It also provides a more precise quantification of criticality using data on the probability of mission loss due to the failure mode, the part failure rate, the ratio that a particular failure mode has to the total part failure rate, and the duration of the applicable mission phase. Commercial FMECA software packages are available from various companies including ReliaSoft Corporation (2011b), Isograph Ltd (2011), and PTC Corp. (2011).

The RPNs are assessed with regard to the risk they express. If corrective actions are decided and implemented, then the respective types and timescales of these actions are also recorded. The effect of these corrective actions is evaluated and the RPN is again calculated, and a decision is made if the new result is acceptable. An FMECA is an interactive, never-ending process and its form is a living document that is regularly updated and modified to reflect the criticality of the items listed therein. Thus, the form offers an insight into the status of each potential failure mode, but shows also the overall risk factor for successfully realizing the product.

6.1.3 Relevant parameters for long-term stability and reliability

The most important parameters of single-mode diode lasers to be monitored during reliability tests include threshold current, thermal rollover optical power, kink-free power, drive current and voltage, slope efficiency, differential resistance, center lasing wavelength, and laser beam far-field angles. These parameters have to be measured before and after each reliability test within the test plan to be discussed in Section 6.1.5 below. Thus, any changes or failures in the internal and external laser structure including the active region, mirror facets, cavity, transverse vertical and lateral confinement, Ohmic contacts, and heat sinks can be detected. Suitable characterization and analysis techniques can then be applied to reveal the actual root causes of these changes. Part III will give more information on this matter.

6.1.4 Test preparations and operation

6.1.4.1 Samples; fixtures; and test equipment

In the following, standard approaches and procedures for sample selection, mounting on test sites, and reliability test equipment are described. In-spec laser chips are selected and visually inspected under an optical microscope at $\sim 1000\times$ magnification for their mechanical integrity: that is, absence of any mechanical defects such as scratches, metal voids/blistering/peeling, chipouts, cracks, facet coating cracks/peeling/voids/discoloration, and contaminants including coating overspray, organics and micrometer-sized debris, and particles in the critical cavity and facet area. Optical microscopy with a Nomarski contrast option is useful for detecting topography changes, poor morphology, growth roughness, and metal lifting. To prevent malfunctions or degradation of the lasers by uncontrolled contaminations, the inspections must be performed in a clean airflow box.

The lasers are mounted junction-side up on AlN submounts using AuSn solder and are subsequently bonded into standard capped TO can header packages (see Section 1.4.3). These packages are then inserted into metal device adapter assemblies before attaching them to the resistively heated metal heat sinks on the boards of the life tester. A high-temperature uniformity of better than 1 °C across the heat sink can be achieved by using a suitable gradation of the resistor values.

Due to the high sensitivity of diode lasers to ESD effects (see Section 3.3), extreme care has to be taken when handling the laser chips into the TO headers and from there into the assemblies. For example, electrical connectors in the TO package adapter are doubly short-circuited by wires and microswitches. The wire short-circuits are removed and the microswitches opened only after the TO header has been carefully inserted into the electrical base of the adapter, and the photodiode/filter for monitoring the laser power has been attached to the assembly package.

A simple computerized life tester consists of driving and measuring electronics partitioned into independent units each serving a certain number of laser channels. Each unit comprises an interface card located in the personal computer, a data acquisition system, laser driver and photodiode receiver circuits. The tester software controls the transmission of control signals and measurement data, which can simply be stored in ASCII files. In contrast, sophisticated, fully automated, and computer-controlled life-test systems are commercially available (Sharetree Ltd, 2011; Reltech Ltd, 2011). These systems are designed for precise static, dynamic, or monitored testing and measurement of all required laser parameters and changes as well as recording and collation of test data. They offer both the devices to be loaded on actively temperature-controlled heat sinks or into trays whose loading sections are thermally insulated from the sections containing the control and monitoring electronics. This allows each tray to be removed independently for device loading/unloading without affecting the environment of the remaining trays. Each tray may be configured for a different device type or test conditions, making these systems ideal for applications with a variety of small batches. The maximum capacity of trays is about 36, with typically 20 test sockets (depending on device under test size) per tray; that is, more than 700 lasers can be life tested in one batch or in individual batches. The flexibility of these systems allows them to be used also in burn-in and environmental stress tests.

6.1.4.2 Sample sizes and test durations

Also in reliability testing, proper experimental planning is an important factor for carrying out experiments successfully. The complex and different forms of life distributions make the planning difficult, for example, to specify sample sizes and test durations.

However, when using the exponential distribution function simple first expressions for these quantities can be written by starting with Equation (5.30) in Chapter 5. To express the confidence bounds on the failure rate the right-hand side of Equation (5.30) has to be multiplied by the χ^2 distribution factor, which depends on the confidence level, number of failures, and one- or two-sided estimations for failure- or time-terminated tests, as discussed in Section 5.4.2. To avoid looking up the values in the tables referenced in Section 5.4.2, we restrict the discussion to the case of zero failures with Equation (5.43) as the relevant expression. From Equation (5.43) we can calculate a minimum test sample size

$$N_0 = \frac{-\ln \alpha}{\lambda_{obj} \times T} = \frac{-\ln \alpha}{T} \times \text{MTTF}_{obj} \tag{6.1}$$

and minimum testing time

$$T = \frac{-\ln \alpha}{\lambda_{obj} \times N_0} = \frac{-\ln \alpha}{N_0} \times \text{MTTF}_{obj} \tag{6.2}$$

where α forms the confidence level $(1 - \alpha)$ (see Section 5.4.2), λ_{obj} is the failure rate objective, and MTTF_{obj} is the MTTF objective. Both expressions work for any α level without the need for tables. However, note that the failure rate objective or MTTF objective will not be confirmed at the desired confidence level if there is even one failure during the test.

Example for illustration: We will give two examples to illustrate the sizes of the numbers of the different parameters involved in the last two equations. In the first example, we want a minimum sample size that will allow us to verify a MTTF_{obj} of 8×10^4 h in an accelerated test with 90% confidence ($\alpha = 0.1$) and a given test time of 1.3×10^4 h ≈ 1.5 years. After rounding down, we obtain from Equation (6.1) the minimum sample size $N_0 = 14$. In the second example, the failure rate objective in an accelerated test is the low value of 10^{-6} h^{-1} = 1000 FITs, which should be confirmed with 95% confidence ($\alpha = 0.05$). The question is: how long has the test to run when 350 devices are put on test? A minimum test time $T \cong 8560$ h ~ 1 year can be calculated from Equation (6.2).

6.1.5   Overview of reliability program building blocks

Figure 6.1 gives a graphical representation of the building blocks of a diode laser reliability program. It shows the three main blocks:

1. Reliability testing.

2. Data collection.

3. Data analysis, reporting, and failure reporting analysis corrective action system (FRACAS).

It also shows the various links between individual elements of the three building blocks. The next three sections give a short description of the main building blocks. Details on the different reliability tests will be discussed from Section 6.1.6 onwards.

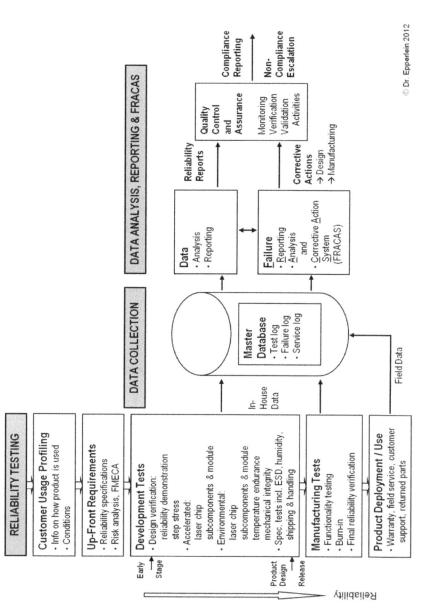

Figure 6.1 Reliability engineering program overview: building blocks.

© Dr. Epperlein 2012

6.1.5.1   Reliability tests and conditions

The first two elements of the first building block deal with topics that have already been discussed in detail in the up-front requirements of Section 6.1.2. The reliability tests are designed to uncover particular suspected failure modes, and the type of testing a laser device undergoes will change with the different phases of its life cycle. The more the reliability of the product grows, the more tests the product design passes in the course of the various types of in-house tests.

Tests during the earlier design stages in the development phase can identify early failures to provide useful information on issues in design, certain parts of the laser device, manufacturing, or other areas leading to nonconformance. The primary goal of these *design verification tests* is to verify the reliability of the design before committing major resources to production. *Accelerated tests* are performed to get meaningful data within a reasonable time and these data are used to predict the lifetime of the laser product under use conditions. *Environmental stress tests* simulate the operating conditions that the product may experience in the field. At this stage, the product design is usually released for production. We refer the reader to Section 5.5.2.5, where we have already discussed various aspects of selecting proper accelerated test conditions, in particular the constraints on temperature.

Subsequent tests mainly measure the *manufacturing process* and the impact any potential *post-release design changes* may have on the reliability. Properly designed, so-called *burn-in* procedures have the purpose to weed out those units that would experience early-life failures in the field. *Reliability verification tests* are performed at regular intervals during production or immediately following a post-release design change or a change in the manufacturing process. In this way, the reliability can be kept to specifications. These tests are also called *lot qualification tests* and comprise accelerated life tests on a sample of laser devices from the many devices on a single laser wafer (lot).

The last phase in building up knowledge of a product's reliability happens when the product is in use. *Field data*, usually available in vast amounts, could in principle contribute to reliability analyses if efficient field data collection and reporting systems are in place and the actual operating conditions of the product are known.

6.1.5.2   Data collection and master database

There are two types of reliability data, *in-house* and *field data*. In-house data result from the different types of reliability testing, which is in general carefully set up and executed by trained staff. One important aspect is to have certain common elements including similarly tested parameters that extend across all of the tests. This helps to perform similar analyses across the different tests and provides a high consistency and synergy in test analysis, which is beneficial to the entire REP. Field data result from a range of sources including warranty activities, field services, customer support, and returned parts analyses. In principle, field data are the true measure of product performance and reliability. However, in contrast to in-house data, field data carry a much higher degree of error, because their sources are much less well defined and there may be also more operator-induced failures. A clear tracking and reporting system is required, providing such data as time to failure and the operating and environmental conditions needed to perform a reliable field-data reliability analysis. Only then can the potential discrepancy between in-house and field data be minimized and a better "connect" between these two established.

Concise and comprehensive data collection is required for a REP to deliver useful results to improve the quality of a product. Modern data collection and recording systems are fully automated and computer controlled, thus decreasing drastically the potential for implementing errors. In setting up a database system, there are at least three data types that need to be included in the database structure: (i) a test log, to include test characterizing data and test data inputs; (ii) a failure log, to contain the information required to characterize and identify all the failures; and (iii) a service log, to record all the information necessary to track any service actions or modifications performed on test equipment. As discussed above, access to field data and therefore their collection for reliability analyses can be quite a difficult endeavor. If there is no formal field data collection system in place, then there is no other way than to gather the necessary reliability information in a laborious and time-consuming effort and enter the data manually into the database, which may be not error-free. Further difficulties arise from the fact that the field data may reside in different locations and in very different forms.

A solution to this is to employ a Web-based reporting system that allows all of the pertinent data to be gathered from the various sources and databases, and to enter it into the master database, where it can easily be processed and analyzed. These functions can be performed automatically using ReliaSoft's XFRACAS software tool (ReliaSoft Corporation, 2011b), a Web-based, closed-loop, enterprise-wide failure reporting, data analysis, and corrective action system. This user-configurable software is designed for the acquisition, management, and analysis of product reliability, quality, safety, and other data from multiple locations. It supports the entire incident management process, from the initial development stages to complete tracking of field serialized units. Bringing the different sources of information and key players together may result in a beneficial synergy by identifying untapped existing resources, improving existing processes, developing fresh perspectives, generating financial rewards through better product designs, enhanced control of product warranties, and more efficient customer support.

6.1.5.3 Data analysis and reporting

This section deals predominantly with the parametric statistical analysis of reliability data, understanding, interpretation, and reporting of the findings, and the provision of quality control and assurance activities aimed at demonstrating that the product is fit-for-use. The analysis mainly relies on in-house data generated in detailed tests designed for specific purposes (see the following sections). These hard-core reliability data with all the various types of evaluation described in Chapter 5 are used to predict the behavior of the product in the field. The data should be cross-referenced against the test log and service log so that irrelevant information can be excluded from the evaluations. Furthermore, in-house and field data should not be intermixed to avoid confusion and to ensure that the conclusions are precise and unambiguous.

The role of field data in this respect is only a secondary role due to the availability and quality issues described in the previous section. However, they will be used to support the results based on evaluations of in-house data. For example, there have been many occasions when the failure rates of pump lasers operated in optical data communication links over hundreds of millions of accumulated device hours have shown similar, very low levels of random failures of <100 FITs to a 95% upper confidence limit, as derived from the in-house data. This is excellent evidence of the supporting role of field data.

The various reports should contain not only numerical and graphical evaluations and conclusions, but also interpretations of the findings. This includes a discussion of the individual failure mechanisms and their likely root causes based on an elaborate failure analysis, but also recommendations for suitable necessary corrective actions to be applied in design and manufacturing with the goal to prevent failures in future runs.

The status of the actual reliability performance is regularly evaluated and compared to the given goals by the manufacturer's quality department. It will act on any deviations. Quality assurance provides both the protection against quality problems and proof to the customer that the laser product is fit-for-use. Readers interested in further detailed information on the role of the quality function in industry are referred to the classic, standard textbooks by Juran and Gryna (1988), Juran and Gryna (1980), Feigenbaum (1991), Caplan (1990), and Montgomery (1991).

6.1.6 Development tests

Development testing comprises a range of tests starting from the early product design phase and extending to final product design release. The severity of the tests increases as the product progresses through the different tests aimed at assuring that the product meets the required reliability specifications before the design is released for production.

6.1.6.1 Design verification tests

These tests are based on the principle of verifying the reliability of the product design before major resources are committed to more elaborate and stringent tests. Two types of early verification tests have gained popularity with many diode laser manufacturers.

Reliability demonstration tests

In these tests a series of stress factors such as high temperature and optical power are defined and if the laser passes these tests over a sufficiently long time, it is considered worthwhile to undertake more elaborate and expensive reliability qualification tests. The challenge is to apply stresses that are high enough for the results to justify proceeding with the more detailed tests but not exceedingly high enough to cause failures. Even if the laser passes the test, this test type leaves open questions such as: were the stresses high enough to get sufficient information on the robustness of the laser or would the laser have failed if slightly higher stresses had been applied; or were the most potential failure mechanisms activated; and what is the real reliability of the laser? Some of these questions are better addressed by step stress testing, to be discussed next.

Step stress testing

The goal of step stress tests is to determine in a short time the design margins with respect to operating and environmental stresses. Usually these tests are performed on a small sample of devices to identify the realistic stress conditions, which are required to set up accelerated tests for delivering conclusive predictions of use lifetimes. They also provide a measure of design robustness. Typical stress factors for diode lasers are temperature and drive current, and these are applied to separate sets of samples. The following list shows a typical step stress program for high-power single-mode pump lasers:

- *Temperature step stress:*

  Devices – 40 laser chips on submounts from five different wafers.
  Initial step condition – heat sink temperature $T_h = 70\,°C \triangleq\, \sim 98\,°C$ junction temperature at a constant drive current $I_{dr} = 300$ mA = constant.
  Steps – increment 10 °C, duration 168 h, minimum number six.

- *Current step stress:*

  Devices – 40 laser chips on submounts from five different wafers.
  Initial step condition – $I_{dr} = 300$ mA, $T_h = 70\,°C$ = constant.
  Steps – increment 100 mA, duration 168 h, minimum number six.

Figure 6.2a shows the principle of step stress testing. The first step is usually below the specification limits. During the tests, the optical power P is monitored, which allows the time to failure to be obtained for each failed device (Figure 6.2b). After each stress step, failed devices are removed and analyzed, and the remaining devices are successively stressed at higher levels until all devices have failed or irrelevant failures begin to occur. Failed parts are analyzed to verify any possible failure mechanisms.

Standard procedures include using visual inspection and optical microscopy to check for fingerprints of COMD events or signatures in the optical power versus time characteristics. More advanced failure analysis techniques include scanning electron microscopy, transmission electron microscopy, electron beam-based techniques such as electron beam-induced current and cathodoluminescence, and optical spectroscopy including electro- and photoluminescence techniques. Deep-level transient spectroscopy, which requires an electrically controllable depletion zone for the generation of the relevant signal, is capable of detecting small concentrations of deep

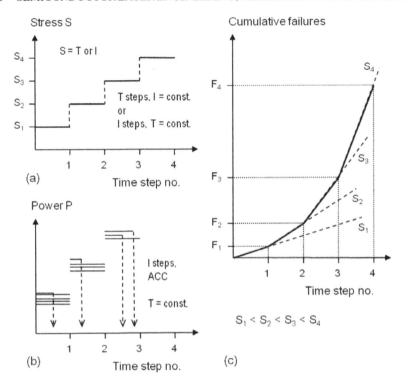

Figure 6.2   Step stress tests within a diode laser chip qualification including current (power) and temperature as stress factors. (a) General illustration with temperature T or current I as stress factors. (b) Schematic representation of failed devices within a test current (power) steps versus time at constant temperature. Constant current operation within each step. Optical power monitored versus time. Power change or drop as indicator of failure. (c) Schematic illustration of the cumulative failure distribution composed of the individual failures F in the entire step stress test. The slope of the individual linear segments gives the strength of a stress level (failure rate) within a step.

traps and nonradiative recombination centers in thin layers and at interfaces. In Part III, we will discuss some of these techniques employed for the advanced diagnosis and characterization of diode lasers in the development phase.

Based on the results of the physical analyses, it is usually possible to identify root causes of the failures. The aim is to improve reliability with suitable corrective actions in design, materials, and manufacturing processes. Alternatively, adjusted maximum limits on the operation or environmental conditions can assure the specified reliability figures. It is this positive feedback loop formed by design, manufacturing, testing, and corrective actions that leads to the maturity of the existing product and development of new product generations with enhanced performance.

Figure 6.2c shows schematically the cumulative failure distribution composed of segments from stress testing. The failure rate curves are steeper for higher stress levels, which means a higher percentage of device failures in the same step duration. In principle, it should be possible to determine both the failure rate at any stress level, by plotting the failure data from the various steps, and the failure rate distribution for each failure mechanism. This would then allow the prediction of the product reliability, which could be considered an additional benefit of step stress testing (Iuculano and Zanini, 1986; Bai *et al.*, 1989; Nelson, 1975, 1980; Miller and Nelson, 1983).

6.1.6.2   Accelerated life tests

As discussed in Section 5.5, the purpose of accelerated testing is to get within reasonable test times sufficient failure data to then allow prediction of the lifetime of the product under use conditions. Common stress factors applied in diode laser accelerated lifetime tests are temperature, current, or optical power in either an automatic current control (ACC) or automatic power control (APC) configuration. In the following, we describe typical accelerated tests for pump laser chips and laser modules.

Laser chip

Laser chips are soldered to a heat sink mounted on a submount assembly, which includes a reverse-biased protection diode. The submount assembly also has a pin photodiode for monitoring the power and a thermistor for temperature control in the module (see Section 1.4.3). The submounted lasers undergo a burn-in process and are screened for threshold current, slope efficiency, kink power, forward voltage, peak wavelength, spectral width, and far-field angles before they are mounted onto the test sites. Every submount receives an ACC burn-in typically at 350 mA, 70 °C for 96 h and with a failure criterion of > 10% the threshold current increase.

A typical matrix life-test program is shown in Table 6.2 aimed at identifying and modeling acceleration factors for wear-out and steady-state reliability. Test conditions

Table 6.2 Typical matrix accelerated life-test plan for submounted single-mode high-power diode laser chips. Applicable standards: MIL-STD-883(D), Method 1005 (IHS, Inc., 2008); TR-NWT-000468 (Telcordia Inc., 2011).

Cell	Heat sink temperature [° C]	Drive current [mA]	Optical power [mW]	Sample size / wafer	Minimum test time [h]	Failure criterion
A	70	500	~250	80 / 5	4000	
B	70	250	~130	80 / 5	4000	5% drop in peak power ex-facet
C	100	500	~200	80 / 5	4000	
D	100	250	~100	80 / 5	4000	
E	120	500	~150	80 / 5	4000	
F	120	250	~80	80 / 5	4000	

and sample sizes are more severe and higher, respectively, than those recommended in the relevant standards MIL-STD-883D Method 1005 Condition B (IHS, Inc., 2008) and Bellcore TR-NWT-000468 (Telcordia Inc., 2011).

The temperatures T_h are the temperatures measured on the heat sink of the laser submount assembly. The junction temperatures T_j are much higher and depend on the drive current and/or output power and can be either calculated using the actual thermal resistance R_{th} (see Equation 2.37) or measured via the wavelength shift determined in a spatially resolved electroluminescence experiment (see Part III).

A typical value for T_j is 145 °C using $R_{th} = 42$ °C/W, $I_{dr} = 500$ mA, and $P = 200$ mW at $T_h = 100$ °C in the calculation. It is anticipated that failures arise in cells A, C, and E within 2000 h test time, yielding a target sudden failure rate figure of <100 FITs at 2000 h.

Typical results of highly accelerated lifetime tests on submounted InGaAs/AlGaAs pump diode lasers aged at high temperatures and drive currents (common practice) have been discussed in Sections 5.3 and 5.5. Depending on the values of the thermal activation energy and current exponent used for calculating the temperature and current acceleration factors, lifetimes can be expected to be above 1×10^6 h > 110 years at use conditions of 25 °C and 250 mA (for 200 mW ex-facet) for both the sudden failure rate and wear-out regimes.

In the accelerated tests, defined in Table 6.2, the optical power is monitored as a function of stress time and failures are counted when the power drops by $\geq$5% of the rated peak power level, for example, 200 mW at $T_h = 100$ °C. Other important operating parameters to be measured periodically during the life tests include threshold current, slope efficiency, kink power, and peak wavelength. In general, the following trends can be expected at stress conditions: for example, $T_h = 100$ °C, $I_{dr} = 500$ mA, $P \sim 200$ mW. The changes (Epperlein *et al.*, 2001, unpublished) usually occur gradually in the first few thousand hours of aging and saturate after $\sim$5000 h:

- Threshold current: average increase $\sim$5%.

- Slope efficiency: average reduction $\sim -5\%$.

- Kink: typical changes in kink currents $< \pm 10\%$.

- Wavelength: maximum increase $\sim +6$ nm.

In addition to the standard accelerated lifetime tests, so-called *benign life tests* should be executed on diode lasers for applications in demanding areas such as submarine optical communication networks that require the utmost product reliability to avoid the extremely high repair/replacement costs in case of product failure. The aim of these tests is to simulate real-life conditions and demonstrate that no unexpected behavior occurs under such conditions. Thus, an upper confidence limit can be placed on the correctness of the models derived from the matrix life tests. Typical test conditions are 180 devices driven at maximum rated optical power at 50 °C for 4000 h.

Laser module

Submounted lasers chips with the photodiode and thermistor attached undergo a burn-in and screening process, as described above, before they are inserted into a fibered module. The module is housed in a 14-pin butterfly hermetic package (see Figure 1.37), which contains a thermoelectric-cooled optical platform. This platform includes the diode laser on a heat sink controlled by the photocurrent from the back-facet monitor photodiode. The laser beam is tightly coupled to a highly stable fiber alignment system. The platform temperature is monitored by a thermistor whose output is used to control the thermoelectric cooler current via an external feedback circuit. The fiber exits the package via a hermetically sealed feedthrough tube (see Section 1.4.3 and Figure 1.37).

The goal of accelerated aging tests of pump laser modules is to demonstrate their endurance (in combination with environmental tests) and to evaluate median lifetimes, failure rates, and thermal activation energies. The test methods and failure criteria are based on the standards Bellcore Advisory TA-NWT-001312 and Technical Reference TR-NWT-000468. The former is now being replaced by Telcordia GR-1312-CORE, the latter by GR-468-CORE (Telcordia Inc., 2011). The accelerated aging test conditions are listed in Table 6.3.

The modules are aged under ACC conditions in special package test systems, where the ovens are adjusted to maintain the selected package temperature T_p monitored with a thermocouple outside the package on the test board. The module thermistor controls the cooler to maintain a constant submount temperature T_s. The systems provide in situ monitoring of photocurrent, laser current, and cooler current and voltage. The modules are removed periodically, for example, every 120 h, from the test ovens for measurements of the *P/I* characteristics, from which key operating parameters such as threshold current, slope efficiency, and kink power are determined. Alternatively, there are systems that provide for in situ measurements of the *P/I*.

The failure criterion is given by the end of life (EOL), which is determined by a 50% increase in laser current at the specified fiber-coupled power as recommended in the Bellcore standards. To derive wear-out time to failure, the criterion is applied for

Table 6.3 Typical matrix accelerated life-test plan for single-mode high-power diode laser modules. T_s = submount temperature, T_p = package temperature. Applicable standards: TR-NWT-000468, TA-NWT-001312 (Telcordia Inc., 2011).

Cell	T_s [°C]	T_p [°C]	I_{dr} [mA]	Sample / wafer size	Minimum test time [h]	Failure criterion
A	25	70	400	15 / 3	5000	
B	25	85	400	15 / 3	5000	End of life 50% drop
C	30	85	400	15 / 3	5000	in in-fiber
D	55	85	400	15 / 3	5000	power

each module by using a linear extrapolation of the time dependence of the drive current $I_{dr}(t)$ during the life test. However, the results of this analysis depend very strongly on the period of time over which the linear extrapolations are made. Therefore, it is important to use the same aging time period, typically 5–10 kh, for this process for comparing different module groups.

Data of long-term projections confirm that the pump laser modules of all leading manufacturers are inherently mature and extremely reliable products. Thus, median lives $>8 \times 10^5$ h > 90 years at 200 mW in-fiber power, 25 °C, and maximum wear-out failure rates <1000 FITs are standard figures. The thermal activation for wear-out relative to the package temperature T_p is usually negligibly small ($E_a < 0.1$ eV). Root causes for degradation in the module can be attributed both to the movement of the fiber tip relative to the laser emission area occurring in the early failure regime with some saturating effect, and to the submounted laser itself, in particular at longer times.

6.1.6.3 Environmental stress testing – laser chip

The purpose of these tests is to examine the robustness of the submounted laser chip with an attached wire bond and to expose weaknesses and latent defects that may result in field failures if corrective action is not taken. In general, the various tests can be divided into three groups: temperature endurance, mechanical integrity, and special tests. These different types of tests along with typical results are discussed below by means of single-mode high-power InGaAs/AlGaAs SQW laser devices, but are in general also valid for other diode laser types as required.

Temperature endurance

Table 6.4 lists the temperature endurance test comprising a temperature cycling test and its conditions and effects according to Bellcore TR-NWT-000468 recommendations. The requirements are that the in-spec devices maintain their specified characteristics within the allowable change limits after completion of the test.

Table 6.4 Temperature endurance test plan for submounted single-mode high-power diode laser chips.

Test	Conditions	References	SS	Potential effects
Temperature cycling (unbiased)	−40 to +70 °C; 50 cycles with fast 12 min ramps and short 15 min dwells	MIL-STD-883D method 1010. Bellcore TR-NWT-000468 Issue 1	11	Distortion due to expansion and contraction, peeling, cracking due to liquid evaporation, fatigue, cracks in finished surfaces. Changes in electro-optical characteristics

Notes:
SS = sample size
MIL-STD-883D method 1010 (IHS, Inc., 2008)
Bellcore TR-NWT-000468 (Telcordia Inc., 2011)

In-spec and defect-free, single-mode InGaAs/AlGaAs pump laser devices from state-of-the-art manufacturing lines in general survive the temperature cycling tests; that is, the changes, for example, in threshold current and slope efficiency, are far below the acceptable limits of $\pm 5\%$ and $\pm 10\%$, respectively.

Mechanical integrity

The tests investigating the mechanical integrity of the submounted laser chip are detailed in Table 6.5 and comprise mechanical shock, vibration, wire bond pull, and die/heat sink shear.

Also here, in-spec and defect-free, single-mode high-power InGaAs/AlGaAs laser devices from state-of-the-art manufacturing lines in general survive the mechanical

Table 6.5   Standard mechanical integrity test plan for submounted single-mode high-power diode laser chips.

Test	Conditions	References	SS	Potential effects/ requirements
Mechanical shock (unbiased)	$5-1500g \cong 50-15\,000$ m/s^2 half sine wave of 0.5 ms duration, 5 shocks per axis, 6 axes	MIL-STD-883D method 2002. Bellcore TR-NWT-000468 Issue 1	11	Mechanical looseness, fatigue destruction, wire disconnection, destruction due to harmonics, solder destruction, cracking
Vibration (unbiased)	$20g \cong 200$ m/s^2, 20–2000 Hz, 4 min per sweep, 4 sweeps per axis, 3 axes	MIL-STD-883D method 2007. Bellcore TR-NWT-000468 Issue 1	11	Same as above
Wire bond pull (no prior conditioning)	1.0 mil gold bond wire diameter. Bond pulled to destruction at 4 mm/s	MIL-STD-883D method 2011. Bellcore TR-NWT-000468 Issue 1	11	Minimum required bond strength 3.0 gram force (gF)
Die shear (no prior conditioning)	Laser die shear from heat sink. Heat sink shear from submount	MIL-STD-883D method 2019. Bellcore TR-NWT-000468 Issue 1	11	Minimum laser die shear force for die area 0.75 $\times$ 0.3 mm^2 with $\geq 50\%$ solder adhesion area: 100 gF Minimum heat sink shear force: 300 gF

Notes:
SS = sample size
MIL-STD-883D method 2002, 2007, 2011, 2019 (IHS, Inc., 2008)
Bellcore TR-NWT-000468 (Telcordia Inc., 2011)

Table 6.6   Special tests including electrostatic discharge robustness testing of submounted single-mode high-power diode laser chips.

Test	Conditions	References	SS	Requirements/ failure definitions
Electrostatic discharge (ESD) based on human body model (HBM) and charged device model (CDM)	Voltage steps 250 V, 1000 V; 5 discharge pulses per voltage step. Pulses in forward and reverse bias	MIL-STD-883D method 3015. Telcordia TR-NWT-000870	12	Minimum 500 V in both bias directions. Failure definitions: 50% threshold current increase. Facet degradation. P/I characteristic distortions.

Notes:
SS = sample size
MIL-STD-883D, method 3015 (IHS, Inc., 2008)
Telcordia TR-NWT-000870 (Telcordia Inc., 2011)

shock and vibration tests. The devices maintain their specifications within the change limits after the tests; for example, the changes in threshold current and slope efficiency are well below the limits of ±5% and ±10%, respectively.

Moreover, wire bonds are usually very reliable, provided the bonding process was carefully carried out on stable and clean metal layer surfaces. Typical bond pull forces can be above 20 grams force (gF), which fully meets the specification of 3 gF. Similarly, die shear does not cause a problem if the solder wetting degree is more than 50%. In this case, die shear strengths can be higher by a factor of at least 3 compared to the failure criterion of 100 gF. The same is true for the heat sink shear strength, which can exceed the specification by more than a factor of 10.

Special tests

The conditions for investigating the robustness of in-spec submounted laser devices to electrostatic discharge (ESD) exposure are given in Table 6.6.

High-quality devices can pass the 500 V objective without any severe failure event and usually withstand forward-biased pulses up to 10 kV. In contrast, reverse-biased ESD can be limited to about 5 kV for failure-free operation.

6.1.6.4   Environmental stress testing – subcomponents and module

These tests focus on the qualification of subcomponents of the module and the module itself. Main subcomponents include the fiber pigtail, package lid, and anti-reflection coating of bulk lenses or lensed fiber. The pigtail comprises the fiber with a lensed tip, built-in grating for wavelength stabilization, and the fiber tube assembly attaching the lensed fiber hermetically sealed to a module package such as the 14-pin butterfly type one. The lid comprises the getter material to trap water and organic compounds.

Key factors for a reliable module package include extreme power coupling stability, which requires submicrometer mechanical stability, and an atmosphere inside the module, which prevents package-induced failures, for example, by keeping the water content and organic pollution at the lowest possible levels and by establishing an oxygen-rich atmosphere inside the module (see Section 1.4.3.5). The various tests to verify the performance and reliability of the components and module are summarized and listed in Tables 6.7–6.9 along with the relevant effects, conditions, and failure criteria.

Temperature endurance

Temperature endurance testing is made up of high/low-temperature storage and fast/slow temperature cycling of nonoperating modules. Table 6.7 gives the conditions for these tests and the failure criteria recommended by Bellcore. The failure

Table 6.7   Typical temperature endurance test plan for single-mode high-power diode laser modules.

Test	Conditions	References	SS	Failure criteria
High-temperature storage	85 °C for 2000–5000 h	Bellcore TA-NWT-001312, Section 8.2.3	11	Laser/fiber: ±10% or ±0.5 dB change in coupling efficiency
				Photodiode: ±10% change in monitor current
Low-temperature storage	−40 °C for 2000–5000 h	Additional; not required by Bellcore	11	Thermoelectric cooler: ±5% change in ac resistance of cooler
Temperature cycling – fast (unbiased)	−40 to +85 °C; 100 pass/fail cycles with fast 12 min ramps and short 15 min dwells.	MIL-STD-883C, method 1010.7 MIL-STD-883D, method 1014.9	11	Laser/fiber: ±10% or ±0.5 dB change in coupling efficiency
				Photodiode: ±10% change in monitor current
Temperature cycling – slow (unbiased)	−40 to +85 °C; 100 pass/fail cycles with slow 30 min ramps and long 30 min	Bellcore TR-NWT-000468 Issue 1 Bellcore TA-NWT-001312	11	Thermoelectric cooler: ±5% change in ac resistance of cooler Package: loss of hermeticity

Notes:
SS = sample size
MIL-STD-883C, method 1010.7; MIL-STD-883D, method 1014.9 (IHS, Inc., 2008)
Bellcore TR-NWT-000468; TA-NWT-001312 (Telcordia Inc., 2011)

criteria define the acceptable upper changes in fiber-coupled power, monitor current, thermoelectric cooler current, and hermeticity.

The stability of the fiber-coupled power and monitor current is sensitively dependent on the specific module design and the stability of its optical train. For example, the glass seal design of the fiber tube assembly including its thickness and length as well as the degree of eccentricity of the fiber in the metal tube are decisive factors for the stability and reliability of the assembly. Good control of these parameters usually leads to a reliable performance of the module conforming to the 10% Bellcore requirements in coupling efficiency and monitor current. Regarding the monitor current, we have to consider that this parameter is also sensitive to internal reflections along the optical axis. Finally, according to common experience, butterfly modules remain hermetic after the various types of thermal stress tests.

Mechanical integrity

Table 6.8 lists the conditions and failure criteria of the various mechanical tests including mechanical shock, vibration, fiber pull, solderability, and thermal shock.

A solid design of the mechanical structure of the module including an optimized design of the fiber tube assembly realized, for example, by a glass seal typically 2 mm thick and a quartz tube approximately 12 mm long, give high assurance of encountering no fiber breaks after the mechanical shock and vibration tests. In addition, changes in fiber-coupled power and monitor current are then very likely to be within the Bellcore 10% upper limits. Similarly, a highly effective fiber retention design can withstand the external forces applied to the fiber pigtail in the fiber pull test leading to no damage to the fiber and coupling shifts well within the allowed limits. This demonstrates secure fiber attachment and alignment. The purpose of the thermal shock test is to verify package integrity and hermeticity. In general, thorough design and well-controlled, optimized manufacturing assure high mechanical ruggedness of the modules, meeting the Bellcore requirements with regard to mechanical integrity.

Special tests

This category comprises tests to evaluate the internal moisture and robustness of the module to ESD exposure (Table 6.9).

The MIL Standard MIL-STD-883 Method 1018 describes the measurement of the water vapor content inside a hermetic package and sets the upper limit to 5000 ppm. State-of-the-art quadrupole or time-of-flight mass spectrometer systems are suitable techniques to be used for this purpose. A low level of moisture content of a few 100 ppm can be achieved by using a stringent level of piece part and material control, materials totally free of flux, adhesive, and solvents, applying a high standard of package cleanliness enforced throughout manufacture, and using assembly processes designed to minimize ingress of moisture and other contaminants.

MIL-STD-883 Method 3015 defines the ESD pass requirement of at least 500 V in both bias directions and based on the human body model (HBM). Many laser module manufacturers carry out additional ESD tests based on the charge device

Table 6.8   Extended mechanical integrity test plan for single-mode high-power diode laser modules.

Test	Conditions	References	SS	Failure criteria
Mechanical shock (devices non-operating)	$500g \cong 5000$ m/s², 0.5 ms, 5 times per axis. Mech. shock & vibration on same sample in sequence	MIL-STD-883C method 2002.3-B. Bellcore TR-NWT-000468 Issue 1, TA-NWT-001312 Sects. 8.1.3/5	11	Laser/fiber: ±10% or ±0.5 dB change in coupling efficiency Thermoelectric cooler: ±5% change in ac resistance of cooler Package: loss of hermeticity
Vibration (devices non-operating)	$20g \cong 200$ m/s² peak acceleration, 20–2000 Hz, 4 min per cycle, 4 cycles per axis	MIL-STD-883C, method 2007.1-A. Bellcore TR-NWT-000468 Issue 1	11	
Fiber pull	1 kg, 3 times for ≥ 10 s in direction coaxial with fibre	Bellcore TA-NWT-001312, Sects. 8.1.3, 8.1.5	11	Laser/fiber: ±10% or ±0.5 dB change in coupling efficiency
Solderability	60/40 tin–lead solder, type R flux, solder temperature 230 °C, 10× visual inspection	MIL-STD-883, method 2003-A	11	
Thermal shock (devices non-operating)	$\Delta T = 100$ °C from 0 to 100 °C within 10 s, soak time 5 min, 15 times	MIL-STD-883, method 1011.9-A. Bellcore TA-NWT-001312 Section 8.1.3.2-10. MIL-STD-883D, method 1014.9-A	11	Package: loss of hermeticity
Hermeticity test afterward	45 psia He bombing, ≥ 2 h, baking time 25 min, 5 min cool-down			He leak rate: $< 1 \times 10^{-8}$ atm cm³/s He

Notes:
SS = sample size
MIL-STD-883C method 2002, 2007, 2003, 1011, MIL-STD-883D, method 1014.9-A (IHS, Inc., 2008)
Bellcore TR-NWT-000468, TA-NWT-001312 (Telcordia Inc., 2011)

Table 6.9 Special tests including electrostatic discharge and internal moisture robustness testing of single-mode high-power diode laser modules.

Test	Conditions	References	SS	Requirements/ failure definitions
Internal moisture	Water outgassing in hermetically sealed module package	MIL-STD-883, method 1018	11	Maximum allowable water vapor content: 5000 ppm
Electrostatic discharge (ESD) based on human body model (HBM) and charged device model (CDM)	Voltage steps 250 V, 1000 V; 5 discharge pulses per voltage step; pulses in forward and reverse bias	MIL-STD-883D method 3015. Telcordia TR-NWT-000870	12	Minimum 500 V in both bias directions. Failure definitions: 50% threshold current increase. Facet degradation. P/I characteristic distortions.

Notes:
SS = sample size
MIL-STD-883D method 3015, 1018 (IHS, Inc., 2008)
Telcordia TR-NWT-000870 (Telcordia Inc., 2011)

model (CDM), because they simulate more realistically the conditions occurring during the fabrication and packaging processes. Usually one can expect the modules to pass the tests well above the 500 V limit and up to a few thousand volts in both forward- and reverse-biased directions without any physical damage and impact on the laser characteristics.

6.1.7 Manufacturing tests

Manufacturing testing addresses any post-release design changes, and measures the manufacturing process. It also involves verification tests of design and process changes.

6.1.7.1 Functionality tests and burn-in

This phase of the product reliability development is particularly dedicated to testing and elaborating the extent that any change after the design release might have on the reliability performance. It also measures the quality of so-called pre-production trials to discover deficiencies in the manufacturing process and to remedy them before going into full-scale production. In order to obtain only the intrinsic reliability capability of the product, which is determined by the quality of the design and manufacturing, so-called *burn-in* processes are applied for a predetermined time to weed out those units that would experience infant failures in the field. As mentioned in Section

6.1.6.2, a typical burn-in program for high-power single-mode diode lasers includes the following conditions: 350 mA ACC, 70 °C for 96 h and with a failure criterion of $>10\%$ threshold current increase.

6.1.7.2   Final reliability verification tests

The final *design verification* includes final *product qualification testing*, after design maturity and test stability have been proved and assurance of process capability and controllability has been provided. The qualification test program is intended to qualify the approved *engineering configuration* prior to its final implementation, aimed at insuring that the changed product and/or process specifications had not degraded the laser product's reliability. It is useful to conduct this type of testing also at regular intervals during production and with adequate duration and sample size to detect any deviations.

Pre-production samples are also sent to the customer to gain further information on how closely the needs of the customer are being met, and to secure the customer's *verification approval*. The *final design review* includes the identification of any outstanding issues that would impinge on quality assurance and affect customer satisfaction. The approval of the final product design constitutes both the formal *full-scale production release* on the new design and the new design baseline.

6.2   Reliability growth program

The concept of reliability growth assumes that the product is undergoing continuous improvement during the development stage and that the improvement can continue well into production. It is crucial that any weaknesses, failure modes resulting from design, manufacturing processes, and relevant changes are uncovered from the very early stages on, their root causes identified, and effective corrective actions taken before testing is resumed. This reliability growth *test–analyze–and–fix concept* is documented in MIL-HDBK-189 "Reliability Growth Management" and MIL-STD-1635 "Reliability Growth Testing" (IHS, Inc., 2008).

A commonly used model for reliability growth quantifying the reliability improvement is the Duane model (Duane, 1964). At each failure, the accumulated test time t_{cum} is calculated and the cumulative mean time between failures (MTBF$_{cum}$) determined. The simple relationship can be written as

$$\text{MTBF}_{cum} = \frac{1}{K} t_{cum}^{\alpha} \qquad (6.3)$$

where K is a constant determined by the initial MTBF, design margins, and design factors, α is the growth rate defined as change in the MTBF or time interval over which change occurred. The instantaneous MTBF$_{inst}$ is calculated using

$$\text{MTBF}_{inst} = \frac{\text{MTBF}_{cum}}{(1 - \alpha)} \qquad (6.4)$$

and an expression for t_{cum} can be derived from Equations (6.3) and (6.4) as

$$t_{cum} = [K \times \mathrm{MTBF}_{inst} \times (1 - \alpha)]^{1/\alpha} . \tag{6.5}$$

The usual approach is to begin plotting MTBF_{cum} early in the program, and to continue plotting it throughout the process. The growth rate α is the slope of a log–log plot of MTBF_{cum} versus t_{cum} by fitting the data points to a straight line by least-squares analysis. The extrapolation of the line provides a prediction of future reliability performance. This has several uses, including the monitoring of actual progress toward meeting specific reliability goals.

6.3   Reliability benefits and costs

6.3.1   Types of benefit

The implementation of an efficient reliability engineering program (REP) can give rise to a series of advantages leading to improved products and business of a company (ReliaSoft Corporation, 2007). We discuss them briefly in the following.

6.3.1.1   Optimum reliability-level determination

Designing the optimal reliability goal into a product provides in a multiple way improved information about, for example, the types of failures that support R&D efforts to minimize these failures, or which failures occur at what time in the life of the product, and hence imply better preparation to cope with them. Further advantages include guidance regarding corrective action decisions to minimize failures and reduce maintenance and repair times, which will eliminate over- and under-designs. In addition, total costs to own, operate, and maintain a product for its lifetime can be minimized.

6.3.1.2   Optimum product burn-in time

This covers the elimination of early-life failures and stabilization, thereby reducing inventory-carrying costs, tooling cost, labor costs, and energy requirements.

6.3.1.3   Effective supplier evaluation

An REP provides guidelines for evaluating suppliers from the point of view of product reliability. This avoids failures in reliability qualification due to issues in vendor materials, products, and supplier-provided data.

6.3.1.4   Well-founded quality control

An REP including an effective FRACAS help provide guidelines for quality control practices, hence improving the effectiveness and efficiency of quality control, since each of the quality control (QC) jobs includes important reliability-related activities, such as new design control, incoming material control, product control, verification of corrective action, and quality improvement.

6.3.1.5   Optimum warranty costs and period

If a product fails to perform its function within the warranty period, the costs for replacing or repairing it will impact profits, but also, what is worse, attract negative attention. Introducing reliability analysis is an important step in taking corrective action, ultimately leading to a more reliable product and hence to a reduction in warranty costs or, for the same cost, an increase in length and coverage of warranty.

6.3.1.6   Improved life-cycle cost-effectiveness

A sound REP leads to both minimization of the manufacturer's overall lifetime costs of a product (fewer repairs, less maintenance) without compromising product reliability and quality, and minimum life-cycle costs for the user, with a long-term positive impact on the company's future business.

6.3.1.7   Promotion of positive image and reputation

The reputation of a company is very closely linked to the reliability of its products. The more reliable these are, the more likely the company is to have a good reputation and positive image.

6.3.1.8   Increase in customer satisfaction

A reliable product may not dramatically affect customer satisfaction in a positive manner, because products of high quality and reliability are just taken for granted. In contrast, however, an unreliable product is more likely to cause strong customer dissatisfaction. Therefore, it is mandatory to offer highly reliable products for achieving great customer satisfaction. Customer satisfaction and goodwill have a strong positive impact on future business.

6.3.1.9   Promotion of sales and future business

Based on reliability indices and metrics, advantages over competitors can be gained who either do not publish their reliability data or simply have noncompetitive data. This advantage can act also as a deterrent to new competitors entering into their markets.

A dedicated effort toward continuously improving product reliability shows customers that the manufacturer is serious about its products, and is committed to customer satisfaction. This type of attitude has a positive impact on future business.

6.3.2   Reliability–cost tradeoffs

In minimizing the manufacturer's reliability costs, we can take a similar approach, which is widely practiced in industry for balancing quality costs, see, for example,

Juran and Gryna (1988). Here an important aspect of quality costs is the possibility of reducing internal failure costs due to rework or repair, fault finding, downtime or downgrading, and external failure costs caused by, for example, returns, free replacements, warranty, and transport damage by investing in prevention and appraisal. Prevention includes process optimization, quality training, evaluation of vendors, quality planning in design and manufacturing, and implementation of reliability methods, and appraisal activities include process control, quality audits, testing, maintenance, calibrations, and inspections. As a result, the total quality costs reduce while the quality level improves and an economic balance can be achieved where the quality costs are at a minimum.

This concept can be transferred to the task of finding the optimum level of reliability performance. Designing a product with inexpensive and unreliable parts and using low-quality manufacturing processes will result in a product with low initial costs but high post-shipment support and warranty costs. In terms of the *quality costs model* discussed above, the prevention and appraisal costs are then low and the internal and external failure costs high. On the other hand, by increasing product reliability, one may increase the initial product costs (*pre-sales costs*), but decrease the *post-shipment costs*. In other words, using costly, highly reliable designs, parts, and processes will result in low *after-sales costs* including costs due to product returns, warranty and replacement costs, costs incurred by loss of customers, and subsequent sales.

The overall cost of the product is the sum of the pre-shipment design and production costs and the various post-shipment cost factors. As Figure 6.3 illustrates, an *optimum reliability point* can be found where the total cost is at a minimum. This optimum is not fixed in time, but shifts to higher reliability levels and lower total costs figures as the manufacturing process is improved over time.

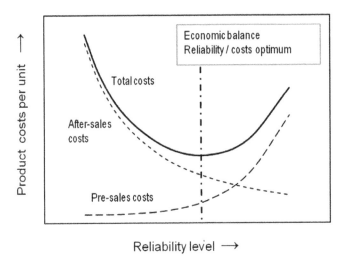

Figure 6.3 Schematic representation of the reliability costs model to determine optimum reliability by balancing pre-sales and after-sales costs.

References

Bai, D. S., Kim, M. S., and Lee, S. H. (1989). *IEEE Trans. Reliab.*, **R-38**, 528.

Caplan, F. (1990). *The Quality System*, 2nd edn, Chilton, Radnor, PA.

Duane, J. T. (1964). *IEEE Trans. Aerosp.*, **2**, 563.

Epperlein, P. W., Hawkridge, A. R., and Skeats, A. (2001). For further information see www.pwe-photonicselectronics-issueresolution.com.

Feigenbaum, A. V. (1991). *Total Quality Control*, 3rd edn, McGraw-Hill, New York.

IHS, Inc. (2008). *IHS Standards Store.* http://global.ihs.com.

Isograph Ltd (2011). *Reliability Software Products.* www.isograph-software.com.

Iuculano, G. and Zanini, A. (1986). *IEEE Trans. Reliab.*, **R-35**, 409.

Juran, J. M. and Gryna, F. M. (1980). *Quality Planning and Analysis*, 2nd edn, McGraw-Hill, New York.

Juran, J. M. and Gryna, F. M. (1988). *Juran's Quality Control Handbook*, 4th edn, McGraw-Hill, New York.

Miller, R. and Nelson, W. (1983). *IEEE Trans. Reliab.*, **R-32**, 59.

Montgomery, D. C. (1991). *Statistical Quality Control*, 2nd edn, John Wiley & Sons, Inc., New York.

Nelson, W. (1975). *IEEE Trans. Reliab.*, **R-24**, 230.

Nelson, W. (1980). *IEEE Trans. Reliab.*, **R-29**, 103.

PTC Corp. (2011). *Windchill Quality Solutions.* www.ptc.com/products.

ReliaSoft Corporation (2007). *About Reliability Engineering.* www.reliasoftindia.com.

ReliaSoft Corporation (2011a). *Military Directives, Handbooks and Standards Related to Reliability.* www.weibull.com/knowledge/milhdbk.htm.

ReliaSoft Corporation (2011b). *Reliability Software Product List.* www.reliasoft.com.

Reltech Ltd (2011). *Products.* www.reltechtestlab.co.uk.

Sharetree Ltd (2011). *Products.* www.sharetree.com.

Telcordia, Inc. (2011). *Documents.* http://telecom-info.telcordia.com.

Part III

DIODE LASER DIAGNOSTICS

Overview

As diode laser applications continue to demand increasingly higher optical output power and longer lifetime data, the stress on the material, particularly at the susceptible mirror facets, poses an ever-growing technological challenge. That is, it has become both increasingly vital and more intricate to identify and analyze the real root causes of specific performance degradation modes at the ultimate laser operating conditions, at which laser performance and reliability are affected at much lower defect densities. In addition, the growing demand for the highest laser functionality and yield makes it indispensable to develop superior materials and device technologies so that they can excel the current and meet the next generation of laser performance requirements. In this sense, material and device characterization employing powerful diagnostic and analytical techniques within the laser product development process will play an increasingly important role and thus continue to make vital contributions to the advancement of the state-of-the-art.

These diagnostic investigations should be implemented as an integral part of the diode laser development plan. The objective is to reveal potential causes for failures in performance and reliability *before* the devices are put through life tests and *before* they are deployed in the field. This proactive and preventive action is in contrast to failure analysis (FA) activities, which are executed only when a device *has failed*.

It is *not the intention* of this part of the text to elaborate on FA techniques and approaches. Although many books and numerous articles in scientific journals have been published on FA of diode lasers, we want to mention briefly in the following some commonly used FA techniques for information and comparison. These can be divided into the general categories of destructive versus nondestructive, contactless versus with contacts, and plan view (view from the top) versus cross-section view detection (view from the side).

Optical microscopic inspection can very quickly detect the first fingerprints of device failure, for example, caused by mechanical damage or contamination. Epitaxial defects, cracks, topography changes, poor morphology, growth roughness, and metal lifting can be easily detected when the microscope is equipped with differential interference contrast (Nomarski).

Electroluminescence (EL) images can be generated by forward biasing the laser to below-threshold operation and detect the integral or spectrally resolved spontaneous emission. This can be achieved in plan view by detecting the light through an opening in the top metal contact or at the device back side with a thinned substrate and without metal contact. Side view EL images can be obtained via a lateral window realized by cleaving the laser in close proximity along the cavity. This approach has been applied for measuring spatially resolved operating temperatures in the cavity and will be described in Chapter 9. EL imaging is particularly suited to study the development of dark-line defects (DLDs) caused by luminescence killing dislocation networks in the laser cavity (see Chapter 3). EL line scans across a defect structure provide useful quantitative information.

Useful information on the extent of device degradation can be obtained quickly from P/I and V/I characteristics. The former can provide information on the location of the failure due to changes, for example, in threshold current, slope efficiency, and peak optical power. From the latter we can derive information on potential changes in contact integrity resulting in changes in device resistance and turn-on voltage. The leakage current of reverse-biased devices is an indicator of the size of impact caused, for example, by DLD networks and electrostatic discharge (ESD) events. In general, leakage currents due to ESD are much higher due to the damage to the depletion zone in the active region.

Scanning electron microscopy (SEM) is well established for imaging the microstructure of solid surfaces and detecting failures such as delamination, oxidation, contamination, and melting showing up at the surface. It combines high spatial resolution (max. $\sim$1 nm) with depth of field in the same image, requires minimal sample preparation, but cannot detect root causes hidden below the surface. SEM on cross-sections obtained by cleaving through the laser structure provides useful information on the thickness and position of the individual layers in the vertical structure. The contrast in these SEM images is caused by the secondary electron emission intensity, which is sensitive to the different compositions of the layers, that is, chemical makeup and surface work function. Using a suitable composition- and doping-sensitive stain etch before taking the SEM image is a further option to highlight relevant features such as layer cross-sections or defect structures.

Electron beam-induced current (EBIC) is a technique that can be run in an SEM by generating electron–hole pairs within the effective electron beam volume of excitation. Electrons and holes are collected within the depletion zone of the p–n junction of the diode laser where they are separated by the electric field, resulting in a small current, which flows through an external circuit, and after amplification is used to modulate the intensity of the viewing screen of the microscope. EBIC is very suitable for evaluating the electrical properties of the semiconductor and in particular electrically active defects such as dislocations. Carriers recombine at such sites before they can be collected at the n-region and p-region, which leads to a lower current level. This then make these areas appear dark on the EBIC screen. Cross-sectional EBIC is useful in locating the p–n junction in the growth direction whereas plan view EBIC shows the lateral distribution of active defects in the junction plane. In Chapter 8, we discuss the application of EBIC in evaluating the effects of stress-enhanced defect formation and migration in diode lasers.

The cathodoluminescence (CL) signal is generated by the radiative recombination of minority carriers excited by the electron beam in an SEM. CL requires no bias, electrical connectivity, or junction, and emission emerges from all excited layers that have a direct bandgap. Nonradiative carrier recombination at electrically active defect sites dramatically reduces the CL efficiency. A common setup for light excitation and collection consists of a polished ellipsoidal mirror where at one focal point the electron beam enters through a hole in the mirror and impinges on the sample. For light collection, usually a light pipe sits at the other focal point. Alternatively, arrays of photodiodes can be used in place of the light pipe. The detector signal is used to modulate the intensity of the viewing screen to generate the CL image. CL images can be detected in plan view and cross-sectional view. Spectrally resolved CL images require cooling of the sample on a low-temperature stage in the SEM system to liquid nitrogen or helium temperatures depending on the specific requirements, and to spectrally detect the CL signal in a monochromator. Cooling the sample reduces both detrimental phonon broadening and thermally-activated nonradiative recombination events via deep-trap defect centers. In general, both effects narrow the spectral lines and dramatically increase the signal of any luminescence spectrum. In addition, many weak transitions can only be detected at very low temperatures below the boiling point of liquid helium at 4.2 K.

Transmission electron microscopy (TEM) is employed to image physical features of structures such as spatial uniformity of quantum wells, integrity of interfaces, defects including precipitates, inclusions, stacking faults, microloops, and dislocations at a very high spatial resolution in the subnanometer regime, if required. TEM requires sample thicknesses well below 1 μm, which is a challenging task for preparation when applying the conventional methods including grinding and ablating. However, preparation is eased and produces more reproducible and satisfying results by using a modern focused ion-beam (FIB) technique, where a gallium beam cuts away the unwanted material to finally leave a thin membrane. TEM can be applied on cross-sectional and plan view samples. Cross-sectional TEM requires knowledge of the spatial origin of device failure so that it is included when cutting the sample. Practical lateral limits are some tens of micrometers for plan view samples, which are cut along the laser cavity and include the active region, usually the dominant part in the development of bulk failure mechanisms.

Scanning laser techniques are useful in evaluating the uniformity of epitaxial layers and semiconductor wafers regarding the two-dimensional distribution of composition, dopants, and defects. Section 7.1 below describes a scanning laser technique producing within minutes a digital photoluminescence (PL) image of a full 2 inch silicon-doped GaAs wafer. By removing the contact electrode layers and thinning the top and/or back side of the laser device so that the information of interest is within the total depth given by the sum of light penetration and carrier diffusion length, the technique can be applied successfully to analyze failed laser devices. Apart from collecting the PL signal, the carriers generated by the excitation laser with an above-bandgap energy can be collected and used in a manner similar to EBIC. This mode is called optical beam-induced current (OBIC) and is less demanding and costly in its realization but yields effectively the same results within a resolution of a couple of micrometers. By rastering the laser beam with a below-bandgap energy across the sample a so-called thermally induced voltage alteration (TIVA) image can be obtained. In this case, the laser beam merely heats the sample and localized heating can be pronounced at extended defect sites causing a change in resistance, which in turn causes a change in voltage of a constant current biased device. The change in voltage is used to plot the image as a function of the position of the exciting laser beam spot. This technique and similar laser probe techniques are suitable for detecting and isolating faults such as open junctions and shorts, in particular in integrated circuits (ICs).

The following description gives an overview of the topics to be dealt with in Chapters 7, 8, and 9. Many of the new diagnostic approaches and techniques described here were pioneered and adjusted by the author specifically for applications in diode laser research and optimization. They have yielded numerous world's first results and many have been adopted by other researchers in academia and industry.

The various diagnostic data on parameters such as impurity trapping in active layers, deep traps at active layer interfaces, laser operating temperatures, stress fields and crystallographic material instabilities, as well as the various root causes and correlations have provided invaluable insights into the significance of these parameters in laser operation. They have contributed significantly to a more detailed understanding of potential degradation mechanisms of laser functionality and thus have furnished the knowledge for effectively optimizing laser design, fabrication processes, laser performance, and reliability. The following chapters report the immense endeavors in pursuit of these goals and also describe the novel techniques, approaches, and diagnostic data in detail.

We will discuss the vast experimental data set obtained from various types of single transverse mode diode lasers regarding the most different material-oriented and device-linked effects including:

- wafer substrate optical uniformity;
- impurity trapping in active layers;
- deep-level defects at active layer interfaces;
- local operating temperatures at mirror facets of different technologies and vertical device structures;
- local operating temperatures along laser cavities;
- mirror temperature topographs;
- temperature-monitored degradation processes;
- mechanical stress in ridge waveguide structures;

- detection of "weak spots" in diode laser mirror facets;

- stress-induced formation, migration and separation of electrically active defects;

- structural and compositional disorder in mirror facets; and

- recrystallization effects in mirror coatings.

Correlations of these parameters with laser performance and reliability data are also discussed. The basics of the employed measurement techniques and approaches are described, which include:

- photoluminescence (PL) scanning;

- low-temperature PL spectroscopy;

- EL spectroscopy with high spatial resolution;

- laser microprobe Raman scattering spectroscopy;

- microspot reflectance modulation or thermoreflectance;

- EBIC; and

- deep-level transient spectroscopy.

Chapter 7

Novel diagnostic laser data for active layer material integrity; impurity trapping effects; and mirror temperatures

Semiconductor Laser Engineering, Reliability and Diagnostics: A Practical Approach to High Power and Single Mode Devices, First Edition. Peter W. Epperlein.
© 2013 John Wiley & Sons, Ltd. Published 2013 by John Wiley & Sons, Ltd.

Introduction

This chapter discusses optical uniformity of doped laser wafer substrates, impurity trapping effects in active laser quantum wells (QWs), deep-trap accumulation at active layer interfaces, and local laser mirror temperatures as a function of output power, laser material, vertical structure, number of active quantum wells, mirror coating, heat spreader, and the laser die mounting technique. Fundamentals and experimental setups of the relevant measurement techniques will also be discussed and include photoluminescence (PL) mapping, low-temperature PL spectroscopy, deep-level transient spectroscopy (DLTS) and Raman microprobe inelastic light scattering spectroscopy, respectively.

7.1   Optical integrity of laser wafer substrates

7.1.1   Motivation

Improving the performance and reliability of optical devices such as diode lasers requires that the electrical and optical material parameters be precisely controlled. This is true not only for the epitaxial growth of the individual layers of the laser structure, but also for the underlying wafer substrate these layers are grown on. Wafers used for diode lasers are typically n-type doped to yield an electron concentration of $\sim 1.5 \times 10^{18}$ cm^{-3}, and standard wafers have a dislocation etch pit density of $\sim 2 \times 10^3$ cm^{-2}.

Any nonuniform distribution of dopants and defects can impact the performance and reliability characteristics of the laser and consequently the yield. The increasingly demanding requirements for materials with a high degree of purity and structural homogeneity place concomitantly higher demands on semiconductor growth, processing technologies, and the characterization techniques, which support these technologies. As mentioned above, in particular, knowledge of the spatial variation of the material parameters is vital because of possible correlations with device properties.

Optical techniques are particularly suited for two-dimensional imaging because they are fast, nondestructive, and noncontact. PL is a simple, but particularly sensitive and informative technique for detecting concentrations of imperfections as low as 10^{11} cm^{-3} in semiconductors. PL intensity maps, for example, of as-grown and processed GaAs or InP wafers contain information on material uniformity, dislocation density, density of deep traps, distribution of residual impurities, uniformity of epitaxial buffer layers, implant and anneal uniformity, and the presence of process-induced defects (Epperlein, 1990; Guidotti *et al.*, 1987; Hovel and Guidotti, 1985).

In the following, we describe two PL scanning techniques, one for fast scanning of full 2 or 3 inch wafers at room temperature and the other for scanning at specific wafer locations with high spatial resolution of ∼2 μm at 300 K and for spectrally resolved scans with a spatial accuracy of ∼30 μm at 2 K. As an example, the PL mapping results of an n-type silicon-doped GaAs wafer will be discussed.

7.1.2   Experimental details

Figure 7.1 shows the setup of a simple "galvo" mirror scanner capable of producing within minutes digital maps of the dominating near band edge (NBE) PL at room temperature of full 2 (3) inch wafers with a resolution of ∼50–100 μm. The wafer placed on a dc motor-driven stage is moved in one direction, while a focused laser beam is scanned in the orthogonal direction. The integrated PL light is detected by a long (typically 4 inch) Si p–n junction photodiode placed close to the wafer surface and parallel to the scanning direction of the focused laser beam. An edge filter in front of the photodiode blocks the laser light and transmits the PL signal for detection in the range ∼1.1–1.6 eV. A personal computer correlates the lock-in amplified signal with the position on the wafer and displays the digitized signal on a monitor in high-resolution color or 8-bit grayscale contrast images. The minimum scanning field with this setup is ∼9 × 9 mm² corresponding to a ∼30 μm resolution.

The second optical mapping system is more sophisticated, capable of much higher spatial resolution and of producing spectrally resolved PL scans. Figure 7.2 shows the setup of the system, which has an operating principle similar to the one described in the literature (Luciano and Kingston, 1978; Yokokawa *et al.*, 1984).

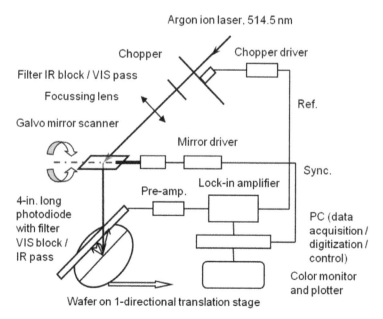

Figure 7.1   Schematics of the photoluminescence (PL) mapping system using a galvo mirror scanner.

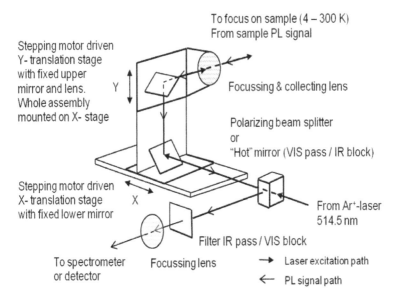

Figure 7.2 Schematic illustration of the high spatial resolution PL scanner for room temperature and cryogenic temperature integral intensity or spectrally resolved mapping applications.

The system consists essentially of two mirrors attached to two orthogonally mounted, high-precision (0.1 μm step size) stepping motor-driven translation stages. The horizontal X-stage moves the vertical Y-stage. Laser light is guided via a dichroic filter or polarizing beamsplitter along the axis of the horizontal stage onto the first mirror, directed vertically upward where the second mirror deflects the beam 90° perpendicular to the laser plane through the focusing lens onto the sample, which can be at room temperature or at low temperature in a helium bath cryostat. The lens fixed to the vertical stage is used both to focus the beam and to collect the PL emission. The way the mirrors and lens are mounted on the two stages insures that the beam always hits the same spot on the upper mirror and passes through the center of the lens during translation. The PL light follows the same optical path as the laser beam. After decoupling from the incident light by the dichroic filter or polarizing beamsplitter, the PL light is directed to a suitable detector or into a spectrometer for spectral analysis. The PL data are processed for evaluation and display in a computer system similar to that described above. By using a lens with a large numerical aperture (0.5), such as a 20 × microscope objective, PL maps with high spatial resolution of ~2 μm and high collection efficiency can be recorded at 300 K. The spatial resolution of low-temperature (2 K) spectral maps is typically 30 μm.

7.1.3 Discussion of wafer photoluminescence (PL) maps

Figure 7.3a shows the grayscale contrast image of the 300 K NBE PL intensity of a full 2 inch Si-doped (100) GaAs wafer where the bright areas correspond to high PL signal. The GaAs is doped to a concentration of the shallow Si donors to yield an

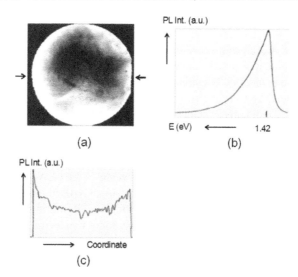

Figure 7.3 Photoluminescence (PL) of a 2-inch, Si-doped n-type (100) GaAs wafer. (a) Grayscale contrast image of the near-band edge, integrated PL intensity distribution at room temperature. Bright areas: high intensity; dark areas: low intensity. (b) Room temperature PL spectrum. (c) Line scan of the integrated PL intensity across the wafer at the arrowed positions in (a).

electron concentration in the range of 1.2 to 1.7×10^{18} cm^{-3} and has a standard etch pit density (EPD) of 2×10^3 cm^{-2}. The 300 K PL spectrum is depicted in Figure 7.3b and consists of a broad asymmetric peak at 1.42 eV with a FWHM of 34 meV primarily due to thermal broadening. The peak can be ascribed to a free-to-bound recombination, in this case from an electron in the neutral Si donor level to a hole in the valence band (D^0,h), while a small shoulder ~6 meV off the main peak in the upper part of the signal and on the high-energy side can be attributed to a band–band transition considering the fact that the binding energy of Si in GaAs is shallow at ~6 meV. In the next section we will discuss the major radiative recombination processes in a semiconductor.

The PL intensity map has an overall bowl-like pattern with high PL intensity at the wafer edges dropping toward the wafer center, which can also be clearly verified by the line scan in Figure 7.3c. The uneven and cloud-like PL distributions may be caused by Si striations and streaks during the GaAs ingot crystal growth.

These PL maps are totally different to the PL maps of semi-insulating (s.i.) (100) GaAs wafers grown by the so-called liquid-encapsulated Czochralski (LEC) method. The macroscale pattern of these maps shows the typical fourfold azimuthal symmetry superimposed on a radial W-type variation (Epperlein, 1990; Guidotti *et al.*, 1987) predicted by a thermoelastic analysis of dislocation generation in s.i. (100) LEC-grown GaAs crystals (Jordan *et al.*, 1980). The small-scale structure in these maps forms the well-known cellular network with cell sizes of a few hundred micrometers and can be fully correlated with the dislocation EPD (Epperlein, 1990).

This effect can be explained in terms of intrinsic gettering of optically and electrically active defects to the dislocation cores during ingot formation and cooling, such that the dislocation network becomes visible. It is generally understood that the causes of NBE PL enhancement near dislocations are manifold, including lifetime variations of free carriers (Leo *et al.*, 1987), spatial variations of deep-levels and shallow acceptor concentrations leading to spatially varying compensation (Wakefield *et al.*, 1984), or a decrease of nonradiative centers forming a denuded zone (Cottrell effect) near dislocations (Chin *et al.*, 1984).

In contrast, however, the local PL fluctuations observed in Figure 7.3a and in particular in the line scan of Figure 7.3c are by far less pronounced, and also show no regular pattern compared to those in maps and scans of s.i. GaAs wafers (Epperlein, 1990). We conclude therefore that the two-dimensional NBE PL distribution is caused primarily by a nonuniform distribution of Si dopants during ingot growth and not by dislocations. The PL intensity and hence the donor concentration can vary by up to a factor of two across the wafer with the consequence that the required electron concentration would be out of spec and too low for most of the wafer area. Laser devices grown on such wafer substrates would show deteriorated performance and reliability figures due to the increased series resistance and linked heating effects. The yield of in-spec devices would be drastically reduced.

7.2 Integrity of laser active layers

7.2.1 Motivation

It is known that the uncontrolled incorporation of impurities such as shallow acceptors in the nominally undoped active layer of molecular beam epitaxy (MBE) grown double-heterostructure (DH) and quantum well (QW) diode lasers and electrical devices such as heterojunction field-effect transistors (FETs) has a degrading effect on the performance and reliability of these devices. In particular, the impurity buildup at the inverted heterointerface (GaAs on AlGaAs) as compared to the normal AlGaAs on GaAs interface of AlGaAs/GaAs devices and the linked formation of point defects, traps, and interface roughness have been identified as potential root causes (Masselink *et al.*, 1984a; Epperlein and Meier, 1990).

Intrinsic and impurity-related PL from nonintentionally doped AlGaAs/GaAs SQW and MQW structures has been studied systematically as a function of well width L_z, bottom AlGaAs cladding layer thickness, and various prelayer sequences (Epperlein and Meier, 1990). Detailed investigations of the dependence of the impurity binding energy on L_z identified carbon acceptors as the dominating impurity species. Carbon acceptors are trapped during growth at the inverted interface of the first in-a-sequence-grown QW. These inhibit the growth, for example, by preventing the lateral propagation of the atomic layers due to pinning steps on the surface and consequently leading to increased roughening of the interface and interface recombination velocity. This detrimental effect increases with increasing bottom cladding thickness but it can be drastically suppressed by growing so-called prelayers before the actual active QW layer.

In the next sections we describe, first, both the fundamental radiative transitions allowed in a semiconductor and the experimental PL setup capable of detecting efficiently and reliably the impurity-related PL emissions. For comparison, we also discuss detection of the intrinsic PL, in particular the faint and narrow lines of free and bound exciton recombinations only detectable at very low temperatures, such as 2 K, for example. We then discuss the various impurity-related effects and a procedure for effectively suppressing the gettering of background impurity atoms at the active layer during MBE growth, and thereby improve the performance of the laser device.

7.2.2   Experimental details

7.2.2.1   Radiative transitions

Figure 7.4 shows schematically the basic radiative recombination processes in a semiconductor following, first, a generation (I) of excess, nonequilibrium electron–hole carriers, for example, by an external light source with an above-bandgap energy E_g, and then a fast relaxation or thermalization process of the excited carriers to the band edges. The transitions can be classified and described as follows:

- (1): Band-to-band recombination, usually difficult to observe in weakly doped samples. More likely in less pure or less perfect crystals and at thermal energies $k_B T > E_x$, the exciton binding energy leaving free carriers available to recombine radiatively in a band-to-band transition.

- (2): Free exciton (FE) recombination in a sufficiently pure material where electrons and holes pair off into excitons, which then recombine emitting a narrow spectral line. Energy of emitted photon in a direct gap semiconductor with momentum conservation in a simple radiative transition is $h\nu = E_g - (1/n^2)E_{x1}$ with n the quantum number of the exciton excited states where $E_{x1} \cong 4.4$ meV (GaAs) is the exciton ground state ($n = 1$) energy. Probability of $n > 1$ exciton transitions decreases with n^{-3} (Elliott, 1957), difficult to observe in the presence of other radiative processes, electric fields (dopants),

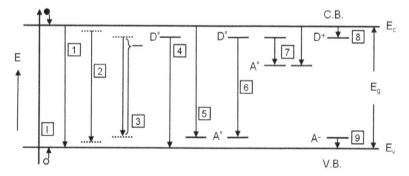

Figure 7.4   Schematic diagram of the fundamental radiative recombination processes in a semiconductor. C.B. is the conduction band, V.B. the valence band.

and temperatures >4 K (exciton breaking up into free carriers). Involvement of a phonon in an FE transition in an indirect gap material to conserve momentum; emitted photon energy reduced by the phonon energy. Narrow FE linewidths $\cong 1$ meV for GaAs at 1 K.

- (3): Bound exciton (BE) recombination in the presence of impurities. Emission with a narrow spectral width of typically 0.1 meV and at a lower photon energy than that of FE, for example, $h\nu(D^\circ, X) = E_g - E(D^\circ, X) = E_g - E_x - 0.13E_D$ (GaAs) with donor binding energy in GaAs $E_D \cong 6$ meV. Bound exciton transitions may be from neutral and ionized donors and acceptors in the forms (D°, X), (D^+, X), (A°, X), (A^+, X) where X designates the exciton bound to an impurity state. Process (3) in Figure 7.4 shows a donor bound exciton transition.

- (4) and (5): Band-to-impurity (free-to-bound (FB)) recombinations involving shallow impurities (hydrogen-like centers) characterized by (h, D°) and (e, A°) where a neutral donor state interacts radiatively with a hole in the valence band and an electron from the conduction band transfers to a neutral acceptor state, respectively. The photon energy is $h\nu = E_g - E_i - E_p$ (in case phonons with energy E_p are involved in transition; phonon replicas) where E_i is 34 meV and 5.2 meV for the ionization energy for acceptors and donors in GaAs, respectively, calculated within a hydrogen model approximation. Occurs in samples with relatively low dopant concentrations, but is dominated in purer material by exciton recombination processes. From the spectra one can deduce the acceptor binding energies but less reliably donor binding energies due to the very small chemical shift of the donor atoms involved.

- (6): Donor–acceptor pair recombinations involve both donor and acceptor impurities and are characterized by (D°, A°). The energy separating the paired donor and acceptor states is $h\nu = E_g - (E_A + E_D) + q^2/\varepsilon r$ where the last term is the Coulomb interaction energy, which is very small for distant pairs (r large). The pair recombination is assisted by a tunneling process for donor and acceptor atoms separated by distances r greater than the effective Bohr radius. Distant pair recombinations are less likely than short-distance pair transitions leading to an increase of the emission intensity as r decreases, which in turn decreases the number of pairs. This means that the intensity must go through a maximum as r is changed in a discrete fashion resulting in an emission spectrum with discrete lines. The fine structure can be resolved for pair separations typically in the range of 1 to 4 nm for GaP; however, above this value the lines overlap and form a broad spectrum. In GaAs where $E_A + E_D$ is small, only the large r-pairs contribute to the emission spectrum, because the Coulomb energy for small r-pairs drives the donor and acceptor levels into the conduction and valence bands, respectively. There, a pair recombination would have to compete with the band-to-band transitions, which involve high density of states, and hence a pair recombination would have to be very efficient to generate a line spectrum (Pankove, 1975). Typical of pair recombinations is the high-energy shift and narrowing of the spectrum with increasing excitation intensity, which can be ascribed to the saturation of long-distance pairs and dominance of short-distance donor–acceptor pairs.

- (7): Deep transitions involving the transition of an electron from the conduction band or from a donor to a deep-level with a large ionization energy $E_i > 0.2$ eV (GaAs). Formation of such deep-levels, for example, by complex centers, vacancy–impurity complex defects such as Ga vacancy–donor complex band structures near 1.2 eV in n-type GaAs, or by 3-d transition metals in III–V compounds. The figure shows the transitions to a deep acceptor level with its energy deep in the bandgap. These transitions yield very low photon energies.

- (8) and (9): Shallow transitions to ionized donors or acceptors. These transitions could be radiative in the far infrared. Although radiative transitions from the conduction band to the donor have been reported in GaAs (Melngailis *et al.*, 1969), we consider a further discussion of this topic beyond the scope of this section.

7.2.2.2 The samples

Different $Al_xGa_{1-x}As$/GaAs QW structures were grown by conventional MBE on (100) s.i. GaAs substrates. All structures were deposited on a superlattice (SL) buffer layer 0.6 μm thick followed by a bottom $Al_xGa_{1-x}As$ cladding layer of varying thickness $L_B \cong 0.05$–0.5 μm, the prelayer sequence consisting either of an SL or a GaAs SQW with thickness $L_{z1} \cong 0.8$–12 nm. Further layers include an $Al_xGa_{1-x}As$ barrier layer with thickness $L_b \cong 0.03$–0.5 μm, the test QW with $L_{z2} \cong 4$–15 nm, and finally the upper $Al_xGa_{1-x}As$ cladding $\cong 0.1$ μm thick. The top test QW probes the efficiency of the prelayers as impurity trapping centers. In addition, several MQW samples were grown with different single-well widths in the range of 0.8 to 12 nm and sequences of the wells and with constant $L_b = 30$ nm. All layers are unintentionally doped with a p-type doping in the GaAs of about 2×10^{14} cm^{-3}. The nominal AlAs mole fraction x of all AlGaAs layers was 0.3.

7.2.2.3 Low-temperature PL spectroscopy setup

The PL spectra were recorded at $T = 2$ K and higher temperatures at various laser excitation power densities in the range $\sim 10^{-2}$–10^2 W/cm^2. The samples were mounted stress-free in a variable temperature cryostat immersed in superfluid liquid helium or in a cold helium gas flow. After standard excitation by the 514.5 nm line of a cw argon-ion laser and appropriate dispersion in a 0.85 m, f/7.8 dual-holographic grating monochromator, the PL light was detected by a thermoelectrically cooled, selected GaAs photomultiplier tube (PMT) with low dark count rate and conventional single-photon pulse counting electronics or an optical multichannel analyzer (OMA) with a cooled charge-coupled device (CCD) detector array.

The state-of-the-art optical spectroscopy setup schematically illustrated in Figure 7.5 was a multipurpose system used by the author for many applications on different semiconductor materials and devices requiring high spatial resolution of ~ 1 μm and

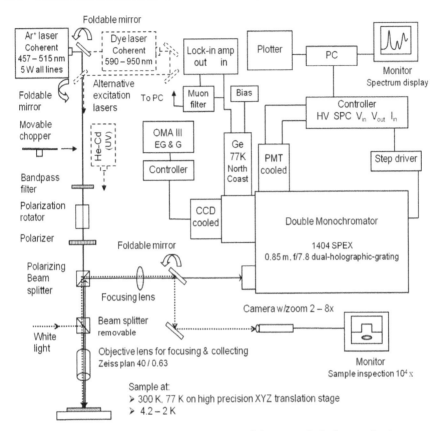

Figure 7.5 Schematics of a multipurpose, state-of-the-art optical characterization system for applications requiring techniques such as low-temperature PL spectroscopy, PL excitation (PLE) spectroscopy, Raman spectroscopy, electroluminescence (EL), and thermoreflectance (TR) with capabilities including high spatial and spectral resolution, high signal detection sensitivity, and variable temperatures in the range of 2 to 300 K. The Ar$^+$ laser is used for PL, Raman and TR, the Ar$^+$ laser-pumped dye laser for PLE, the He–Cd UV laser for III–nitride semiconductor PL, the mechanical chopper for light detection using a photodiode, and the various polarization components are used for Raman spectroscopy.

different temperatures in the range of 2 to 300 K and included PL spectroscopy, PL excitation (PLE) spectroscopy, Raman spectroscopy, electroluminescence (EL), thermoreflectance, and other suitable techniques (see Chapters 7–9). The cw dye laser was used as a wavelength-tunable excitation source for PLE spectroscopy measurements to determine the energies of the various free exciton transitions in the QW. The mechanical chopper was used for all applications not requiring signal detection by the PMT or CCD but only by a photodiode. The various polarization-dependent components in the laser excitation path were used in Raman spectroscopy measurements, see next sections below.

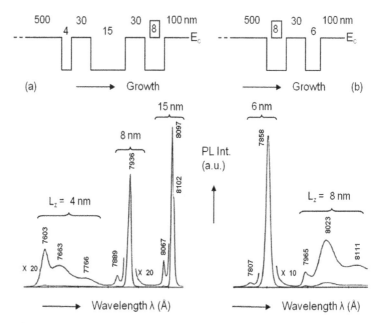

Figure 7.6 Photoluminescence (PL) spectra of AlGaAs/GaAs multi-quantum well (MQW) structures at low temperatures of 2 K. For clarity only the simplified conduction band portion of the MQW structures is shown. (a) MQW with 4, 15, 8 nm wide in-a-sequence MBE-grown QWs. (b) MQW with 8, 6 nm wide in-a-sequence grown QWs. Focus is on the shape of the PL spectrum of the 8 nm QW in (a) and (b).

7.2.3   Discussion of quantum well PL spectra

7.2.3.1   Exciton and impurity-related recombinations

A typical 2 K PL spectrum of a MQW sample with SQWs 4, 15, and 8 nm wide is shown in Figure 7.6a where the 4 nm QW is the first in-a-sequence-grown well after the bottom AlGaAs cladding layer 500 nm thick. The spectrum of this 4 nm well shows a sharper peak on the high-energy side end and two broader peaks on the low-energy side (only visible with signal expansion by 20×), whereas the spectra of the subsequently grown 15 and 8 nm wells with 30 nm barriers each exhibit a dominating, high-intensity sharp emission, without signal expansion.

The peak on the high-energy side of the 4 nm QW spectrum and the strong narrow lines in the 8 and 15 nm well spectra can be attributed to the intrinsic $n = 1$ free heavy-hole exciton recombination, FE(e1–hh1), with a photon energy E_{1h}. We can exclude impurity (acceptor) bound exciton luminescence, which occurs only occasionally with an almost unobservable emission, Stokes-shifted from E_{1h} by about 4 meV.

The lower intensity, sharp peaks on the high-energy sides (signal expanded 20×) of the dominating peaks in the 15 and 8 nm well spectra can be assigned to the $n = 1$ free light-hole exciton recombination, FE(e1–lh1), with a photon energy E_{1l}.

These E_{1h} and E_{1l} assignments have been concluded from the dependence of the PL intensity on excitation power and lattice temperature. The intrinsic PL intensity is nearly linear in excitation power, as expected for a monomolecular (excitonic) recombination process, and decreases by about 20% when the sample temperature is increased from 2 to $\sim$30 K. For details see Epperlein and Meier (1990). PLE spectroscopy measurements and subband calculations confirm these assignments.

Figure 7.6b shows the spectrum of a MQW where an 8 nm SQW is grown first with a subsequent 6 nm well growth. In contrast, the 8 nm well now shows the reduced signal emission with two broader peaks on the low-energy side and the 6 nm well the dominating, high-intensity sharp emission with a narrow peak again on the high-energy side after signal expansion 10×. The various narrow emissions on the high-energy sides of the sub-spectra can again be ascribed to free exciton recombinations.

On the other hand, the extrinsic PL, that is, the intensity of the luminescence on the low-energy side of the intrinsic peak of the 4 nm well in Figure 7.6a and the 8 nm well in Figure 7.6b, tends to saturate at higher excitation powers (Epperlein and Meier, 1990). The presence of a limited number of impurities can plausibly account for such a saturation. This effect and the energy of the transition strongly suggest that it can be ascribed to the recombination of free electrons in the $n = 1$ quantum confined state with neutral acceptors in the QW, marked as (e1,A°). Donor–acceptor pair recombination can be ruled out for three reasons: (i) the intensity of the extrinsic peak is not very sensitive to temperature in the range of 2 to 20 K; (ii) the energy of the peak responds only slightly to the excitation intensity; and (iii) to temperature. Further evidence of the involvement of acceptors in the formation of the broad extrinsic PL emissions and their identification will be discussed in the next section.

Moreover, the presence of extrinsic QW PL correlates with the linewidth of the intrinsic exciton PL emission and with the thickness of the bottom cladding and barrier layers.

Figure 7.7a shows the intrinsic FWHM of the dominating intrinsic heavy-hole exciton recombination of different, nominally 8 nm wide wells plotted as a function of the extrinsic PL intensity normalized to the intrinsic intensity. These QWs have exciton wavefunctions, which only weakly penetrate into the barrier layers and therefore are sensitive to interfacial disorder. The figure shows that samples with practically no extrinsic PL have small linewidths of typically 2.5 meV indicating smooth and abrupt interfaces. The linewidth increases with increasing extrinsic PL signal and tends to saturate.

Figure 7.7b describes a simple model for the average interface roughness $\overline{\Delta L_z}$ based on the L_z dependence of the quantum confined energy (cf. Equation 1.19) and by using the maximum observed FWHM of 8 meV. The model calculates $\overline{\Delta L_z} \cong$ 0.02 nm and evaluates the lateral extension of the roughness expressed by the ratio between the island-covered area A_{with} and the island-free area $A_{w/o}$ to roughly 10%. It is well known that monolayer fluctuation steps of the well thickness can be the source of formation of deep-levels and nonradiative recombination centers, which can strongly impact the efficiency and threshold of the diode laser and its long-term

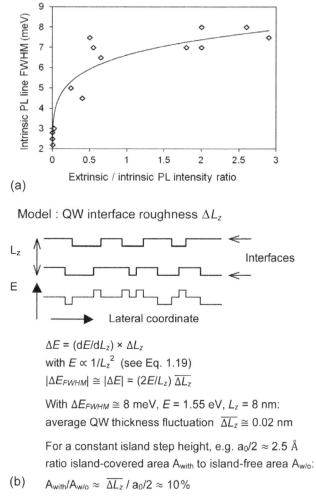

(a)

Model : QW interface roughness ΔL_z

(b)

$\Delta E = (dE/dL_z) \times \Delta L_z$

with $E \propto 1/L_z^2$ (see Eq. 1.19)

$|\Delta E_{FWHM}| \cong |\Delta E| = (2E/L_z)\,\overline{\Delta L_z}$

With $\Delta E_{FWHM} \cong 8$ meV, $E = 1.55$ eV, $L_z = 8$ nm:

average QW thickness fluctuation $\overline{\Delta L_z} \cong 0.02$ nm

For a constant island step height, e.g. $a_0/2 \approx 2.5$ Å

ratio island-covered area A_{with} to island-free area $A_{w/o}$:

$A_{with}/A_{w/o} \approx \overline{\Delta L_z}\,/\,a_0/2 \approx 10\%$

Figure 7.7   (a) Linewidth of the 2 K PL emission of the 8 nm well plotted as a function of the extrinsic acceptor-related PL intensity normalized to the intrinsic exciton-related PL intensity of the 8 nm well at low excitations of 1 W/cm². (b) Simple interface roughness model estimating an average QW thickness fluctuation and ratio between island-covered area and island-free area. Solid line: calculated trendline.

lifetime. Line broadening due to band filling via charge transfer from the acceptor impurities at the interfaces was estimated to play only a minor role.

7.2.3.2   Dependence on thickness of well and barrier layer

More information on the acceptors involved in the (e1,A°) transition, in particular on their energetic states and identification, has been obtained by determining the acceptor binding energy $E(A^\circ)$ as a function of L_z, which varied in the range $\sim$1–12 nm in many samples grown at the residual MBE background pressure.

$E(A^\circ)$ was evaluated in the usual way (Miller *et al.*, 1982) from $E(A^\circ) = E_{1h} - E(e1,A^\circ) + B(1h) = E_g - E(e1,A^\circ)$, where E_{1h} is the $n = 1$ free heavy-hole (hh) exciton transition energy, $E(e1,A^\circ)$ the energy of the broad peak(s) of the extrinsic PL, $B(1h)$ the binding energy of the $n = 1$ ground state hh exciton, and E_g the $n = 1$ QW energy gap. $E(A^\circ)$ is the energy necessary to transfer a hole from A° to the $n = 1$ hh level in the valence band of the well. The actual L_z values of the various QWs were determined from calculations of the transition energies and by using the experimental E_{1h} values.

The experimental data plotted as $E(A^\circ)$ versus L_z fall distinctly into a low-energy and high-energy branch and increase with decreasing L_z. The interested reader is referred for further details to the original publication of Epperlein and Meier (1990). This splitting into two components can be expected, since the acceptor density of states is strongly enhanced at the well center and interface positions (Masselink *et al.*, 1984b; Bastard, 1981). The lowering of the binding energy for on-edge impurities compared to on-center impurities is a direct consequence of the repulsive interface potential, which tends to push the hole charge distribution away from the attractive ionized acceptor center leading to a reduced effective Coulomb attraction.

It was demonstrated for the first time that the experimental data in the upper and lower branch are in good agreement with calculations (Masselink *et al.*, 1984b) for the ground state of neutral carbon acceptors (C°) at the center and at the interfaces of $Al_{0.3}Ga_{0.7}As/GaAs$ QWs, with typical binding energies $E(C^\circ)$ of 37 and 23 meV for a 6 nm QW, respectively (Epperlein and Meier, 1990). In addition, control experiments on samples grown in a beryllium background atmosphere leading to higher binding energies of $\sim$27 meV at the interface of a 6 nm well confirmed the involvement of neutral carbon acceptors in the formation of the broad extrinsic PL peaks shown in Figure 7.6.

The extrinsic $(e1,A^\circ)$ PL intensity increases with increasing thickness of the bottom AlGaAs cladding layer or AlGaAs barrier layer in the MQW as clearly demonstrated in Figure 7.8. This dependence shows that AlGaAs >100 nm thick underlying the GaAs well is required to build up a sufficiently high impurity level for detection. From the above observations and experimental results, that is, that prelayers grown near the normal interface (AlGaAs on GaAs) do not have a measurable effect on the extrinsic PL of the test QW, the following model of the incorporation of carbon acceptors in the QW structure can be derived.

Due to the lower solubility of impurities in AlGaAs than in GaAs, they stay afloat on the AlGaAs growth surface and are progressively trapped in a thin layer (Meynadier *et al.*, 1985) at the inverted interface upon deposition of the GaAs. The atomic scale interface roughness deduced from the linewidth measurements (Figure 7.7) are most likely due to the growth-inhibiting nature of carbon (Phillips, 1981), such as, for example, by preventing the lateral propagation of the atomic layers due to the pinning steps on the surface. This irregular structure of the interface is the source for the formation of performance and reliability-impacting defect centers. A

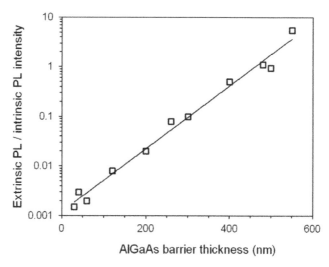

Figure 7.8 Normalized extrinsic PL intensity versus $Al_{0.3}Ga_{0.7}As$ lower cladding or barrier layer thickness of QWs nominally 8 nm wide at 2 K and 1 W/cm^2. The ratio is corrected for its excitation intensity dependence; $\gtrsim 1\%$ normalized extrinsic, acceptor-related PL intensity obtainable for 150 nm and thicker AlGaAs barriers. Solid line: calculated trendline.

saturation of the intrinsic FWHM PL (see Figure 7.7), equivalent to a maximum roughness detectable with the excitonic PL as an optical probe in QWs, appears to be conclusive from this model.

7.2.3.3 Prelayers for improving active layer integrity

The trapping of detrimental impurity centers in the nominally undoped test QW or active QW in a diode laser can be efficiently suppressed by growing thin GaAs prelayers before the actual QW. These undesired impurities can be found in the residual MBE background atmosphere released for example during outgassing of the Al oven and shutter.

Figure 7.9 shows that the extrinsic impurity-related QW PL intensity normalized to the intrinsic exciton-related QW PL intensity can be suppressed below 1% by a GaAs SQW prelayer with a thickness of at least 5 nm. This dependence is similar to the one shown for the first time by Epperlein and Meier (1990). Similarly strong suppression can be obtained by using SL prelayers with the same total GaAs thickness. The results in Figure 7.9 are in agreement with the model of impurity incorporation described above.

It has been demonstrated that InGaAs/AlGaAs QW GRIN-SCH lasers with GaAs prelayers positioned in the lower AlGaAs cladding layer before the GRIN-SCH layer showed a higher efficiency and lower threshold current and, in particular, a positive effect on laser reliability compared to devices without such prelayers gettering the segregating impurities during growth.

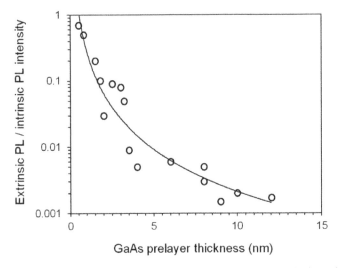

Figure 7.9 Extrinsic acceptor-related PL intensity normalized to the intrinsic exciton-related PL intensity of 8 nm QWs at 2 K and 1 W/cm² and for lower AlGaAs claddings 500 nm thick and AlGaAs barriers 30 nm thick versus the thickness of GaAs prelayers acting as impurity trapping centers. The ratio is corrected for its excitation intensity dependence. Suppression of extrinsic PL to below 1% for 5 nm and thicker GaAs prelayers. Solid line: calculated trendline.

7.3   Deep-level defects at interfaces of active regions

7.3.1   Motivation

Impurities and native defects in the crystal lattice of a semiconductor disturb the proper periodicity, which causes the formation of localized, electrically active energy states within the forbidden bandgap. Selective incorporation of dopants such as silicon or beryllium added to the semiconductor, for example, GaAs, as donors or acceptors to control the conductivity as n-type or p-type, form shallow energy levels close to the conduction or valence band, respectively.

In contrast, point defects, clusters, complexes, lattice defects, and impurities such as transition metals produce energy levels further away and are called deep-levels, traps, or nonradiative recombination centers. Deep-levels can be incorporated during device growth or processing, or both, but they can also be present in the incoming wafer material. They can have two major effects on materials and devices. Clearly, deep-levels can be either beneficial or detrimental in device manufacture. Without elaborating further, an example of the first category is the deliberate addition of deep chromium acceptors to GaAs in order to compensate the residual shallow donors to produce very high-resistivity material known as semi-insulating GaAs, which is used as the wafer substrate for electronic devices such as field-effect transistors.

More importantly in the context of this text, however, are the undesirable deep states unintentionally introduced into the laser device. There they can reduce the minority carrier lifetime by acting directly as recombination centers or by increasing the probability of Auger processes, or they can form the seed for the growth of extended dislocation networks. These effects impact the quantum efficiency and threshold current and the degradation properties, which can occur already at very low defect concentrations well below 10^{15} cm^{-3} depending on the efficiency of the trap carrier capture rate. Therefore, the need to detect and monitor deep-levels is indispensable and will grow as devices become more sophisticated and as manufacturing becomes increasingly dependent on removing the last traces of undesirable deep-levels.

It has long been established that the localized states of deep-level imperfections in the depletion region of semiconductor devices can be characterized by measuring the capacitance (or current) transients after excitation with a bias pulse or light pulse. A technique, which detects and characterizes very low concentrations of electrically active defects down to 10^9 atoms/cm^3 is deep-level transient spectroscopy (DLTS). It has developed from the laborious single-shot transient recordings into the ingenious automation of the transient measurements and spectroscopic survey technique with inherent higher resolution due to signal averaging (Lang, 1974). Compact, commercial turnkey instruments (e.g., SEMILAB Ltd, 2012) available today are designed for measuring deep-levels including their parameters such as energy level, capture cross-section, concentration, and for profiling their spatial density in the depletion zone of a p–n junction or Schottky barrier.

In the following sections, the operating principle of the DLTS technique including a block diagram and two measurement examples are discussed. One is on the formation of deep electron traps at the upper interface of n-AlGaAs/GaAs SQWs and the other deals with the generation of a deep-level when etching the ridge waveguide of a diode laser down to close proximity to the active layer.

7.3.2   Experimental details

Figure 7.10 contains a schematic representation of the development of the capacitance transient due to a majority carrier trap in a one-sided p$^+$–n diode. A majority (minority) carrier trap can be an electron trap in an n(p)-type material with the electron emission rate e_n much greater than the hole emission rate e_p or a hole trap in a p(n)-type material with the hole emission rate e_p much greater than the electron emission rate e_n. To establish conditions for unique interpretations of DLTS spectra, the junction has to be one-sided, so that the depletion zone is located predominantly at one side of the junction. Therefore, the p$^+$–n diode configuration in the figure can detect majority electron traps and minority hole traps in the n-layer of the junction.

The DLTS measurement sequence starts with reverse biasing the junction to generate the depletion zone (Figure 7.10a). It then works by processing the capacitance transient signal resulting from the presence of deep-levels in the depletion layer. The transient is excited by applying repetitive bias pulses in the forward direction. Alternatively, light pulses can be applied. During the forward bias pulse the deep-levels (here electron traps) will fill with electrons, because the depletion layer width shrinks

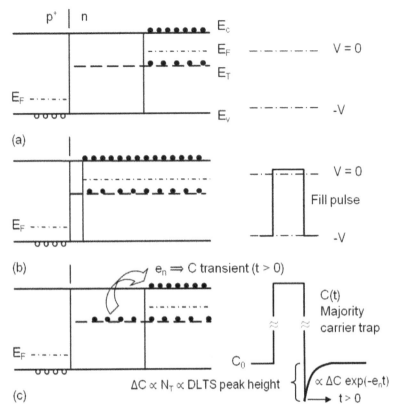

Figure 7.10 Schematic representation of the DLTS principle of operation for a majority electron trap in a one-sided p^+-n diode. (a) Reversed-biased diode leads to depletion zone mainly in the n-layer and in this space-charge region trap states are empty of carriers (electrons) ($t < 0$ quiescent state). (b) During forward-biased majority carrier pulse the space-charge region collapses and traps capture carriers (electrons) ($t = 0$ state). (c) Upon turning off the filling pulse quickly (compared to trap emission time) change in capacitance is negative (positive for minority carrier traps), capacitance transient begins ($t = 0^+$), reverse bias re-establishes, and capacitance transient decays due to thermal carrier emission from traps ($t > 0$) and proceeds until traps are empty.

(Figure 7.10b). Upon turning off quickly the forward filling pulse (compared to electron emission time), the depletion layer expands by thermal emission of electrons from the traps leaving behind the positively ionized defects (Figure 7.10c), which increases the space-charge density and is key to the DLTS measurement. Under the constant reverse bias voltage condition, this causes the diode capacitance to increase. Defect parameters such as energy level and concentration can be deduced from this capacitance transient. Compared to the velocity of free electrons, the emission rate of the electrons from the traps is relatively slow, depending on the specific nature of the trap, but most importantly it is very sensitive to temperature.

As already discussed in the previous section, the major strength of the DLTS technique is that it processes the capacitance transients in a way that yields a spectral output, that is, one capacitance peak for each deep level. This is accomplished by sampling the capacitance transient at two points in time t_1 and t_2 at least and by averaging the capacitance difference ΔC between these two points over many repetitions of the pulse, which finally is the basic output of the DLTS system. The capacitance meter has to respond faster than the deep-level and it has to be gated-off during the pulse to avoid overloading. By slowly varying the sample temperature while maintaining the same pulse sequence and sampling points, the output ΔC sweeps through a peak. This occurs at the temperature where the time constant set by the sampling times t_1 and t_2

$$\tau_{max} = \frac{t_1 - t_2}{\ln(t_1/t_2)} \qquad (7.1)$$

matches the emission time constant of the deep-levels, $1/e_n$. Or equivalently, when the electron trap emission rate toward the conduction band

$$e_n = \sigma_n N_c v_{n,th} \exp\{-(E_c - E_T)/k_B T\} \qquad (7.2)$$

is equal to the electronically set and adjustable rate window τ_{max}^{-1}. Equation (7.2) is written in a simplified form where σ_n is the electron capture cross-section of the trap, N_c the free electron density in the conduction band, $v_{n,th}$ the average electron thermal drift velocity, E_c the conduction band edge energy, E_T the energy level of the trap, and $E_{a,T} = E_c - E_T$ is the activation energy of the trap. Figure 7.11 shows schematically the various stages in the development of the final DLTS spectral signal.

The emission prefactor in Equation (7.2) is temperature dependent and in a simplified form is proportional to T^2 because $v_{n,th} \propto T^{1/2}$ and $N_c \propto T^{3/2}$ where $\sigma_n = \sigma_n(T)$ is dependent on the type of capture, which is usually unknown and therefore omitted here. By plotting $\ln(e_n/T^2)$ versus $1/T$ (Arrhenius plot) we can determine the activation energy of the trap from the slope.

A capacitance DLTS temperature scan can distinguish between the types of carrier traps involved. For example, the spectral peaks are directed downward for majority electron traps in n-type material with $\Delta C(t) \propto -n_T(t)$ the density of filled electron traps, and directed upward for minority hole traps in n-type material with $\Delta C(t) \propto p_T(t)$ the density of filled hole traps.

Further types of the DLTS technique along with their relevant features are as follows:

- Isothermal capacitance DLTS: Frequency (rate window) scanned at constant temperature; for details and advantages see paragraph below.

- Current DLTS: Transient current signals; suitable for very small capacitance devices (<1 pF); more sensitive than capacitance DLTS for faster transients; no distinction between majority and minority carrier traps, $I(t) \propto e_n n_T(t)$, $I(t) \propto e_p p_T(t)$.

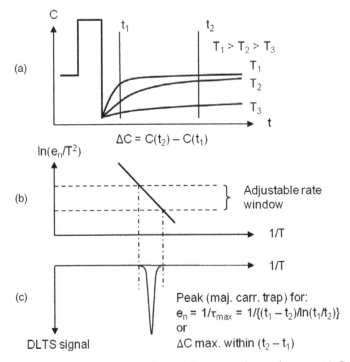

Figure 7.11 Development of the DLTS signal for a majority carrier trap. (a) Capacitance transients for different temperatures. (b) Arrhenius plot illustrating the meaning of the user-defined and adjustable rate window to match the thermal emission rate at the DLTS signal peak (c).

- Scanning DLTS: Pulsed electron beam (SEM) fills deep levels with carriers; two-dimensional mapping of deep levels with high spatial resolution.

- Optical DLTS: Uses pulsed bandgap light to inject electron–hole pairs; can detect minority carrier traps with Schottky diodes; extends DLTS to high-resistivity semiconductors; use of sub-bandgap light (cw) to photoionize mid-gap levels to prevent carrier freeze-out at low temperatures.

- Constant capacitance DLTS: Voltage transient; particularly applicable for high trap densities > shallow dopant concentrations.

- Double-correlated DLTS (DDLTS): Uses two filling pulses of different forward voltages; signal difference yields trap concentration in a thin layer of depletion layer; for spatial trap concentration profiling.

In practical, commercial systems the averaging is done by a double box-car integrator or by a broadband lock-in amplifier. The box-car signal averaging principle uses only a small portion of the actual capacitance transients, hence the signal to noise is far from the optimum value. Another disadvantage is that the DLTS peak amplitude will remain constant only at selected discrete values of t_1 and t_2. Since physical reasons might lead to changing peak heights, for example, strongly temperature-dependent capture cross-sections, to assure proper evaluation, only preselected rate windows, typically 8 to 12, are provided in box-car systems.

The lock-in averaging method (SEMILAB Ltd, 2012), however, overcomes these difficulties. It is the generalization of the box-car averaging with gate widths corresponding to half of the repetition time. However, to realize this generalization special lock-in phase-setting procedures and signal gating are necessary. A certain gate-off period has to be applied to avoid overloading the capacitance meter and the lock-in amplifier. During the excitation pulse the input to the capacitance meter is temporarily blocked. After termination of the pulse, the input is reopened, but the capacitance meter now requires a certain settling time. This is achieved in the second portion of the gate period. It is a constant fraction of the repetition time, hence the phase of the lock-in amplifier is independent of the repetition rate. This fundamentally new idea made it possible to use the theoretically more favorable lock-in amplifier in practice (Ferenczi *et al.*, 1984). Due to the repetition rate independent phase-setting principle, the repetition rate of the lock-in amplifier can be selected continuously over four orders of magnitude instead of a few discrete rate-window settings allowed by the box-car method. It can be shown (Ferenczi *et al.*, 1986) that the output signal of the lock-in amplifier is a function of temperature and repetition time. The function for the lock-in output signal has a maximum repetition time at fixed temperature just as it has a maximum temperature at fixed repetition time. The latter case represents the well-known temperature scan DLTS, whereas the former is the isothermal DLTS approach called frequency scan DLTS (sweeping the rate window or repetition rate). It should be noted that the temperature dependence is exponential and the frequency dependence is linear, and hence a much wider range should be scanned in frequency than in temperature to cover the same activation energy interval. With an automatic lock-in reference frequency sweep feature built into commercial systems (SEMILAB Ltd, 2012) the DLTS signal is measured typically at 1024 different frequencies covering four decimal orders.

Frequency scan DLTS has some fundamental advantages over the temperature scan approach: it avoids thermal annealing of the sample; it has better establishment of the accurate sample temperature and hence more accurate peak position determination; temperature-dependent phenomena such as basic sample capacitance or capture process do not interfere with the measurement of the trap emission rate; it is much faster because it is done electrically; and it is suitable for wafer mapping by coupling the system to an automatic wafer mapping setup.

Figure 7.12 shows a block diagram of the essential components of a standard temperature-scanned DLTS system. Most of the components have already been introduced above and their functions will be summarized briefly as follows: the temperature of the reverse-biased device under test (DUT) is ramped up typically between 77 and 420 K; the pulse generator supplies the train of voltage pulses to the sample; the capacitance meter measures the static capacitance and dynamic capacitance transient response of the sample; the capacitance transients are fed after amplification and digitization in an analog/digital (A/D) converter into the computer for signal

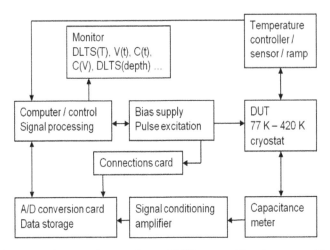

Figure 7.12 Block diagram of a capacitance DLTS system. To enhance sensitivity an automatic capacitance compensator is provided in the SEMILAB system, which compensates the variation of the static sample capacitance with temperature and hence keeps the fast capacitance bridge with a response time <5 μs in balance.

processing; the temperature controller measures and controls the temperature in the cryostat and supplies the digitized data to the computer; the pulse generator triggers through a connections card the digitization process of the transient; and all functions including DLTS vs. *T*, *V* vs. *t*, *C* vs. *t*, *C* vs. *V*, etc., can be displayed on the monitor.

7.3.3 Discussion of deep-level transient spectroscopy results

In the following, we want to discuss two typical examples demonstrating the formation of growth-induced and processing-induced deep levels. The first case deals with the generation and accumulation of deep-levels under nonoptimum growth temperature conditions at the upper interface of AlGaAs/GaAs SQW structures, whereas the second case discusses the formation of a deep-level in the active region of a ridge laser, when the ridge waveguide is etched down too close to the active layer.

Nonradiative carrier recombination in the active region, at the interfaces, and in the wide-bandgap cladding regions of diode lasers is one of the major limiting factors to achieving low threshold current densities and high quantum efficiencies. To identify the responsible deep-levels and laser degradation mechanisms, DLTS measurements have been performed both on DH and SQW lasers (Uji *et al.*, 1980; Hamilton *et al.*, 1985; Debbar and Bhattacharya, 1987; As *et al.*, 1987, 1988).

Figure 7.13 shows a capacitance DLTS spectrum recorded from the upper AlGaAs cladding layer of an n-AlGaAs/GaAs SQW device. The samples were originally designed to study deep-level emission effects in QW structures. They consisted of a GaAs SQW 3–40 nm wide embedded in an upper n-$Al_xGa_{1-x}As$ layer 100 nm

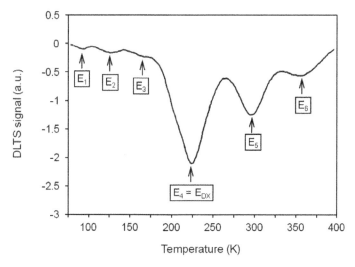

Figure 7.13   DLTS spectrum recorded from the upper AlGaAs cladding layer of an n-AlGaAs/GaAs SQW device sampled by a Schottky diode at a reverse bias of -3.5 V, filling pulse amplitude of $+0.5$ V, and a rate window of 278 s^{-1}. (Data adapted in amended form from As *et al.*, 1988.) E_1–E_6: majority electron traps.

thick and a lower n-Al$_x$Ga$_{1-x}$As layer ($x = 0.24$–0.39) 350 nm thick Si-doped to 4×10^{17} cm^{-3} and were MBE grown on an s.i. GaAs substrate with a suitable buffer consisting of a GaAs layer and AlGaAs SL (As *et al.*, 1988). The depletion zone was generated by using a Schottky diode formed on top of the device structure and consisted of a (Ti, Pt, Au) Schottky contact and an annealed (Au, Ge, Ni) Ohmic contact. Since the zero-voltage depletion width was smaller than 100 nm, the distance to the QW, it was possible to sweep the depletion edge through the QW. The spatial resolution was determined by the Debye length and was typically 8 nm.

The spectrum in Figure 7.13 was taken at a reverse bias of -3.5 V, forward voltage pulses 10 µs wide of $+0.5$ V height, and a rate window of 278 s^{-1}. It shows six peaks directed downward, which indicates, as discussed above, that they are generated from the emission of majority electron traps. Their T^2-corrected thermal activation energies were obtained from $\ln(e_n/T^2)$ versus $1/T$ plots and range between 0.12 and 0.63 eV.

At lower (0 V) and higher (-5.5 V) reverse voltages and the same 0.5 V forward voltage pulses, there is only the dominating center peak, which can be ascribed to the well-known, so-called DX center, a relatively deep donor state related to the presence of a simple substitutional donor impurity in Al$_x$Ga$_{1-x}$As ($x > 0.22$) and that controls the electrical properties of this material (Mooney, 1990; Ghezzi *et al.*, 1991). This feature is characteristic of spatially localized defects. The deep levels (E_1–E_3, E_5, and E_6) in addition to the DX center have been extensively investigated by Yamanaka *et al.* (1987) and are in general very dependent on the growth ambient, growth temperature, and group V/III flux ratio. By determining the depth distribution of these traps the actual root cause for the formation of these traps could be revealed.

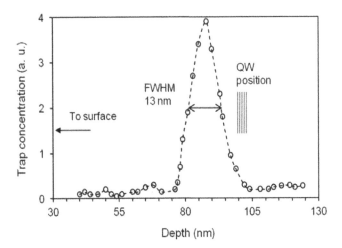

Figure 7.14 Depth profile of trap concentration of trap E_6 (see Figure 7.13) obtained by DDLTS with amplitude difference between the two excitation pulses $\Delta V = +0.1$ V, pulse duration 300 μs, rate window 2787 s^{-1}. The same distribution was obtained for the other traps E_1–E_3 and E_5. Position of the 3 nm wide QW is indicated for comparison. (Data adapted in amended form from As *et al.*, 1988.)

Figure 7.14 shows the depth profile of the E_6 trap concentration obtained from DDLTS measurements performed at a rate window of 2787 s^{-1} and 0.1 V voltage difference between the two trap filling pulses. It clearly exhibits a maximum concentration in the upper AlGaAs layer at a depth of 88 nm with a FWHM of 13 nm just above the QW 3 nm wide located 100 nm below the surface. Similar distributions have been obtained for all the other additional traps with FWHM values between 12 and 26 nm. Maximum trap concentrations are high, in the order of 5×10^{15} cm^{-3}, as calculated by convoluting the DDLTS depth signal with the *C/V* profile carrier concentration measured during the majority carrier pulse (Lang, 1974; As *et al.*, 1988).

The origin of the defects may be due to the nonoptimum growth condition for the first 10–20 nm of the AlGaAs layer. In this region the substrate temperature of 680 °C is too low for defect-free growth. Yamanaka *et al.* (1987) have shown that deep traps with comparable activation energies are observed if the growth temperature is below 715 °C. The time to stabilize to the optimum temperature of 715 °C is equal to the time to grow about 25 nm, which agrees well with the FWHM of the trap distributions observed. Since for the lower AlGaAs growth the optimum substrate temperature was kept constant up to the start of the GaAs growth, no defects were observed (cooling-down time for GaAs growth is much quicker than heating-up time for upper AlGaAs).

An interrupted growth after the GaAs QW would allow the resumption of optimal AlGaAs growth conditions and would most likely prevent the formation of these detrimental deep states, which are in contrast to the DX centers mainly responsible for the degradation of the PL efficiency in MBE-grown AlGaAs layers and the performance of AlGaAs/GaAs QW lasers (Yamanaka *et al.*, 1984, 1987; Bhattacharya *et al.*, 1984).

The second example describes the formation of a deep trap in the etching process of ridge waveguide InGaAs/AlGaAs SQW GRIN-SCH lasers. The DLTS spectrum of a laser device with an etched ridge depth of 1.5 μm, equivalent to a 0.5 μm residual thickness, which is the distance from the bottom of the etch to the active layer, is displayed in Figure 7.15a and shows only the dominating DX center complex

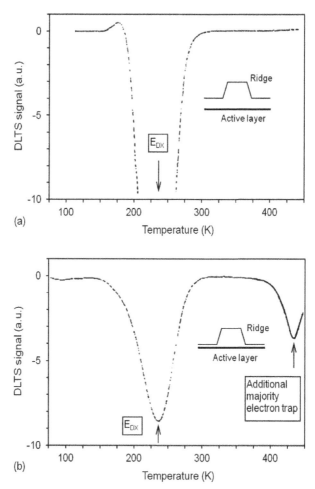

Figure 7.15 DLTS spectra of ridge waveguide InGaAs/AlGaAs GRIN-SCH SQW diode lasers recorded at a reverse bias of −3 V, filling pulse width of 20 μs, and rate window of 250 s^{-1}. (a) Laser with large residual thickness between bottom of wet-etched ridge and active layer. (b) Laser with low residual thickness inducing deep majority carrier trap in addition to DX defect center, which is known to be inherent in Si-doped bulk n-AlGaAs.

(typically 10^{18} cm^{-3}), which is known to have no impact on optical properties (Yamanaka *et al.*, 1984). In this case, the etching is not interfering with the p–n junction of the diode and these lasers show no abnormal degradation behavior.

In contrast, ridge lasers deeply etched down to about 1.9 µm leaving a residual thickness of $\cong 0.1$ µm show a distinct additional peak in the DLTS spectrum, which can be attributed to majority carrier trap emission (Figure 7.15b). In principle, this additional majority carrier trap could be either an electron trap in the n-type material or a hole trap in the p-type material of the junction. However, the presence of the DX peak in the same spectrum indicates that it is more likely an electron trap in the n-layer.

The symmetry of the p–n junction leading to an extension of the depletion zone into the n- and p-layers in principle did not allow unique identification of the traps (apart from DX which is linked to n-type Si-doped AlGaAs). As the doping levels in the symmetrical p–n junction of the diode are high, the depletion zone could be extended to a maximum width of only ± 0.15 µm with the available maximum reverse bias voltage. However, it was sufficient to probe the GRIN-SCH region width and DDLTS showed that the trap is located at the interfaces of the GRIN-SCH layer to the surrounding cladding layers, laterally predominantly effective at the lower corners of the ridge waveguide where it can have a strong impact, particularly on laser reliability. It has been demonstrated that the trap, which is very efficient due to its high activation energy of $\cong 0.8$ eV and concentration, can lead to rapid failure of the diode laser.

7.4 Micro-Raman spectroscopy for diode laser diagnostics

7.4.1 Motivation

Raman spectroscopy is a very effective spectroscopic technique for studying the interaction between monochromatic light and matter, in which the light is inelastically scattered (Raman, 1928; see, e.g., Balkanski, 1971; Cardona and Güntherodt, 1975–91). It has become a very important and highly versatile analytical and research tool, which can be used for fast, nondestructive chemical and structural analyses of solid, liquid, and gaseous matter in industrial and research applications as wide ranging as pharmaceuticals, forensics, chemistry, material science, and, above all, semiconductors. As the inelastically scattered light has interacted with the fundamental atomic and molecular excitations of the matter, it carries a wealth of information and all the crucial data necessary to characterize the microscopic properties and status of the matter, which finally show up, in effect as a fingerprint, in a Raman spectrum.

Figure 7.16 shows a schematic overview of the Raman effect including relevant Raman signal parameters and Raman interactions in semiconductors. It shows the Raman backscattering or near-backscattering configuration, which is widely used on opaque materials such as semiconductors. The small penetration depth of light of

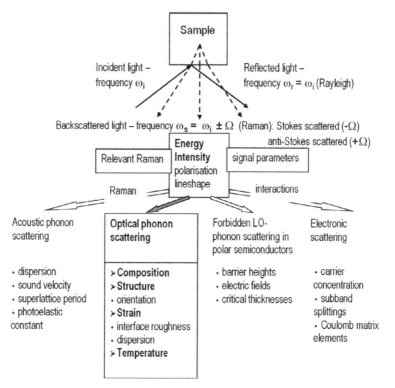

Figure 7.16   Schematic overview of the Raman inelastic light scattering process on the opaque semiconductor sample in backscattering geometry. Also shown are relevant signal parameters and interactions with elementary excitations in a semiconductor. Terms highlighted in bold are major subject topics in this text. (Parts adapted in modified form from Brugger, 1990.)

typically 100 nm in semiconductors such as GaAs permits the study of very thin layers. Within the topic of this text, we are especially interested in the signal parameters of energy and intensity and in the inelastic light scattering process with optical phonons yielding information on composition, structure, lattice disorder, strain, and temperature of the material and device.

This section deals predominantly with local temperatures at laser facets measured in dependence on laser output power, applied mirror technology, laser material, vertical laser structure, and laser chip mounting technique. In Chapter 8 next, we use Raman to reveal and characterize the microscopic root causes of enhanced mirror temperatures, strain along the active layer on the facet of ridge laser devices, and the instability of the mirror coating material by recrystallization under high laser output power.

But first we discuss briefly the physical fundamentals of the Raman effect, the Raman scattering configuration and phonon modes on a standard laser facet, the expression used to calculate the mirror temperature from the Raman spectrum, and finally some experimental details and results.

7.4.2   Basics of Raman inelastic light scattering

The Raman effect is based on the molecular polarizability of the material which is determined by deformations of the molecules in the material by the oscillating electromagnetic wave of the incident light. An electric dipole moment equal to the product of polarizability and electric field is subsequently induced, and the molecules start vibrating (and rotating in a gas or liquid) with a characteristic frequency Ω due to the periodic deformation. Monochromatic laser light with frequency ω_i excites molecules and transforms them into oscillating (and rotational) dipoles, which can emit light at three different frequencies.

As already indicated in Figure 7.16, photons interacting with the molecules of the material scatter elastically and inelastically. The former process is called elastic Rayleigh scattering where the scattered photons have the same frequency $\omega_r = \omega_i$ as the incident light. In this case, the molecule has no Raman-active mode and absorbs a photon with frequency ω_i, and the excited molecule returns to the same basic excitation, which in a solid can be, for example, a vibrational (phonon) mode.

However, only one out of about 10^{12} photons is inelastically scattered. This can be obtained by assuming a typical Raman scattering efficiency per unit length x and unit solid angle Θ as low as

$$\partial^2 I/(\partial\Theta\partial x) \cong 10^{-7}\mathrm{sr}^{-1}\mathrm{cm}^{-1} \tag{7.3a}$$

where I is the Raman scattering efficiency of the Stokes or anti-Stokes Raman line in the spectrum (see below), Θ is the solid angle, and x the coordinate of the light into the sample. By using a typical penetration depth of 10^{-5} cm for the excitation light, for example, in GaAs, the total efficiency can be written as

$$10^{-7}\mathrm{sr}^{-1}\mathrm{cm}^{-1} \times 10^{-5}\mathrm{cm} = 10^{-12}\mathrm{sr}^{-1} \tag{7.3b}$$

which means that only one photon out of 10^{12} is inelastically scattered! This extremely low generation probability of Raman photons requires the use of an extremely efficient detection system and the removal of any stray light and Rayleigh background light (see next sections), which might hinder greatly the Raman light detection. The Raman signal is proportional to

$$\partial I/\partial\Theta \propto \sigma \times n(\Omega) \times V \tag{7.3c}$$

with σ the scattering cross-section, $n(\Omega)$ the scatterer concentration, and V the scattering volume.

The energy of the Raman light is shifted either lower or higher (red or blue shift, respectively). It is the change in energy which provides the chemical and structural information.

Red-shifted photons are the most common, because at room temperature the population is principally in the molecule's ground state (vibrational, rotational, or electronic level), and therefore the Raman scattering effect is larger. Upon interaction with the photon, the molecule is excited into a higher virtual state, part of the incident photon's energy is transferred to the Raman active mode with frequency Ω, and the resulting frequency of the scattered light is reduced to $\omega_i - \Omega$, called the *Stokes shift*. The loss of energy is directly related to the functional group bonds, the structure of the molecule to which it is attached, the types of atoms in that molecule, and its environment. Factors such as the polarization state of the molecule determine the Raman scattering intensity, which increases with increasing change in polarizability. This means that some vibrational (*phonon*) or rotational transitions or transitions of any other fundamental excitations, which show low polarizability, will not be Raman active and therefore do not appear in a Raman spectrum.

On the other hand, blue-shifted photons are much less common. In this case, a photon with frequency ω_i is absorbed by a Raman active molecule, which at the time of interaction is already in an excited state with higher energy. Excess energy of the excited Raman active mode is released, the molecule returns to the basic energetic state, and the resulting frequency of scattered light will be shifted to a higher frequency $\omega_i + \Omega$, which is designated as the *anti-Stokes shift*. The processes discussed here are schematically illustrated in Figure 7.17.

Figure 7.18 shows schematically the Stokes and anti-Stokes Raman spectra in connection with the PL spectrum, which can be excited in the same process. The

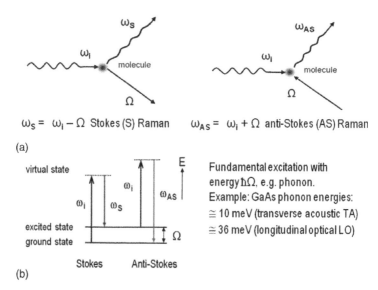

Figure 7.17   Schematics showing fundamental Raman scattering processes. (a) Graph of Stokes and anti-Stokes scattering process. (b) Energy-level diagram for Raman scattering. Gray tone intensity indicates strength of Raman line.

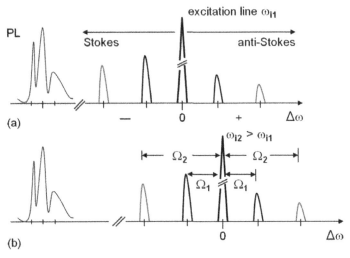

Figure 7.18 Schematic illustration of Raman spectra at two different light excitation frequencies $\omega_{i2} > \omega_{i1}$ in relation to the PL spectrum excited simultaneously.

Raman effect, which is nonresonant, differs in general from a luminescence process, which is resonant, by exciting the material to a discrete (not virtual) energy level. This means that the Raman effect can take place actually for any energy of incident light. And the luminescence spectrum is anchored at a specific excitation energy, whereas a Raman spectral line maintains a constant separation from the excitation energy. The correct selection of the exciting laser wavelength is crucial so that the luminescence spectrum does not interfere with the Raman spectrum to such an extent that it is no longer detectable.

By using Bose–Einstein statistics for the occupation number of the elementary excitations, the intensity of the temperature-dependent Stokes (I_S) and anti-Stokes (I_{AS}) scattered light can be estimated as follows (Cardona and Güntherodt, 1975–91):

$$I_S \propto 1 + n(\Omega) = (1 - \exp\{-\hbar\Omega/k_B T\})^{-1} \tag{7.4a}$$

$$I_{AS} \propto n(\Omega) = (\exp\{\hbar\Omega/k_B T\} - 1)^{-1} \tag{7.4b}$$

$$I_S/I_{AS} \propto \exp\{\hbar\Omega/k_B T\} \tag{7.4c}$$

where $n(\Omega)$ is the density of the fundamental vibrational, rotational, or electronic excitations with frequency Ω taking part in the Raman inelastic light scattering process. As we will show further below, the ratio between the integrated intensities of the Stokes and anti-Stokes optical phonon lines in the Raman spectrum will be used for calculating the sample temperature, which is already indicated in Equation (7.4c). In all further discussions, we will use *phonons* as the fundamental excitations taking part in the Raman process.

7.4.3   Experimental details

The extremely low Raman efficiencies (see above) require experimental equipment optimized in at least the following four major components:

- Monochromatic and collimated excitation source, usually a laser with selectable wavelengths.

- Optical system for producing a magnified image of microscopic areas of the sample, for focusing the laser beam onto the sample and for collecting efficiently the weak inelastically scattered radiation.

- Spectrometer system with high stray and Rayleigh light rejection ratio and high spectral resolving power for analyzing the spectral content of the inelastically scattered laser light.

- Detector with high sensitivity for detecting the extremely weak, spectrally resolved Raman light.

In Figure 7.5 we have already introduced the multipurpose optical system, which comprises also the essential components needed for generating Raman spectra on diode lasers, in particular on the laser mirrors, in a very efficient way with high sensitivity, high spectral and spatial resolution, and high stray and laser light rejection. In the following, we shall give some details of the system, which was assembled by the author from individual components.

The Ar^+ laser line 457.9 nm with a low intensity of typically 1 mW and below was used throughout the Raman experiments on diode laser facets. This ensured, in combination with a high-quality 40/0.63 objective lens and xyx-translation stage with 0.1 μm resolution, the achievement of the required spatial resolution of ≤ 1 μm with negligible self-heating ($\Delta T < 10$ K) of the laser spot on the laser die, which was attached to a heat sink. The mirror structure was observed with a video camera and monitor under high magnification of $10^4\times$. This was made possible by coupling simple lamplight from a fiber bundle via a beamsplitter into the optical path through the objective lens and onto the sample. The sample was then imaged in reverse path at a large image distance onto a camera with zoom. This procedure was used to align the sample to the laser focus. The beamsplitter was removed for the actual measurement.

To block out the laser light from detection usually commercially available interference (notch) filters are employed, which cut-off a spectral range of typically ± 80–120 cm^{-1} from the laser line. This method may be efficient in stray light elimination, but it does not allow detection of low-frequency Raman modes in the range below 100 cm^{-1}. However, these filters have additional severe drawbacks related to quantitative Stokes and anti-Stokes Raman spectra measurements, for example, for determining the absolute temperature at a laser facet. The drawbacks include: (i) the rejection of Rayleigh scattered light in the notch region is not sufficient; (ii) the spectral bandwidth of the notch region is not sufficiently narrow; and (iii) the edges of the notch region are not sufficiently sharp and steep. The consequences are that only

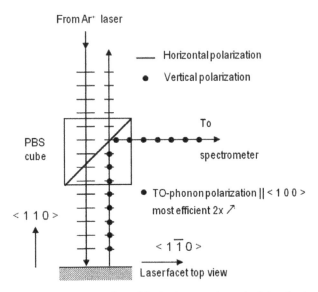

Figure 7.19 Schematic representation of the basic operating principle of using a polarizing beamsplitter (PBS) for effectively separating the elastically scattered Rayleigh light from the inelastically scattered Raman light before entering the spectrometer. Top view: (100) orientation perpendicular to figure plane.

Raman signals with larger energy shifts are accessible, which means that essential parts of the Raman spectrum in the low-energy regime cannot be revealed. Another drawback is: (iv) the transmission outside the notch (stopband) is not uniform and different in the Stokes and anti-Stokes energy regime of the Raman spectrum. Therefore, the spectra have to be corrected with the exact filter transmission curves for the quantitative measurements of facet temperature, which is cumbersome and not error-free. Further drawbacks are: (v) each filter is designed for only one specific probe laser wavelength; and (vi) the filters are expensive, of limited use, and not easy to handle.

The author has developed and successfully used a novel and efficient method to reject the interfering Rayleigh light from detection in Raman microprobe spectroscopy on usually (110)-oriented laser facets (Epperlein, 1992). It employs crossed polarizations of incident and scattered light in conjunction with a commercial polarizing beamsplitter (PBS), which has polarization-dependent transmissions. This new method avoids all the drawbacks listed above for notch filters.

Figure 7.19 demonstrates the basic function by using a simple PBS, which consists of a pair of right-angled prisms, cemented together, hypotenuse-face to hypotenuse-face, with a special multilayer dielectric film in between. The incident light from the excitation laser is linearly polarized (*p-polarization*) such that it passes straight through the cube onto the laser facet surface with the correct polarization for symmetry-allowed Raman scattering (see Section 7.4.4 and Figure 7.20 below).

Upon scattering, the Rayleigh light with unchanged polarization is transmitted back through the cube. In contrast, the interesting Raman light with a polarization orthogonal to that of the incident beam (*s-polarization*) is reflected by the multilayer dielectric film at a 90° angle and fed into the spectrometer for spectral detection.

The extinction ratio (ER) between Rayleigh and Raman light of this basic configuration is in the range of 10^{-2} to 10^{-3} at the detection input and the cut-off spectral range from the laser line is now dramatically reduced to about ± 25 cm^{-1}. By using a crossed pair of PBS cubes or a high-quality polarizing Thompson calcite prism ER can be further increased to 10^{-4}–10^{-6} and Raman signals as close as ~ 10 cm^{-1} to the laser line can be detected, which is never possible with notch filters.

The requirements of the spectral detection system are enormously high considering the facts that, first, the Raman signal is many orders of magnitude weaker than the elastically scattered Rayleigh light and that, second, the difference between the Raman signal energy and laser light energy is extremely low, in the region of 1% of the laser energy. For example, 1% of the energy of the Ar$^+$-laser 457.9 nm line amounts to 0.01×2.7 eV $= 27$ meV, which is a typical phonon energy in GaAs.

The spectrometer must meet stringent conditions to observe this weak sideband close to the strong laser line. It must have a high spectral resolving power $\Delta\lambda/\lambda$, which is for the double spectrometer in Figure 7.5 with a specified focus length of 0.85 m (effectively 0.9 m due to the detector column added) and dual-holographic grating greater than 10^4. This means that spectral structures separated by ≤ 0.05 nm can be resolved at an excitation laser line in the visible region.

Another stringent condition is that the stray light rejection ratio, that is, the ratio of the integrated background stray light to the signal light intensity, is sufficiently small. The specified stray light rejection ratio of the double spectrometer in Figure 7.5 is $< 10^{-14}$, which in combination with the use of the external PBS discussed above fully meets the requirements to perform high-performance reliable Raman spectroscopy measurements. Stray light, which is produced by imperfections in the optics and by the scattering of light at walls and dust particles inside the spectrometer, is about an order of magnitude less intense from holographic gratings than from ruled gratings. It can be further minimized by using a smooth sample surface and by flushing the spectrometer with a purified and inert purge gas such as nitrogen. This avoids negative effects such as water vapor absorption and oxygen/ozone contamination of the optics, although ozone buildup occurs only in the presence of deep UV radiation.

Finally, modern Raman systems employ single-photon counting electronics and multichannel detection systems. In the former, a discriminator selects and counts only photocurrent pulses arriving at the photomultiplier tube (PMT) anode with large enough amplitude and originating at the photocathode. The background noise or dark count rate originating from thermionic emission of electrons at the photocathode has to be kept low. This can be achieved by cooling the PMT to ~ -20 °C by using a thermoelectric cooler or even to liquid nitrogen temperature of -196 °C and by selecting a PMT with a Cs-activated GaAs photocathode. The system in Figure 7.5 used a selected GaAs:Cs PMT with a low dark count rate of ~ 5 counts/s at the maximum anode/cathode voltage of 1800 V and had a dynamic range of $> 10^6$.

The second system is an optical multichannel analyzer (OMA) using a cooled charge-coupled device (CCD) array detector. This method has the advantage of detecting and imaging a certain width of the spectrum after a given signal integration time without having to scan the spectrometer output as in the case of the PMT method. The achievable width of the spectrum depends on the spectral resolving power of the spectrometer, the exit slit width, and the width of the array. The spectral resolution, which depends also on the pixel size of the detector array, is generally lower than that in a scanning system with the same spectrometer specifications. Finally, the sensitivity of the CCD detector can be enhanced by adding an image intensifier tube. A discussion of these systems is, however, beyond the scope of this text, but we refer the interested reader to the relevant literature, for example, Chang and Long (1983).

7.4.4 Raman on standard diode laser facets

The Raman scattering event has to meet the energy and wavevector conservation laws

$$\hbar\omega_S = \hbar\omega_i \mp \hbar\Omega \tag{7.5}$$

$$\hbar k_s = \hbar k_i \mp \hbar K \tag{7.6}$$

where K is the crystal momentum of the elementary excitation (e.g., phonon), the minus sign stands for the Stokes processes, and the plus sign for the anti-Stokes processes. The wavevectors for k_i and k_s of the incident (i) and scattered (s) photon, respectively, are given by $2\pi/\lambda$ where λ is the wavelength of the light typically in the region of 5000 Å. These wavevectors are very small compared to the large wavevectors of the phonon estimated by $2\pi/a_0$ where a_0 is the lattice constant typically in the order of 5 Å. This means that only phonons very close to the center of the Brillouin zone ($K = 0$) can participate in the inelastic light scattering process. However, any destruction of the translational symmetry, which leads to a breakdown of the k-conservation law, for example, by lattice disorder effects, allows the participation of phonons of any K vector ($K \neq 0$) in the Raman scattering process.

Figure 7.20 shows the relevant crystal orientations of a (zinc-blende semiconductor) diode laser mirror with a typical (110)-oriented facet. Raman microprobe spectroscopy is done in a backscattering configuration from the (110)-oriented facet

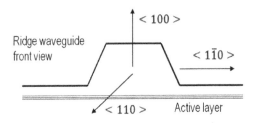

Figure 7.20 Crystal orientations in a typical (110) ridge diode laser facet.

of the opaque semiconductor surface, which means that the k vectors of the incident and scattered light are anti-parallel to each other or, in other words, the scattered light is collected by the same lens as that through which the incident light is focused.

Raman selection rules determine the existence of Raman-active phonons for certain choices of the incident and scattered polarizations and scattering geometries (for details see, e.g., Yu and Cardona, 2001). According to these, at an ideal (110) facet, scattering at transverse optical (TO) phonons is allowed by symmetry, whereas scattering at longitudinal optical (LO) phonons is symmetry forbidden. The symmetry-allowed scattering configurations written in the so-called Porto notation $k_i(e_i, e_s)k_s$, where k_i and e_i are the wavevector and polarization vector of the incident light and k_s and e_s the wavevector and polarization vector of the scattered light, respectively, can be written as

$$\bar{1}\bar{1}0\left(1\bar{1}0, 1\bar{1}0\right)110 \tag{7.7}$$

for the parallel-scattering geometry and

$$\bar{1}\bar{1}0\left(1\bar{1}0, 100\right)110 \tag{7.8}$$

for the crossed-polarization configuration.

As examples, scattering at AlGaAs leads to two modes, GaAs-like and AlAs-like zone center TO phonon modes. In contrast, scattering at AlGaInP shows a one-mode behavior, that is, only one Raman AlGaInP TO phonon mode (Bayramov et al., 1981).

At real (110) facets the translational symmetry can be destroyed by structural and compositional disorder, surface electric fields, strain, or other perturbations, which results in a relaxation of the k-conservation law. As mentioned above, this allows Raman scattering at phonons of any K vector ($K \neq 0$) including first-order transverse optical (TO), longitudinal optical (LO), transverse acoustic (TA), longitudinal acoustic (LA) phonons, and second-order phonons such as 2TO and 2TA.

7.4.5   Raman for facet temperature measurements

By using the ratio I_S/I_{AS} between the integrated intensities of the Stokes (S) and anti-Stokes (AS) Raman lines of the appropriate TO phonon mode in the material system under investigation, the local mirror temperature T can be determined as follows (Compaan and Trodahl, 1984):

$$\frac{I_S}{I_{AS}} = C_1 \left(\frac{1 - R_{AS}}{1 - R_S}\right) \left(\frac{\alpha_{AS} + \alpha_i}{\alpha_S + \alpha_i}\right) \left(\frac{n_{r,AS}}{n_{r,S}}\right) \left(\frac{\omega_i - \Omega}{\omega_i + \Omega}\right)^3 \exp\{\hbar\Omega / k_B T\} \tag{7.9}$$

where R is the reflectivity, α the absorption coefficient, and n_r the refractive index of the facet material at frequencies ω_S and ω_{AS}. C_1 is a correction factor due to the different responses of the detector and gratings at Stokes and anti-Stokes frequencies.

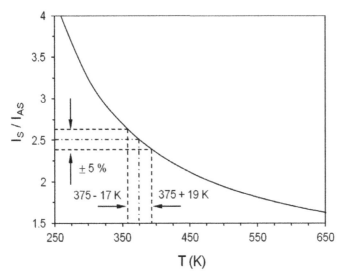

Figure 7.21   Calculated sensitivity of the mirror temperature T on the Stokes/anti-Stokes intensity ratio I_S/I_{AS} of a suitable phonon mode in the Raman spectrum; $\hbar\Omega = 33$ meV is used for the GaAs TO phonon and 0.9 for the total correction factor in the calculation.

For a backscattering geometry, R can be calculated for the Stokes and anti-Stokes light from (see also Equation 1.12)

$$R = \frac{(n_r - 1)^2 + \kappa^2}{(n_r + 1)^2 + \kappa^2} \tag{7.10}$$

where κ is the extinction coefficient. By using ellipsometry data for n_r and κ (Aspnes *et al.*, 1986) of the major materials GaAs, AlGaAs, InGaAs, and AlGaInP used in the Raman studies of this text, a typical TO phonon energy $\hbar\Omega \cong 33$ meV, and the energy of the incident laser $\hbar\omega_i \cong 2.7$ eV, the total correction factor in front of the exponential term can be calculated as $C_1 \times C_2 \cong 0.9$. The temperature dependence of the S- and AS-Raman susceptibilities has been neglected in the calculations.

The sensitivity of T on I_S/I_{AS} was calculated on the basis of Equation (7.9) by using the values of the phonon energy and correction factor in the previous paragraph. Figure 7.21 shows that a $\pm 5\%$ error in a typical value 2.5 for I_S/I_{AS} leads to a roughly ± 20 K error in temperature of 375 K corresponding to $I_S/I_{AS} = 2.5$. Of course, at higher I_S/I_{AS} values the error in temperature is lower, whereas at lower values it is much higher. The I_S/I_{AS} intensity ratio is very insensitive to temperature changes at high-temperature levels, which hampers accurate temperature measurements.

7.4.5.1   Typical examples of Stokes- and anti-Stokes Raman spectra

Figure 7.22 shows typical examples of AS- and S-Raman spectra of a cleaved, uncoated facet of an AlGaAs/GaAs GRIN-SCH SQW laser 3 μm wide (Brugger and

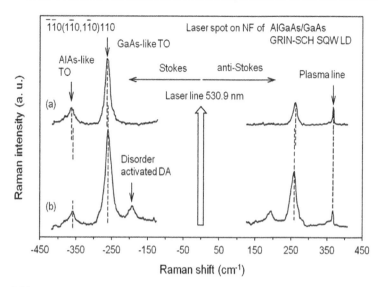

Figure 7.22 Raman spectra of a cleaved, uncoated (110) facet in the near-field (NF) of an (Al)GaAs GRIN-SCH SQW ridge laser 3 μm wide during operation: (a) 5 mW per facet, laser not degraded; (b) 18 mW per facet, laser degraded. (Parts adapted in modified form from Brugger and Epperlein, 1990.)

Epperlein, 1990). The probe laser was focused on the optical near-field spot. The upper curves were taken from the nondegraded laser at a cw power of 5 mW per facet. The spectra, detected in symmetry-allowed, parallel-scattering configuration, show the first-order GaAs-like and AlAs-like TO phonon modes, as expected. The spectra show the typical two-mode behavior in contrast to AlGaInP with a one-mode spectrum, as we will show in the next example below. The Ar^+-ion laser plasma line recorded in the spectra can be used as a very useful energy reference for measurements of very accurate frequency shifts and to obtain qualitative information on changes in surface roughness via intensity fluctuations of the elastically scattered Rayleigh light (see also Section 8.1).

The lower curves in Figure 7.22 were recorded from the same mirror region and at the same low excitation intensity of $\lesssim 100$ kW/cm^2 (leading to a temperature rise <10 K), but after heavy degradation of the laser with a strong increase in threshold current. The laser was here operated at 18 mW/facet resulting in a significant heating of the facet. This can be seen in a shift of the phonon lines to lower energies and broadening of the linewidths caused by anharmonic effects in the crystalline material. However, most importantly, the S/AS intensity ratio is decreased with the higher temperature. The absolute intensities of the AS and S signals are also higher due to the temperature-dependent change of the Bose–Einstein occupation number (Equations 7.4a, 7.4b), optical material constants, and the resonance curve of the Raman susceptibility (Compaan and Trodahl, 1984). The line at 195 cm^{-1} can be attributed to a disorder-activated (DA) phonon mode. Its existence demonstrates

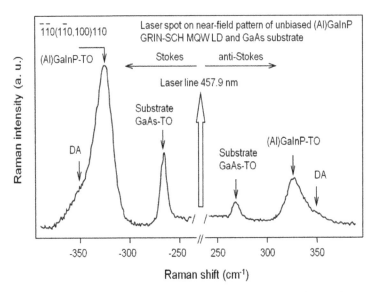

Figure 7.23 Raman spectra of a cleaved, uncoated (110) mirror facet of an unbiased (Al)GaInP GRIN-SCH MQW ridge laser 5 μm wide on a GaAs substrate. Excitation laser spot in parts on near-field pattern and in parts on GaAs substrate. (Parts adapted in modified form from Epperlein *et al.*, 1992.)

strong crystal damage at the facets of the degraded laser. We discuss this mode in detail in the context of revealing the microscopic root causes of enhanced facet temperatures and linked COMD probability (see Section 8.1).

In contrast, Figure 7.23 illustrates the one-mode Raman spectrum of (Al)GaInP. It shows the first-order Stokes and anti-Stokes spectra of a nondegraded facet of an unbiased AlGaInP/GaInP GRIN-SCH MQW ridge laser 5 μm wide detected in symmetry-allowed, crossed-polarization geometry (Epperlein *et al.*, 1992). The focused (∼1 μm), low-power (1–2 mW) probe laser beam irradiated part of the near-field region and part of the GaAs substrate. With this probing condition the temperature gradient expected between the hot (Al)GaInP regions and the GaAs substrate could be evaluated in one spectrum recorded at different drive currents and power levels (see next section). The spectra show the GaAs substrate TO phonon mode well separated from the (Al)GaInP TO phonon peak exhibiting the widely accepted one-mode behavior of AlGaInP. The peak appearing as a shoulder at ± 350 cm^{-1} on the high-energy side of this peak can be attributed to one-phonon optical transitions due to disorder in the crystal structure (Bayramov *et al.*, 1981).

7.4.5.2 First laser mirror temperatures by Raman

Figure 7.24 shows mirror temperature ΔT versus power P calculated from Stokes- and anti-Stokes GaAs-like TO phonon modes of cleaved uncoated and coated

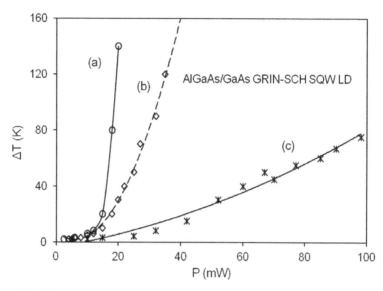

Figure 7.24 Measured temperature rises ΔT at mirror facets inside the near-field spot of (Al)GaAs GRIN-SCH SQW lasers as a function of cw optical output power P: (a) 3 µm wide ridge, cleaved and uncoated mirror; (b) 8 µm wide ridge, cleaved and uncoated mirror; (c) 8 µm wide ridge, cleaved and passivation-coated Al_2O_3 mirror $\lambda/2$-thick. Solid/dashed lines are guides to the eye. (Data adapted in modified form from Brugger and Epperlein, 1990.)

AlGaAs/GaAs lasers with different ridge widths (Brugger and Epperlein, 1990; Brugger *et al.*, 1990). The temperature data consider the small temperature increases due to self-heating of the low-power excitation laser (see previous section). The temperature increase ΔT is relative to the temperature at threshold current with $P \cong$ 0. The narrower the ridge, the higher are the temperatures due to the higher power densities and the more pronounced is the nonlinear $\Delta T/P$ dependence.

Typical values for cleaved uncoated mirrors are $\Delta T > 100$ K for $P > 5$ mW per 1 µm ridge width. Once degradation has occurred, a continuous increase of ΔT can be observed as a function of time at constant P (not shown; see also critical facet temperature to COMD event in Chapter 9). ΔT can go up to 10^3 K on highly degraded uncoated mirrors. On the other hand, passivation-coated mirrors with transparent Al_2O_3 layers $\lambda/2$-thick show temperatures strongly reduced by about a factor of 10 compared to cleaved uncoated devices with the same widths (Figure 7.24c). These optical thicknesses of $\lambda/2$ lead to no interference losses and leave the reflectivity at about 30%, the value for cleaved uncoated GaAs mirrors. The degradation threshold of these laser facets is about five times higher than for uncoated ones, which can be ascribed to the passivation of dangling bonds at the cleaved surface and therefore a reduction of effective nonradiative recombination centers and surface recombination velocity, as discussed in Chapter 4.

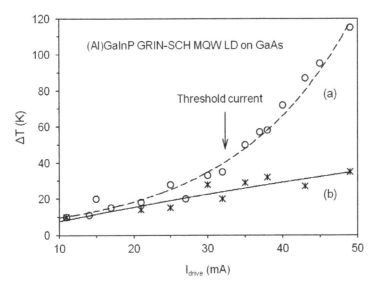

Figure 7.25   Measured temperature rises ΔT at cleaved uncoated facets of two ridge (Al)GaInP GRIN-SCH MQW lasers 5 μm wide as a function of drive current I_{drive}: (a) temperature in (Al)GaInP near-field spot; (b) temperature in GaAs substrate near GaAs/AlGaInP interface. Solid/dashed lines are guides to the eye. (Data adapted in modified form from Epperlein *et al.*, 1992.)

The next two figures show ΔT as a function of the drive current I_d and power P for cleaved uncoated facets of (Al)GaInP GRIN-SCH MQW ridge lasers 5 μm wide (Epperlein *et al.*, 1992).

The temperatures in Figure 7.25 were obtained from the S/AS intensity ratios of the (Al)GaInP TO phonon peak and GaAs substrate TO phonon mode of the spectra as described in Figure 7.23. The ΔT data comprise data obtained for two nearly identical lasers and allow for the low heating effect ($\sim$10 K) of the low probe laser power $\sim$1 mW.

There are clearly two regimes indicating different heating mechanisms. Below threshold in the spontaneous emission region, ΔT increases to a maximum of $\sim$35 K, a value which agrees well with the bulk temperature rise of the active p–n junction region along the cavity axis as found from the comparison between P/I_d characteristics recorded under pulsed and dc conditions. In this regime, ΔT is caused by Joule heating of I_d. Above threshold, ΔT rapidly increases with I_d reaching $\sim$100 K at $1.5I_{th}$. Here, the Joule heating effect is superimposed by heating due to absorption of laser radiation, which leads to an enhanced nonradiative surface recombination.

The low ΔT values in the lower branch of Figure 7.25, measured in the GaAs substrate close ($<$1 μm) to the epilayer/substrate interface, indicate a steep temperature gradient in the direction perpendicular to the active layer. This is more pronounced than what is known from temperature maps of AlGaAs/GaAs laser mirrors

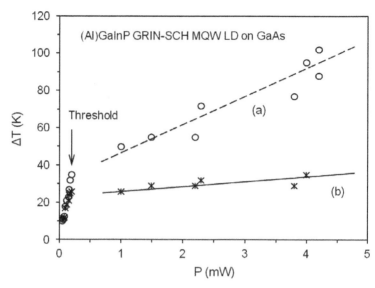

Figure 7.26 Measured temperature rises ΔT at cleaved uncoated facets of two ridge (Al)GaInP GRIN-SCH MQW lasers 5 µm wide as a function of cw power P: (a) temperature in (Al)GaInP near-field spot; (b) temperature in GaAs substrate near GaAs/AlGaInP interface. Solid/dashed lines are guides to the eye. (Data adapted in modified form from Epperlein *et al.*, 1992.)

(Epperlein, 1993; see Chapter 9) due to the higher thermal resistivity of AlGaInP (Adachi, 1983).

This thermal behavior is more distinct in the ΔT versus P plots in Figure 7.26 where the slope of the curves in the lasing regime is indicative of the thermal resistivity. The slope ratio of $\sim$6 is in reasonable agreement with the known ratio between the thermal resistivities of the used composition $(Al_{0.7}Ga_{0.3})_{0.52}In_{0.48}P$ and GaAs. The ΔT data are in good agreement with independent measurements using the Raman phonon line energy shift (Epperlein *et al.*, 1992) and thermoreflectance as a novel approach for laser mirror characterization (Epperlein, 1993; see Chapter 9). The data demonstrate the high heating efficiency of the (Al)GaInP facets with $\Delta T >$ 100 K for 0.8 mW lasing power per 1 µm ridge width and therefore pose a high risk for laser reliability.

7.4.6   Various dependencies of diode laser mirror temperatures

The next three sections deal with the dependence of mirror temperatures on laser material, mirror surface treatment, number of active QWs, cladding layer material, use of a heat spreader, and type of mounting the laser die on a heat sink. Further dependencies such as on a bent-waveguide nonabsorbing mirror structure and the type of mirror passivation technology will be discussed in Chapters 8 and 9, and also in connection with the novel thermoreflectance measurement technique pioneered and introduced successfully by the author as a fast and highly versatile technique for diode laser characterization (Epperlein, 1993).

7.4.6.1 Laser material

The dependence of ΔT on laser material is shown in Figure 7.27. For clarity, the mirror temperatures of cleaved uncoated SQW ridge lasers 5 μm wide are plotted (Epperlein, 1996). They are higher for (Al)GaInP lasers than for (Al)GaAs lasers primarily because of the relatively high thermal and electrical resistivity of AlGaInP cladding layers. The low mirror temperatures of the compressively-strained AlGaAs/InGaAs lasers are predominantly due to low bulk and surface defect densities. It is worth mentioning that EBIC studies (Jakubowicz and Fried, 1993) on strained lasers showed that, in terms of defect formation, the strained active QW can effectively screen the ridge-induced strain fields (Epperlein *et al.*, 1990, 1995). This can reduce the generation and migration of defects at the mirrors during laser operation resulting in lower facet heating and hence higher mirror stability and laser performance. It may also be supported by the fact that we never observed the discontinuity at threshold in the $\Delta T/I_d$ plots of nondegraded AlGaAs/InGaAs lasers. However, upon moderately degrading these lasers at the mirror surface, which appeared as an increase in I_{th}, the kink at I_{th} then appeared and the temperatures increased by a factor of ~ 4 (see Chapter 9). This kink at threshold was always observed for all the other laser

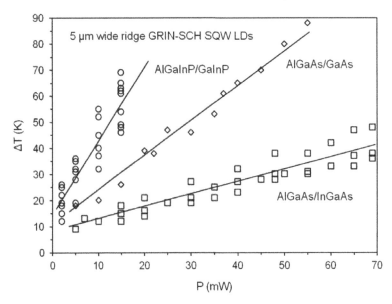

Figure 7.27 Temperature increases ΔT at cleaved uncoated facets inside the near-field of (Al)GaInP, (Al)GaAs, and AlGaAs/InGaAs GRIN-SCH SQW ridge lasers 5 μm wide as a function of cw power P. Solid lines: calculated trendlines.

types investigated. Based on this information, a laser structure can be designed with minimum mirror heating and hence increased power level at COMD.

7.4.6.2 Mirror surface treatment

Figure 7.28 plots the mirror temperature increases ΔT versus the optical power P for different mirror treatments of (Al)GaAs GRIN-SCH SQW ridge lasers 7 μm wide. The temperatures are highest for cleaved, untreated, and uncoated mirrors. They can be reduced by an oxygen-plasma ashing process usually applied to remove residual photoresist from an etched mirror structure and the deposition of a Si_3N_4 coating layer $\lambda/2$ thick. The removal of the damage layer formed after the oxygen-ashing treatment by a wet etch and the deposition of Si_3N_4 $\lambda/2$-thick decreases the temperatures further. The lowest temperatures can be achieved with passivation-coated mirrors realized with transparent Al_2O_3 layers $\lambda/2$-thick.

All the data have been confirmed by independent thermoreflectance measurements. The thermoreflectance technique and its novel applications in diode laser diagnostics will be discussed in Chapter 9. At point A in Figure 7.28c the laser experiences partial COMD, which leads to a much higher mirror temperature at a much reduced power of one-third the power at A. The temperature at A can be interpreted as a critical temperature $\Delta T \cong 125$ K for the onset of the COMD event in (Al)GaAs lasers. Within measurement accuracy of ± 20 K it is practically independent of

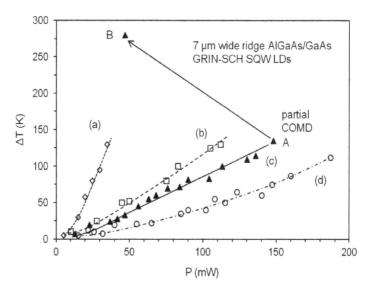

Figure 7.28 Experimental temperature rises ΔT at cleaved mirror facets as function of cw optical power P for different mirror treatments of (Al)GaAs GRIN-SCH SQW ridge lasers 7 μm wide: (a) untreated, uncoated; (b) oxygen-ashed, Si_3N_4-coated, $\lambda/2$-thick; (c) oxygen-ashed, wet-etched, Si_3N_4-coated, $\lambda/2$-thick; (d) Al_2O_3 coated, $\lambda/2$-thick. Various lines: guides to the eye.

surface treatment and coating. As discussed in Chapter 3, at a critical temperature thermal runaway is initiated at the mirror surface leading to partial or even complete COMD of the laser device. We will resume this topic in Chapter 9 and demonstrate the accurate measurement of temperature at the time of onset of the COMD process by thermoreflectance.

7.4.6.3 Cladding layers; mounting of laser die; heat spreader; and number of active quantum wells

The effect of cladding layers, laser mounting, heat spreader, and number of active QWs on the facet temperature was investigated with Raman spectroscopy and thermoreflectance measurements on 5 μm ridge waveguide, uncoated, lattice-matched, and strained GaInP lasers with AlGaInP and mixed AlGaInP–AlGaAs cladding layers, respectively. Selected data from the $\Delta T/I_d$ and $\Delta T/P$ plots are listed in Table 7.1 along with schematic illustrations of the various device configurations.

The major conclusions can be summarized as follows, where points 2 through 4 are of a general nature and apply also to other diode laser types:

1. Temperatures measured up to 5 mW show that mixed claddings consisting of 200 nm AlGaInP and 1.5 μm AlGaAs lead to lower facet heating by a factor of ~2.5 × compared to AlGaInP just 0.8 μm thick, which is caused by the lower electrical and thermal resistivity of AlGaAs in the two-layer cladding stack.

2. Facet heating is lower by ~2 × at 15 mW for junction-side down mounting than for junction-side up mounting. The lasers used in this comparison had a gold heat spreader 5 μm thick electroplated on top of the p-contact and covering the ridge waveguide. The heat spreader was recessed ~10 μm from the cavity ends.

3. A heat spreader layer recessed >10 μm from the mirror edge has no measurable impact on the temperature. However, a heat spreader aligned with the mirror edge within ≤1 μm efficiently cools the mirrors of the junction-side up mounted lasers by a factor of ~2.5×, which is stronger than for junction-side down mounted lasers. This efficient cooling with aligned heat spreader can be expected, considering that the characteristic decay length of temperature from the hot mirror into the cavity is ~6 μm. This value was determined from electroluminescence spectra detected along the cavity with high spatial resolution (see Chapter 9; Epperlein and Bona, 1993a; Epperlein, 1997).

4. Mirror temperatures increase with the number of active QWs. As already discussed in Section 4.4.2, this can be understood in simplified terms by considering that, first, each QW may act as a possible heating source fed by any possible nonradiative recombination of the injected carriers, and, second, the individual heating sources in lasers with more than one QW are cumulative

Table 7.1   Averaged temperature rises at selected power values of uncoated mirrors of (Al)GaInP GRIN-SCH QW ridge lasers 5 μm wide with different vertical structures and in different mounting configurations. Temperature rise at 0 mW corresponds to Joule heating of the drive current. This value has to be subtracted from the value at >0 mW to obtain the temperature rise only due to absorption of the stimulated laser radiation.

Symbol	Laser type	0 mW	5 mW	10 mW	15 mW
AlGaInP — 5-QW, n⁺ - GaAs, H–Sink	5 QW, AlGaInP claddings: No H, JSU	28	65	–	–
H-Spr. Recessed, AlGaInP/ AlGaAs — 2-QW, n⁺ - GaAs, H–Sink	2 QW, AlGaInP / AlGaAs claddings: H recessed, JSU	10	20	31	47
n⁺ - GaAs, AlGaInP/ AlGaAs — 2-QW, H-Spr. Recessed, H–Sink ⁄ n⁺ – GaAs, Solder, H.-Spr. recessed, H.-Sink	H recessed, JSD	10	15	20	25
H-Spr. Aligned, AlGaInP/ AlGaAs — 2-QW, n⁺ - GaAs, H–Sink ⁄ H.-Spr. aligned, n⁺ – GaAs	H aligned, JSU	5	9	13	18
AlGaInP/ AlGaAs — 1-QW, n⁺ - GaAs, H–Sink	1 QW, AlGaInP / AlGaAs claddings: No H, JSU	10	16	22	32

Note: JSU / JSD = junction side up / junction side down mounting. H (H. Spr.) = heat spreader. H. Sink = heat sink

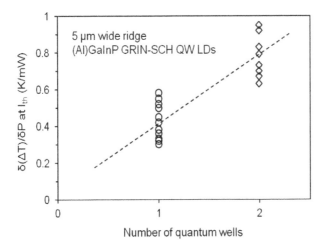

Figure 7.29 Mirror temperature gradient per lasing power unit at threshold current only due to absorption of laser radiation versus number of active quantum wells in (Al)GaInP GRIN-SCH QW ridge lasers 5 μm wide. Dashed line is least-squares fit.

in their effect. Figure 7.29 demonstrates the details of the exact proportionality between facet heating and number of QWs (one and two) (Epperlein and Bona, 1993b). In this case, facet heating is expressed as the temperature gradient per lasing power unit at threshold and the plotted data represent multiple measurements on two different samples at each QW number. This dependence is in full agreement with that in Figure 4.7 showing the averaged temperatures for devices with one, two, and five QWs.

From the viewpoint of obtaining the lowest possible facet heating, lasers with one QW, AlGaAs claddings, and a heat spreader aligned with the mirror edges provide considerable improvements over conventional GaInP laser structures with AlGaInP cladding layers. Thus, the ramped COMD power level could be improved by a factor of 6× (cf. Figure 4.7).

References

Adachi, S. (1983). *J. Appl. Phys.*, **54**, 1844.

As, D. J., Epperlein, P. W., and Mooney, P. M. (1987). *Inst. Phys. Conf. Ser.*, **31**, 561.

As, D. J., Epperlein, P. W., and Mooney, P. M. (1988). *J. Appl. Phys.*, **64**, 2408.

Aspnes, D. E., Kelso, S. M., Logan, R. A., and Bhat, R. (1986). *J. Appl. Phys.*, **60**, 754.

Balkanski, M. (ed.) (1971). *Light Scattering in Solids*, Flammarion, Paris.

Bastard, G. (1981). *Phys. Rev.*, **B24**, 4714.

Bayramov, B. H., Toporov, V. V., Ubaydullaev, S. B., Hildisch, L., and Jahne, E. (1981). *Solid State Commun.*, **37**, 963.

Bhattacharya, P. K., Subramanian, S., and Ludowise, M. J. (1984). *J. Appl. Phys.*, **55**, 3664.

Brugger, H. (1990). *Nato Advanced Research Workshop on Light Scattering in Semiconductor Structures and Superlattices*, Mont Tremblant, Canada.

Brugger, H. and Epperlein, P. W. (1990). *Appl. Phys. Lett.*, **56**, 1049.

Brugger, H., Epperlein, P. W., Beeck, S., and Abstreiter, G. (1990). *Inst. Phys. Conf. Ser.*, **106**, 771.

Cardona, M. and Güntherodt, G. (eds.) (1975–91). *Light Scattering in Solids I–VI, Topics in Applied Physics* **8**, **50**, **51**, **54**, **66**, **68**, Springer-Verlag, Berlin.

Chang, R. K. and Long, M. B. (1983). *Optical multichannel detection.* Chapter 3 in *Light Scattering in Solids II* (eds. Cardona, M. and Güntherodt, G.), *Topics in Applied Physics* **49**, 179, Springer-Verlag, Berlin.

Chin, A. K., Caruso, R., Young, M. S., and von Neida, A. R. (1984). *Appl. Phys. Lett.*, **45**, 552.

Compaan, A. and Trodahl, H. J. (1984). *Phys. Rev.*, **B29**, 793.

Debbar, N. and Bhattacharya, P. (1987). *J. Appl. Phys.*, **62**, 3845.

Elliott, R. J. (1957). *Phys. Rev.*, **108**, 1384.

Epperlein, P. W. (1990). *Photoluminescence topography studies in III-V semiconductors.* In *Defect Control in Semiconductors* (ed. Sumino, K.), *Proceedings of the International Conference on Science and Technology of Defect Control in Semiconductors, The Yokohama 21st Century Forum*, Elsevier Science (North-Holland), Amsterdam, **1**, 699.

Epperlein, P. W. and Meier, H. P. (1990). *Impurity trapping in nominally undoped GaAs/AlGaAs quantum wells.* In *Defect Control in Semiconductors* (ed. Sumino, K.), Elsevier Science (North-Holland), Amsterdam, **2**, 1223.

Epperlein, P. W. (1992). *IBM Tech. Disclosure Bull.*, **35**, 496.

Epperlein, P. W. (1993). *Jpn. J. Appl. Phys.*, **32**, Part 1, No. 12A, 5514.

Epperlein, P. W. (1996). *Proceedings of the Conference on Lasers and Electro-Optics, CLEO'96, OSA Tech. Dig.*, **9**, 108.

Epperlein, P. W. (1997). *Proc. SPIE*, **3001**, 13.

Epperlein, P. W. and Bona, G. L. (1993a). *Appl. Phys. Lett.*, **62**, 3074.

Epperlein, P. W. and Bona, G. L. (1993b). *Proceedings of the Conference on Lasers and Electro-Optics, CLEO'93, OSA Tech. Dig.*, **11**, 478.

Epperlein, P. W., Fried, A., and Jakubowicz, A. (1990). Chapter 8, in *Inst. Phys. Conf. Ser.*, **112**, 567.

Epperlein, P. W., Bona, G. L., and Roentgen, P. (1992). *Appl. Phys. Lett.*, **60**, 680.

Epperlein, P. W., Hunziker, G., Daetwyler, K., Deutsch, U., Dietrich, H. P., and Webb, D. J. (1995). Chapter 5 in *Inst. Phys. Conf. Ser.*, **141**, 483.

Ferenczi, G., Horvath, P., Toth, F., and Boda, J. (1984). *US Patent* No. 4437060.

Ferenczi, G., Boda, J., and Pavelka, T. (1986). *Phys. Status Solidi (a)*, **94**, K119.

Ghezzi, C., Gombia, E., and Mosca, R. (1991). *J. Appl. Phys.*, **70**, 215.

Guidotti, D., Hovel, H. J., Albert, M., and Becker, J. (1987). *Room temperature photoluminescence imaging of dislocations.* In *Review of Progress in Quantitative Nondestructive Evaluation* (eds. Thompson, D. and Chimenti, D.), Plenum Press, New York, **6B**, 1369.

Hamilton, B., Singer, K., and Peaker, A. (1985). *Inst. Phys. Conf. Ser.*, **79**, 241.

Hovel, H. J. and Guidotti, D. (1985). *IEEE Trans. Electron Devices*, **32**, 2331.

Jakubowicz, A. and Fried, A. (1993). *Proceedings of the 182nd Electrochemical Society Meeting*, **93-10**, 212.

Jordan, A. S., Caruso, R., and von Neida, A. R. (1980). *Bell Syst. Tech. J.*, **59**, 593.

Leo, K., Rühle, W., and Hägel, N. (1987). *J. Appl. Phys.*, **62**, 3055.

Lang, D. V. (1974). *J. Appl. Phys.*, **45**, 3023.

Luciano, M. J. and Kingston, D. L. (1978). *Rev. Sci. Instrum.*, **49**, 718.

Masselink, W. T., Klein, M. V., Sun, Y. L., Chang, Y. C., Fischer, R., Drummond, T. J., and Morkoc, H. (1984a). *Appl. Phys. Lett.*, **44**, 435.

Masselink, W. T., Chang, Y. C., and Morkoc, H. (1984b). *J. Vac. Sci. Technol.*, **B2**, 376.

Melngailis, I., Stillman, G. E., Dimmock, J. O., and Wolfe, C. M. (1969). *Phys. Rev. Lett.*, **23**, 1111.

Meynadier, M. H., Brum, J. A., Delalande, C., Voos, M., Alexandre, F., and Lievin, J. L. (1985). *J. Appl. Phys.*, **58**, 4307.

Miller, R. C., Tsang, W. T., and Munteanu, O. (1982). *Appl. Phys. Lett.*, **41**, 374.

Mooney, P. M. (1990). *J. Appl. Phys.*, **67**, R1.

Pankove, J. I. (1975). *Optical Processes in Semiconductors*, Dover, New York.

Phillips, J. C. (1981). *J. Vac. Sci. Technol.*, **19**, 545.

Raman, C. V. (1928). *Ind. J. Phys.*, **2**, 387; *Nature*, **121**, 619.

SEMILAB Ltd (2012). *Products*. http://www.semilab.hu.

Uji, T., Suzuki, T., and Kamejima, T. (1980). *Appl. Phys. Lett.*, **36**, 655.

Wakefield, B., Leigh, P. A., Lyons, M. H., and Elliott, C. R. (1984). *Appl. Phys. Lett.*, **45**, 66.

Yamanaka, K., Naritsuka, S., Mannoh, M., Yuasa, T., Nomura, Y., Mihara, M., and Ishii, M. (1984). *J. Vac. Sci. Technol.*, **B2**, 229.

Yamanaka, K., Naritsuka, S., Kanamoto, K., Mihara, M., and Ishii, M. (1987). *J. Appl. Phys.*, **61**, 5062.

Yokogawa, M., Nishine, S., Matsumoto, K., Akai, S.-I., and Okada, H. (1984). *Jpn. J. Appl. Phys.*, **23**, 663.

Yu, P. Y. and Cardona, M. (2001). *Fundamentals of Semiconductors*, 3rd edn, Springer-Verlag, Berlin.

Chapter 8

Novel diagnostic laser data for mirror facet disorder effects; mechanical stress effects; and facet coating instability

Semiconductor Laser Engineering, Reliability and Diagnostics: A Practical Approach to High Power and Single Mode Devices, First Edition. Peter W. Epperlein.
© 2013 John Wiley & Sons, Ltd. Published 2013 by John Wiley & Sons, Ltd.

Introduction

This chapter investigates three phenomena closely linked to the performance and reliability of diode lasers. The rst phenomenon is related to speci c, microscopic root causes of degradation processes in the susceptible mirror facets. It has been found by Raman microprobe spectroscopy that the strength of lattice disorder is positively correlated with the local facet temperature and negatively correlated with the power level at catastrophic optical mirror damage (COMD) of etched (Al)GaAs laser mirrors. By using different analysis techniques the atomic origin of the lattice disorder effect could be identi ed. The second phenomenon deals with the formation of mechanical stress in the active layer under the ridge waveguide of (Al)GaAs laser devices. Raman and photoluminescence (PL) measurements have yielded camel hump-like stress pro les with maximum compressive stress amplitudes of ~5 kbar at the ridge slopes and reduced compressive or tensile stress components toward the ridge center. The development of high stress gradients between the compressive and tensile stress elds can lead to well-pronounced stress-enhanced defect formation and migration effects, which can have a sensitive impact on laser performance. The third phenomenon concerns the stability of mirror coatings during strong laser radiation exposure. Strong and fast silicon recrystallization effects have been observed in coating layer stacks comprising amorphous ion-beam (IB) deposited silicon layers in contrast to plasma-enhanced chemical-vapor deposited (PECVD) silicon. The recrystallized silicon leads to a reduction in mirror re ectivity due to a reduced refractive index and to a movement of the optical mode and shrinkage of the lateral near- eld pattern.

8.1 Diode laser mirror facet studies by Raman

8.1.1 Motivation

The speci c, microscopic origins for degradation processes in the susceptible mirror facets are usually not well known. Furthermore, a detailed systematic and quantitative investigation of the relationships between facet temperature, microscopic root causes for degradation, and relevant laser performance parameters has to our knowledge not yet been published. In this context, we have probed for the rst time the lattice disorder in laser mirrors by Raman microprobe spectroscopy and correlated the strength of disorder to local facet temperature and power level at COMD.

8.1.2 Raman microprobe spectra

The Stokes Raman spectra in Figure 8.1 were detected in crossed-polarization ge-ometry from a series of six dry-etched, $\lambda/2$-Al_2O_3-coated, (110) facets of (Al)GaAs SQW ridge lasers 5 μm wide on the same laser bar (Epperlein *et al.*, 1993).

The spectra show the symmetry-allowed GaAs-like and AlAs-like TO phonon modes and additional, normally forbidden modes in AlGaAs. These can be ascribed

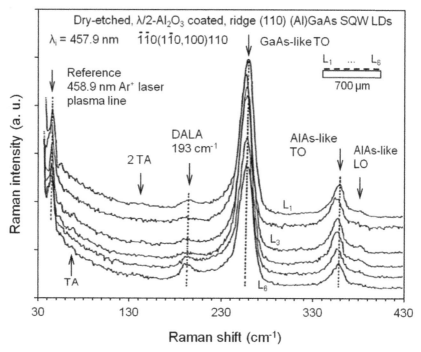

Figure 8.1 Typical Stokes Raman spectra detected in air at room temperature in a crossed-polarization geometry from a series of six dry-etched, $\lambda/2$-Al_2O_3-coated, (110) facets in the near- eld spots of (Al)GaAs SQW ridge lasers 5 μm wide on the same laser bar over a length of 700 μm.

to the rst-order AlAs-like longitudinal optical (LO(Γ)), short-wavelengh ($K > 0$) transverse acoustic (TA(L)) and second-order 2TA(L) phonon modes (Jusserand and Sapriel, 1981), which become Raman active as a consequence of disorder (softening of k-conservation law, see Section 7.4.4).

The spectra generally show the AlAs-like LO and not the GaAs-like LO mode in as-grown facets, which indicates the presence of disorder mainly in the AlAs sublattice. The origins of the disorder-activated mode at 193 cm^{-1} will be elaborated in detail in the next section. The Ar$^+$ laser 458.9 nm plasma line was used for Rayleigh scattering investigations and calibration. The spectra recorded from the six lasers shift up to 3 cm^{-1} across a distance of 700 μm. The stronger the 193 cm^{-1} mode, the stronger is the blueshift of the entire spectrum, which indicates the presence of compressive strain in the mirror region increasing the level of disorder even further by inhomogeneous atomic displacement. The observed peak shifts of up to 3 cm^{-1} correspond to strain amplitudes up to 5 kbar in the mirror surfaces, which may originate from the adhesion of the dielectric overlays varying from sample to sample (see Section 8.2; Epperlein *et al.*, 1990).

8.1.3 Possible origins of the 193 cm^{-1} mode in (Al)GaAs

The strong mode at 193 cm^{-1} can be attributed in principle to disorder-activated longitudinal acoustic (DALA) phonon scattering (Kawamura *et al.*, 1972). However, this intrinsic Raman mode may be superimposed by scattering from elemental arsenic As. The Raman spectrum from elemental As is well known and shows sharp TO modes called E_g and A_{1g} in the crystalline form at 195 cm^{-1} and 257 cm^{-1}, respectively (Farrow *et al.*, 1977; Renucci *et al.*, 1973).

This As–E_g TO phonon mode position, in conjunction with the presence of the 193 cm^{-1} mode in the crossed-polarization geometry between incident and scattered light, suggests that elemental As may be present in the laser facet. Since the 193 cm^{-1} mode is, however, also present in the parallel polarization geometry, which is preferred by the DALA mode, we have also recorded Raman spectra with increased resolution of 1.5 cm^{-1} from a polycrystalline arsenic sample for a de nitive identi cation of the 193 cm^{-1} mode. Figure 8.2 shows Raman spectra from a dry-etched as-grown (110) AlGaAs facet and a polycrystalline As sample under the same conditions for comparison.

Line shapes and line frequencies, but above all the appearance of a shoulder on the low-energy side of the GaAs-like TO peak, which can be ascribed to As–A_{1g}

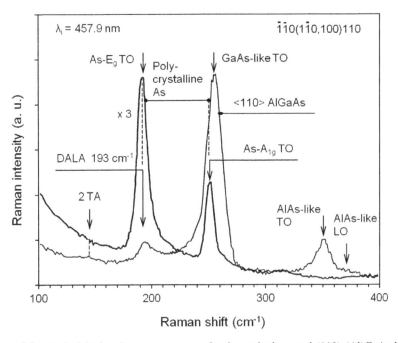

Figure 8.2 Typical Stokes Raman spectrum of a dry-etched, coated (110) (Al)GaAs laser mirror (thin line) compared to Raman spectrum of a polycrystalline arsenic sample (thick line), both detected in backscattering crossed-polarization geometry at ambient conditions. (Parts adapted in modi ed form from Epperlein *et al.*, 1993.)

TO phonon scattering, strongly indicate the presence of excess As in the AlGaAs facet. Furthermore, this nding is supported by energy dispersive x-ray (EDX) and Rayleigh scattering measurements as a function of the 193 cm^{-1} mode/GaAs-like TO mode intensity ratio (Figure 8.3).

Figure 8.3a shows the increase of the As concentration normalized to the Ga concentration with increasing normalized 193 cm^{-1} mode intensity. The existence of elemental As in the dry-etched laser mirrors is most likely due to oxidation during the removal of the etch mask in an oxygen plasma ashing process or due to a disturbed stoichiometry leaving excess As behind. The normalized 193 cm^{-1} Raman mode intensity uctuates on a micrometer scale indicating the presence of As clusters and hence a locally varying enhanced strength of compositional disorder.

Since Rayleigh scattering is sensitive to local mass density uctuations due to alloy disorder and clustering effects, the positive correlation between normalized Rayleigh scattering intensity and relative 193 cm^{-1} mode intensity in Figure 8.3b is further direct proof of the existence of disorder in the laser mirrors studied.

Finally, as mentioned above, the stronger the relative 193 cm^{-1} mode intensity, the stronger the blueshift observed for the entire spectrum. This indicates the presence of compressive strain in the mirror region increasing the level of disorder even further by inhomogeneous atomic displacement (Figure 8.3c and Figure 8.1).

In summary, according to the different measurements, we can now ascribe the 193 cm^{-1} mode to Raman scattering at the E_g–TO phonon of elemental As and DALA phonon in AlGaAs. The relative intensity of the 193 cm^{-1} mode can be used as a useful measure for the degree of compositional and structural disorder in laser mirror facets.

8.1.4   Facet disorder – facet temperature – catastrophic optical mirror damage robustness correlations

For the rst time quantitative correlations between lattice disorder, local mirror temperature rise ΔT, and COMD power level have been established (Epperlein, 1993; Epperlein *et al.*, 1993) (Figure 8.4).

The cw COMD power level decreases, whereas ΔT increases with increasing strength of disorder. The third correlation is the decrease of COMD with increasing ΔT, as expected. This anti-correlation between COMD and ΔT clearly demonstrates for the rst time in a quantitative manner the signi cance of lattice disorder in optical power absorption and thus in nonradiative carrier recombination, which leads to facet surface heating and thus shortens laser lifetimes. In this respect, the contribution of elemental As to heating seems to be signi cant, because of its large complex refractive index (Renucci *et al.*, 1973), particularly a large extinction coef cient which causes a large absorption ($\alpha \cong 4 \times 10^5$ cm^{-1}) of the 830 nm laser radiation. The thickness of the As clusters could be estimated to be in the region of 3 nm. In principle, the relationships in Figure 8.4 can be used to predict laser lifetimes from the strength of disorder measured at virgin mirror facets.

Finally, Figure 8.5 shows the Stokes Raman spectrum after the occurrence of COMD. Compared to the as-grown facet, the relative strength of the integrated

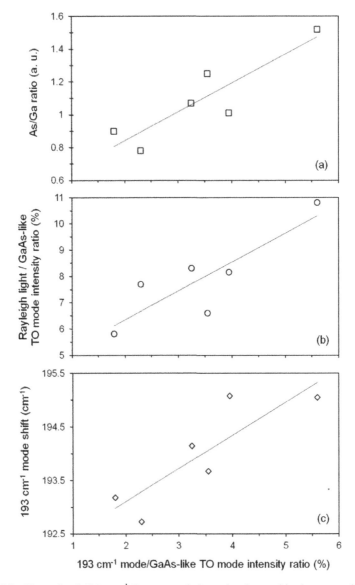

Figure 8.3 Normalized 193 cm^{-1} Raman mode intensity detected in the near- eld pattern of dry-etched (110) (Al)GaAs laser mirrors dependent upon (a) the As/Ga ratio from EDX spectroscopy measurements, (b) the normalized elastic Rayleigh light scattering intensity, and (c) the 193 cm^{-1} mode shift. Solid lines are least-squares ts. (Adapted in extended form from Epperlein *et al.*, 1993.)

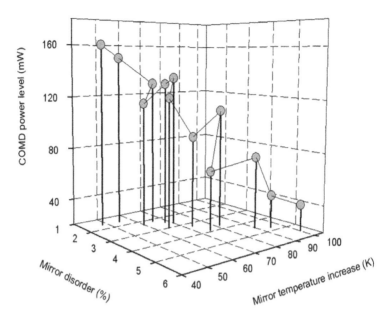

Figure 8.4 COMD cw power levels, mirror temperature rises at 30 mW, and mirror disorder strengths measured as the normalized 193 cm^{-1} Raman mode intensities on dry-etched (Al)GaAs laser mirrors. Significant correlations constituted for the very first time: COMD power level versus lattice disorder strength, mirror temperature rise versus lattice disorder strength, COMD power level versus mirror temperature rise.

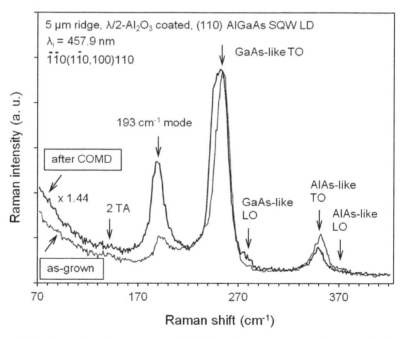

Figure 8.5 Comparison between a typical Stokes Raman spectrum of a dry-etched (110) (Al)GaAs laser mirror before (thin line) and after COMD event (thick dark line).

193 cm^{-1} peak intensity increased by a factor of at least ve. The strong lattice damage is also supported by the additional appearance of the normally forbidden GaAs-like LO phonon mode. In this sample, the 193 cm^{-1} mode is now predominantly due to DALA phonon scattering. Scattering at As–As vibrations can be excluded, because according to microprobe Auger spectroscopy measurements, arsenic As is depleted in the near- eld spot where the COMD event led to localized melting and consequently evaporation of As from the facet region.

8.2 Local mechanical stress in ridge waveguide diode lasers

8.2.1 Motivation

The distribution of mechanical stress in semiconductor laser waveguide structures is a subject of great perennial interest, because stress can have positive and negative effects on laser performance (Thijs, 1992). Thus, QW lasers with elastically built-in stress have performance and reliability characteristics that are substantially superior to those of unstrained devices (see Section 1.1.4.1). On the other hand, stress may be the driving force for a variety of degradation phenomena (Ueda, 1988) and may modify the laser waveguide modes via the pressure dependence of the refractive index. For example, the pressure coef cient of the index is $\partial n_r/\partial p = -4 \times 10^{-6}$ bar^{-1} for GaAs (Samara, 1983). Thus, depending on the type of stress, compressive or tensile, the index may be increased or decreased, respectively. We want to illustrate this effect by means of a ridge (Al)GaAs laser employing Si$_3$N$_4$ for ridge embedding and Ti/Pt/Au for the p-metallization layer. In the case of Si$_3$N$_4$ on GaAs, the nitride is under compressive stress at the interface while the GaAs is under tensile stress. In the case of Ti/Pt/Au on GaAs, the metallization is under tensile stress and GaAs under compressive stress (see Section 3.1.1.4). Stress type and level depend sensitively on the differences in thermal expansion coef cients (thermal stress) and on the microstructure of the deposited layer (intrinsic stress). Moreover, stress varies also sensitively on the applied deposition conditions, such as angle of deposition, deposition pressure, and substrate bias in the case of a sputtering deposition process. In general, Pt dominates the stress level in Ti/Pt/Au, which doubles when the evaporation angle is increased from 45° to 90°.

In the following, we describe microprobe stress measurements made for the rst time on cleaved uncoated (110) facets of ridge (Al)GaAs SQW diode lasers (Epperlein et al., 1990, 1995; Epperlein, 1997). The devices employed Si$_3$N$_4$ for embedding the ridge structure and Ti/Pt/Au for the p-electrode. We will discuss the stress distribution along the active layer under the ridge, the origins of the stress, and the potential in uence of the measured stress elds on the formation and migration of defects. Finally, stress model experiments employing a three-point laser bar bending technique will be described along with the effects of external stress on near- eld pattern and P/I characteristic of narrow ridge (Al)GaAs lasers.

8.2.2   Measurements – Raman shifts and stress profiles

Figure 8.6 shows a series of rst-order Stokes Raman GaAs-like TO phonon modes detected with a high spatial resolution of better than 1 μm on the cleaved (110) facets of ridge (Al)GaAs GRIN-SCH SQW laser devices 5 μm wide at different locations along the active layer. The laser power density for exciting the Raman spectra was very low (<100 kW/cm^2) causing a negligible self-heating effect with no measurable impact on the Raman spectra. The top spectrum was recorded far away from the strained region around the ridge and therefore its peak position of 261 cm^{-1} can be taken as the reference point for the spectral shifts observed when approaching the ridge structure. Stress in the crystal affects the frequency of the phonons and hence the position of the Raman peak. Typically, it is possible to observe and detect stress by analyzing the shift in the Raman mode position, but it can also affect the mode shape and induce broadening and deformation of the Raman peak.

Mode shifts of three different laser devices are shown in detail in Figure 8.7. Typical maximum positive shifts of $+2$ cm^{-1} can be found close to both ridge slopes. Inside the ridge the Raman mode exhibits lower positive shifts or even negative shifts with respect to the reference point as shown by one device in Figure 8.7. The mode shift pro les of the three different samples are similar in the overall camel hump-like shape, but are differently well-pronounced in shape and strength at the ridge slopes, inside and outside the ridge regions. These mode shift pro les indicate the distribution of mechanical stress in the ridge waveguide structure, which may vary qualitatively and quantitatively from sample to sample.

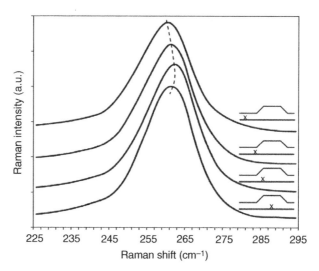

Figure 8.6   Typical rst-order Stokes GaAs-like TO phonon Raman mode spectra at different locations (marked by x) along the active layer of uncoated (110) ridge (Al)GaAs SQW laser mirror facets. (Parts adapted in modi ed form from Epperlein *et al.*, 1995.)

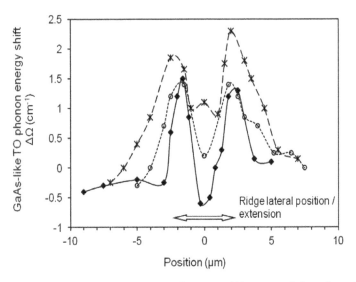

Figure 8.7 GaAs-like TO phonon Raman mode energy shift measured along the active layer of cleaved, uncoated (110) ridge (Al)GaAs SQW laser facets 5 μm wide of three different devices (data marked by different symbols). Measurement points are connected as a guide to the eye.

Using the relationship between compression and pressure (Murnaghan, 1944)

$$P = \frac{B_0}{B_0'} \left\{ \left(\frac{a_0}{a} \right)^{3B_0'} - 1 \right\} \qquad (8.1)$$

and the experimental (hydrostatic) pressure dependence of the GaAs TO phonon mode (Trommer *et al.*, 1980)

$$\Delta\omega_{TO} = 1.12 \times 10^3 \left(\frac{-\Delta a}{a_0} \right) - 2.28 \times 10^3 \left(\frac{-\Delta a}{a_0} \right)^2 \qquad (8.2a)$$

$$\frac{-\Delta a}{a_0} = 1 - \frac{a}{a_0} \qquad (8.2b)$$

where $B_0 = 7.25 \times 10^{11}$ dyn/cm^2 is the (GaAs) bulk modulus, $B_0' = (\partial B_0/\partial P)_T$ the bulk modulus pressure derivative, $B_0' = 4.67$ for GaAs, a_0 the lattice constant, and a the lattice constant under compression, the following conversion between mode shift in wavenumber cm^{-1} units and pressure in kbar units could be derived:

$$1 \text{ cm}^{-1} \text{ GaAs TO phonon mode shift} \stackrel{\wedge}{=} 2.2 \text{ kbar.} \qquad (8.3)$$

Accordingly, the mode shift pro les in Figure 8.7 can be converted to stress pro les. There are high compressive stress amplitudes of up to 5 kbar measured close to the ridge slopes and reduced compressive stress or even low tensile stress elds toward the ridge center. These stress gures are supposed to be much higher when using a probe with a spatial resolution higher than $\cong 1$ μm in the experiments. Crossover regions from compression to tension are located inside and outside the slope regions. These regions will play an important part in defect generation and migration as we will discuss in the next section. PL measurements led to the same camel hump-like stress pro les, but proved to be superior to the Raman measurements because of their higher sensitivity as will be discussed in Section 8.2.4 below.

The experimental stress pro les are supported by calculating the stress distribution in planes parallel to the mirror facet of a ridge-type laser. The calculation involves the solution of the stress function equation on a semi-in nite, continuous, and isotropic medium, and assumes tangential forces originating from the difference in the thermal expansion coef cients between the bulk material and its overlay. The phonon energy shift is then related to the stress via the local variations in the lattice parameter (Epperlein *et al.*, 1990; Fried *et al.*, 1991).

Measurements on special samples with (i) just the ridge structure, (ii) the ridge structure with the overlaid insulating Si_3N_4 ridge-embedding layer, and (iii) the ridge structure with the top Ti/Pt/Au p-metallization layer, revealed that both the nitride and the metallization layer can introduce stress.

8.2.3   Detection of "weak spots"

The susceptibility of laser structures to the formation of electrically active defects can be investigated by exposing the laser mirror to electron irradiation in a scanning electron microscope (SEM) and monitoring the formation of defects and strain-enhanced migration and separation of defects by electron beam induced current (EBIC).

8.2.3.1   Electron irradiation and electron beam induced current (EBIC) images of diode lasers

Figure 8.8 shows for comparison the EBIC images of two identical ridge (Al)GaAs GRIN-SCH SQW lasers taken from the (110) cleaved facets: the top trace was recorded on a new, nondegraded device, the lower one on a device exposed to electron irradiation by multiple electron beam scans along the active region under the conditions of 15 kV and 8×10^{-10} A (Epperlein *et al.*, 1990).

What is remarkable are the two symmetric features close to the lower corners of the ridge in Figure 8.8b showing a reduced EBIC signal which developed during irradiation. This can clearly be seen in the EBIC line scan along the active layer (white stripe). The line scans in Figure 8.8b were taken at two different electron irradiation exposure times, 15 h (curve A) and 39 h (curve B), under the conditions of 29 kV, 5×10^{-11} A, and 0.8 μm/s. From these curves it can be concluded that the regions close to the lower ridge corners demonstrate an enhanced susceptibility to

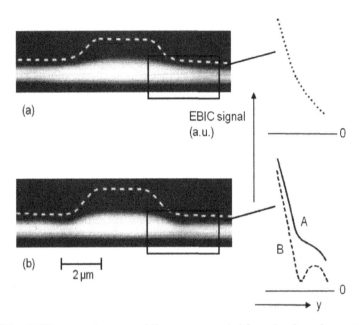

Figure 8.8 EBIC contrast images and line scans recorded from the cleaved uncoated (110) facets of two identical ridge (Al)GaAs SQW lasers. (a) As-grown, nondegraded laser measured before electron irradiation and (b) contrast image obtained after SEM electron irradiation by multiple scans along active layer with 15 kV, 8×10^{-10} A strength; line scans after 15 h (curve A) and 39 h (curve B) for 29 kV, 5×10^{-11} A, 0.8 μm/s electron irradiation conditions. Dashed lines indicate ridge structure. (Parts adapted in modi ed form from Epperlein *et al.*, 1990.)

degradation processes. There is no change of the EBIC signal in the center region of the ridge after 15 h exposure time. However, a clear change can already be seen close to the ridge corners where the signal at the shoulder on curve A is higher than the signal measured at the same position before irradiation (Figure 8.8a). We can therefore conclude that in this region the defect state has been modi ed by the electron irradiation.

Strong stress elds formed by the close proximity of zones under compressive and tensile stress are the driving forces for point defect migration (Robertson *et al.*, 1981). Accordingly, interstitial-type defects might be moved to positions of maximum dilatation, whereas vacancy-type defects might be expected to move to positions with a compressed crystal lattice. The effect of such a migration, resulting in a separation of both defect types, is actually manifested in the local maximum and adjacent minimum of the EBIC line traces in Figure 8.8b. These locations are the weakest points with respect to a high susceptibility to degradation processes.

A further example of the stress-induced defect formation effect is demonstrated in Figure 8.9. Here, defect structures have developed around the lower corners of the ridge of an (Al)GaAs laser after 100 h of operation at 10 mW optical output power (Fried *et al.*, 1991). These weak spots with enhanced susceptibility to degradation

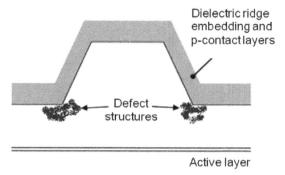

Figure 8.9   Schematic, qualitative illustration of a transmission electron microscope image of an uncoated ridge (Al)GaAs laser mirror 3 μm wide after 100 h operation at 10 mW, and of defect structures (dark spotted areas), which developed at the lower ridge corners due to high stress gradients in the ridge structure. (Adapted in schematic form from Fried et al., 1991.)

correlate with the location of high stress gradients. As already discussed brie y in Section 7.4.6.1, it is worth mentioning again that EBIC studies on compressively-strained InGaAs/AlGaAs SQW lasers (Jakubowicz and Fried, 1993) showed that, in terms of defect formation, the strained active QW layer can effectively screen the ridge-induced stress elds. This screening effect can reduce the generation and migration of defects at the mirrors during laser operation resulting in lower facet heating and hence higher mirror stability and laser performance.

8.2.3.2   EBIC – basic concept

In brief, an electron beam in an SEM is raster scanned over the surface of a p–n junction or Schottky barrier diode comprising a depletion region. High-energy electrons penetrate the specimen and generate electron–hole pairs within the pear-shaped volume of excitation (Figure 8.10).

Before recombination occurs, the carriers diffuse within the diffusion lengths in the semiconductor. If they penetrate the depletion zone they are separated into electrons and holes by the electric eld of the junction. These carriers then drift to the respective contacts where they are collected, which results nally in a small current in an external circuit. The charge collection current is inversely proportional to the formation energy of an electron–hole pair and directly proportional to the beam current, incident beam voltage, and charge collection ef ciency. This current varies in the scanning process depending on the type of electrically active defects (recombination or generation defect center) with a charge collection ef ciency different to that in the surrounding matrix. These variations are used to modulate the intensity of the SEM viewing screen and thus yield information on the distribution of electrically active defects.

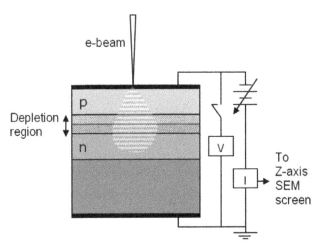

Figure 8.10 Schematics of the experimental setup for realizing the electron beam induced current (EBIC) technique using a p–n junction for generating the required depletion layer.

8.2.4 Stress model experiments

8.2.4.1 Laser bar bending technique and results

To investigate the effect of stress on laser operation in a controlled way, model experiments can be performed, in which stress is applied externally, for example, by bending a laser bar (parallel array of individual diode lasers) in a mechanical device (Figure 8.11). The three-point bar bending device shown in the gure allows an effective length of 6 mm, width of 750 μm, and thickness of 150 μm of the bar.

Figure 8.12 displays the achievable stress magnitudes in ridge (Al)GaAs GRIN-SCH SQW lasers. Here, the relative QW–PL energy shift ΔE is plotted as a function of bending amplitude s with the probe laser spot positioned at the cross-section of the bar in the active layer below the ridge located at maximum s in the middle of the bar. An upward bending with the ridge at the top (upper) side of the bar generates a compressive stress in in the epitaxial structure and tensile stress in in the substrate near the interface between epitaxial layer and substrate.

The energy shift was converted to stress amplitudes by using 10 meV/kbar for the pressure coef cient of the QW–PL energy (Gell *et al.*, 1987). Maximum bending am-plitudes of ∼40 μm (fracture limit), corresponding to ∼0.6 kbar, have been achieved. As demonstrated in Figure 8.12, the bending technique enables the measurement of built-in stress amplitudes (at $s = 0$) as low as only 50 bar, hardly achievable by any other nondestructive and contactless technique.

Stress measurements by using Raman or PL each have their own merits. In the case of Raman, the pros are a de nite local signal origin in contrast to the cons, which include spectral peak overlap leading to loss of de nite local signal origin, small signal levels, small pressure sensitivity (∼500 bar), and small peak shifts (∼1 cm^{-1} $\triangleq$ 0.2 A for 2 kbar). In the case of PL, the pros are spectral peaks with

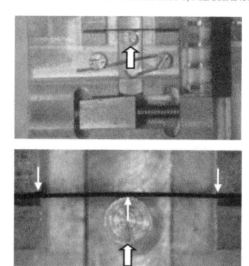

Figure 8.11   Mechanical device for realizing a three-point laser bar bending technique. Upper image shows the top overall view including the cross-section of the bar at the top side of the image. Lower image shows details with the bar pressed from below (arrow upward) to the points of support on both sides (arrows downward) separated by a distance of 6 mm. The circular stub is moved upward against the bar lower side by a simple screw-driven movement of a wedge. Bar can bend freely upward.

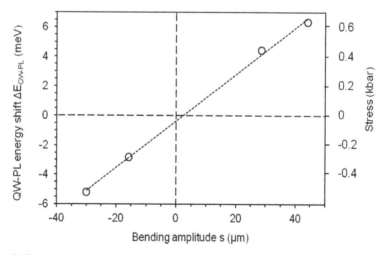

Figure 8.12   Dependence of the photoluminescence (PL) (Al)GaAs QW heavy-hole exciton energy shift ΔE_{QW-PL} on the laser bar bending amplitude s for compressive and tensile stress applied to the ridge lasers. Probe laser spot was positioned on the (110) facet at the center of ridge on the active layer. Dotted line is least-squares t. (Parts adapted in modi ed form from Epperlein *et al.*, 1995.)

high signal levels, whereas the cons are broad peaks at 300 K, relatively small peak shifts ($\sim$10 meV/kbar for GaAs QW at 300 K), and limited lateral resolution (carrier diffusion).

Finally, the bending experiments showed that the near- eld patterns and P/I characteristics are highly sensitive to the external stress (Epperlein *et al.*, 1995). In particular, shrinkage of the lateral mode by an external tensile stress is accompanied by modi cation of the differential ef ciency at a certain power level. This can be understood qualitatively by the fact that the lateral effective index pro le is modi ed asymmetrically in the center regions of the ridge as a result of bending the bar. Therefore, the position and intensity distribution in the two lobes of the rst-order mode, which is present at zero external stress, are laterally shifted toward each other. This leads to an effective reduction in the width of the total intensity distribution, whereby the internal losses of the waveguide are modi ed and the differential quantum ef ciency is changed. Furthermore, under external stress these laser devices showed a reversible increase in the threshold current of up to 15%, which might be ascribed to a shift of the light-hole and heavy-hole valence bands in the bending experiment (Epperlein *et al.*, 1995).

8.3 Diode laser mirror facet coating structural instability

8.3.1 Motivation

The application of diode lasers requires, in general, devices with excellent electrical and optical properties and reliability that do not change with time, in particular in sensitive areas such as ber optic communications. Front and back mirrors are known to be very sensitive elements in laser cavities. In this respect, it is useful to investigate the structural stability of the mirror coatings, in particular that of thin amorphous silicon layers involved in the high-re ectivity coating stacks of back mirrors under high optical power exposure. The effect of any coating instability on mode behavior and laser characteristics is of utmost importance and hence justi es to be investigated thoroughly.

8.3.2 Experimental details

The devices used were ridge compressively-strained InGaAs/AlGaAs GRIN-SCH SQW lasers 4 μm wide with cleaved (110)-oriented facets. Back facets were coated with high-re ective (95%) PECVD Si/Si$_3$N$_4$ or IB-deposited Si/Al$_2$O$_3$ ve-layer stacks with layers $\lambda/4$-thick in the double-pair stacks (see Section 1.3.8). The optical thickness (optical path length) of the thin Si layers is $\lambda/4n_r = 64$ nm for $\lambda = 980$ nm and $n_r = 3.8$. Single PECVD Si$_3$N$_4$ or IB Al$_2$O$_3$ layers, respectively, were deposited on the front facets to obtain a re ectivity of $\sim$10%.

The laser chips were soldered junction-side up onto heat sinks. Raman microprobe spectra were measured in the near- eld spots on the facets at air and room temperature in the backscattering geometry by focusing the 457.9 nm line of an Ar^+ laser with a power ≤ 1 mW onto a spot with diameter ~ 1 μm. The detection and further processing of the Raman signal has been described in detail in Section 7.4.

8.3.3   Silicon recrystallization by internal power exposure

8.3.3.1   Dependence on silicon deposition technique

The Raman spectra of as-deposited PECVD and IB Si layers show a broad asymmetric peak close to 475 cm^{-1} with a typical linewidth of 70 cm^{-1} (Figure 8.13) (Epperlein and Gasser, 1995). The spectra strongly resemble those usually observed for amorphous silicon (a-Si) (Tsang *et al.*, 1985) and thus verify that the as-deposited

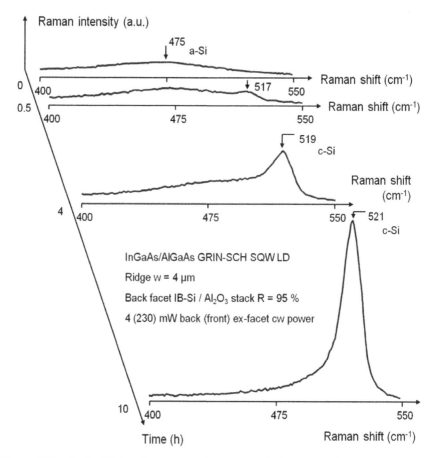

Figure 8.13   Typical Stokes Raman scattering spectra of silicon layers $\lambda/4$-thick in ion beam (IB) deposited Si/Al_2O_3 back-facet coating stacks of 4 μm ridge InGaAs/AlGaAs SQW lasers at 4 (230) mW back- (front-)facet power emission and 980 nm lasing wavelength for different operation times. (Parts adapted in modiﬁed form from Epperlein and Gasser, 1995.)

Si layers were amorphous. The broad peak arises from the relaxation of the normal Raman selection rules for scattering from a crystal as a result of the loss of translational symmetry in the amorphous phase (see Section 7.4.4).

However, the IB a-Si and PECVD a-Si layers behaved quite differently under strong laser emission. Figure 8.13 shows for the first time the change in the Stokes Raman IB a-Si spectrum with increasing exposure time to strong internal laser power (Epperlein and Gasser, 1995). In particular, the hydrogen-free IB a-Si layers show an additional sharp line with a position in the range of 517 to 521 cm^{-1} and a size dependent on the exposure time. This line can be attributed to scattering from the threefold degenerate, $K = 0$, optical phonon of crystalline Si (c-Si) (Tsang et al., 1985). The line clearly exhibits an increase in intensity relative to that of the a-Si mode with time. After $\sim$10 h at 200 mW, the c-Si peak intensity saturates indicating a nearly complete recrystallization of a-Si to c-Si.

By deconvoluting the spectra and using integrated Raman backscattering cross-sections in the two phases, the volume fraction of crystallinity ρ in the two-phase system of Si microcrystallites embedded in the a-Si matrix can be calculated.

Figure 8.14 shows the dependence of ρ on operating time at 4 (230) mW back-(front-)facet cw power emission. It is strongest within the near-field spot and decays rapidly to zero toward the substrate at a distance of $\sim$5 μm. The figure also gives the maximum energy absorbed in the Si layers with increasing recrystallization time. Moreover, the c-Si Raman peak shifts to higher energies with increasing time (crystallinity) (Figure 8.13), which can be interpreted as a particle size effect (Tsu,

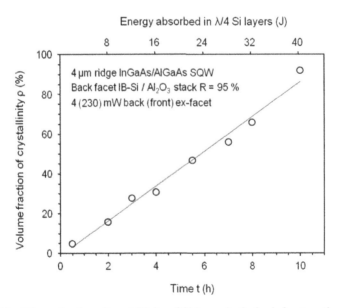

Figure 8.14 Volume fraction of crystallinity of Si layers in the back-facet coating stacks of InGaAs/AlGaAs ridge lasers recrystallized under 4 (230) mW back- (front-)facet cw power emission and at 980 nm lasing wavelength at different operation times. The maximum energy absorbed in the Si layers as a function of time (crystallinity) is indicated in the upper axis.

1981). The Si crystallite size was estimated to be ~5 nm for $\rho \cong 5\%$ and >10 nm for $\rho \cong 100\%$.

In contrast, PECVD a-Si remains amorphous, independent of exposure time and cw optical power emission. This structural stability can be ascribed to the large amount of ~10 at.% of hydrogen in the a-Si lms (Mei *et al.*, 1994) prepared in a glow discharge decomposition of silane, SiH_4. Whereas hydrogen is known to be essential for good electrical performance of a-Si based devices by saturating randomly distributed dangling bonds, it protects the PECVD a-Si from recrystallization.

8.3.3.2 Temperature rises in ion beam- and plasma enhanced chemical vapor-deposited amorphous silicon coatings

Another distinctive feature of IB a-Si and PECVD a-Si is the temperature in the respective silicon layer. Figure 8.15 shows plots of the temperature increases ΔT versus operating current I and optical power P for both IB-Si and PECVD-Si. The data were obtained from the Stokes/anti-Stokes intensity ratios of the a-Si Raman mode and give directly the temperatures in the silicon layers. The ΔT data are with respect to 300 K and allow for the low heating effect (<10 K) of the probe laser intensity. The power scale is valid for both lasers with IB-Si and PECVD-Si coating layers, which have practically the same slope ef ciency. The temperatures in IB-Si are at least higher by a factor of two compared to those in PECVD-Si containing stacks. This is mainly due to the higher absorption coef cient α of IB a-Si, which is

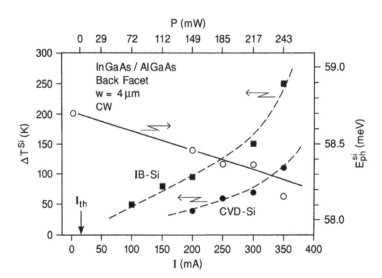

Figure 8.15 Temperature rises ΔT measured in the ion beam (IB) and plasma-enhanced chemical vapor-deposited (PECVD) amorphous silicon (a-Si) layers in back-facet coating stacks of InGaAs/AlGaAs ridge lasers as a function of current I and optical power P by using the Stokes/anti-Stokes Si Raman mode intensity ratio. The temperature (drive current) dependence of the Si phonon energy $E_{ph}{}^{Si}$ is also shown. (Parts adapted in modi ed form from Epperlein and Gasser, 1995.)

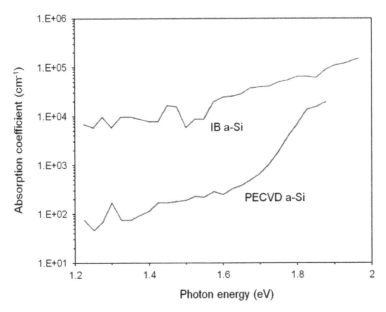

Figure 8.16 Absorption coef cient measured as function of photon energy in ion beam (IB) and plasma-enhanced chemical vapor-deposited (PECVD) amorphous silicon (a-Si) layers by using an optical spectroscopic ellipsometry technique.

$\alpha \cong 8 \times 10^3$ cm^{-1} compared to only $\alpha \cong 60$ cm^{-1} for PECVD a-Si at 980 nm lasing wavelength (1.26 eV).

Furthermore, Figure 8.15 shows also the a-Si phonon energy decreasing with increasing temperature due to the increasing laser current and optical power emission. This weak temperature dependence of the phonon energies is in agreement with the temperature dependence of Si Raman lines (Tsu and Hernandez, 1982) and inelastic neutron scattering measurements (Delaire *et al.*, 2011) and can be ascribed to phonon softening due to thermal expansion and thermal disorder effects. Generally, it is expected that phonon frequencies decrease (soften) with increasing interatomic separations from thermal expansion (Grimvall, 1999; Fultz, 2010).

The absorption data in Figure 8.16 have been achieved by ellipsometric measurements (Webb, private communication). The highest temperatures measured in the IB–Si layers are signi cantly lower than those given in the literature for recrystallizing a-Si exclusively with thermal annealing treatments (Hartstein *et al.*, 1980), and thus suggest a photothermal activation of the crystallization process in the IB a-Si layers.

8.3.4 Silicon recrystallization by external power exposure – control experiments

To quantify the effect of silicon recrystallization on laser characteristics in a controlled way, we produced the crystallization in well-de ned steps by irradiating the back facet in the near- eld area with a focused Ar$^+$ laser beam (Epperlein and Gasser, 1995; Epperlein, 1997). Thus, a volume crystallinity of $\rho \cong 100\%$ could be produced in IB a-Si layers in $\sim$2 h with 30 mW of the 457.9 laser line focused in a spot size of $\sim$1–2 µm diameter. The lasers were characterized by taking near- eld images and *P/I* characteristics at both the back-facet and front-facet for comparison before and after the recrystallization process.

8.3.4.1 Effect on optical mode and *P/I* characteristics

A value of $\rho \cong 100\%$ increases the transmitted power at the back facet by a factor of 1.7. This enhancement of power emission can be accounted for by a decrease in the re ectivity of the as-deposited coating stack from 95% to 91% due to a transition of the refractive index from 3.8 of a-Si to 3.45 of c-Si. An increase in the transmitted power, caused by an absorption coef cient decreasing from 8×10^3 (a-Si) to 5×10^2 cm^{-1} (c-Si), plays only a minor role with a contribution of $\sim$10%.

Figure 8.17 shows the corresponding *P/I* characteristics measured at the front facet. Of particular note is the clear movement of the kink (nonlinearity) to higher

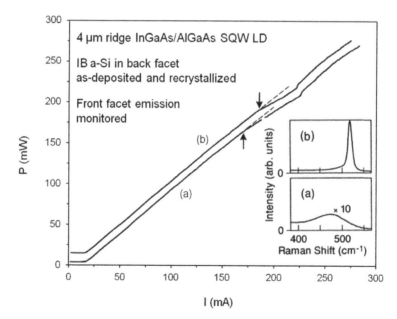

Figure 8.17   Light (*P*)/current (*I*) characteristics taken from the as-grown front facet of an InGaAs/AlGaAs ridge laser with ion-beam (IB) a-Si layers in the high-re ectivity (*R* = 95%) back-facet coating stack, as-deposited (a) and completely recrystallized (b). Inset shows Raman spectra for as-grown amorphous Si (a-Si) (a) and crystalline Si (c-Si) (b). *P/I* plot (b) is displaced vertically for clarity. Arrows indicate the onset of a kink (movement of mode) in the *P/I* characteristics. (Parts adapted in modi ed form from Epperlein and Gasser, 1995.)

power levels observed in the *P/I* and a small shrinkage ($\sim$10%) of the lateral near-eld pattern extension upon Si recrystallization in the back-facet coating. This effect is known to be linked to a movement of the laser mode and usually causes a big problem in achieving a stable coupling of single-mode laser power into a single-mode optical ber.

References

Delaire, O., Marty, K., Stone, M., Kent, P., Lucas, M., Abernathy, D., Mandrus, D., and Sales, B. (2011). *Proc. Natl. Acad. Sci. USA*, **108**, 4725.

Epperlein, P. W. (1993). *Proceedings of the Conference on Lasers and Electro-Optics, CLEO '93, OSA Tech. Dig.*, **11**, 158.

Epperlein, P. W. (1997). *Proc. SPIE*, **3001**, 13.

Epperlein, P. W. and Gasser, M. (1995). Chapter 5 in *Inst. Phys. Conf. Ser.*, **141**, 537.

Epperlein, P. W., Fried, A., and Jakubowicz, A. (1990). Chapter 8 in *Inst. Phys. Conf. Ser.*, **112**, 567.

Epperlein, P. W., Buchmann, P., and Jakubowicz, A. (1993). *Appl. Phys. Lett.*, **62**, 455.

Epperlein, P. W., Hunziker, G., Daetwyler, K., Deutsch, U., Dietrich, H. P., and Webb, D. J. (1995). Chapter 5 in *Inst. Phys. Conf. Ser.*, **141**, 483.

Farrow, R. L., Chang, R. K., Mroczkowski, S., and Pollak, F. H. (1977). *Appl. Phys. Lett.*, **31**, 768.

Fried, A., Jakubowicz, A., Newcomb, S. B., and Stobbs, W. M. (1991). *Inst. Phys. Conf. Ser.*, **117**, 585.

Fultz, B. (2010). *Prog. Mater. Sci.*, **55**, 247.

Gell, M. A., Ninno, D., Jaros, M., Wolford, D. J., Kuech, T. F., and Bradley, J. A. (1987). *Phys. Rev.*, **B35**, 1196.

Grimvall, G. (1999). *Thermophysical Properties of Materials*, North-Holland, Amsterdam.

Hartstein, A., Tsang, J. C., DiMaria, D. J., and Dong, D. W. (1980). *Appl. Phys. Lett.*, **36**, 836.

Jakubowicz, A. and Fried, A. (1993). *Proceedings of the 182nd Electrochemical Society Meeting*, **93-10**, 212.

Jusserand, B. and Sapriel, J. (1981). *Phys. Rev.*, **B24**, 7194.

Kawamura, H., Tsu, R., and Esaki, L. (1972). *Phys. Rev. Lett.*, **29**, 1397.

Mei, P., Boyce, J., Hack, M., Lujan, R., Johnson, R., Anderson, G., Fork, D., and Ready, S. (1994). *Appl. Phys. Lett.*, **64**, 1132.

Murnaghan, F. D. (1944). *Proc. Natl Acad. Sci. USA*, **30**, 244.

Renucci, J. B., Richter, W., Cardona, M., and Schönherr, E. (1973). *Phys. Status Solidi*, **B60**, 299.

Robertson, M. J., Wake eld, B., and Hutchinson, P. (1981). *J. Appl. Phys.*, **52**, 4462.

Samara, G. A. (1983). *Phys. Rev.*, **B27**, 3494.

Thijs, P. J. A. (1992). *Proceedings of the 13th IEEE International Semiconductor Laser Conference*, 2.

Trommer, R., Müller, H., Cardona, M., and Vogl, P. (1980). *Phys. Rev.*, **B21**, 4869.

Tsang, J. C., Oehrlein, G. S., Haller, I., and Custer, J. S. (1985). *Appl. Phys. Lett.*, **46**, 589.

Tsu, R. (1981). *Raman scattering characterization of bonding effects in* silicon. In *Defects in Semiconductors* (eds. Narayan and Tan), North-Holland, Amsterdam, 445.

Tsu, R. and Hernandez, J. G. (1982). *Appl. Phys. Lett.*, **41**, 1016.

Ueda, O. (1988). *J. Electrochem. Soc.*, **135**, 11C.

Chapter 9

Novel diagnostic data for diverse laser temperature effects; dynamic laser degradation effects; and mirror temperature maps

Semiconductor Laser Engineering, Reliability and Diagnostics: A Practical Approach to High Power and Single Mode Devices, First Edition. Peter W. Epperlein.
© 2013 John Wiley & Sons, Ltd. Published 2013 by John Wiley & Sons, Ltd.

Introduction

The main focus of this chapter is on a detailed description of the fundamental concept, physical realization, diverse applications, and results of the novel thermoreflectance technique pioneered and successfully introduced by the author as a powerful, highly versatile, experimental approach for characterizing diode lasers (Epperlein, 1990, 1993, 1997; Epperlein and Martin, 1992; Epperlein and Bona, 1993). The thermoreflectance technique has been adopted by many researchers in academia and industry for temperature monitoring of electronic devices (e.g., Schaub, 2001; de Freitas *et al.*, 2005; Ju *et al.*, 1997) and optical devices in particular diode lasers (e.g., Wawer *et al.*, 2005; Piwoński *et al.*, 2005, 2006; Xi *et al.*, 2005; Mansanares *et al.*, 1994).

We will discuss further mirror temperature measurements including a nonabsorbing mirror structure, devices with different heat spreader configurations, facet treatments, and a line scan perpendicular to the active layer toward the substrate side. In addition, we present a comparison between the properties of both thermoreflectance and optical spectroscopies, such as Raman and photoluminescence, demonstrating the various benefits of the thermoreflectance technique. Moreover, by using a special electroluminescence technique, the sharp decrease in the temperature from the mirror surface into and along the laser cavity could be measured for the first time with a high, submicrometer spatial resolution (Epperlein and Bona, 1993; Epperlein, 1997).

Finally, measurements on real-time temperature-monitored laser degradation processes and two-dimensional mirror temperature distributions, successfully made for the first time by employing the newly developed thermoreflectance technique (Epperlein, 1990, 1993, 1997), will be discussed. The former activity includes processes such as critical facet temperature to the COMD event, development of the facet temperature with increasing operation time, and temperature associated with dark-line defects, whereas the latter activity also compares the experimental temperature maps to numerically modeled ones.

9.1 Thermoreflectance microscopy for diode laser diagnostics

9.1.1 Motivation

In the preceding two chapters, we have described how Raman microprobe spectroscopy can be employed to measure local laser mirror temperatures with sufficient spectral and spatial resolution even though the measurement accuracy at higher temperatures is limited (see Figure 7.21) due to the exponential term involved in the evaluation process (see Equation 7.9) using the Stokes/anti-Stokes phonon mode intensity ratio.

The Raman signal detection sensitivity (see Section 7.4.3) has been dramatically improved since the first application of Raman for measuring laser mirror temperatures (Todoroki, 1986). This implies also the use of much lower laser power densities ($\gtrsim 10^2 \times$ less) for exciting the Raman scattering signal. These are now typically low ($\lesssim 100$ kW/cm^2, corresponds to ~ 1 mW in ~ 1 μm spot size) for a state-of-the-art setup resulting in low temperature rises <10 K of a laser die mounted on a heat sink (see Section 7.4).

Furthermore, to achieve the maximum temperature signal on a laser facet, the probe laser spot has to be aligned within the near-field spot where the laser intensity is highest. Also here, different powerful techniques have been developed, including simultaneous imaging of the probe laser spot and near-field pattern by using a microscope (see Figure 9.3 below) or the approach described in Section 7.4.3 and Figure 7.5. A very efficient and effective approach is given by using the p–n junction of the diode laser itself as a photocell. Centering the laser spot can then be easily achieved by maximizing the photovoltage by moving the laser spot in the near-field pattern.

Further advantages of using Raman spectroscopy in temperature measurements are its inherent strength to probe simultaneously the properties of the laser material at any location of the laser mirror surface (see Sections 8.1, 8.2, and 8.3). This conventional, nondestructive, contactless, and powerful optical characterization technique is well established; however, it can be time consuming and measures the data point-by-point, which may be a drawback, in particular in laser temperature measurements.

Although not discussed in this text, we should mention that the temperature information can also be derived from the energy shift of the photoluminescence (PL) signal offering, however, some advantages over Raman, such as higher signal levels and fewer (one) measurement runs per temperature point. The energy shift of a PL signal, for example, from the bulk or a quantum-confined area of a material, is predominantly determined by the temperature dependence of the bandgap energy.

Figure 9.1 shows the temperature dependence of the energy gap calculated using the empirical relation (Varshni, 1967) given in the figure. The energy shifts are small

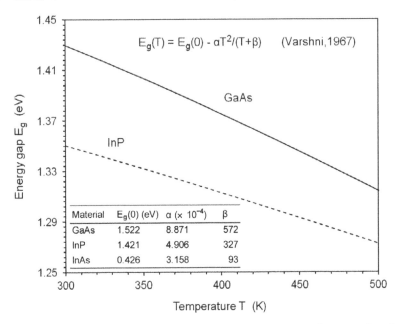

Figure 9.1   Calculated temperature dependence of the bandgap energy of some semiconductor compounds using the empirical expression in the figure.

with 5.4 meV/10 K for GaAs the largest of the three materials listed, followed by InP with 3.7 meV/10 K and InAs with 3.1 meV/10 K. However, given the fact that the PL lines are fairly broad for temperatures $\gtrsim 300$ K, it is very difficult to measure the temperature better than ~ 5 K in a reliable way. Thus, the essential drawbacks of insufficient sensitivity, speed, and amount of data per unit time exist also here.

Another approach is based on analyzing the laser radiation itself emerging from the mirror facet (Sweeney *et al.*, 2003). Under certain conditions the authors derived a simplified expression $I \propto E^2 \times \exp(-E/k_B T)$ for the radiation intensity I of photons with energy E emitted from the facet, which is no longer dependent on the effective bulk absorption coefficient and optical confinement factor. By plotting $\ln(I/E^2)$ versus E the authors claim to deduce the temperature from the slope of the plot in the high-energy Boltzmann tail of the facet emission. However, considering both the huge experimental and data evaluation effort involved in the procedure and the limited validity of the expression derived for the radiation intensity (see above), this technique cannot be regarded as a reliable, practicable, and versatile technique for diode laser mirror characterization. This is also underlined by the fact that the technique did not deliver in cw measurements on InGaAs/AlGaAs QW lasers the critical temperature to COMD at $\Delta T \sim 120$ K, which is now a well-established value.

Indispensable, however, for local laser temperature measurements is a technique capable of delivering the data in a fast and preferentially in a continuous mode, that is, the quasi-instant temperature response to the laser optical output power over a large range. These requirements can only be met by nonspectroscopic techniques that rely on a single easily and promptly measurable parameter. Certainly, commercial specialized thermal imaging cameras using focal plane sensor arrays made of materials such as InSb, InGaAs, or HgCdTe respond to mid- and long-wavelength infrared emission and could comfortably detect the thermal emission from the diode laser in the form of a thermal image. However, these cameras are expensive, and the more sensitive models require cryogenic cooling, but above all these cameras cannot deliver the required spatial resolution of $\lesssim 1$ μm due to the limited size of both pixels and arrays (max. 640×512). Therefore, it would be also technically very difficult to record reliably and efficiently a line plot temperature versus optical output power from the diode laser near-field spot by using such cameras.

We will demonstrate throughout this chapter that reflectance modulation (RM) is an alternative, highly versatile, and powerful new technological approach to characterize diode lasers, in particular their thermal behavior, hence the reason why the new technique is also called thermoreflectance (TR).

9.1.2   Concept and signal interpretation

In optical modulation spectroscopy techniques, the response of the optical constants of a solid is measured against a periodic change of an applied perturbation such as mechanical stress, temperature, or electric field (Cardona, 1969; Matatagui *et al.*, 1968; Seraphin, 1972). This information can then be used to investigate for example electronic band structure properties and effects in semiconductors.

Figure 9.2 illustrates how the principle of reflectance modulation (RM) can be applied to a semiconductor diode laser to determine in a fast way local absolute

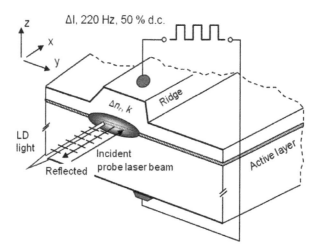

Figure 9.2   Schematics of optical reflectance modulation at the mirror facet of a ridge waveguide diode laser quasi-cw square-pulse operated.

mirror temperatures with high (sub)micrometer spatial resolution (Epperlein, 1990, 1993, 1997; Epperlein and Martin, 1992; Epperlein and Bona, 1993).

The diode laser is square-pulse power modulated with a low repetition frequency of $f = 220$ Hz and a duty cycle (d.c.) of 50%. This ensures that the reflectance response is maximum during operation and turned off between the pulses (quasi-cw operation). A cw Ar^+ laser beam with wavelength $\lambda = 457.9$ nm and low optical power $P_0 < 0.1$ mW (causing negligible $\Delta T < 1$ K) was used for probing the mirror surface within the light penetration depth of $\sim$40 nm. (The Ar^+ laser with its blue line was used to achieve the lowest possible focus spot size with the available equipment.) The probe laser beam focused to a spot size of FWHM $\cong 0.8$ µm impinges on the mirror at normal incidence. Upon reflection the probe laser beam is intensity modulated with the frequency f. The modulation depth of the reflected beam $P_0 \times \Delta R$ with ΔR the change of reflectivity is a measure for the strength of the periodic perturbations on the reflectance. However, both the reflected probe laser light and the diode laser emission are modulated with the same frequency f. To avoid any artifacts in the reflectance measurements from where the temperatures are finally derived, the diode laser emission has to be entirely excluded from the detection process. The original appropriate measures taken by using available components will be described in Section 9.1.4 below.

Below laser threshold, the signal can be ascribed to a modulation of the reflected light by a modulation of the electric field of the space-charge layer in the mirror surface (electroreflectance). The space-charge layer, that is, the surface barrier, is controlled and modulated by carrier injection due to the pulsed power operation of the device. A reduction of the surface barrier with increasing carrier injection may lead to a strong nonradiative recombination via surface states, and therefore is expected to contribute to a temperature rise at the facet. A higher carrier injection density and hence a stronger modulation of the surface potential may account for the observation that the electroreflectance signal $\Delta R/R$ below threshold is higher in narrow lasers than in wider ones at the same operating current (Epperlein, 1990). Since surface states and adsorbed impurities usually determine the surface potential, we conclude that the signal height below threshold is a measure for the local density of these states. The complex electric field distributions in the near-field region of the laser and the nonuniformity of the space-charge field along the penetration path of the probe light, however, complicate a quantitative analysis of the electroreflectance data. A quantitative analysis requires resolving at least two issues: first, the effect of the photoexcited carriers generated by the cw probe laser on the initial band bending; and, second, the contribution of any electric field in the mirror surface plane by a controlled changing of the orientation of the polarization vector of the incident probe laser beam.

Above laser threshold, the temperature modulation ΔT of the optical properties due to the pulsed power operation is practically the major contributor to the intensity modulation of the reflected light (thermoreflectance). As we will see in the following sections, the $\Delta R/R$ signals and thus ΔT are higher in narrow-ridge than in wide-ridge lasers due to the higher lasing power densities, and they are also higher for uncoated than for coated facets.

9.1.3   Reflectance–temperature change relationship

A relationship between the normalized reflectance change $\Delta R/R$ and the size of the temperature modulation ΔT can be derived from the normal incidence reflectance law (see Equation 7.10) of a solid/air interface with n_r the real refractive index and κ the extinction coefficient. Because n_r is usually large and κ small, for example, $n_r \cong 4.1$ and $\kappa \cong 0.38$ for $Al_{0.4}Ga_{0.6}As$ measured at 457.9 nm (Aspnes *et al.*, 1986), Equation (7.10) simplifies to $R \cong (n_r - 1)^2/(n_r + 1)^2$. By applying calculus of variations to this simplified R expression, the following expression for ΔT is obtained:

$$\Delta T \cong \left(\frac{n_r^2 - 1}{4}\right) \left(\frac{dn_r}{dT}\right)^{-1} \left(\frac{\Delta R}{R}\right) = C \left(\frac{\Delta R}{R}\right) \qquad (9.1)$$

with dn_r/dT the temperature coefficient of the refractive index (Epperlein, 1990). Using appropriate data for n_r at $\lambda = 457.9$ nm and dn_r/dT (e.g., 4.1 and 1×10^{-3} K^{-1}, respectively, for $Al_{0.4}Ga_{0.6}As$) the constant C is obtained as $C = (4 \pm 0.5) \times 10^3$ K and $C = (5.5 \pm 0.5) \times 10^3$ K yielding the final expressions

$$\Delta T \cong 4 \times 10^3 \left(\frac{\Delta R}{R}\right) \text{[K]} \qquad (9.2a)$$

and

$$\Delta T \cong 5.5 \times 10^3 \left(\frac{\Delta R}{R}\right) \text{[K]} \qquad (9.2b)$$

for GaAs-based and GaInP-based diode lasers, respectively (Epperlein, 1993).

9.1.4   Experimental details

Figure 9.3 shows the schematics of the experimental setup for realizing the thermoreflectance (TR) technique.

This scheme is in principle the same as that first published by Epperlein (1990), but slightly changed to adjust to the special requirements of a Navitar microscope (Navitar Inc., 2011). The essential part is to retrieve only the reflected (modulated) probe laser light, which carries the information by passing the total radiation (diode laser and probe laser light both modulated with the same frequency) through a very effective and efficient filter system:

- The cold mirror reflects only the 457.9 nm of the probe laser light and passes the diode laser light (usually IR).

- The bandpass interference filter (IF) stack and notch filter pass only the 457.9 nm light, which finally enters a grating spectrometer.

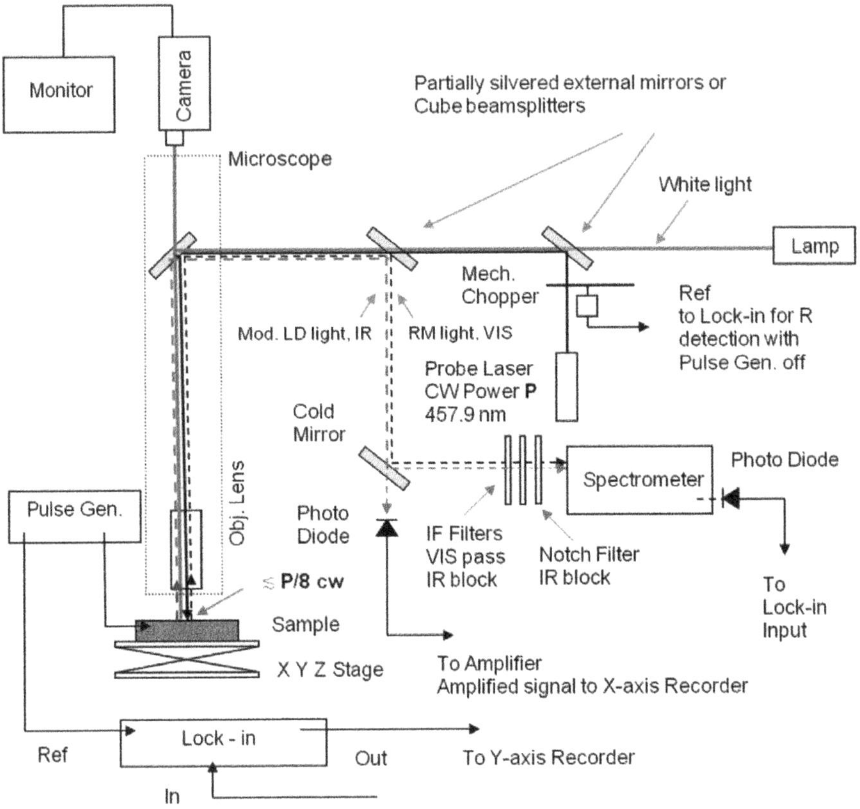

Figure 9.3 Scheme of the experimental setup for reflectance modulation measurements on diode lasers emitting in the infrared (IR) wavelength regime.

- The spectrometer performs an ultimate discrimination of the signal light from any residual diode laser radiation (<10 pW measured, which is about three orders of magnitude below the detection limit of the setup and therefore negligible with no impact on the TR measurements).

At the spectrometer exit the spectrally clean 457.9 nm reflected, modulated probe laser light is directed to a silicon photodiode, and the periodic signal with modulation depth $P_0 \times \Delta R$ is phase-sensitive detected by a lock-in amplifier tuned to modulation frequency f.

The dc reflected power $P_0 \times R$ is measured in the same way but with the diode laser switched off and the mechanical chopper tuned to the same frequency f and 50% d.c. The lock-in signal proportional to $P_0 \times \Delta R$ is normalized to the $P_0 \times R$ signal and displayed on the y axis of an x,y recorder. On the other hand, the diode laser beam passing through the cold mirror is detected by a silicon photodetector and after amplification and calibration displayed on the x axis of the recorder in absolute linearized optical power values.

The alignment of the probe laser spot to the center of the near-field pattern and hence maximization of the $\Delta R/R \propto (\Delta T)$ response can be achieved either by simultaneously imaging the probe laser spot and near-field pattern using a microscope (Figure 9.3) or the approach described in Figure 7.5, or by using the p–n junction of the diode laser itself as a photocell and maximizing its photovoltage by moving the laser spot in the near-field pattern (see also Section 9.1.1).

9.1.5   Potential perturbation effects on reflectance

In principle, there are some effects that can impact the reflectance results. However, as these effects are effective on both the actual modulated reflected probe laser power $P_0 \times \Delta R$ and also the dc reflected probe laser power $P_0 \times R$ measured at the same modulation frequency, the ratio of both quantities relevant for deriving ΔT cancel out such effects to a first approximation. Nevertheless, we briefly discuss such potential effects.

First, protective Al_2O_3 or Si_3N_4 mirror coatings $\lambda/2$- or $\lambda/4$-thick do not interfere with the reflectance process as long as their optical thicknesses are equal to an integral multiple of one-half of the probe laser wavelength λ_0: $n_r \times t = m \times \lambda_0/2$. Two examples for illustration are:

- Al_2O_3 or Si_3N_4 coatings $\lambda/2$-thick:

  $t = 265.6$ nm $\Rightarrow n_r \times t = \underline{451.5\ \text{nm}} \cong 2 \times (\lambda_0/2)\ (= \underline{457.9\ \text{nm}})$

- Al_2O_3 or Si_3N_4 coatings $\lambda/4$-thick:

  $t = 132.8$ nm $\Rightarrow n_r \times t = \underline{225.8\ \text{nm}} \cong 1 \times (\lambda_0/2)\ (= \underline{228.9\ \text{nm}})$

with $\lambda \cong 840$ nm the AlGaAs/GaAs diode laser wavelength, $\lambda_0 = 457.9$ nm the probe laser wavelength, t the coating thickness, n_r the refractive index of coating at λ (for calculating t, see Section 1.3.8) and at λ_0 (for calculating $n_r \times t$). In the cases above, the reflectance is to a good approximation unchanged and a maximum, that is, the coatings do not impact the reflectance process because their optical thicknesses are practically an integral multiple of half of the probe laser wavelength λ_0, which can be adjusted sufficiently well to meet this condition, if required.

Second, temperature changes generally shift the energy of the absorption edge and alter its shape leading to a change in the refractive index and finally in a reflectivity change which represents the fundamental physics of measuring temperatures using the reflectance modulation (thermoreflectance) technique. However, it may be affected by any change in the number of free carriers generated, for example, by the above-bandgap probe laser energy or by current injection. The excess energy of the carriers is redistributed among the carriers and the lattice via carrier–carrier and carrier–phonon interactions. Refractive index and hence reflectance changes can then be caused by this free-carrier plasma (Auston et al., 1978). Carrier-induced index changes have been estimated (Epperlein, 1993) by using the expressions for intra- and interband absorptions for free carriers derived from dispersion relations (Stern, 1964). It turned out, however, that these changes are very small; $\Delta n_r < 1 \times 10^{-4}$ at $\lambda_0 = 457.9$ nm, leading to a reflectance change $<1 \times 10^{-5}$, which is more than one order of magnitude below the detection limit of the setup. This is more valid the stronger the probe laser energy E_p differs from the fundamental bandgap energy E_0 (Stern, 1964). In all the measurements on the various diode lasers, E_p was about half-way between E_0 and the energy E_1 of the next higher critical point in the band structure (Aspnes *et al.*, 1986). All in all, the free carrier changes induced by the above-bandgap probe laser energy or by current injection have no measurable impact on the $\Delta R/R$ ratio because they affect both ΔR and R measurements roughly with the same strength.

Third, the cw photoreflectance effect due to a constant offset in surface potential by carrier injection through the cw probe laser has only a second-order effect on the value of $\Delta R/R$ above diode laser threshold and can be considered as negligible in the temperature measurements using TR.

9.2 Thermoreflectance versus optical spectroscopies

9.2.1 General

Diode laser mirror temperature measurements offer a series of benefits, which are very useful for developing high-quality diode lasers. These include:

- Evaluating mirror surface quality; effect of coatings.
- Supporting studies on degradation mechanisms.
- Optimizing coating material/process; increase of output power.
- Detecting "hot spots" impacting reliability.
- Designing reliable waveguide structures.
- Assessing thermal crosstalk in laser arrays.
- Monitoring mirror quality on a laser production line.

All of these activities can be carried out most effectively and efficiently by the TR technique, which has further relevant properties superior to alternative techniques such as Raman and PL spectroscopy as will be demonstrated in the next section.

9.2.2 Comparison

The most important parameters for evaluating the potential of a laser temperature measurement technique are listed in Table 9.1 to compare TR with optical spectroscopies, such as Raman and PL. The comparison shows the superiority of the TR technique.

Table 9.1 Comparison of the thermoreflectance (TR; reflectance modulation) technique with optical spectroscopies of Raman scattering and photoluminescence (PL) with respect to relevant operating parameters to deduce the temperature information from the sample.

	Optical micro-thermometer			
Feature	TR	Bonus	Spectroscopy Raman vs. PL	Bonus
Speed	Fast, $\cong 1$ s/pt.	+	V. slow (Raman), $\cong 1000$ s/pt. Slow (PL), $\cong 5$ s/pt.	
Data display	Continuous	+	Point by point	
Data flux	Very large	+	Very small (Raman) Small (PL)	
ΔT vs. P curve	Complete. Fine structure. $\cong 1$ min	+	Fragmentary. Fewer details, even for 100× time (TR)	
Sensitivity	≤ 1 K	+	≥ 10 K (Raman) ≥ 5 K (PL)	
Accuracy	$\cong \pm 1$ K	+	$\geq \pm 10$ K ($\Delta T < 100$ K), $\geq$ ± 20 K ($\Delta T > 100$ K) (Raman), $\geq \pm 5$ K (PL)	
Spatial resolution	≤ 1 μm	+	≤ 1 μm	+
Contactless	Yes	+	Yes	+
Nondestructive	Yes	+	Limited (Raman) Yes (PL)	+
Stress on diode laser	Low	+	V. high at high P (Raman), measuring time limited by time to COMD	(+)
Maximization of signal	Easy, lock-in signal	+	V. tedious, integral peak intensity (Raman). Simpler for PL	(+)
Use	Simple	+	Demanding, complicated (Raman). Simpler for PL	(+)
	Advanced features			
Temperature mapping	Very practicable	+	Practically not feasible	
Degradation processes	Real-time detection feasible	+	Real-time detection not feasible	
Surface states	Accessible via electroreflectance	+	Accessible via electric field-induced Raman scattering	+

9.3   Lowest detectable temperature rise

An expression for the lowest detectable $\Delta R/R$ and therefore ΔT can be derived from the following simple assumptions. We approximate the square-pulse modulation of the diode laser by a sinusoidal modulation with angular frequency ω and the reflected probe laser beam power can then be written as

$$(P + \Delta P) = P_0(R + \Delta R\cos(\omega t)) \qquad (9.3a)$$

where P_0 is the incident power and R the reflectivity.

The current in the photodiode detector can be expressed as

$$(I + \Delta I) = \eta(P + \Delta P)q/h\nu \qquad (9.3b)$$

where η is the quantum efficiency of the photodiode with a value in the range of 0 to 1, q the electron charge, and $h\nu$ the incident photon energy. The noise power of the photodiode is given by

$$P_N = \left(4k_B T + 2qIR_L\right) \times \Delta f \qquad (9.3c)$$

where R_L is the photodiode load resistance and Δf the detection bandwidth. For $R_L I \gg 2k_B T/q$ the thermal noise can be neglected and Equation (9.3c) becomes

$$P_N = 2qIR_L \times \Delta f. \qquad (9.3d)$$

By assuming the noise power equals the minimum detectable signal power

$$P_S = (\Delta I)^2 R_L = 2qIR_L \times \Delta f \qquad (9.3e)$$

and using Equation (9.3a) and Equation (9.3b) we get

$$\left(\frac{\Delta R}{R}\right)^2 \overline{\cos^2(\omega t)} = \left(\frac{\Delta R}{R}\right)^2 \frac{1}{2} = \frac{2h\nu \times \Delta f}{\eta P_0 R} \qquad (9.3f)$$

where the average of the cosine function has been taken over a complete period. Finally, the expression for a minimum detectable $\Delta R/R$ is

$$\frac{\Delta R}{R} = \sqrt{\frac{4h\nu \times \Delta f}{R P_0 \eta}}. \qquad (9.3g)$$

The detection limit of the original experimental setup used for the first TR measurements (Epperlein, 1990, 1993) was $\Delta R/R \cong 1 \times 10^{-4}$ yielding a lowest detectable $\Delta T \cong 0.5$ K.

9.4 Diode laser mirror temperatures by micro-thermoreflectance

9.4.1 Motivation

In this section, further mirror temperature measurements are discussed by using the novel micro-thermoreflectance technique. On the one hand, these measurements are on laser structures different from those used in the Raman measurements discussed in Chapters 7 and 8, but on the other hand they confirm also the results of these Raman measurements and thus demonstrate the potential and many benefits of the TR technique.

9.4.2 Dependence on number of active quantum wells

What is striking in Figure 9.4 are the continuous curves in the mirror temperature increase ΔT versus drive current I_d plots recorded by TR, in contrast to the single data points of Raman measurements. The temperatures were measured in the near-field patterns of mirror facets of junction-side up mounted ridge (Al)GaInP lasers 5 μm wide with one, two, and five active quantum wells (QWs). The ΔT data are relative to ambient temperature and allow for the small heating effect of the low intensity (<1 mW) of the 457.9 nm Ar^+ probe laser line focused to a spot size of $\lesssim 1$ μm

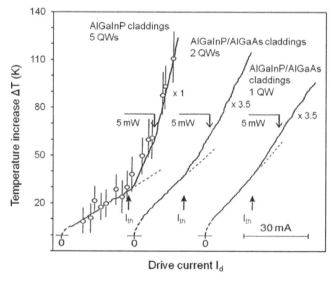

Figure 9.4 Typical temperature rise ΔT versus drive current I_d plots from thermoreflectance (TR) measurements on cleaved uncoated ridge GaInP lasers 5 μm wide with one and two active quantum wells (QWs) and mixed AlGaInP–AlGaAs claddings on the one side, and five QWs and AlGaInP claddings on the other side. The $\Delta T/I_d$ characteristic measured by using Raman spectroscopy is also shown for comparison for the 5-QW device.

FWHM (Epperlein and Bona, 1993). To verify the temperatures derived from the TR measurements, the figure also includes Raman data obtained from the 5-QW device for comparison, which demonstrates excellent agreement between the data from the two totally different measurement techniques.

The ΔT versus I_d plots clearly show two typical regimes indicating different heating mechanisms. Below threshold current I_{th}, ΔT increases with I_d and is mainly due to Joule heating of the drive current. This is supported by the fact that the ΔT data at I_{th} agree well with the bulk temperature rises of the active p–n junction region along the cavity, as found from the comparison between P/I_d characteristics recorded under pulsed and dc conditions. However, a distinct onset in ΔT is observed at I_{th} for all three lasers, even for the SQW laser, and ΔT increases rapidly with I_d. As discussed already in Section 7.4.5.2, in this regime the Joule heating effect is superimposed by heating due to the absorption of laser radiation leading to an enhanced nonradiative carrier recombination rate in the mirror facets. The slope of the curve in this regime with respect to the dashed line of the extrapolated current heating is a sensitive measure for assessing the heating efficiency only under lasing conditions (current heating subtracted) and can be expressed by the differential temperature increase $\delta(\Delta T)/\delta P$ per laser output power unit at a reasonably low value of, say, 5 mW.

The temperatures in Figure 9.4 are linearly proportional to the number of QWs (see also Figure 7.29) up to two, whereas the dependence is superlinear up to five QWs (cf. Figure 4.7). The latter can be accounted for by the use of AlGaInP cladding layers in the 5-QW device, which have a higher electrical and thermal resistivity than the mixed AlGaInP–AlGaAs claddings used in the SQW and DQW devices (cf. Sections 4.4.2.2 and 7.4.6.3) and therefore a higher heating power.

In terms of a simplified picture, the QW number dependence of the mirror temperature can be understood by taking into account, first, that each QW may act as a possible heating source fed by any possible nonradiative recombination of the injected carriers, and, second, that the individual heating sources in lasers with more than one QW are cumulative in their effect. See also the discussion in Section 1.3.4.3 on the dependence of the threshold current on the number of QWs involving different loss mechanisms such as free-carrier absorption leading to an increased optical loss coefficient α_i and thus higher internal heating.

9.4.3   Dependence on heat spreader

Figure 9.5 shows ΔT versus I plots of 980 nm compressively-strained InGaAs/ AlGaAs GRIN-SCH SQW ridge uncoated lasers 5 µm wide recorded under different conditions. First, the dependence on the use of a thick Au heat spreader (HS) layer on top of the p-contact (Figure 9.5b) is shown in (A) and (B) of Figure 9.5a. The graphs clearly show a decrease in the temperature for devices with the HS lined up with the mirror edge (B), in contrast to devices with the HS recessed (A) by ~20 µm (Figure 9.5b). The averaged drop in temperature is by a factor of two measured at the 60 mW mark of the three devices in each group. This difference can be understood from the fact that the temperature decreases exponentially

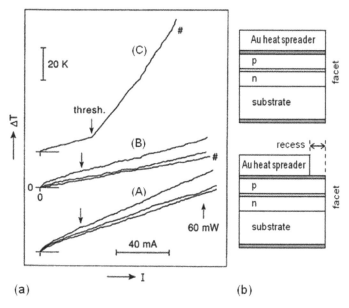

Figure 9.5 Maximum temperature rise ΔT versus drive current I plots measured by TR on uncoated mirrors of 5 µm ridge InGaAs/AlGaAs SQW lasers depending on a heat spreader (HS) on top of the structure (b). (a): (A) HS $\cong$ 20 µm recessed; (B) HS lined up with mirror edge; (C) after COMD of device marked # in comparison to as-grown device in (B).

from the mirror surface toward the laser cavity, which will be discussed in detail in Section 9.6 below.

The second effect deals with the emergence of the kink at I_{th} in the $\Delta T/I$ characteristic after a COMD event occurred. The characteristics of as-grown lasers in (A) and (B) of the figure show no kink at threshold and thus are dominated by the Joule heating of the drive current. However, after COMD occurred a distinct kink at threshold appeared, as demonstrated by the characteristic marked # in (B) for the as-grown facet and (C) for the facet after COMD. This demonstrates in a distinct manner the superposition of laser radiation heating caused by surface absorption effects as argued already on many occasions in the text, in particular in Sections 7.4.6.1 and 8.2.3.1.

9.4.4   Dependence on mirror treatment and coating

A similar topic has been discussed in Section 7.4.6.2 on ridge (Al)GaAs lasers 7 µm wide in a ΔT versus P dependence representation by using Raman spectroscopy for measuring the mirror temperature rises ΔT. Here, we discuss ΔT versus I measurements on (Al)GaAs devices 10 µm wide with similar mirror treatments and coatings achieved, however, using the TR technique (Epperlein, 1993). The objective is to quantify the mirror surface quality/heating efficiency by the slope of signal rise at threshold relative to the current-heating signal slope below threshold.

10 μm wide ridge (Al)GaAs SQW lasers. Cleaved mirrors.

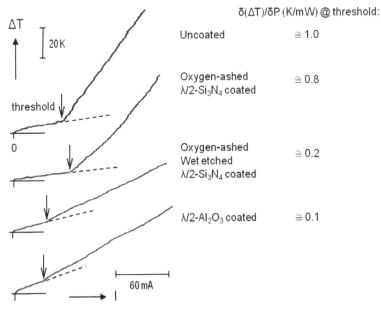

Figure 9.6   TR facet temperature rises ΔT measured on (Al)GaAs SQW lasers 10 μm wide with different mirror treatments and coatings. Differential temperature increases per power unit at threshold are listed for the various mirror technologies to compare the respective mirror surface qualities/heating efficiencies.

Figure 9.6 shows the graphs for (i) uncoated, (ii) oxygen-ashed, $\lambda/2$–Al_2O_3-coated, (iii) oxygen-ashed, wet-etched, $\lambda/2$–Si_3N_4-coated, and (iv) $\lambda/2$–Al_2O_3-coated mirrors with the differential temperature increases $\delta(\Delta T)/\delta P \cong 1.0$, 0.8, 0.2, and 0.1 K/mW, respectively.

What is remarkable is the reduction of heating due to optical power absorption by a factor of 4 due to the removal of the damage layer after an oxygen-ashing treatment by a wet etch. In addition, measurements on mirrors with the same surface treatment prior to coating showed no significant difference in ΔT between $\lambda/2$–Al_2O_3 and $\lambda/2$–Si_3N_4 passivation layers.

Finally, the signal of the slightly convex characteristics near the zero-point $I \gtrsim 0$ can be ascribed to electroreflectance caused by a surface potential modulated by the pulsed carrier injection during the TR measurement. This interpretation is in agreement with results from electric field-induced Raman scattering (EFIRS) measurements (Beeck *et al.*, 1989) on the band bending of laser mirror surfaces.

9.4.5   Bent-waveguide nonabsorbing mirror

A nonabsorbing mirror (NAM) structure based on a bent-waveguide concept has been discussed in Section 4.3.4 to enhance the optical strength of the laser facet.

In this approach, the optical beam is decoupled from the active waveguide layer into the nonabsorbing cladding layer with a higher bandgap energy (Gfeller *et al.*, 1992a, 1992b). As described previously, the bent-waveguide structure in the mirror regions of ridge (Al)GaAs lasers 3.5 µm wide is formed by overgrowing a wet-etched channel 1 µm deep with 22° sloped sidewalls in the n$^+$-doped (100) GaAs substrate. Exact wafer orientation is necessary to obtain symmetrical growth conditions on both channel slopes.

To assess the effectiveness of the NAMs with respect to a reduction of optical absorption, the rise of mirror temperatures ΔT of lasers with dry-etched (CL$_2$ reactive ion etching) and $\lambda/2$–Al$_2$O$_3$-coated mirrors has been measured for the first time for progressive NAM section lengths x in the range $-1 < x < 16$ µm by spatially resolved ($\lesssim 1$ µm) TR.

Figure 9.7 shows the schematics of the NAM structure and, for illustration, the $\Delta T/I$ characteristics at three typical positions $x = -1$, 5, and 15 µm with high, low, and high temperatures measured for example at 30 mW for each device, respectively. This temperature behavior is also clearly reflected in the differential temperature increase at threshold, which is lowest for a device with a NAM section length $x = 5$ µm. The characteristics can be accounted for by the properties of the conventional mirror at $x = -1$ µm, a NAM with a beam fully in the nonabsorbing cladding at $x = 5$ µm, and a NAM with a partially recoupled beam into the absorbing waveguide at $x = 15$ µm, respectively.

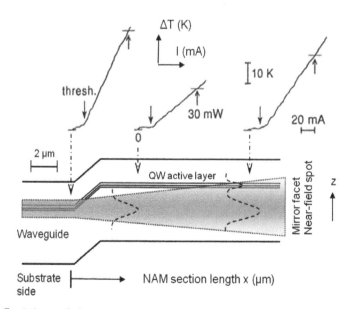

Figure 9.7 Scheme of a bent-waveguide nonabsorbing mirror (NAM) structure of dry-etched, coated (Al)GaAs lasers. Schematic side view of the beam propagation in the structure. Selected temperature ΔT versus current I characteristics recorded on mirror facets at different NAM section lengths $x = -1$, 5, and 15 µm. Scale valid along cavity in x direction.

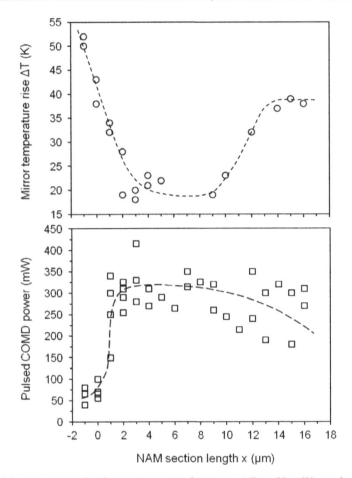

Figure 9.8 Experimental mirror temperature increases ΔT at 30 mW quasi-cw power (top) and pulsed COMD power levels (bottom) for etched and passivation-coated mirrors of (Al)GaAs lasers 5 µm wide as a function of progressive nonabsorbing mirror (NAM) section length x. (Adapted in modified form from Gfeller *et al.*, 1992a.) Dashed lines are guides to the eye.

A more detailed $\Delta T/x$ dependence is plotted in Figure 9.8, which can be divided into three distinct regions.

In the region $-1 < x < 2$ µm (transition from conventional mirror to sloped waveguide section) the temperature measured at 30 mW optical output power falls sharply from typically 50 to 20 K.

In the range $2 < x < 10$ µm (optical beam propagating fully in nonabsorbing cladding), the temperature remains at this low level of 20 K, and increases again for $x > 10$ µm (partial beam recoupling into the absorbing waveguide leads to increased heating; recoupling is evidenced by monitoring the optical near-field patterns, see Figure 4.5).

The finite temperature of 20 K at the minimum may be due to nonradiative recombination of diffused carriers from the pumped laser section and to residual optical power absorption at surface defects of the etched mirrors as supported by the well-pronounced onset at threshold in the $\Delta T/I$ plots. Further improvements might be achieved by introducing a current blocking layer and by improved mirror etching and coating techniques.

The temperature profile in Figure 9.8 can be interpreted as a demonstration of nonradiative carrier recombination due to optical power absorption depending on the degree of overlap of the optical field intensity with the absorbing waveguide profile. The $\Delta T(x)$ profile is anti-correlated with the COMD(x) profile, as expected (see also Figure 8.4 for comparison). For lasers with observed partial beam recoupling ($x >$ 10 µm), the COMD power shows a downward trend due to an increasing overlap of the optical beam with the absorbing waveguide. The effective length of the NAM section for this particular geometry is in the range $2 < x < 10$ µm. Lasers with NAM lengths in this range show a fourfold improvement in the pulsed (500 ns at 0.1% duty cycle) COMD power levels compared to lasers with conventional mirrors ($x < 0$). A similar or even better improvement has also been achieved for cw-measured COMD power levels on lasers with $x = 4$ µm NAM section lengths.

9.5 Diode laser mirror studies by micro-thermoreflectance

9.5.1 Motivation

For many applications it would be useful to apply a nondestructive, contactless, sensitive, and fast microprobe to monitor local variations of material and laser device properties with high spatial resolution in the (sub)micrometer regime and in real time. These requirements can be fully met by the RM or TR techniques. Application examples include real-time temperature-monitored laser degradation effects, threshold and heating distribution within the near-field pattern, which will be discussed in this section, and laser facet temperature mapping investigations, which will be dealt with in the last section of this chapter.

9.5.2 Real-time temperature-monitored laser degradation

The discussion in this section deals with three effects first investigated by Epperlein (1990, 1992, 1993, 1997). They comprise the temperature leading to the COMD event, temperature during time to COMD, and temperature associated with dark-spot defect formation.

9.5.2.1 Critical temperature to catastrophic optical mirror damage

Typical examples of $\Delta T/I_d$ characteristics of passivation-coated (Al)GaAs lasers with ridge waveguide widths of 5–10 µm are shown in Figure 9.9a, c, d. In all

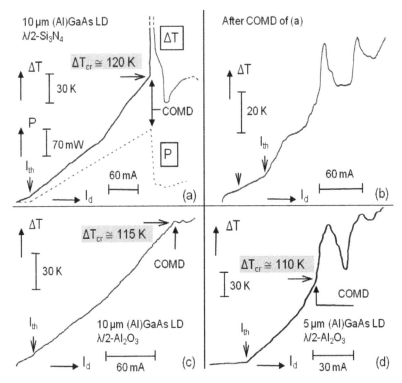

Figure 9.9   Mirror temperature rises ΔT as function of drive current I_d for ridge (Al)GaAs lasers with different widths and passivation coatings (a), (c), (d) demonstrating the onset of COMD at a critical temperature increase ΔT_{cr}. (a) P/I plot with partial COMD event shown for comparison. (b) $\Delta T/I_d$ plot for device in (a) after (partial) COMD event with increased threshold current and reduced power and a quasi-periodic structure at large currents due to an optical power periodically increasing and decreasing with I_d (not shown).

three cases depicted in the figure, the lasers incurred COMD at a critical temperature increase ΔT_{cr} in the range of 110 to 120 K obtained in a quasi-cw operation (see Section 9.1.2). The existence of a critical temperature is consistent with the thermal runaway model for COMD discussed in Section 3.2.2.3. The heating of the mirror surface, together with the rapid exponential increase in the absorption coefficient with temperature, causes the surface to switch into a highly absorbing state for the laser radiation. Bandgap shrinkage and absorption increase form a positive feedback cycle at the critical temperature, ultimately leading to thermal runaway and catastrophic self-destruction of the mirror.

Figure 9.9a shows for illustration the critical temperature in the $\Delta T/I_d$ graph in correlation with a drop in power P due to partial COMD shown in the P/I_d characteristic. The critical temperature obtained for an (Al)GaAs laser in Figure 7.28 by Raman measurements can be considered to be in reasonable agreement with the data obtained here. This conclusion takes into account an overall measurement error of at least ± 10 K and the fact that the samples were different in ridge geometry and mirror technology, which might have an effect on the critical temperature. Figure 9.9b shows the $\Delta T/I_d$ after partial COMD (Figure 9.9a) with an increased threshold current and a quasi-periodic structure at higher drive currents I_d with power values periodically increasing and decreasing with I_d (not shown).

The quasi-cw critical temperature values are compatible with cw and quasi-cw measurements of COMD powers of broad-waveguide, 100 µm stripe InGaAs lasers (Botez, 1999). The critical temperature is practically independent of facet treatment and coating, but is dependent on the semiconductor material system used. The advantage of the TR technique is that the critical temperature can be determined at the very moment the COMD event is occurring by simply ramping up the current (power).

9.5.2.2 Development of facet temperature with operation time

The time development of mirror temperature is demonstrated on a cleaved uncoated InGaP/AlGaInP–AlGaAs DQW 5 µm ridge laser operated at 30 mW.

Figure 9.10 shows a gradual increase of the temperature up to the time t where the mirror rapidly switches at $\Delta T_{cr} \cong 90$ K into partial COMD (see sharp drop of optical power in P/t characteristic and residual power after COMD event). It reaches

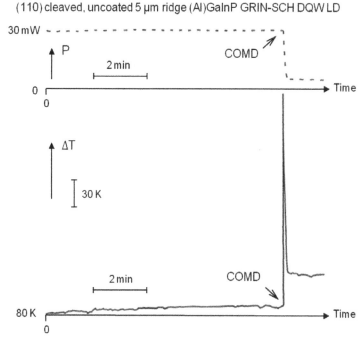

Figure 9.10 TR mirror temperature rise ΔT as function of time of an uncoated (Al)GaInP laser 5 µm wide operated at 30 mW. Demonstration of the onset of COMD at a critical temperature increase (~90 K) after a certain elapsed time. Indicates the formation of active defects with a concentration sufficient to initiate the COMD process. Power versus time plot is shown for comparison.

a maximum of ~320 K within ~1 s (detection system limited). The degraded output power then causes a rapid decay of the temperature, which finally settles at a steady state value. The actual temperature in the thermal runaway phase is much higher than the 320 K and can only be determined by using a sufficiently short time constant in the lock-in measurement of the TR signal. Furthermore, a maximum temperature response can only be expected when the probe laser spot truly hits the center of the hot spot at the facet. However, this cannot be predicted with sufficient precision, because the location where the hot spot will develop depends sensitively on the local heat and mechanical strain distributions under the ridge waveguide structure. Nevertheless, this experiment showed that COMD occurred in the (Al)GaInP laser system at a lower critical temperature increase of ~90 K compared to that in an (Al)GaAs laser material. It also demonstrated that, in principle, the dynamic behavior of thermal laser properties can be measured with the TR technique.

9.5.2.3   Temperature associated with dark-spot defects in mirror facets

The temperature associated with the gradual formation of a dark-spot defect (accumulation of localized nonradiative recombination defects, see Section 3.2.2.2) has been measured on a cleaved uncoated ridge (Al)GaAs laser facet 15 μm wide (Figure 9.11).

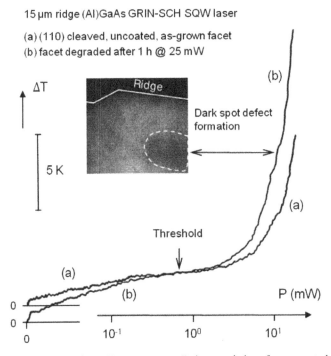

Figure 9.11 Temperature rise ΔT versus power P characteristics of an uncoated mirror of a ridge (Al)GaAs SQW laser 15 μm wide: (a) as-grown mirror without degradation; (b) facet degraded with dark-spot defect formation in center of near-field pattern (see inset).

Temperature and dark spot have been monitored or imaged simultaneously as a function of time. The figure shows the temperature increase ΔT as a function of the logarithm of optical power P for an as-grown laser and a degraded laser with dark-spot formation at the mirror facet. The temperature in the dark spot grew at a rate of ~5 K/h of operation at 25 mW quasi-cw optical output power. The growth rate depends on the effective local mirror temperature.

9.5.3   Local optical probe

This section demonstrates for the first time how the electro-optical and thermal properties within the near-field pattern of a diode laser can be probed by using the reflectance modulation technique.

9.5.3.1   Threshold and heating distribution within near-field spot

Figure 9.12 shows the $\Delta R/R$ versus I_d characteristics around laser threshold at different random locations within the near-field pattern of a slightly degraded (Al)GaAs laser exhibiting two-mode behavior.

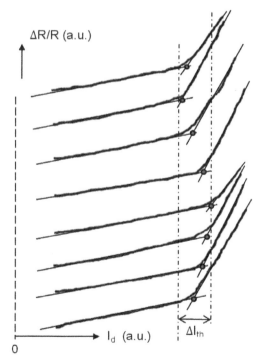

Figure 9.12 Local normalized reflectance change $\Delta R/R$ versus I_d characteristics around laser threshold at different random locations within the near-field pattern of a slightly degraded (Al)GaAs laser exhibiting two-mode behavior. Characteristics displaced vertically for clarity. Linear approximations (thin lines) of the curves. Demonstration of local threshold and heating strength variations within filament structure of lateral mode profile.

The characteristics are displaced vertically for clarity. Below the local thresholds, the characteristics have practically the same shape and slope, whereas the local threshold points and slopes of the characteristics beyond the thresholds vary strongly depending on the location within the near-field spot. In a first attempt, both effects can be made plausible within the concept of the filamentation phenomenon (discussed in various sections of Chapter 2), where the individual filaments within the lateral mode profile along the cavity have different thresholds and heating strengths. A partial correlation between local optical density within the near field and local threshold current could be found in the first measurements, but more detailed and systematic investigations are required to verify the effect.

9.6 Diode laser cavity temperatures by micro-electroluminescence

9.6.1 Motivation

The maximum laser optical output power is limited either by COMD or output power saturation due to heating of the active layer. High junction temperatures have important consequences for laser efficiency and threshold current, since both of these quantities are strongly temperature dependent (see Sections 1.3.7.1, 2.4.5.3, and 1.3.7.3). Therefore, it is crucial to measure the junction temperature under operating conditions. In addition, knowing the temperature profile along the cavity can give useful information about facet heating mechanisms, the effectiveness of a heat spreader, and mounting the laser chip onto a heat sink, and can reveal hot spots caused by heating through nonradiative recombination at high concentrations of defect centers.

Different methods have been used to measure the temperature inside the laser cavity including measuring the spectral shift of electroluminescence (EL) through a window in the epitaxial-side p-contact metallization (Brugger and Epperlein, 1990; Dyment *et al.*, 1975) and GaAs substrate-side n-contact metallization (Dabkowski *et al.*, 1994). Here, we discuss laser cavity temperatures obtained from spectral shifts of the active layer EL emitted on the side of a standard ridge laser chip cleaved close (<5 µm) to the ridge waveguide all along the cavity (Epperlein and Bona, 1993; Epperlein, 1997).

9.6.2 Experimental details – sample and setup

Figure 9.13a shows a sample prepared by cleaving a standard ridge (Al)GaAs laser chip in close proximity (<5 µm) to the ridge along the cavity of total length 750 µm with a maximum deviation of no more than 1 µm. The cleaved edge of total length 750 µm was magnified 350× and projected by appropriate optics perpendicular to the entrance slit of a 0.85 m spectrometer (Figure 9.13b). The magnified EL light line was moved perpendicular to the spectrometer slit with a typical width of $\lesssim$300 µm and thus a lateral resolution of $\lesssim$1 µm could be achieved.

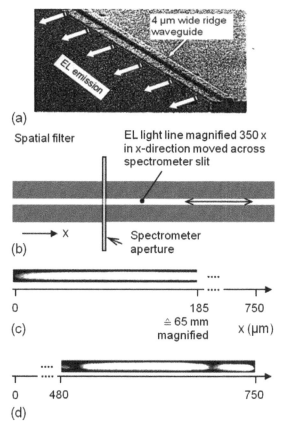

Figure 9.13   Laser cavity temperatures measured by spatially resolved electroluminescence (EL): (a) ridge waveguide laser sample preparation; (b) measurement principle; (c) and (d) partial EL light line images toward both cavity ends with two dark-spot areas in (d) due to local defect formation after thermal rollover of laser. Bright areas indicate high luminescence signal.

The laser was driven by current pulses at a low repetition rate of 200 Hz and d.c. of 50% and the EL emission after dispersion was detected by a suitable detector and phase-sensitive lock-in system. The radiation emitted laterally through the cleaved face originates in the active QW region and escapes from the sample without reabsorption in the higher energy layers of the laser structure. Figure 9.13c shows for illustration part of the uniform EL light line near one end of a laser cavity, whereas part of the EL light with two dark spots (hot spots) that occurred after thermal rollover operation of the laser can be seen close to the other end of the cavity (Figure 9.13d).

9.6.3   Temperature profiles along laser cavity

Typical EL spectra of the GRIN-SCH QW emission of a 5 μm ridge (Al)GaAs laser mounted onto a heat sink are plotted in Figure 9.14 for a cw laser output power of

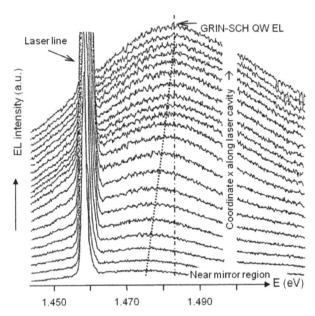

Figure 9.14   EL spectra of a 5 μm ridge (Al)GaAs laser with dominant GRIN-SCH SQW emission recorded at 300 K and 13 mW cw in 1 μm steps from the front mirror along the cavity. Spectra are displaced vertically for clarity.

13 mW. The spectra were recorded from close to the front mirror along the cavity in 1 μm steps and included the scattered laser light which was not blocked to be used as a reference. The spectral position of the EL QW peak shifts clearly to lower energies on approaching the mirror position, which indicates an increase in temperature due to the dominating temperature dependence of the bandgap energy.

By using an experimentally determined temperature coefficient of the peak energy of $dE/dT \cong -0.54$ meV/K (corresponding to $d\lambda/dT \cong 3$ Å/K), which is in excellent agreement with the calculated temperature coefficient of the bandgap energy for GaAs derived from Figure 9.1, the temperature profile $\Delta T(x)$ along the cavity can be determined. Figure 9.15a shows the plot of log ΔT versus x for a ridge InGaAs/AlGaAs and GaInP/AlGaAnP-AlGaAs diode laser.

Both profiles exhibit in the first part an exponential decay of ΔT toward the cavity with a characteristic length of ∼6 μm at the 1/e point. Beyond $x > 10$ μm the ΔT decays are also approximately exponential but slower and different for the two material systems. In this regime, the temperature slowly approaches a constant value effective for $x > 25$ μm. The overall $\Delta T(x)$ profile has a bathtub-like shape with high temperatures at the mirror sides and low temperatures in the center regions of the cavity. To achieve an efficient cooling effect by using a heat spreader on top of the structure, it has to be lined up with the mirror edge; a heat spreader recessed by the characteristic decay length or more has no measurable cooling effect on the mirror.

Finally, Figure 9.15b shows the dependence of ΔT on drive current I_d in a log–log graph measured at a half-length position in the cavity. The cavity temperature

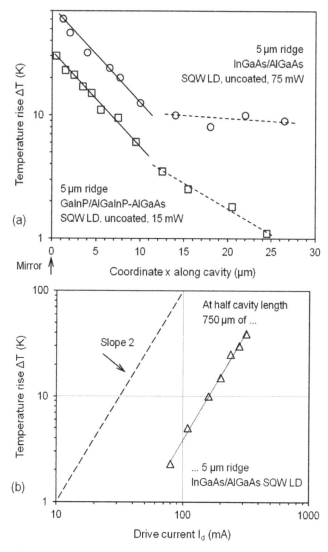

Figure 9.15 Cavity temperatures measured by spatially resolved EL for an InGaAs/AlGaAs and GaInP/AlGaInP–AlGaAs SQW laser. (a) Log of temperature rise ΔT versus coordinate x along cavity plots show for both lasers the same steep exponential decay of ΔT with a characteristic length of ~6 µm (1/e point) followed by a much slower but different decay toward the bulk of the cavity. (b) Temperature rises ΔT taken as a function of drive current I_d near the center of the cavity are proportional to the square of I_d in line with the expected Joule heating power of the drive current.

is strongly proportional to the Joule heating of the drive current given by I_d^2, which is illustrated in the figure. The temperature rise in the cavity center of a 5 μm ridge uncoated InGaAs/AlGaAs laser is $\Delta T \sim 40$ K at 300 mA (150 mW).

9.7 Diode laser facet temperature – two-dimensional mapping

9.7.1 Motivation

Knowledge of the two-dimensional temperature distribution on a laser facet can give useful information about local hot spots, the quality of facet passivation treatment and reflectivity coating layers, the bond between coatings and the semiconductor surface, and can contribute to the optimization of the vertical structure to achieve the goal of lowest possible absolute operating temperatures and temperature gradients.

The first ever mirror temperature map of an operating diode laser was obtained by the author (Epperlein, 1990) and reported in subsequent papers (Epperlein and Martin, 1992; Epperlein, 1993, 1997).

9.7.2 Experimental concept

Temperature maps have been obtained by raster scanning the focus spot of the low-power probe laser across the mirror surface of the quasi-cw square-pulse power-modulated diode laser (see Section 9.1.2). The high spatial resolution (~0.8 μm FWHM optical spot size) two-dimensional scanner with two orthogonally mounted high-precision (0.1 μm mechanical step size) stepping motor-driven translation stages, discussed in Section 7.1.2 and Figure 7.2, was employed in this application. The $\Delta R/R$ signal was detected as a function of location and, after conversion to temperature, a ΔT map was generated.

9.7.3 First temperature maps ever

A contour map of a typical temperature rise (relative to 300 K) distribution on an as-cleaved and uncoated mirror of a 15 μm wide ridge and 750 μm long InGaAs/AlGaAs GRIN-SCH SQW laser, quasi-cw square-pulse operated at a power $P = 40$ mW, is displayed in Figure 9.16.

The scanned area is 35 μm × 15 μm with the first scan aligned along the active layer approximately 0.5 μm (residual thickness, see Section 2.3) away from the lower edge of the etched ridge profile. Note that the scale in the y direction is smaller than that in the z direction.

There are two striking features: first, a very localized hot spot with a strong temperature rise $\Delta T \sim 60$ K within the near-field pattern just below the ridge characterized by the dense 2 K equidistant contour lines; and, second, a much lower temperature regime characterized by the wider contour lines indicating a slower decay of temperature. This regime is in the GaAs substrate as indicated by the arrow

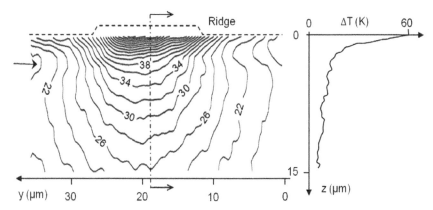

Figure 9.16   Contour map of the temperature rise ΔT distribution recorded by scanning TR on a cleaved uncoated mirror facet of an (Al)GaAs diode laser 15 µm wide operated at 40 mW quasi-cw. The scan was not exactly parallel in the horizontal y direction, which caused a slight displacement in the z direction of the contour indents on both sides due to the epitaxial layer/substrate interface (see arrow) with different thermal conductivities in both regions. The effect of the latter is demonstrated by the different widths of the isothermal contour lines and the different slopes of the two sections in the ΔT versus z line scan. Note that the scale in the y direction is smaller than that in the z direction.

marking the boundary line between the epitaxial (mainly) AlGaAs layers and the GaAs substrate. The ratio of the temperature decay strength in the GaAs and AlGaAs is fully consistent with the thermal conductivity that is about 5–6 times larger in GaAs than in AlGaAs. The two different temperature zones are also clearly demonstrated in the line scan shown in the figure. The contour map exhibits particularly well the epilayer/substrate interface with the indent of the isothermal lines becoming more pronounced in the horizontal y direction and the bending of the contour lines away from the heat source due to the strong heat-spreading effect of the top p-metallization.

9.7.4   Independent temperature line scans perpendicular to the active layer

A high-precision temperature line scan perpendicular to the near-field pattern of an E2-passivated (see Section 4.2.2.1), Al_2O_3-coated front mirror ($R = 0.1$) of a compressively-strained InGaAs/AlGaAs GRIN-SCH SQW single-mode high-power diode laser with a ridge 4 µm wide is shown in Figure 9.17. The temperature measurements were recorded at an output power of 110 mW for two separate runs and with a spatial resolution of 0.8 µm FWHM.

The figure shows the hot spot area within the near-field pattern with a spot size of ~2 µm FWHM in the vertical direction, which is ~3× higher than the effective extension of the near field. The decay of the temperature toward the substrate side is divided into the two zones comprising the epitaxial layers and the substrate as demonstrated in an impressive manner by the log ΔT versus z coordinate graph of

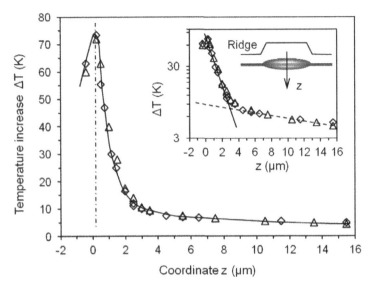

Figure 9.17 Temperature line scan ΔT versus z direction perpendicular across the near-field spot center of an InGaAs/AlGaAs SQW laser 4 μm wide at high power of 110 mW cw toward the GaAs substrate in two separate scans. The log ΔT versus z graph in the inset shows an exponential decay of ΔT in both epitaxial and substrate areas in agreement with the respective thermal conductivities in these areas.

the inset confirming, for the first time, exponential decay in both areas. The slope ratio of the two experimental log curves is in excellent agreement with the ratio of the relevant thermal conductivities in the two regions confirming the result obtained in the section above.

9.7.5 Temperature modeling

The experimental ΔT map can be compared to heat flow calculations. From a theoretical point of view, the temperature distribution in the laser can be obtained by solving the three-dimensional heat equation in the device. For simplicity, however, we can neglect any variation of heat production that may occur along the laser structure. Therefore, we can limit the calculations to a cross-section of the laser where a solution of the two-dimensional heat equation must be found:

$$-\mathrm{div}\,(k \times \mathrm{grad}(T)) = Q + h + (T_{ext} - T) \tag{9.4}$$

where k is the heat conduction coefficient, Q the heat source, h the convective heat transfer coefficient, and T_{ext} the external temperature. We further neglect particular thermal effects at the mirror including laser radiation absorption and nonradiative recombination effects and any convective heat transfer ($h = 0$) and external temperature ($T_{ext} = 0$) in the numerical computation process. The heat flow calculations make use only of the different thermal conductivities in the various sections of the structure and thus can give only a qualitative picture of the temperature distribution in the laser facet. We used the Partial Differential Equation (PDE) Toolbox within MATLAB (The MathWorks Inc., 1997) to solve the elliptic PDE in Equation (9.4), in order to determine the steady state temperature distribution.

9.7.5.1   Modeling procedure

We consider that heat is produced in the active layer, in a region that corresponds approximately to the near-field pattern of the diode laser. Neglecting radiation and convection losses at the surfaces of the device, we assume that heat can flow only through the substrate to a heat sink. The temperature at the bottom of the heat sink is held constant at 300 K. The PDE solving process can be divided into the following steps:

1. Define the geometry (2-D domain). Draw the geometry of the laser structure. The total chip thickness is 102 μm including 97 μm for the substrate. The total chip width is 150 μm with the ridge centered at the middle line of 75 μm (centerline of ridge) (Figure 9.18a). These close-to-reality dimensions are important among others for computing a temperature map which is qualitatively

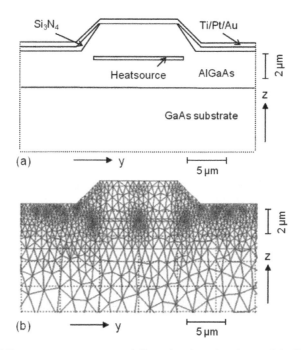

Figure 9.18   Mirror temperature map modeling using the MATLAB Partial Differential Equation Toolbox. (a) Simplified ridge structure – zoomed in. (b) Mesh pattern – zoomed in, used in finite element method to discretize the two-dimensional heat partial differential equation and produce approximation to solution.

as close as possible to the experimental one. For the direct comparison a reduced plot area is generated, which is similar to the 35 μm × 15 μm area of
the experimental map.

2. Define the boundary conditions. Specify the temperature on boundaries
   (Dirichlet). Specify the heat flux ($k \times \nabla T$) across boundaries (Neumann).

3. Define the PDE. Enter the required PDE coefficients (k, Q, (h), (T_{ext})) in each
   subdomain independently. Thus, the various material properties in the PDE
   model can be easily specified.

4. Create the triangular mesh to be used in the finite element method (FEM)
   to discretize the PDE and produce approximate solution (Figure 9.18b). Automated mesh generation and plot of the mesh. Adjusts automatically to the
   thickness of the various materials and refines accordingly.

5. Solve the PDE. In solve mode the nonlinear and adaptive solver for elliptic
   problems is invoked and controlled by using the FEM for problems defined
   on bounded domains in the plane. The PDE is discretized on the mesh and an
   equation is built as a discrete approximation of the solution.

6. Plot the solution and other physical properties calculated from the solution
   (post processing).

The laser device parameters including dimensional, heat conduction, and heat
source data used for the temperature modeling are listed in Table 9.2.

Table 9.2   Laser device parameters and data used in the mirror
temperature map modeling. The various thermal conductivity values
are from the Ioffe Physico-Technical Institute (2001), Carlson et al.
(1965), Afromowitz (1973), and Adachi (1983).

Dimensions [μm]:	
Chip length	≅750
Effective ridge width	≅ 12
Heat conduction of layers k [W/(K × μm)]:	
GaAs substrate	≅ 0.55×10^{-4}
AlGaAs cladding	≅ 0.11×10^{-4}
InGaAs active layer	≅ 0.08×10^{-4}
Si_3N_4 dielectric ridge embedding	≅ 0.14×10^{-5}
Ti/Pt/Au p-contact	≅ 0.30×10^{-3}
Heat source (InGaAs/AlGaAs):	
Volume = $W \times H \times L$ [μm³]	≅ 900
Net power = Input P at 40 mW − optical P [mW]	≅ 113
Heat source Q = net power/volume [W/μm³]	≅ 1.26×10^{-4}

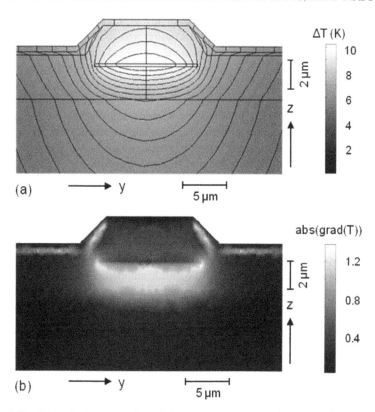

Figure 9.19 Numerical computation of the temperature map $\Delta T(y, z)$ of the laser used in Figure 9.16 by considering only the different thermal conductivities of the various layers in the vertical laser structure. The modeling does not consider mirror heating effects due to laser radiation absorption and nonradiative recombination. (a) Contrast and contour image in good qualitative agreement with experimental temperature map of Figure 9.16. (b) First temperature gradient map of a diode laser mirror with strong gradients below the active layer as heat source toward the epitaxial layer/substrate interface, and along the silicon nitride layer.

9.7.5.2 Modeling results and discussion

Figure 9.19a shows the calculated mirror temperature map $\Delta T(y, z)$ of the laser used in Section 9.7.3 in excellent qualitative agreement with the experimental contour plot of Figure 9.16. In particular, the contrast and contour images show the hot spot area with its highest temperatures at the heat source of the active layer and with a steep temperature decay in the epitaxial region toward the substrate side due to a lower thermal conductivity in the epitaxial $Al_{0.35}Ga_{0.65}As$ layers than in the GaAs substrate. The wider contour lines towards the p-contact indicate the strong heat spreading effect of the metallization.

The maximum temperature of $\Delta T \sim 10$ K is in agreement with the cavity temperature measured toward the center of the cavity caused mainly by the distribution of the different thermal conductivities in the structure. The difference between maximum ΔT values of the calculated map (Figure 9.19a) and the experimental map (Figure 9.16) is due to laser radiation absorption and nonradiative recombination effects at the facet not taken into account in the numerical computation. The ratio of six between both values is in reasonable agreement with temperature measurements made from the mirror edge along the laser cavity (Figure 9.15).

Finally, Figure 9.19b exhibits for the first time a calculated temperature gradient map across the mirror facet. There are two striking areas where the gradients are highest. One is, as expected, below the active layer to the boundary with the substrate, whereas the other is in the dielectric Si_3N_4 layer along the lower edges of the ridge profile, but most importantly also along both ridge slope sides.

Considering the different thermal expansion coefficients of $\sim 14 \times 10^{-6}$ K^{-1}, $\sim 3 \times 10^{-6}$ K^{-1} (Suganuma et al., 1985) and $\sim 5.5 \times 10^{-6}$ K^{-1} (Ioffe Physico-Technical Institute, 2001) for the dominating Au in the p-contact, Si_3N_4, and $Al_{0.35}Ga_{0.65}As$, respectively, these areas are potential sources for the formation of high thermal stress fields leading to high local stress amplitudes. The latter may trigger the formation of detrimental structural defects and cracks in the materials and may cause problems in the adhesion of the nitride and metallization layers.

These thermal stress fields superpose the intrinsic mechanical stress fields (see Section 8.2), caused by the deposition of the dielectric nitride and p-metallization layers at temperatures of $\sim 120°C$ on ridge waveguide structures. As discussed in Section 8.2, local mechanical stress fields, which may now be enhanced through the superposition of both stress fields, are responsible for the formation and migration of defects (see Figures 8.8 and 8.9) resulting in an even stronger negative effect on diode laser performance and reliability.

References

Adachi, S. (1983). *J. Appl. Phys.*, **54**, 1844.

Afromowitz, M. A. (1973). *J. Appl. Phys.*, **44**, 1292.

Aspnes, D. E., Kelso, S. M., Logan, R. A., and Bhat, R. (1986). *J. Appl. Phys.*, **60**, 754.

Auston, D. H., McAfee, S., Shank, C. V., Ippen, E. P., and Teschke, O. (1978). *Solid-State Electron.*, **21**, 147.

Beeck, S., Egeler, T., Abstreiter, G., Brugger, H., Epperlein, P. W., Webb, D. J., Hanke, C., Hoyler, C., and Korte, L. (1989). *Proceedings of ESSDERC'89*, 508.

Botez, D. (1999). *Appl. Phys. Lett.*, **74**, 3102.

Brugger, H. and Epperlein, P. W. (1990). *Appl. Phys. Lett.*, **56**, 1049.

Cardona, M. (1969). *Modulation spectroscopy*. In *Solid State Physics* (eds. Seitz, F. *et al.*), Suppl. 11, Academic Press, New York.

Carlson, R. O., Slack, G. A., and Silverman, S. J. (1965). *J. Appl. Phys.*, **36**, 505.

Dabkowski, F. P., Chin, A. K., Gavrilovic, P., Alie, S., and Beyea, D. M. (1994). *Appl. Phys. Lett.*, **64**, 13.

De Freitas, L., da Silva, E., Mansanares, A., Pimentel, M., Filho, S., and Batista, J. (2005). *J. Appl. Phys.*, **97**, 104510.

Dyment, J. C., Cheng, Y. C., and SpringThorpe, A. J. (1975). *J. Appl. Phys.*, **46**, 1739.

Epperlein, P. W. (1990). *Inst. Phys. Conf. Ser.*, **112**, 633.

Epperlein, P. W. (1993). *Jpn. J. Appl. Phys.*, **32**, Part 1, No. 12A, 5514.

Epperlein, P. W. (1997). *Proc. SPIE*, **3001**, 13.

Epperlein, P. W. and Bona, G. L. (1993). *Appl. Phys. Lett.*, **62**, 3074.

Epperlein, P. W. and Martin, O. J. (1992). *Inst. Phys. Conf. Ser.*, **120**, 353.

Gfeller, F. R., Buchmann, P., Epperlein, P. W., Meier, H. P., and Reithmaier, J. P. (1992a). *J. Appl. Phys.*, **72**, 2131.

Gfeller, F. R., Buchmann, P., Epperlein, P. W., Meier, H. P., and Reithmaier, J. P. (1992b). *Proceedings of the IEEE 13th International Semicondoctor Laser Conference, Conf. Dig.*, 222.

Ioffe Physico-Technical Institute (2001). *New Semiconductor Materials: Physical Properties.* www.ioffe.rssi.ru/SVA/NSM.

Ju, Y., Käding, O., Leung, Y., Wong, S., and Goodson, K. (1997). *IEEE Electron Device Lett.*, **18**, 169.

Mansanares, A. M., Roger, J. P., Fournier, D., and Boccara, A. C. (1994). *Appl. Phys. Lett.*, **64**, 4.

Matatagui, E., Thompson, A., and Cardona, M. (1968). *Phys. Rev.*, **176**, 950.

Navitar, Inc. (2011). *Products*. http://navitar.com.

Piwoński, T., Wawer, D., Szymański, M., Ochalski, T., and Bugajski, M. (2005). *Optica Applicata*, **XXXV**, 611.

Piwoński, T., Pierscinska, D., Pierscinski, K., Malag, A., Jasik, A., Kozlowska, A., and Bugajski, M. (2006). *Proc. Mater. Res. Soc. Symp.*, **916**, DD06-01.

Schaub, E. (2001). *Analytical Sciences* (The Japan Society for Analytical Chemistry), **17** (Special issue), s443.

Seraphin, B. O. (1972). Chapter 1 in *Semiconductor and Semimetals* (eds. Willardson, R. K. and Beer, A. C.), **9**, 1, Academic Press, New York.

Stern, F. (1964). *Phys. Rev.*, **133**, A1653.

Suganuma, K., Okamoto, T., Koizumi, M., and Shimada, M. (1985). *J. Mater. Sci. Lett.*, **4**, 648.

Sweeney, S., Lyons, L., Adams, A., and Lock, D. (2003). *IEEE J. Sel. Top. Quantum Electron.*, **9**, 1325.

The MathWorks, Inc. (1997). *Products*: MATLAB. www.mathworks.com.

Todoroki, S. (1986). *J. Appl. Phys.*, **60**, 61.

Varshni, Y. P. (1967). *Physica*, **34**, 149.

Wawer, D., Ochalski, T. J., Piwoński, T., Wójcik-Jedlińska, A., Bugajski, M., and Page, H. (2005). *phys. stat. sol.* (a), **202**, 1227.

Xi, Y., Xi, J., Gessmann, Th., Shah, J., Kim, J., Schubert, E., Fischer, A., Crawford, M., Bogart, K., and Allerman, A. (2005). *Appl. Phys. Lett.*, **86**, 031907.

Index

For major topics within diode laser engineering (DLE), diode laser reliability (DLR), diode laser diagnostics (DLD) see under the following letters:
DLE: B, C, D, E, F, G, H, L, M, O, Q, R, S, T, U, W
DLR: A, C, D, E, F, L, O, R, S
DLD: A, D, E, L, M, P, R, S, T

Semiconductor Laser Engineering, Reliability and Diagnostics: A Practical Approach to High Power and Single Mode Devices, First Edition. Peter W. Epperlein.
© 2013 John Wiley & Sons, Ltd. Published 2013 by John Wiley & Sons, Ltd.

Printed and bound by CPI Group (UK) Ltd, Croydon, CR0 4YY

16/04/2025

14658369-0004